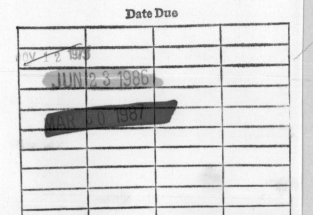

CLASSICAL DESCRIPTIONS OF MOTION

A SERIES OF BOOKS IN PHYSICS

Editors: Henry M. Foley
Malvin A. Ruderman

CLASSICAL DESCRIPTIONS OF MOTION

THE DYNAMICS OF PARTICLE TRAJECTORIES RIGID ROTATIONS, AND ELASTIC WAVES

Emil Jan Konopinski
Indiana University

W. H. FREEMAN AND COMPANY
San Francisco

Printed in the United States of America.
Library of Congress Catalog Card Number: 71–75626
Standard Book Number: 7167 0323–8

9 8 7 6 5 4 3 2 1

PREFACE

This book is addressed to students who wish to arrive at the frontiers of present-day Physical Theory as expeditiously as possible. It is designed to be quite comprehensive as to the fundamentals of Classical Mechanics, but is highly selective as to applications and the mathematical languages introduced, these being chosen for their special pertinence to modern developments.

October 1968 E. J. Konopinski

CONTENTS

CLASSICAL DESCRIPTIONS OF MOTION

INTRODUCTION: ABOUT PHYSICAL THEORIZING

Science or poetry or philosophy each imbues some class of men with heady convictions of having attained insight and truth. Yet beliefs so inspired often contradict each other unless the minds holding them have been subjected to some kind of uniform, orthodox development. And even the most cherished beliefs may only be temporary, being forced to change by the harsh discipline of fact. Drastic upsets occur particularly often in the minds of scientists, since they most explicitly profess a commitment to correspondences between their ideas and factual experiences. A scientist finally tends to abandon all beliefs, to think only as he finds himself forced to think whenever he wants his ideas to work.

Description and Prediction

A scientist naturally considers an idea to be working satisfactorily when it leads him to solutions of his problems. What constitutes a solution of a scientific problem is the formulation of an *expectation*, a *prediction* of what will happen in a given type of circumstance, without the necessity of repeated trial and error. Thus it is its verifiable predictive value that lends worth to a scientific idea, and a scientist feels compelled to accept the idea when he deems it essential to making predictions which will turn out to correspond successfully to observed facts. This seems to be about the only criterion for "truth" that he finds himself really obliged to accept.

At their most rudimentary level, predictions are based on the expectation that whatever has been observed to happen in a given set of circumstances will happen again whenever at least the essential parts of the circumstances are repeated. The word "essential" is a very important one here. Very little predictive power is displayed when the circumstances must be replicated in every discernible respect before a prediction can be ventured. The most impressive predictions have been made by discerning in ostensibly new circumstances subtle relationships to ones already experienced, suggesting the presence of the conditions that are to be regarded as the really essential ones for the expected consequences.

Even the recognition of the essentials in circumstances already requires accurate descriptions of these. Scientific theories consist of the requisite *descriptions* of observed phenomena, accurate enough to enable some success in *predictions* of future phenomena. The greater the precision with which the essentials are discriminated and described, the more accurate can the predictions be made. As these statements imply, theories undergo continuing development as experiences accumulate; the reliability and breadth of the predictions made at any given stage depend on the quantitative accuracy and comprehensiveness of the information available for the descriptions.

The Development of Physical Theories

The generalizations that are made by discerning and describing features common to otherwise diverse physical circumstances can often be fitted into broad logical schemes from which verifiable consequences can be deduced. This introduction of *deductive* procedures accounts for the spectacular progress made in physical theory, as compared to sciences more restricted to immediate and individual instances of *induction* from empirical facts. As a consequence, consistency with a deductive scheme has come to contribute at least as much to confidence in the predictions from a given set of theoretical ideas as do numerous cases of agreement with observations.

However, questions always remain whether a given theory, or some idea in it, is really essential after all to the predictions deduced from it. Sometimes a formula is found which yields amazingly accurate quantitative agreements with corresponding measured numbers. The theoretical formula must usually be attended by words and pictures that at least *help* make clear how it applies in the laboratory. A given set of words and pictures may seem essential to the making of the prediction, yet has quite often been transmogrified without spoiling the success of the formula.* Moreover, a given formula is sometimes derivable from several, apparently distinct, logical schemes,† calling for different verbalizations and visualizations. There is always room for the suspicion that whenever a physicist decides at some point that he has arrived at a uniquely necessary theory, it is only because his ability to imagine alternative schemes is limited.

Nevertheless, closer approaches to uniqueness are continually being achieved: through establishing isomorphisms which reduce apparent differences of formulation to triviality, as well as through increasing the comprehensiveness of the theory and thus bringing ever wider ranges of facts to bear on its determination. A unified, all-embracing physical theory may never be attained, but there have accumulated several quite impressively comprehensive bodies of theory. They are often worth study even after they have been superseded by still more comprehensive ones, since they may retain wide ranges of usefulness and become even more reliable when the limits of their applicability are understood. Thus, some "classical" theories should be given as much attention as the more comprehensive relativistic and quantum-mechanical ones. In many cases such attention is even a necessity, since the more modern theories frequently require reinterpretation in classical terms, over common ranges of applicability, before they can be brought into touch with the laboratory.

Physical theory has to be presented with a residue of uncertainty that must remain as long as men keep themselves open to the possibility of new experience. It is not difficult to learn to live with such uncertainty when it is just the readiness to discard old ideas, whenever new facts demand it, that leads to more efficacious and comprehensive concepts. An evolutionary veiw of theories must apparently be accepted; even the "classical" ones keep growing, at least in particular aspects, under severe tests of their fitness to survive. This affects their exposition profoundly, since it calls for reviews of stages in their development, rather than presenting them in a Euclidean manner, as completely defined and closed systems. Indeed, when physical theories become definable with mathematical rigor, it may indicate that rigor mortis is setting in!

---

* The "wave versus particle duality" was a sensational instance of this.

† Witness Newton's and Einstein's derivations of the planetary orbits, or Bohr's, de Broglie's, Schrödinger's, Heisenberg's and Dirac's derivations of the Balmer formula.

Mathematical Description

The greater the precision with which physical circumstances are discriminated and described, the more accurately can predictions be made. What is really essential to a successful prediction can even escape notice entirely unless subtle distinctions, requiring a refined precision of expression, are searched out. To achieve the requisite accuracy of expression, a language with far more finesse than the ordinary vernacular may become necessary. It usually takes a "special purpose," *mathematical* language, which must sometimes even be invented for the occasion. (That is why Newton invented the calculus!)

A mathematical vocabulary is first adopted by putting physical conceptions, abstracted from the experience to be described, into correspondence with suitable mathematical conceptions. The latter are defined by the roles they play in logical operations,* and the operations form an *algebra* of the mathematical concepts. Algebraic operations yield results still couched in the mathematical language being used. Physical predictions can emerge only after the adopted correspondences are again utilized to translate mathematical results into physical expectations.

The prime example of a mathematical vocabulary invented to help with physical description consists of the positive integers, which can be put into correspondence with any distinguishable physical objects. The relations among them implicit in the conception† of the integers give rise to the operation of addition and its inverse, subtraction. To enforce a logical uniformity—that any integer may be subtracted from any other—negative integers and zero had to be introduced. These also have found use in physical descriptions. Then an enumeration of repeated additions of the same integer leads to the conceptions of multiplication and of powers, thence to division and root-finding. To enable the exceptionless performance of the latter operations, the rational fractions and the irrational and complex numbers were invented. Finally, the problem of establishing correspondence with the physical conception of continuity in space led to the introduction of transcendental numbers such as π. The continuum of just the real numbers, which can be put into one-to-one correspondence with the points of any trajectory in space, or to the moments of time, pervades all quantitative physical description.

---

* It is such a consideration that belies the comment that science has merely substituted the word "atoms" for the word "angels" of the medieval theologian. A word, like a more strictly mathematical concept, carries with it a context of logic in which it can take part.

†A great discovery of the mathematicians is that the statement "two plus two equals four" owes its certainty to its not being a physical assertion but a logical one, "empty" in the sense that it is already implicit in the *definition* of the integers. Physical assertions cannot have such certainty, since they are always at the mercy of future experience.

Because physical conceptions can rarely be adequately understood unless they are expressed in a mathematical language, it is necessary to direct a great deal of attention to the mathematics in any presentation of physical theory. The real number algebra is so widely familiar that this will be taken for granted here, as will familiarity with much of trigonometry, analytic geometry, and the calculus. More detailed attention will be paid to further proliferations of the mathematical languages, starting at a level as elementary as the vector algebra and the vector calculus, because these provide simple examples of how definitions of mathematical languages can be bent to serve the needs of physical description. Most purely mathematical problems will be ignored, since the interest here will be primarily in the correspondent physical descriptions. Suitable extensions of the mathematics will be introduced as they become necessary. These will have the advantages that accrue from defining mathematical conceptions "operationally" (that is, by the uses they are put to).

Two Aims of the Book

The exposition will begin with the classical "Newtonian" description of motion, since this has served as the prototype of all physical theorizing. Some of the initial considerations will be quite elementary but are explicitly needed in preparing for the more highly developed reaches of the subject. Intensive dissection into "axioms" is avoided* since it can be as endless as extensions of the subject, and that it leads any closer to some "absolute truth" may be doubted. In its entirety the exposition will be aimed at (1) a review of the basic principles which have guided the development of classical descriptions of motion, together with those applications which best exhibit the part played by Classical Mechanics in modern physical theory, and (2) an "operational" review of some mathematical languages which are used throughout modern physical theory: vector algebra and calculus, complex numbers, transformation and operator matrices, eigenvectors and orthogonal function sets—all illustrated by applications in the Classical Mechanics.

* On the contrary, use of "intuition" is encouraged, although not for final proof. Intuitive thinking seems to be the way discoveries are initiated, even by mathematicians bent on eliminating intuition altogether.

CHAPTER 1

THE KINEMATICS OF A PARTICLE

The classical theory of mechanics was developed from observations on the ordinary motions of ordinary bodies.*

In making the observations, attention may be concentrated on the positions of *points* of the moving bodies, and each point may be specified by putting it into correspondence with a position "vector," with denotations: $\mathbf{r}_1, \mathbf{r}_2, \ldots, \mathbf{r}_i, \ldots$. How many points of a body are given attention depends upon the purposes for which the description is being constructed, or upon the completeness and accuracy of the information which can be brought to bear. When predictions about just the orbit followed by a planet are desired, or if the planet is so distant that it is observable only as a point, then the whole planet may be represented by a single point: a "point-model" of the planet is employed. When predictions about the planet's diurnal motion are to be

* In the sense meant here, extraordinary motions are ones with velocities approaching the speed of light. Extraordinary bodies are of atomic dimensions or less. Observations on these extremes have led to modifications of the theory about to be presented; however, the modifications are without detectable effect in the "ordinary" motions to be dealt with for the present.

made—that is, rotations about an axis within it are observable—then at least a second point of it must be watched, or a "rigid-body model" of the planet used. To treat tidal motions of the whole body may require paying attention to a whole continuum of points, in an "elastic-body model." To adopt a model is part of discerning just the essentials in circumstances, as needed for some type of prediction.

A motion may be described by supplying the variable positions, $\mathbf{r}_i(t)$, for a continuum of times, t. Such space-time interrelationships are the characteristic concern of *kinematics*. A typical predictive scheme formed by a purely kinematical description was provided by Kepler when he showed that the planets follow elliptical orbits in a certain way. A more comprehensive scheme was developed by Newton after further circumstances influencing motion were discriminated and described. Such further circumstances (such as mass and force) are the concern of *dynamics*. The first two chapters here will survey essential features of just the kinematics and dynamics of point-particles.

1.1 VECTOR ALGEBRA

Men have the intuitive habit of imagining physical phenomena to be distributed in a three-dimensional space. The habit is presumably a product of their evolution, and its value is attested by their survival.

The primary element of description in space is the location of a point in it, and this must be done relative to some reference frame. It may be done by giving some three numbers; cartesian coordinates (x, y, z) and spherical coordinates (r, ϑ, φ) are perhaps the most important examples. Whatever triplet of numbers is used, they stand for a single entity: a point-position, which may be symbolized as $\mathbf{r}$ (relative to the chosen origin) or, in a diagram, by an arrow extending from the origin to the point. It is clearly advantageous to define an algebra for operating with symbols like $\mathbf{r}$; it may enable the forming of a description that is independent of the trio of numbers chosen in any particular evaluation of a point-position. The algebra is defined with an eye for its usefulness in physical descriptions (and is developed in consistency with the conventionally accepted rules of logic).

The mathematical concept symbolized by $\mathbf{r}$, or by the arrow in diagrammatic representations, is called a *vector*. The position-vector discussed here is only a prototype for more general uses of the vector concept; velocity vectors $\mathbf{v}$ and acceleration vectors $\mathbf{a}$ will be introduced in the next sections, and momentum vectors $\mathbf{p}$ and force-vectors $\mathbf{F}$ in the next chapter.

In relation to the position-vector, $\mathbf{r}$, numbers like x, y, z are called

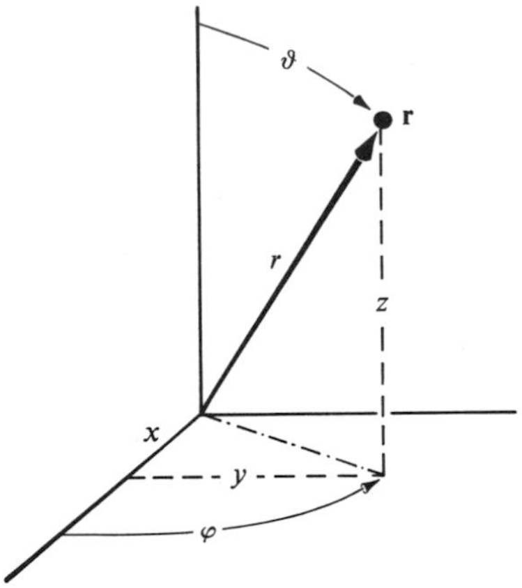

Figure 1.1

cartesian *components*. The number $r \equiv |\mathbf{r}|$ is called the *magnitude* of the vector and ϑ, φ specify its *direction*. A vector is said to differ from a *scalar* (number) in that it has direction, as well as magnitude.

Adding Vectors

A function for which the addition of vectors is useful to define is indicated by Figure 1.2. A point is moved from a position $\mathbf{r}$ to $\mathbf{r}'$ through a displacement represented by a vector $\mathbf{s}$. The *resultant* $\mathbf{r}'$ may be represented as the sum

$$\mathbf{r}' = \mathbf{r} + \mathbf{s}. \tag{1.1}$$

For the components this implies

$$x' = x + s_x, \qquad y' = y + s_y, \qquad z' = z + s_z. \tag{1.2}$$

Any vector equation, like (1.1), can be regarded as standing for a trio of component equations.

The displacement vector $\mathbf{s}$ is a result of a subtraction,

$$\mathbf{s} = \mathbf{r}' - \mathbf{r}. \tag{1.3}$$

Subtraction is plainly equivalent to the addition of the negative of a vector (that is, with its direction reversed).

The Scalar Product

The multiplication of two vectors may be defined so as to give a scalar result and is then denoted $\mathbf{r} \cdot \mathbf{s}$ or $(\mathbf{r}, \mathbf{s})$. The first denotation has led to the name "dot product" for the scalar product. The useful definition may be

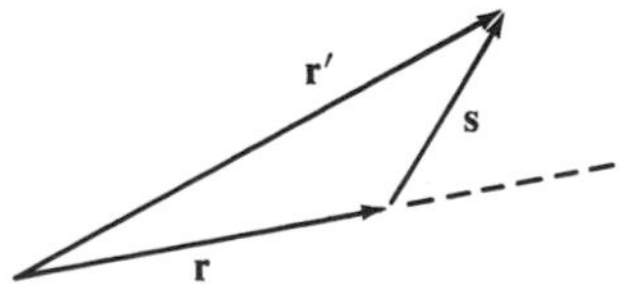

Figure 1.2

made to emerge from the scalar product, distributively applied, of the sum equation (1.1) with itself,

$$\mathbf{r}' \cdot \mathbf{r}' = \mathbf{r} \cdot \mathbf{r} + 2\mathbf{r} \cdot \mathbf{s} + \mathbf{s} \cdot \mathbf{s}. \tag{1.4}$$

This yields the trigonometric "cosine rule" for the magnitude,

$$r' = (r^2 + 2rs \cos \alpha + s^2)^{1/2}, \tag{1.5}$$

if each scalar product is defined on the model

$$\mathbf{r} \cdot \mathbf{s} \equiv rs \cos \alpha, \tag{1.6}$$

where α is the angle (Figure 1.2) between the direction of the two vector factors. Notice that $rs \cos \alpha$ can be regarded as the product of the magnitude of **r** with the projection $s \cos \alpha$ of **s** on **r**, or, equivalently, as the product of s with the projection $r \cos \alpha$ of **r** on **s**. A scalar product of two different vectors has a meaning by itself in such an example as that of the "work," $W = \mathbf{F} \cdot \mathbf{s}$, done by a force **F** during a displacement **s** (to be examined in the next chapter).

It is useful to notice that $\mathbf{r} \cdot \mathbf{s} = 0$ need not imply that either **r** or **s** is zero; it may merely mean that **r** and **s** are perpendicular to each other.

The relation of the cartesian components x, y, z and s_x, s_y, s_z to the product $\mathbf{r} \cdot \mathbf{s}$ will be explored below.

The Vector Product

Another way of multiplying two vectors,* this time with a vector as a result, is useful to define. An illustration is provided by the area of the parallelogram in Figure 1.3. The area has a magnitude,

$$A = rs \sin \alpha, \tag{1.7}$$

and also a definable direction: *facing* into the direction of the vector **A** of the diagram. The vector having such a magnitude and direction is symbolized by

$$\mathbf{A} = \mathbf{r} \times \mathbf{s} \qquad (\text{or } [\mathbf{r} \wedge \mathbf{s}]). \tag{1.8}$$

---

* A logical exhaustion of all possible ways will be discussed in §9.2.

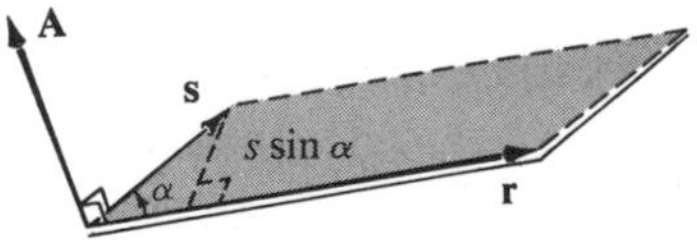

Figure 1.3

The first notation accounts for the name "cross-product," frequently used. The direction of **A** may be described as that in which a right-handed screw would progress if it were turned so as to bring the factor vector **r** into the direction of **s**, through the angle α between them. Notice that $\mathbf{A} \to -\mathbf{A}$ for $\alpha \to \alpha + \pi$. The definition is such that the order of the factors matters:

$$\mathbf{s} \times \mathbf{r} = -\mathbf{r} \times \mathbf{s}. \tag{1.9}$$

The multiplication here may therefore be called "anticommutative."

The vector product of a vector with itself obviously vanishes: $\mathbf{r} \times \mathbf{r} = \mathbf{s} \times \mathbf{s} = 0$. A vanishing of the product $\mathbf{r} \times \mathbf{s}$ need not imply that either **r** or **s** vanishes; it may merely mean that **r** and **s** are parallel or antiparallel to each other.

The result $\mathbf{A} = \mathbf{r} \times \mathbf{s}$ is perpendicular to the plane defined by **r** and **s**, and hence to each of these vector factors; consequently

$$\mathbf{r} \cdot (\mathbf{r} \times \mathbf{s}) = 0 \qquad \text{and} \qquad \mathbf{s} \cdot (\mathbf{r} \times \mathbf{s}) = 0 \tag{1.10}$$

are valid conclusions.

The relation of the components x, y, z and s_x, s_y, s_z to $A_x = (\mathbf{r} \times \mathbf{s})_x$, $A_y = (\mathbf{r} \times \mathbf{s})_y$, $A_z = (\mathbf{r} \times \mathbf{s})_z$ will be explored after the following.

Unit Vector Bases

Vectors of unit magnitude are used for pointing out directions. Thus, if $\hat{\mathbf{r}}$ is a vector in the same direction as that of **r** but with unit magnitude ($|\hat{\mathbf{r}}| = 1$), then

$$\mathbf{r} = r\hat{\mathbf{r}}, \tag{1.11}$$

and there is a separation into a magnitude and a direction factor, which is frequently convenient.

Unit vectors **i**, **j**, **k** are widely used to point out the three orthogonal directions of a cartesian reference frame. Because each has unit magnitude and each pair is perpendicular,

$$i^2 = j^2 = k^2 = 1, \qquad \mathbf{i} \cdot \mathbf{j} = \mathbf{j} \cdot \mathbf{k} = \mathbf{k} \cdot \mathbf{i} = 0. \tag{1.12}$$

Moreover, $\mathbf{i} \times \mathbf{j}$ is a unit vector in the direction of $\mathbf{k}$, and so on; hence

$$\mathbf{i} \times \mathbf{j} = \mathbf{k}, \qquad \mathbf{j} \times \mathbf{k} = \mathbf{i}, \qquad \mathbf{k} \times \mathbf{i} = \mathbf{j}. \tag{1.13}$$

Notice the "cyclic order" in which $\mathbf{i}, \mathbf{j}, \mathbf{k}$ must stand if no minus signs are to be introduced.

If $\mathbf{i}$ points out the direction of the x-axis, $\mathbf{j}$ that of y, and $\mathbf{k}$ that of z, then it is clear from Figure 1.1 that

$$\mathbf{r} = \mathbf{i}x + \mathbf{j}y + \mathbf{k}z. \tag{1.14}$$

Any vector can be similarly resolved, such as

$$\mathbf{F} = \mathbf{i}F_x + \mathbf{j}F_y + \mathbf{k}F_z, \tag{1.15}$$

to make its components explicit. The results,

$$\mathbf{i} \cdot \mathbf{F} = F_x, \qquad \mathbf{j} \cdot \mathbf{F} = F_y, \qquad \mathbf{k} \cdot \mathbf{F} = F_z, \tag{1.16}$$

are significant expressions for the projections.

The usefulness of the resolved forms, like (1.14) and (1.15), is illustrated by the development

$$\begin{aligned} \mathbf{F} \cdot \mathbf{s} &= (\mathbf{i}F_x + \mathbf{j}F_y + \mathbf{k}F_z) \cdot (\mathbf{i}s_x + \mathbf{j}s_y + \mathbf{k}s_z) \\ &= F_x s_x + F_y s_y + F_z s_z, \end{aligned} \tag{1.17}$$

which is the relation of the scalar product to the components of its vector factors promised above. Similarly,

$$\begin{aligned} \mathbf{F} \times \mathbf{s} &= \mathbf{i} \times \mathbf{i}F_x s_x + \mathbf{i} \times \mathbf{j}F_x s_y + \mathbf{i} \times \mathbf{k}F_x s_z \\ &\quad + \mathbf{j} \times \mathbf{i}F_y s_x + \mathbf{j} \times \mathbf{j}F_y s_y + \cdots \\ &= \mathbf{i}(F_y s_z - F_z s_y) + \mathbf{j}(F_z s_x - F_x s_z) + \mathbf{k}(F_x s_y - F_y s_x). \end{aligned} \tag{1.18}$$

Thus, for example, the x-component of the area vector (1.8) is $A_x = ys_z - zs_y$.

Multiple Products

The unit vector resolutions provide one way to prove (Exercise 1.5) useful results for products of three vectors like $\mathbf{A}, \mathbf{B}, \mathbf{C}$:

$$\mathbf{A} \cdot (\mathbf{B} \times \mathbf{C}) = (\mathbf{A} \times \mathbf{B}) \cdot \mathbf{C} = \mathbf{B} \cdot (\mathbf{C} \times \mathbf{A}), \tag{1.19}$$

$$\mathbf{A} \times (\mathbf{B} \times \mathbf{C}) = \mathbf{B}(\mathbf{A} \cdot \mathbf{C}) - \mathbf{C}(\mathbf{A} \cdot \mathbf{B}). \tag{1.20}$$

The first of these is sometimes described by the statement that "a dot and cross may be interchanged." In the last, it should be clear why there is no term on the right which has the direction of **A**.

1.2 THE VELOCITY

Contributions to the kinematics of a moving point begin when a position-vector is considered as a function of time: $\mathbf{r}(t)$. To correspond to the physical conception of a trajectory, this should be a continuous succession of points, forming some curve such as is indicated in Figure 1.4, but not necessarily confined to a plane. Given $\mathbf{r}(t)$, the instantaneous velocity should be calculable at any point of the trajectory.

Derivative of a Vector with Respect to a Scalar

To conform to a definition of velocity $\mathbf{v}(t)$ as a time rate of change of the position, the derivative should be defined by

$$\frac{d\mathbf{r}}{dt} \equiv \lim_{\Delta t \to 0} \frac{\Delta \mathbf{r}}{\Delta t} \equiv \lim \frac{\mathbf{r}(t + \Delta t) - \mathbf{r}(t)}{\Delta t}. \tag{1.21}$$

In the special case of time as the scalar variable, a customary notation is $d\mathbf{r}/dt \equiv \dot{\mathbf{r}}(t)$; hence $\mathbf{v}(t) = \dot{\mathbf{r}}(t)$.

The derivative of a vector with respect to a scalar is itself a vector; in the situation here, $\dot{\mathbf{r}}$ has the direction taken by $\Delta\mathbf{r}$ as it approaches zero, with $\Delta t \to 0$. This direction is *not* the same as the direction of $\mathbf{r}$ itself; rather, $\dot{\mathbf{r}}(t)$ has the direction of the tangent to the path at $\mathbf{r}(t)$.

The different direction of $\dot{\mathbf{r}}$ from that of $\mathbf{r}$ is obviously a consequence of the fact that a vector may change not only in magnitude but also in direction. The two aspects may be isolated by separating off the directional property as in (1.11), but now the notation $\mathbf{r} = \mathbf{e}r$, with $\mathbf{e} \equiv \hat{\mathbf{r}} \equiv \mathbf{r}/r$, is convenient to use. Then $\dot{\mathbf{r}} = \mathbf{e}\dot{r} + \dot{\mathbf{e}}r$. The two terms here are at right angles to each other since, like any vector defined to have a constant magnitude, $\mathbf{e}$ can change only perpendicularly to itself: from $\mathbf{e} \cdot \mathbf{e} = 1$, the result $\dot{\mathbf{e}} \cdot \mathbf{e} + \mathbf{e} \cdot \dot{\mathbf{e}} = 2\mathbf{e} \cdot \dot{\mathbf{e}} = 0$ follows.

The Radial and Angular Velocities

The rate of change $\dot{r}$ of the distance r is the magnitude of the *radial* velocity,

$$\mathbf{v}_r \equiv \mathbf{e}\dot{r}, \tag{1.22}$$

directed "along the line of sight," $\mathbf{e} \equiv \hat{\mathbf{r}}$.

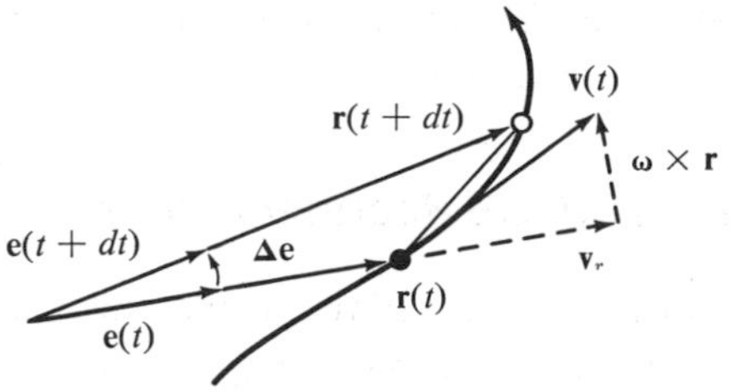

Figure 1.4

Since a unit vector by definition changes at most in direction, $\dot{\mathbf{e}}$ must amount to an angular velocity in magnitude. Referring to Figure 1.4, the magnitude $|\Delta\mathbf{e}| = \Delta\alpha$ if $\Delta\alpha$ is the angle between the unit vectors $\mathbf{e}(t)$ and $\mathbf{e}(t + \Delta t)$; then $|\dot{\mathbf{e}}| = \lim \Delta\alpha/\Delta t = \dot{\alpha}$. It is customary to represent an angular velocity by a vector $\boldsymbol{\omega}$ which has the magnitude $|\boldsymbol{\omega}| = \dot{\alpha}$, but the direction $\hat{\boldsymbol{\omega}} \equiv \boldsymbol{\omega}/\dot{\alpha}$ of the instantaneous axis about which the rotation takes place (as $\Delta t \to 0$). Hence $\hat{\boldsymbol{\omega}}$ is perpendicular to the plane defined by $\mathbf{r}(t)$ and $\mathbf{r}(t + \Delta t)$ as $\Delta t \to 0$, in which the tangent and $\mathbf{v}$ itself lie. The conclusion is that

$$\dot{\mathbf{e}} = \boldsymbol{\omega} \times \mathbf{e}, \tag{1.23}$$

since this has the direction of $\Delta\mathbf{e} \to 0$ and the proper magnitude $|\boldsymbol{\omega} \times \mathbf{e}| = \omega = \dot{\alpha}$.

In terms of the angular velocity $\boldsymbol{\omega}$ associated with the "scanning" of the trajectory from the chosen viewpoint O, the velocity expression $\mathbf{v} = \mathbf{e}\dot{r} + \dot{\mathbf{e}}r$ becomes

$$\mathbf{v} = \mathbf{v}_r + \boldsymbol{\omega} \times \mathbf{r}, \tag{1.24}$$

with $\mathbf{v}_r$ along the "line of sight" and $\boldsymbol{\omega} \times \mathbf{r}$ transverse to it (Figure 1.4). The radial and angular velocities here should be individually calculable when $\mathbf{r}(t)$ is given. The expression $v_r = dr(t)/dt$ is explicit enough for the radial velocity but $\boldsymbol{\omega}$ is only implicit in (1.23). However, it is equally plain that

$$\mathbf{e} \times \mathbf{v} = \mathbf{e} \times (\boldsymbol{\omega} \times \mathbf{r}) = r\boldsymbol{\omega}, \tag{1.25}$$

which gives $\boldsymbol{\omega}$ explicitly for a given $\mathbf{v} \equiv \dot{\mathbf{r}}$. It is essential for this relation that $\mathbf{r}$ be perpendicular to $\boldsymbol{\omega}$.

Cartesian Representation of the Velocity

The cartesian components of the velocity are easily deducible from taking the time-derivative of $\mathbf{r} = \mathbf{i}x + \mathbf{j}y + \mathbf{k}z$, which yields

$$\mathbf{v} \equiv \dot{\mathbf{r}} = \mathbf{i}\dot{x} + \mathbf{j}\dot{y} + \mathbf{k}\dot{z}, \tag{1.26}$$

since **i**, **j**, **k** are constant in direction as well as magnitude for a fixed frame. Thus, simply, $v_x = \dot{x}$, $v_y = \dot{y}$, and $v_z = \dot{z}$.

Polar and Cylindrical Representations

When the motion is restricted to a plane, this may be taken to be the $z = 0$ plane of Figure 1.1, and the point may be located in it by giving the "plane polar coordinates" r, φ. In effect, a moving reference frame is now being used, pointed out by orthogonal unit vectors $\mathbf{e}_r$, $\mathbf{e}_\varphi$, with $\mathbf{e}_r$ always directed toward the point as it moves. Reference to Fig. 1.5 makes it easy to understand that

$$\dot{\mathbf{e}}_r = \mathbf{e}_\varphi \dot{\varphi} \qquad \text{and} \qquad \dot{\mathbf{e}}_\varphi = -\mathbf{e}_r \dot{\varphi}, \tag{1.27}$$

and then

$$\begin{aligned} \mathbf{r}(t) &= \mathbf{e}_r r, \\ \mathbf{v}(t) \equiv \dot{\mathbf{r}} &= \mathbf{e}_r \dot{r} + \mathbf{e}_\varphi r\dot{\varphi}. \end{aligned} \tag{1.28}$$

Thus the "polar" components of the velocity are $v_r = \dot{r}$ and $v_\varphi = r\dot{\varphi}$, quite in consistency with (1.24).

The planar coordinates r, φ may also be retained for describing a spatial motion, merely by supplementation with a z-coordinate of the point perpendicular to the plane of r, φ. (Now $r = (x^2 + y^2)^{1/2}$ rather than $r = (x^2 + y^2 + z^2)^{1/2}$ as for the spherical coordinate trio of Figure 1.1). The results are the "cylindrical coordinates" in which

$$\begin{aligned} \mathbf{r} &= \mathbf{e}_r r + \mathbf{k}z, \\ \mathbf{v} \equiv \dot{\mathbf{r}} &= \mathbf{e}_r \dot{r} + \mathbf{e}_\varphi r\dot{\varphi} + \mathbf{k}\dot{z}. \end{aligned} \tag{1.29}$$

Notice that $r \neq |\mathbf{r}|$ in the notation here. These expressions are most useful in

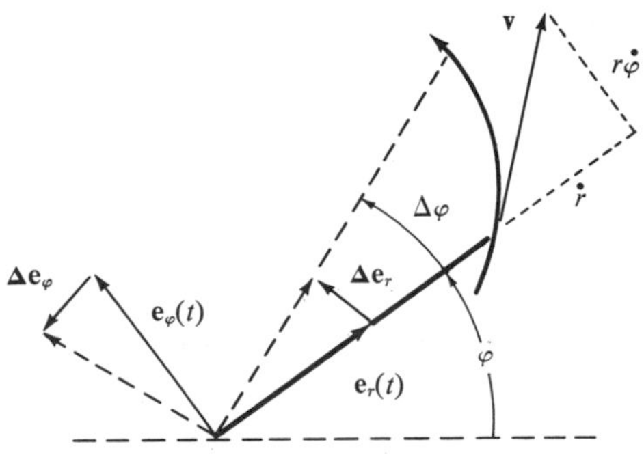

Figure 1.5

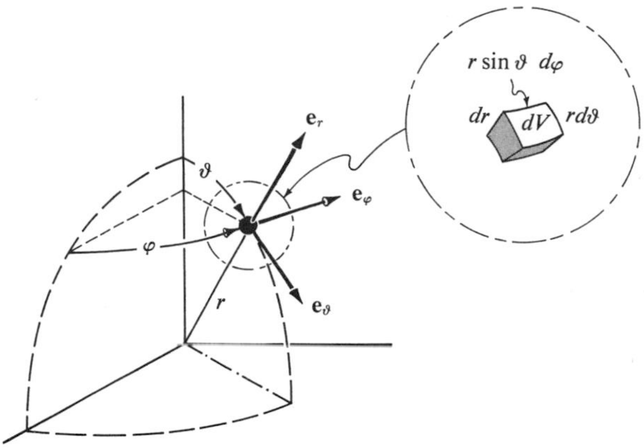

Figure 1.6

problems of axial "symmetry" when the z-direction is taken along the symmetry axis.

Spherical Representation

Use of the spherical coordinates r, ϑ, φ is tantamount to adopting a rotating reference frame determined by orthogonal unit vectors $\mathbf{e}_r$, $\mathbf{e}_\vartheta$, $\mathbf{e}_\varphi$, which respectively point out the directions of increasing r, ϑ, and φ. For anyone familiar with the spherical volume element,

$$dV \equiv r^2\, dr\, d(\cos\vartheta)\, d\varphi, \tag{1.30}$$

it is not difficult to find that the line-element vector is

$$\mathbf{dr} = \mathbf{e}_r\, dr + \mathbf{e}_\vartheta\, r\, d\vartheta + \mathbf{e}_\varphi\, r \sin\vartheta\, d\varphi, \tag{1.31}$$

and hence

$$\mathbf{v} \equiv \dot{\mathbf{r}} = \mathbf{e}_r \dot{r} + \mathbf{e}_\vartheta r\dot{\vartheta} + \mathbf{e}_\varphi r \sin\vartheta\, \dot{\varphi}. \tag{1.32}$$

This follows from taking the derivative of $\mathbf{r} = \mathbf{e}_r r$ if it is recognized that

$$\begin{aligned} \dot{\mathbf{e}}_r &= \mathbf{e}_\vartheta \dot{\vartheta} + \mathbf{e}_\varphi \dot{\varphi} \sin\vartheta, \\ \dot{\mathbf{e}}_\vartheta &= -\mathbf{e}_r \dot{\vartheta} + \mathbf{e}_\varphi \dot{\varphi} \cos\vartheta, \\ \dot{\mathbf{e}}_\varphi &= -\mathbf{e}_r \dot{\varphi} \sin\vartheta - \mathbf{e}_\vartheta \dot{\varphi} \cos\vartheta, \end{aligned} \tag{1.33}$$

which are more intricate than (1.27) but not difficult to derive. A particularly significant derivation will be given in §1.4.

The most rudimentary vector calculus, the taking of a first derivative of a vector function of a scalar has been illustrated here. The next section will illustrate second derivatives; the differentiation and integration of functions of a *vector* variable will be taken up in §2.5, where they will be put to immediate use.

1.3 ACCELERATION

Higher and higher derivatives of a trajectory, $\mathbf{r}(t)$, could be investigated as measures of more and more subtle variations in it. However, interest generally stops at the second derivative, the acceleration

$$\mathbf{a}(t) \equiv \dot{\mathbf{v}}(t) \equiv \ddot{\mathbf{r}}(t), \tag{1.34}$$

since this is given directly by the force imposed, as will be seen in the next chapter.

"Intrinsic" Representation

To discuss the velocity (1.24), it was resolved into a radial component along the "line of sight" and a transverse rotation of the latter. For the acceleration, it may be more illuminating to resolve it into components tangential and normal to the trajectory, an "intrinsic" representation. Let s be distance measured along the curve of the trajectory, so that $v = \dot{s}$ is the magnitude of the velocity and

$$\mathbf{v} = \boldsymbol{\tau}\dot{s}, \tag{1.35}$$

where $\boldsymbol{\tau}$ is a unit vector tangential to the trajectory at the point of $\mathbf{v}$. Then $\dot{\mathbf{v}} = \boldsymbol{\tau}\ddot{s} + \dot{\boldsymbol{\tau}}\dot{s}$ will plainly represent a separation into tangential and normal components of acceleration, since $\dot{\boldsymbol{\tau}}$ can only be perpendicular to $\boldsymbol{\tau}$, as for the rate of change of any unit vector. Actually, there is a whole plane of normals to the tangent at any point of the curve. A unique one, the "principal normal" with direction denoted by a unit vector $\mathbf{n}$, lies in a "plane of successive tangents" (that is, a plane defined by $\boldsymbol{\tau}(t)$ and $\boldsymbol{\tau}(t + \Delta t)$ as $\Delta t \to 0$). The corresponding "successive principal normals" $\mathbf{n}(t)$, $\mathbf{n}(t + \Delta t)$ will intersect in some point C as $\Delta t \to 0$, and the distance ϱ from the "center of curvature" C to the moving point is called the "radius of curvature." Now, the displacement $\Delta s = \dot{s}\Delta t$ subtends an angle $\Delta\psi = \Delta s/\varrho$ at the center of curvature, and $|\Delta\boldsymbol{\tau}| = \Delta\psi$, as Figure 1.7 helps make clear. Thus $\dot{\boldsymbol{\tau}} = \lim \Delta\boldsymbol{\tau}/\Delta t$ has the direction of $\mathbf{n}$ and the magnitude $|\dot{\boldsymbol{\tau}}| = (\Delta\psi/\Delta s)\dot{s} = \dot{s}/\varrho$. As a consequence,

$$\mathbf{a} = \boldsymbol{\tau}\dot{v} + \mathbf{n}v^2/\varrho, \tag{1.36}$$

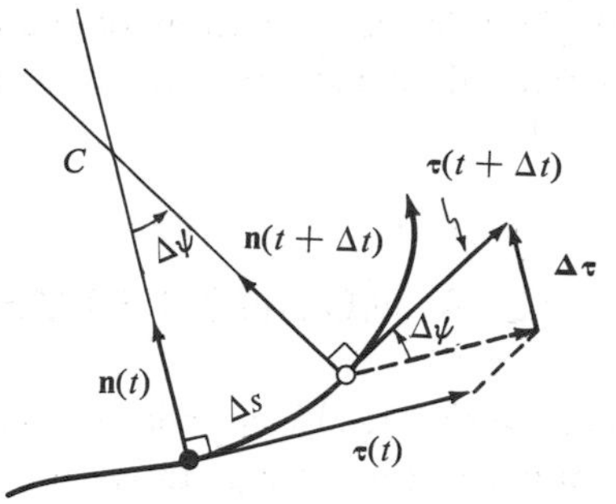

Figure 1.7

where $\dot{v} \equiv \ddot{s}$. This makes specific that, to give a point nonuniform motion along a curved trajectory, not only must an acceleration $\ddot{s}$ along the curve be provided, but also a so-called "centripetal acceleration," v^2/ϱ, normal to the curve, when its radius of curvature is finite.

Cartesian Representation

In the same way that the cartesian components of velocity $v_x = \dot{x}$, $v_y = \dot{y}$, $v_z = \dot{z}$ were derived (1.26), it is easy to see that

$$a_x = \ddot{x}, \qquad a_y = \ddot{y}, \qquad a_z = \ddot{z} \tag{1.37}$$

are the cartesian components of acceleration, relative to a fixed frame.

Cylindrical Representation

This can be obtained formally as the time-derivative of the velocity resolution (1.29), by making use of (1.27):

$$a_r = \ddot{r} - r\dot{\varphi}^2, \qquad a_\varphi = r\ddot{\varphi} + 2\dot{r}\dot{\varphi}, \qquad a_z = \ddot{z}. \tag{1.38}$$

Only the first two of these components are needed in a plane-polar representation. The occurrence of the term $-r\dot{\varphi}^2 = -v_\varphi^2/r$ is perhaps not surprising in view of (1.36), but the significance of the term $2\dot{r}\dot{\varphi}$, called a "Coriolis acceleration," needs further exploration (§1.4).

Spherical Representation

From the velocity expression (1.32) and the angular velocities (1.33) it follows that

$$\begin{aligned} a_r &= \ddot{r} & &- r\dot{\vartheta}^2 - r\sin^2\vartheta\,\dot{\varphi}^2, \\ a_\vartheta &= r\ddot{\vartheta} & &- r\sin\vartheta\cos\vartheta\,\dot{\varphi}^2 + 2\dot{r}\dot{\vartheta}, \\ a_\varphi &= r\sin\vartheta\,\ddot{\varphi} + 2\dot{r}\dot{\varphi}\sin\vartheta + 2r\dot{\vartheta}\,\dot{\varphi}\cos\vartheta. & & \end{aligned} \tag{1.39}$$

The significance of all these terms, as well as of the expressions (1.36) and (1.38), is clarified by referral to rotating frames of reference (in the next section).

1.4 REFERENCE TO A MOVING FRAME

Suppose an observer using a fixed (relative to himself) frame $O, \mathbf{i}, \mathbf{j}, \mathbf{k}$ considers the relation of his observations to those of another observer using a frame $O', \mathbf{I}, \mathbf{J}, \mathbf{K}$, which has some motion relative to the "fixed" frame.

The motion of the frame can be specified by giving the trajectory

$$\mathbf{r}_0(t) = \mathbf{i}x_0(t) + \mathbf{j}y_0(t) + \mathbf{k}z_0(t) \tag{1.40}$$

of the origin O' and the angular velocity $\mathbf{\Omega}(t)$ of the frame about O'. From the experience with the rotation of a unit vector (1.23), it is to be expected that

$$\dot{\mathbf{I}} = \mathbf{\Omega} \times \mathbf{I}, \qquad \dot{\mathbf{J}} = \mathbf{\Omega} \times \mathbf{J}, \qquad \dot{\mathbf{K}} = \mathbf{\Omega} \times \mathbf{K}. \tag{1.41}$$

However, (1.23) presented a simpler situation in that the angular velocity vector $\boldsymbol{\omega}$ was perpendicular to the rotating vector $\mathbf{e}$. Figure 1.8 makes it clear that for any vector $\mathbf{R}_0$ fixed to the rotating frame

$$\dot{\mathbf{R}}_0 = \mathbf{\Omega} \times \mathbf{R}_0, \tag{1.42}$$

and (1.41) is merely a collection of special cases: for the unit vectors pointing out the rotating frame.

It is fairly obvious, but should be checked specifically, that a rotation is the only motion possible to any rigid frame fixed to O', and that such a motion maintains the orthogonality,

$$\mathbf{I} \cdot \mathbf{J} = \mathbf{J} \cdot \mathbf{K} = \mathbf{K} \cdot \mathbf{I} = 0, \tag{1.43}$$

of the unit vectors. For this purpose, consider any point $\mathbf{R}_0$ fixed to the frame,

$$\begin{aligned} \mathbf{R}_0(t) &= \mathbf{I}(t)X_0 + \mathbf{J}(t)Y_0 + \mathbf{K}(t)Z_0, \\ \dot{\mathbf{R}}_0 &= \dot{\mathbf{I}}X_0 + \dot{\mathbf{J}}Y_0 + \dot{\mathbf{K}}Z_0, \end{aligned} \tag{1.44}$$

where X_0, Y_0, Z_0 are the constant projections to be expected of a vector affixed to the frame. Because $\mathbf{I} \cdot \mathbf{I} = 1$, $\dot{\mathbf{I}} \cdot \mathbf{I} = 0$, signifying that $\dot{\mathbf{I}}$ can have no projection on $\mathbf{I}$ if $\mathbf{I}$ is to be a unit vector. Call the remaining, generally non-vanishing, components $(\dot{\mathbf{I}})_Y = \Omega_Z(t)$ and $(\dot{\mathbf{I}})_Z = -\Omega_Y(t)$, without prejudice as

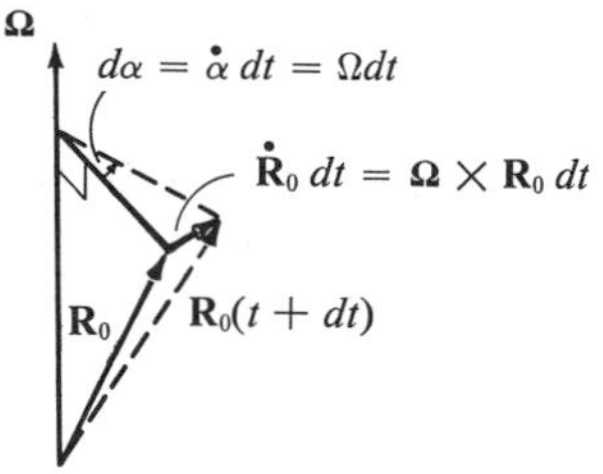

Figure 1.8

to the interpretation of these quantities. Thus,

$$\dot{\mathbf{I}} = \mathbf{J}\Omega_Z - \mathbf{K}\Omega_Y.$$

Next, because $\mathbf{I}\cdot\mathbf{J}=0$, $\mathbf{I}\cdot\dot{\mathbf{J}} = -\dot{\mathbf{I}}\cdot\mathbf{J} = -\Omega_Z$; hence $(\dot{\mathbf{J}})_X = -\Omega_Z$ is a quantity already introduced. The yet unnamed Z-component of $\dot{\mathbf{J}}$ will be called Ω_X, so that

$$\dot{\mathbf{J}} = -\mathbf{I}\Omega_Z + \mathbf{K}\Omega_X.$$

From $\mathbf{I}\cdot\mathbf{K} = \mathbf{J}\cdot\mathbf{K} = 0$ it follows similarly that

$$\dot{\mathbf{K}} = \mathbf{I}\Omega_Y - \mathbf{J}\Omega_X.$$

This exhausts the consequences of the rigid orthogonality and converts (1.44) into

$$\dot{\mathbf{R}}_0 = \mathbf{I}(\Omega_Y Z_0 - \Omega_Z Y_0) + \mathbf{J}(\Omega_Z X_0 - \Omega_X Z_0) + \mathbf{K}(\Omega_X Y_0 - \Omega_Y X_0),$$

which is just (1.42) if the $\Omega_{X,Y,Z}$ are taken to be components of a vector $\mathbf{\Omega}(t)$. Such a relation may always be interpreted as a rotation with an instantaneous angular velocity $\mathbf{\Omega}(t)$.

The effect of the frame rotation on any vector **A**, not necessarily fixed to the frame but resolvable on it, may be seen from:

$$\begin{aligned}\mathbf{A} &= \mathbf{I}A_X + \mathbf{J}A_Y + \mathbf{K}A_Z,\\ \dot{\mathbf{A}} &= \mathbf{I}\dot{A}_X + \mathbf{J}\dot{A}_Y + \mathbf{K}\dot{A}_Z + \dot{\mathbf{I}}A_X + \dot{\mathbf{J}}A_Y + \dot{\mathbf{K}}A_Z.\end{aligned} \tag{1.45}$$

The last three terms may be written

$$\mathbf{\Omega} \times (\mathbf{I}A_X + \mathbf{J}A_Y + \mathbf{K}A_Z) = \mathbf{\Omega} \times \mathbf{A},$$

while the preceding three terms are appropriate to represent as a *partial* time-derivative, $\partial\mathbf{A}/\partial t$, "holding **I**, **J**, **K** fixed." Thus

$$\dot{\mathbf{A}} = \left[\frac{\partial}{\partial t} + \mathbf{\Omega}\times\mathbf{A}\right] \tag{1.46}$$

is the operation which separates the effect of the frame rotation from changes relative to the frame.

For the instantaneous angular velocity vector $\mathbf{\Omega}(t)$ itself,

$$\dot{\mathbf{\Omega}} \equiv \frac{d\mathbf{\Omega}}{dt} = \frac{\partial \mathbf{\Omega}}{\partial t}, \tag{1.47}$$

since $\mathbf{\Omega} \times \mathbf{\Omega} \equiv 0$. The angular acceleration is the same relative to either the fixed or the moving frame. This is not surprising, in view of the fact that an axis by definition does not share in the rotational motion about itself. (A rotation axis may indeed rotate, as is representable by $\partial\hat{\Omega}/\partial t \neq 0$, but none of this is "imparted" by the rotation about itself.)

"Absolute" and "Internal" Velocities

Consider a point ("particle") that moves relative to the moving frame, in which it has the position vector from the origin O': $\mathbf{R} = \mathbf{I}X + \mathbf{J}Y + \mathbf{K}Z$. Rates of change $\dot{X}$, $\dot{Y}$, $\dot{Z}$ may exist, and

$$\mathbf{V} = \mathbf{I}\dot{X} + \mathbf{J}\dot{Y} + \mathbf{K}\dot{Z} \equiv \frac{\partial \mathbf{R}}{\partial t} \tag{1.48}$$

may be regarded as the velocity relative to or "internal to" the moving frame (that is, as though O', $\mathbf{I}$, $\mathbf{J}$, $\mathbf{K}$ were fixed). Relative to the "absolute" frame —the one in which O', $\mathbf{I}$, $\mathbf{J}$, $\mathbf{K}$ have the motions given by $\dot{\mathbf{r}}_0 = \mathbf{v}_0$ and $\mathbf{\Omega}$—the same particle has

$$\begin{aligned} \mathbf{r} &= \mathbf{i}x + \mathbf{j}y + \mathbf{k}z, \\ \mathbf{v} \equiv \dot{\mathbf{r}} &= \mathbf{i}\dot{x} + \mathbf{j}\dot{y} + \mathbf{k}\dot{z}. \end{aligned} \tag{1.49}$$

The relation between the internal velocity $\mathbf{V}$ and the "absolute" velocity $\mathbf{v}$ is desired.

The relation is easy to obtain as the time-derivative of

$$\mathbf{r} = \mathbf{r}_0 + \mathbf{R} \rightarrow \dot{\mathbf{r}} = \dot{\mathbf{r}}_0 + \dot{\mathbf{R}}, \tag{1.50}$$

bearing in mind the operation (1.46):

$$\mathbf{v} = \mathbf{V} + \mathbf{v}_0 + \mathbf{\Omega} \times \mathbf{R}. \tag{1.51}$$

Thus the total velocity consists not only of the internal one, $\mathbf{V}$, but has added to it the translational velocity, $\mathbf{v}_0 \equiv \dot{\mathbf{r}}_0$ of the moving origin, plus a part $\mathbf{\Omega} \times \mathbf{R}$ imparted by the rotation of the frame.

The Accelerations

Let the acceleration of the particle in its trajectory as viewed from O' **I**, **J**, **K** be

$$\mathbf{A} = \mathbf{I}\ddot{X} + \mathbf{J}\ddot{Y} + \mathbf{K}\ddot{Z}. \tag{1.52}$$

The relation between this and the "absolute" acceleration $\mathbf{a} = \mathbf{i}\ddot{x} + \mathbf{j}\ddot{y} + \mathbf{k}\ddot{z}$ is desired.

The time-derivative of the total velocity (1.51) yields

$$\mathbf{a} \equiv \dot{\mathbf{v}} = \dot{\mathbf{v}}_0 + \left(\frac{d}{dt}\right)[\mathbf{V} + \boldsymbol{\Omega} \times \mathbf{R}].$$

After resolving **V** and $\boldsymbol{\Omega} \times \mathbf{R}$ on the moving axes **I**, **J**, **K**, the operation (1.46) may be applied to give

$$\frac{d\mathbf{V}}{dt} = \mathbf{A} + \boldsymbol{\Omega} \times \mathbf{V},$$

$$\left(\frac{d}{dt}\right)\boldsymbol{\Omega} \times \mathbf{R} = \dot{\boldsymbol{\Omega}} \times \mathbf{R} + \boldsymbol{\Omega} \times \mathbf{V} + \boldsymbol{\Omega} \times [\boldsymbol{\Omega} \times \mathbf{R}],$$

since the instantaneous axis does not share in the rotation about itself (1.47). Thus,

$$\mathbf{a} = \mathbf{A} + \dot{\mathbf{v}}_0 + \dot{\boldsymbol{\Omega}} \times \mathbf{R} + \boldsymbol{\Omega} \times [\boldsymbol{\Omega} \times \mathbf{R}] + 2\boldsymbol{\Omega} \times \mathbf{V}. \tag{1.53}$$

There are superposed on the "internal" acceleration **A**, the several components: (1) $\dot{\mathbf{v}}_0$, the acceleration of the moving reference origin. (2) $\dot{\boldsymbol{\Omega}} \times \mathbf{R}$, arising from any angular accelerations of the moving frame, (3) $\boldsymbol{\Omega} \times (\boldsymbol{\Omega} \times \mathbf{R})$, a centripetal acceleration, (4) $2\boldsymbol{\Omega} \times \mathbf{V}$, a Coriolis acceleration.

The Centripetal Acceleration

This is directed toward the rotation axis $\hat{\boldsymbol{\Omega}}$, as Figure 1.9 should help make clear. It is an acceleration which must be supplied to the particle to maintain its distance from the axis as for a rotation. That distance, normal to the axis, is $\mathbf{R} - \hat{\boldsymbol{\Omega}}(\hat{\boldsymbol{\Omega}} \cdot \mathbf{R})$ if $\hat{\boldsymbol{\Omega}}$ is the unit vector giving the direction of the axis, and

$$\boldsymbol{\Omega} \times (\boldsymbol{\Omega} \times \mathbf{R}) = -\Omega^2[\mathbf{R} - \hat{\boldsymbol{\Omega}}(\hat{\boldsymbol{\Omega}} \cdot \mathbf{R})]$$

is the acceleration needed.

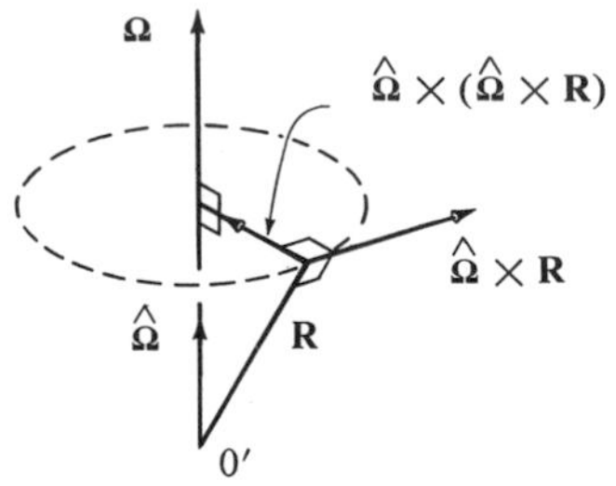

Figure 1.9

The Coriolis Acceleration

This acceleration must be supplied to the particle when it is desired to maintain it at a given velocity **V** relative to a rotating frame. For example, to keep the particle moving steadily on a normal outward from the axis, a Coriolis acceleration $2\Omega V$ into the direction of the rotational displacements must be supplied. When walking outward on the platform of a merry-go-round, a man must "lean" backwards to provide centripetal acceleration (counteract "centrifugal force") and also to his left (if that is the direction of the rotational displacements) in order to help rotate himself with the platform.

The origin of the factor 2 in $2\boldsymbol{\Omega} \times \mathbf{V}$ is particularly clear in the above example (Figure 1.10). First, in a time Δt, the radial distance is increased by $V\,\Delta t$; hence the transversal velocity $\boldsymbol{\Omega} \times \mathbf{R}$ by $\boldsymbol{\Omega} \times \mathbf{V}\,\Delta t$. Second, the *direction* of the radius changes by the angle $\Omega\,\Delta t$; hence a radial **V** undergoes an additional transversal change $\boldsymbol{\Omega} \times \mathbf{V}\,\Delta t$.

The "Intrinsic" Representation

The resolution (1.36) of the acceleration refers to a reference frame $\boldsymbol{\tau}, \mathbf{n}, \boldsymbol{\tau} \times \mathbf{n}$ centered on the moving point so that $\mathbf{V} = 0$ in it. The frame rotates with $\boldsymbol{\Omega} = \boldsymbol{\tau} \times \mathbf{n}v/\varrho = \mathbf{v} \times \mathbf{n}/\rho$, but $\mathbf{R} = 0$, and hence $\mathbf{v} = \mathbf{v}_0 = \boldsymbol{\tau}\dot{s}$. The expression for the "absolute acceleration" (1.53) is replaced by $\mathbf{a} = \dot{\mathbf{v}}_0 = \boldsymbol{\tau}\ddot{s} + \boldsymbol{\Omega} \times \boldsymbol{\tau}\dot{s} = \boldsymbol{\tau}\dot{v} + \boldsymbol{\tau} \times (\mathbf{n} \times \boldsymbol{\tau})v^2/\varrho$, according to the operation (1.46), and the result reduces to (1.36), as was expected. Notice that a Coriolis acceleration is not separated out in a frame following the trajectory as it does in the "intrinsic" representation, as it also is not in the cartesian representations.

Polar Representation

The directions of increase $\mathbf{e}_r$, $\mathbf{e}_\varphi$ of the plane polar coordinates r, φ define a frame (Figure 1.5), which may be centered at the origin of **r** and rotating with $\boldsymbol{\Omega} = \mathbf{k}\dot{\varphi}$. In it, $\mathbf{V} = \mathbf{e}_r\dot{r}$ and $\mathbf{A} = \mathbf{e}_r\ddot{r}$, since $\mathbf{e}_r$ persistently points at the

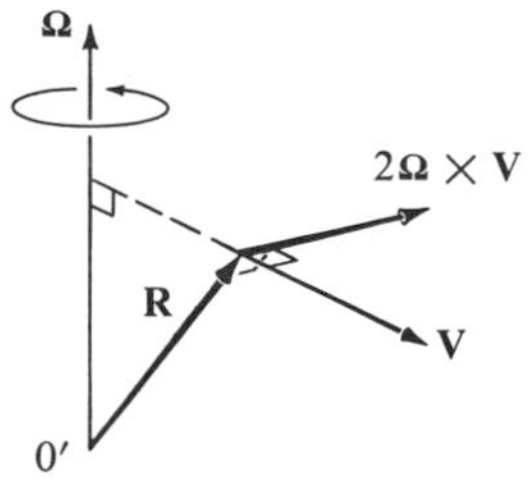

Figure 1.10

moving particle. With $\mathbf{r}_0 = 0$ and $\mathbf{v}_0 = 0$, expressions (1.51) and (1.53) become

$$\mathbf{v} = \mathbf{e}_r \dot{r} + \mathbf{k}\dot{\varphi} \times \mathbf{e}_r r = \mathbf{e}_r \dot{r} + \mathbf{e}_\varphi r\dot{\varphi},$$

$$\mathbf{a} = \mathbf{e}_r \ddot{r} + \mathbf{k} \times \mathbf{e}_r r\ddot{\varphi} + \mathbf{k} \times (\mathbf{k} \times \mathbf{e}_r) r\dot{\varphi}^2 + 2(\mathbf{k} \times \mathbf{e}_r)\dot{r}\dot{\varphi},$$

equivalent to results (1.28) and (1.38). The origins of the terms $-\mathbf{e}_r r\dot{\varphi}^2$ as centripetal acceleration and of $\mathbf{e}_\varphi 2\dot{r}\dot{\varphi}$ in Coriolis acceleration are now plain. It is also plain that the rotational effects in (1.51) and (1.53) cannot influence a component $\mathbf{k}\ddot{z}$ of acceleration parallel to the rotation axis, arising in the cylindrical representation.

Spherical Representation

The spherical coordinate frame $\mathbf{e}_r$, $\mathbf{e}_\vartheta$, $\mathbf{e}_\varphi$ of (1.32) and (1.33) rotates with the angular velocity

$$\begin{aligned}\mathbf{\Omega} &= \mathbf{e}_\varphi \dot{\vartheta} + \mathbf{k}\dot{\varphi} \\ &= \mathbf{e}_\varphi \dot{\vartheta} + (\mathbf{e}_r \cos\vartheta - \mathbf{e}_\vartheta \sin\vartheta)\dot{\varphi}. \qquad (1.54)\end{aligned}$$

This, together with $\mathbf{V} = \mathbf{e}_r \dot{r}$, $\mathbf{A} = \mathbf{e}_r \ddot{r}$, as befits taking $\mathbf{r}_0 = \mathbf{v}_0 = 0$, makes (1.51) and (1.53) yield just the results (1.32) and (1.39). Of course, the relations (1.33) are just the special cases of the velocities

$$\dot{\mathbf{e}}_r = \mathbf{\Omega} \times \mathbf{e}_r, \qquad \dot{\mathbf{e}}_\vartheta = \mathbf{\Omega} \times \mathbf{e}_\vartheta, \qquad \dot{\mathbf{e}}_\varphi = \mathbf{\Omega} \times \mathbf{e}_\varphi .$$

1.5 TRAJECTORIES OR ORBITS

When a motion is described by a function $\mathbf{r}(t)$, this may be regarded as a parametric description of a geometric object: the trajectory. That is already evident from the example of uniform motion in a straight line (Figure 1.11). In this case

$$\mathbf{r} = \mathbf{r}_0 + \mathbf{v}_0 t, \qquad (1.55)$$

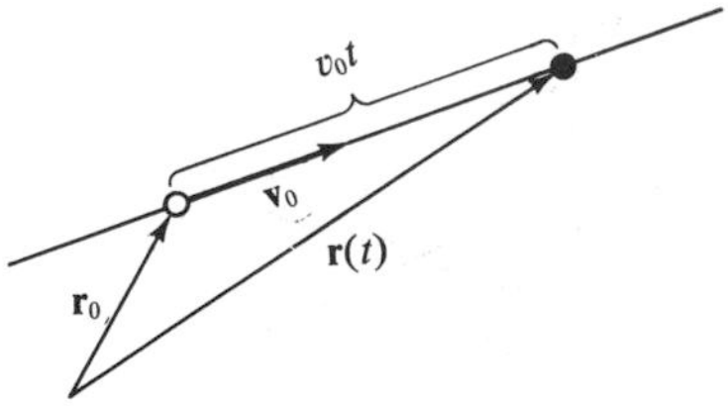

Figure 1.11

where $\mathbf{r}_0$ is the initial ($t = 0$) position and $\mathbf{v}_0$ is the constant velocity.

Trajectories under Constant Acceleration

Studies of motion under constant acceleration, $\mathbf{a} \equiv \mathbf{g}$, acquired importance when Galileo found that this seems to characterize any otherwise undisturbed, but only "gravitationally accelerated," motion near the surface of the earth. Now $\ddot{\mathbf{r}} = \mathbf{g}$, a constant, implies $\dot{\mathbf{r}} = \mathbf{g}t +$ an arbitrary constant vector, $\mathbf{v}_0$. In turn, $\mathbf{r} = \frac{1}{2}\mathbf{g}t^2 + \mathbf{v}_0 t +$a second arbitrary constant vector, $\mathbf{r}_0$:

$$\mathbf{r}(t) = \mathbf{r}_0 + \mathbf{v}_0 t + \tfrac{1}{2}\mathbf{g}t^2 \tag{1.56}$$

is the modification of the uniform motion (1.55) generated by a constant acceleration.

To get a description of the trajectory more familiar in analytic geometry, adopt a cartesian frame with $\mathbf{r}_0$ as origin (hence $\mathbf{r}_0 = 0$) and having its z-direction opposite to that of $\mathbf{g}$ (hence $g_z = -g$). Moreover, rotate the cartesian x, z plane about the z-axis until it contains $\mathbf{v}_0$ (hence $v_{0y} = 0$). Now the vector equation (1.56) has only x and z components, these being

$$\begin{aligned} x &= (v_0 \cos \alpha)t, \\ z &= (v_0 \sin \alpha)t - \tfrac{1}{2}gt^2, \end{aligned} \tag{1.57}$$

with α the "angle of elevation" of $\mathbf{v}_0$ from the horizontal. Such an analysis leads to regarding the motion as a superposition of two independent (orthogonal) linear motions, a uniform horizontal one and a uniformly accelerated (and decelerated) vertical one. The time serves as a parameter for a trajectory defined by (1.57); it can be eliminated between the two equations to give

$$\begin{aligned} z(x) &= x \tan \alpha - \tfrac{1}{2}g(x/v_0 \cos \alpha)^2 \\ &= (\tan \alpha)x[1 - (x/R)], \end{aligned} \tag{1.58}$$

if $R \equiv 2(v_0^2/g) \sin \alpha \cos \alpha = (v_0^2/g) \sin 2\alpha$. This should be recognized as defining a parabola with "range" (where the trajectory comes down to the initial height) R, and a maximum at $x = \frac{1}{2}R$. Many predictions about the motion can

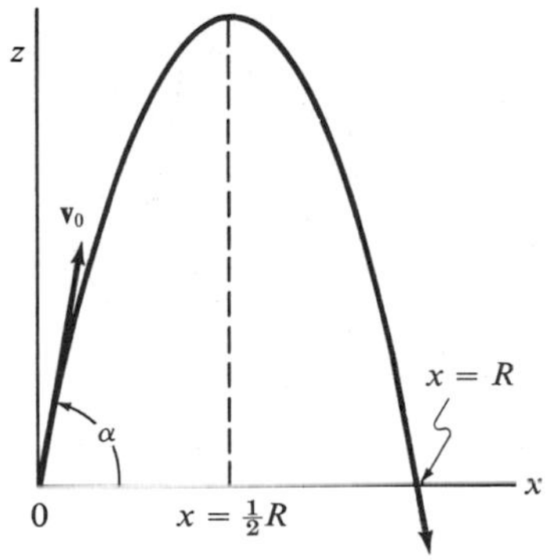

Figure 1.12

be derived from this expression: the maximum height reached will be $z(\frac{1}{2}R) = \frac{1}{4}R \tan \alpha$, and the elevation angle needed for maximum range is $\alpha = \pi/4$.

Kepler Orbits

The achievement which perhaps gave the greatest impetus to the theory of classical mechanics was Newton's demonstration of a relationship between the gravitational acceleration **g** and the Laws of Planetary Motion found by Kepler to describe Tycho Brahe's observations on the planets. The account of Newton's achievement will be left to §4.3; here only Kepler's geometrical and kinematical laws will be formulated.

According to Kepler's first law, each planet traces out an ellipse in space, with the sun at one elliptic focus. To express this mathematically, a precise definition of an ellipse must be used. One definition describes it as a locus traced by a point which maintains a constant sum of distances, $r + r' = 2a$, from two "focal points" separated by a fixed distance $2c$. In terms of plane polar coordinates centered on the focus occupied by the sun, $r' = (r^2 + 4c^2 - 4cr \cos \varphi)^{1/2}$. Solving $r + r' = 2a$ for $r(\varphi)$ yields

$$r(\varphi) = b^2/(a - c \cos \varphi), \tag{1.59}$$

where $b^2 = a^2 - c^2$. Clearly, a and b are the major and minor (maximum and minimum) radii from the center half-way between the foci; $r(0) = a + c$ and $r(\pi) = a - c$ correctly. If cartesian coordinates with origin at the center of the ellipse are used, $x = r \cos \varphi - c$ and $y = r \sin \varphi$. Substitution into (1.59) then gives

$$\frac{x^2}{a^2} + \frac{y^2}{b^2} = 1, \tag{1.60}$$

which may be a more familiar description of an ellipse.

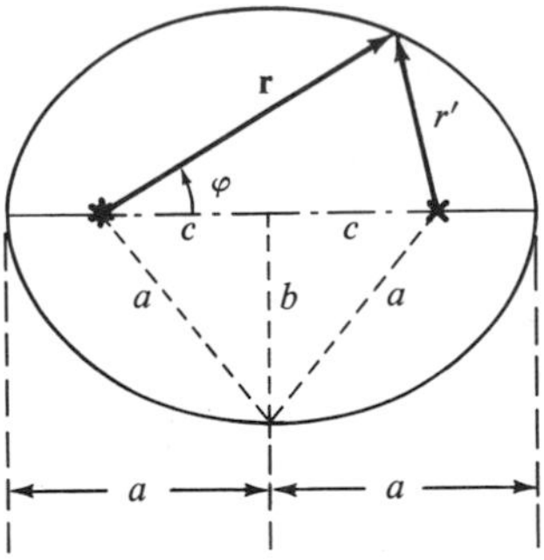

Figure 1.13

Kepler's second law states that the line from sun to planet sweeps over equal areas in equal times; that is, the so-called "sectorial velocity" is constant. Figure 1.14 makes it plain that the triangular sector element of area subtended by an increment of angle $d\varphi$ is just $dA = \frac{1}{2}r^2\, d\varphi$, and hence the law may be expressed as

$$\dot{A} = \tfrac{1}{2}r^2\dot{\varphi}, \text{ constant.} \tag{1.61}$$

This can be used to obtain a kinematical relation between the coordinates and time through integrating

$$dt = d\varphi/\dot{\varphi} = (\dot{A})^{-1}\tfrac{1}{2}r^2\, d\varphi = (b^4/2\dot{A})\, d\varphi/(a - c\cos\varphi)^2. \tag{1.62}$$

Since the entire area of the ellipse is $A = \int_0^{2\pi} \frac{1}{2}r^2\, d\varphi = \pi ab$, the period of one revolution (the "year") is

$$T = \pi ab/\dot{A}, \tag{1.63}$$

giving the sectorial velocity $\dot{A}$ in terms of the year, T, and the dimensions of the orbit a, b.

A third law found by Kepler was that a proportionality

$$T^2 \sim a^3 \tag{1.64}$$

exists between the indicated powers of the periods and the major orbital radii

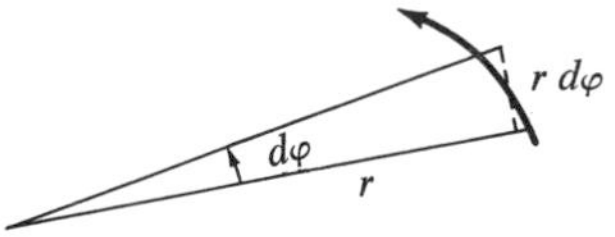

Figure 1.14

of the various planets. Giving explanations of these laws on a basis which also comprehends gravitation to the earth was Newton's great achievement (§4.3).

EXERCISES*

1.1. Investigate the significance of the two vectors $\mathbf{e}(\mathbf{a} \cdot \mathbf{e})$ and $\mathbf{e} \times (\mathbf{a} \times \mathbf{e})$ in relation to their sum, when $\mathbf{e}$ is a unit vector and $\mathbf{a}$ an arbitrary one.

1.2. Suppose it is known about a vector $\mathbf{a}$ only that its projection on a given direction $\mathbf{e}$ is $s = \mathbf{e} \cdot \mathbf{a}$, and that $\mathbf{e} \times \mathbf{a} = \mathbf{c}$. Show that this information is sufficient to determine $\mathbf{a}$ by expressing it in terms of $\mathbf{e}$, s, and $\mathbf{c}$.

1.3. Besides the "cosine rule" (1.5), one of the principal formulas of plane trigonometry is the "sine rule,"

$$\frac{\sin \alpha}{a} = \frac{\sin \beta}{b} = \frac{\sin \gamma}{c},$$

where α, β, γ are interior angles of a triangle, opposite to sides of length a, b, c, respectively. Show how this follows from properties of vectors such that $\mathbf{a} + \mathbf{b} + \mathbf{c} = 0$.

1.4. By the "centroid" $\bar{\mathbf{r}}$ of a set of points $\mathbf{r}_1, \mathbf{r}_2, \mathbf{r}_3, \ldots$ is meant

$$\bar{\mathbf{r}} = \frac{m_1\mathbf{r}_1 + m_2\mathbf{r}_2 + m_3\mathbf{r}_3 + \cdots}{m_1 + m_2 + m_3 + \cdots},$$

where $m_1, m_2, m_3, \ldots$ are "weights" attached to the respective points.

(a) Investigate the positions of the centroid of just *two* points, as a function of their weight ratio, m_1/m_2.

(b) Show that the centroid of *three equally weighted* points lies at the intersection of the "medians" of the triangle formed by lines joining the points. (A median of a triangle is a line from any vertex to the midpoint of the side opposite it.) Argue from your result that the medians of any triangle "trisect" each other.

(c) Let $\mathbf{r}_1, \mathbf{r}_2, \mathbf{r}_3, \mathbf{r}_4, \ldots$ be the position vectors of *equally weighted* points relative to their centroid itself as origin. Show that those vectors can be made to form a closed polygon (by displacement without changes of direction or magnitudes).

1.5. Derive the multiple products (1.19) and (1.20), and from them an expression for $(\mathbf{A} \times \mathbf{B}) \cdot (\mathbf{C} \times \mathbf{D})$ in terms of just scalar products among the four vectors.

---

*All Exercises in this book are meant to suggest investigations as extensive as the reader may wish to make them.

1.6. Suppose that the angle between each pair of the four unit vectors $\mathbf{e}_1$, $\mathbf{e}_2$, $\mathbf{e}_3$, $\mathbf{e}_4$ is known. Express the angle between the vectors $\mathbf{e}_1 \times \mathbf{e}_2$ and $\mathbf{e}_3 \times \mathbf{e}_4$ in terms of the six known angles.

1.7. Although "a dot and a cross may be interchanged" as in (1.19), the factors of the double cross-product (1.20) cannot be. However, show that

$$\mathbf{A} \times (\mathbf{B} \times \mathbf{C}) + \mathbf{B} \times (\mathbf{C} \times \mathbf{A}) + \mathbf{C} \times (\mathbf{A} \times \mathbf{B}) = 0.$$

This is known as a "Jacobi identity."

1.8. Show that three points with position vectors related by

$$\mathbf{r}_1 \times \mathbf{r}_2 + \mathbf{r}_2 \times \mathbf{r}_3 + \mathbf{r}_3 \times \mathbf{r}_1 = 0$$

lie on a straight line.

1.9. (a) Suppose that ϑ, φ are the spherical coordinates of the unit vector $\hat{\mathbf{r}}$, and that $\hat{\mathbf{r}}'(\vartheta', \varphi')$ is a second point on the unit sphere (see Figure 13.8). If Θ is the angle between the two unit vectors, show that

$$\cos \Theta = \cos \vartheta \cos \vartheta' + \sin \vartheta \sin \vartheta' \cos(\varphi - \varphi'),$$

and that this reduces to familiar results for $\varphi = \varphi'$, and so on.

(b) The arcs ϑ, ϑ', and Θ of (a) form a "spherical triangle" on the unit sphere, and the result (a) is one of the principal formulas of spherical trigonometry. Another one is

$$\frac{\sin \vartheta}{\sin \alpha} = \frac{\sin \vartheta'}{\sin \alpha'} = \frac{\sin \Theta}{\sin(\varphi - \varphi')},$$

where α and α' are interior angles of the sperical triangle opposite the arcs ϑ and ϑ', respectively. Derive this result. When does it effectively reduce to the plane trigonometric "sine rule" of Ex. 1.3?

1.10. Let the apex of this triangle represent the position of a lamp post, $\mathbf{c}$ a fence, and $\mathbf{d}$ a path that bisects $\mathbf{b}$.

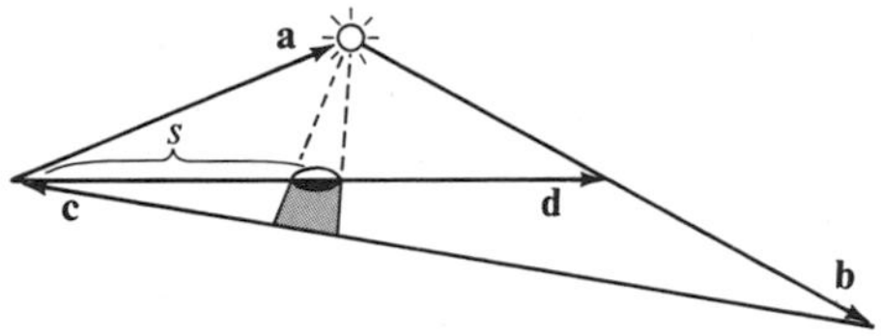

(a) A man walks along the path. At what position, s, will he be when his shadow reaches the midpoint of the fence? (See Exercise 1.4(b).)

(b) If the man walks at a uniform speed v, what is the velocity of his shadow, expressed in terms of v, s, and the lengths c and d? Why does the shadow's velocity approach infinity as the man's path is extended to $s = 2d$, and the fence is also extended?

(The reader may want to construct other problems like this for himself. For example, the path might be a circle centered on the lamp post and the fence a tangent to the circle, in which case the shadow's velocity may be given expression as $v/\cos^2(vt/a)$.)

1.11. Accelerations may be associated with ostensibly uniform motions. For example, suppose a barge is being dragged toward a dock by means of a line from a winch on its bow to a ring at the edge of a dock which is at some height h higher than the winch. Even if the winch reels the line in at a *uniform* speed v, the barge has a horizontal acceleration. Evaluate it at the point the barge reaches the horizontal distance x from the dock $(-h^2v^2/x^3)$.

1.12. Two ships converge toward a crossing point on perpendicular courses, each maintaining the same, uniform speed. Show that their distance apart, $r(t)$, varies with an acceleration $\ddot{r} \neq 0$ unless they are headed for bow-to-bow collision. If they are a distance r_0 apart when closest to each other, how far apart are they as one crosses the other's wake? ($\sqrt{2}r_0$.) (The thoughtful reader will want to reconcile the result $\ddot{r} \neq 0$ with the vanishing relative acceleration, $\ddot{\mathbf{r}}_1 - \ddot{\mathbf{r}}_2 = 0$, of any two uniform motions by showing that, relative to either ship, there is a centripetal acceleration just canceling the radial acceleration $\ddot{r}$. He will also want to show that an observed angular acceleration is similarly canceled by a Coriolis effect.)

1.13. A small cylinder of radius a rests in a smooth radial groove in the platform of a turntable, at the distance r_0 from the rotation axis. The groove has a circular-sector cross-section, of the same radius as that of the cylinder, and a depth $h < a$. The turntable begins revolving at a uniform angular speed, Ω. Show that the cylinder will jump the groove at the radius

$$R = \left[r_0^2 + \frac{g^2}{4\Omega^4}\frac{h(2a-h)}{(a-h)^2}\right]^{1/2}.$$

Give arguments that the way this varies, with any valid changes made in the quantities involved, is in a direction to be expected.

1.14. Calculate in detail how (1.33), (1.32), and (1.39) follow from (1.54).

1.15. A subway train is capable of a steady maximum acceleration, and can decelerate at a higher uniform rate. Use is made of these properties for covering distances between stations in the least time possible, from rest to rest. Show that in each such trip, the train's speed averages to just half the maximum speed it can afford to attain in the stretch.

1.16. For the parabolic path of Figure 1.12, find the maximum range (for a given initial velocity) along a slope making the angle ϑ with the horizontal.

1.17. Show explicitly that the area of an ellipse is πab, as assumed for (1.63). Notice that it properly reduces to an area of a circle for $a = b$.

1.18. The expression $(x^2/a^2) - (y^2/b^2) = 1$ defines two branches of a hyperbola. Find definitions for $r(x, y)$ and $\varphi(x, y)$ which allow representing one branch of the hyperbola as in (1.59), with only the replacement of $-c$ by $+(a^2 + b^2)^{1/2}$. Then show that just a reversal of the sign of a in the resulting expression yields the second branch.

CHAPTER 2

ELEMENTS OF PARTICLE DYNAMICS

The kinematics just reviewed dealt with the concepts of space and time; primary ones in the sense that others—velocity and acceleration—were derived from them. Another concept that might be "derived" is that of a "point-particle." What amounts to an analysis into point-particles is adopted whenever a motion is described by giving a discrete set of position-vectors: $\mathbf{r}_1(t)$, $\mathbf{r}_2(t)$, . . . , $\mathbf{r}_i(t)$, ... Thus, in the kinematical context, a particle is effectively defined by two properties.

1. *Localizability*: a "configuration" which is fully specified by giving a single position, $\mathbf{r}$, of a point at any given moment.

2. *Duration*: a continuity of $\mathbf{r}(t)$ as a function of time.

For a dynamical context, the concept is generalized to that of a "mass-particle," by associating with the point a third property:

3. *Mass*, to be specified by giving a scalar magnitude, m. This concept is an abstraction from experience with what is sometimes called "inertia" to acceleration.

A fourth primary concept must be introduced, besides space, time and mass, before what is called a *dynamical* description of the circumstances of a motion can be completed. In the Newtonian formulation, the additional concept is represented by a vector **F** and is called "force." It is roughly an abstraction from experience with pushing and pulling masses about. Better definitions of both the mass and force concepts are needed before they can be made part of a predictive procedure with some semblance of accuracy.

A theorist might consider a concept like mass to be adequately defined for his purposes when the assertion that it is representable by a scalar number has been made, and its role in theoretical descriptions demonstrated. However, an experimentalist has the far more difficult task of identifying it with something in his observations, before an adequate correspondence between theory and experience can be established. This calls for the specification of procedures by which a number can be assigned to the mass (that is, a measurement obtained). Such specifications constitute what have come to be called "operational" definitions.*

To present an operational definition without dragging the reader into a laboratory requires quoting results which are found to issue from appropriate experimental operations. Thus, the definitions of mass and force will be natural by-products of reviewing the experience from which the theory as a whole is developed, as done in the following sections.

Meanwhile, operational definitions of length and time could also be demanded. Such definitions have been provided by Einstein, but it seems preferable to reserve their discussion to a stage at which substantial problems arise, as when speeds comparable to that of light are considered. For the present, especially in view of the impending discussion of relativity, it should be pointed out that the space and time concepts were, in the preceding chapter, treated as mental constructs imposed on experience for the purpose of providing terms and frames of reference.

2.1 GALILEAN RELATIVITY

The first type of experience to be drawn upon has implications for the way a dynamical theory should be developed, and is embodied in what is called a "special" Principle of Relativity. The phenomenon in question has been experienced by anyone who has been aboard some vehicle capable of smooth uniform motion and has found it difficult to tell whether the vehicle was moving or not. The eventual conclusion from that type of experience is

* Thus, a physical conception can be said to require two definitions, or two aspects of one, before its full meaning can be communicated. One establishes correspondence to physical processes, the other to mental ones.

that no test confined to the reference frame presented by the vehicle itself can reveal a really uniform motion of it. A much-quoted example is provided by the test of dropping an object from the masthead of a vessel afloat on a smoothly flowing river; insofar as air friction is negligible, the object falls to the foot of the mast even when the shore can be seen to be gliding by. When an explicit statement of the conclusion from such experience is attempted, it is said that only *relative* motion has meaning, and that all frames of reference in uniform relative motion form equally valid bases for judging motion. The restriction to *uniform* relative motions of the frames is responsible for the qualification "special" attached to this relativity principle.

The Galilean Transformation

The relativity principle may be formalized by referring to specific descriptions—for example, to the position-vector functions of time, $\mathbf{r}_i(t)$, defined in one reference frame, and to an alternative set, $\mathbf{R}_i(t)$, defined in a second frame moving uniformly with respect to the first. When the connection between the two frames is taken to be

$$\mathbf{v}_i(t) = \mathbf{V}_i(t) + \mathbf{v}_0 \text{ (a constant)}, \tag{2.1a}$$

where $\mathbf{v}_i \equiv \dot{\mathbf{r}}_i$ and $\mathbf{V}_i \equiv \dot{\mathbf{R}}_i$, so that

$$\mathbf{r}_i = \mathbf{R}_i + \mathbf{v}_0 t + \mathbf{r}_0 \text{ (a constant)}, \tag{2.1b}$$

then the substitutions effected by these relations are said to constitute a "Galilean transformation." They are relations appropriate for transforming expressions referring to one frame so that they refer to the other, instead. In such terms, the implication of the relativity principle is that any relation like $f(\mathbf{r}_1, \mathbf{r}_2, \ldots, t) = 0$, which helps to describe circumstances of a motion as viewed from one frame of reference, may with equal validity be expressed in the other as $f(\mathbf{R}_1, \mathbf{R}_2, \ldots, t) = 0$:

$$f(\mathbf{R}_1 + \mathbf{v}_0 t + \mathbf{r}_0, \mathbf{R}_2 + \mathbf{v}_0 t + \mathbf{r}_0, \ldots, t) = f(\mathbf{R}_1, \mathbf{R}_2, \ldots, t) = 0, \tag{2.2}$$

and all reference to the frame motion cancels out of relations which are explicitly consistent with relativity. The relations may be understood to contain a given $\mathbf{r}_i(t)$ more than once, evaluated at various times; derivatives $\dot{\mathbf{r}}_i, \ddot{\mathbf{r}}_i, \ldots$ may also be included.

Descriptions with properties like (2.2) are said to have *manifest* Galilean invariance. Examples of manifest invariance, (having explicit consistency with Galilean relativity) will be exhibited after relations descriptive of motion have been developed (as in (2.10) and the equations following it).

Newton's First Law

The conclusion from the special relativity which immediately affects the way dynamical theory is developed is this: assigning a velocity $\mathbf{v}_i \neq 0$ to a particle, and giving it none, $\mathbf{v}_i = 0$, differ only in point of view. There is no more to explain about a state of uniform motion than about a state of rest. The Laws of Motion should be constructed on the basis that uniform motion is "natural," and that only lapses from it need to be accounted for.*

Newton proposed to account for disturbances of uniform motion as effects of "forces" and his First Law,

> *A body remains at rest, or in a state of uniform motion, until it is compelled to change that state by an impressed force,*

may be regarded as an initial, qualitative definition of force. This Law is consistent with special relativity by holding that uniform motion does not require maintenance by forces any more than does rest.

2.2 MASS AND THE CONSERVATION OF MOMENTUM

In view of the foregoing, the development of the theory is continued by drawing on experience with velocity *changes.* An elementary way to induce such changes is to allow a pair of uniformly moving bodies to collide, with each afterwards resuming a different uniform motion. Some of the experience with such interactions is embodied in Newton's Third Law:

> *To every action there is an equal and opposite reaction.*

However, the meaning of "action" and "reaction" which permits the quantitative assertion here presumes that masses are already defined. Without this, only a kinematical statement can be made,

> *Corresponding to every acceleration of a body interacting with another, there is some oppositely directed acceleration of the second body.*

---

* It seems that cats agree, since they seem to jump at the spot where a fleeing mouse will be if it continues in uniform motion.

Operational Definition of Mass

The initial definition will be concerned with the mass to be associated with the motion of a point. There will be a minimal ambiguity in the identification and measurement of accelerations if bodies so small are considered that the motion of each is adequately representable as that of a single point-center. The mass to be measured is "isolated" together with a second mass, which will have the role of a standard, at least temporarily. Isolation is achieved by experimenting with arrangements of the two bodies until one is found in which all observable reactions to the acceleration of either seems to be confined to the other of the pair. The acceleration may attend collisions of the bodies or arise from circumstances otherwise described, as "gravitational attraction," "electrostatic repulsion," and so on. The latter types of interaction could be viewed as "collisions at a distance."

The implication of Newton's Law of Reactions is that any acceleration $\mathbf{a}_{12}$, exhibited by body #1 interacting with #2, has a ratio to the reacting acceleration $\mathbf{a}_{21}$ which is some negative scalar number, $\mathbf{a}_{12}/\mathbf{a}_{21} = -c_{21}$. The experiences embodied in the Law show that this number is independent of the particular magnitudes of the individual accelerations found in a given trial. Repetition of the experiment using a body #3 in isolated partnership with #1, and then, in turn with #2, would yield: $-\mathbf{a}_{13}/\mathbf{a}_{31} = c_{31}$ and $-\mathbf{a}_{23}/\mathbf{a}_{32} = c_{32}$. Quantitative measurements of "classical" accuracy would show that $c_{32} = c_{31}/c_{21}$. As a result, $c_{21}\,\mathbf{a}_{23} = -c_{31}\,\mathbf{a}_{32}$, and this characterizes the experiment without the presence of #1. Thus #1 has merely served as a standard unit for measuring what may now be called the "masses," $m_2 = c_{21}$ and $m_3 = c_{31}$ of #2 and #3, respectively. (The unit mass #1 may then be put away into a safe at the Bureau of Standards.)

The relationships found between pairs of reacting accelerations have the form

$$m_i\,\mathbf{a}_{ij} = -m_j\,\mathbf{a}_{ji}. \tag{2.3}$$

This can be taken as the quantitative expression of the Law of Reactions. "Action" and "reaction" are identified with products of mass and acceleration; such products will shortly be equated to "forces."

Extensions of the experiments to cases in which three or more particles are isolated together would show that the individual accelerations $\mathbf{a}_1, \mathbf{a}_2, \mathbf{a}_3, \ldots$ exhibited by the particles are related as

$$m_1\mathbf{a}_1 = -m_2\,\mathbf{a}_2 - m_3\,\mathbf{a}_3 - \cdots \equiv -\sum_{i\neq 1} m_i\,\mathbf{a}_i, \tag{2.4}$$

with masses as determined from two-body experiments. Thus "reactions" or "forces" may be superposed vectorially to produce resultants.

Finding (2.4) can serve as a basis for attributing mass to extended bodies, treated as accumulations of many mass-particles (see Chapters 6, 10, and 15). It is characteristic of any initial operational definition that it gives only a start to the effective meaning of a concept as its use is extended to fresh situations. As the area of application grows, so does the concept. It acquires a continuum of "operational definitions," comprised of its uses, beyond the bounds of any one set down at any given stage of development. Moreover, the successful extensions in the use of a concept play a greater part in giving confidence in it than does any isolated operational definition.

The remark may be added that the purely mathematical concepts also gain their effective meaning by "operational" definition, as through the algebraic uses they validly acquire.

The Principle of Momentum Conservation

More directly measurable than the accelerations, in the types of observations just reviewed, are the velocity changes that take place over some *finite* time interval: $\int \mathbf{a}_i \, dt = \mathbf{v}_i - \mathbf{v}_i^0$. The measurements become particularly unambiguous when collisions are observed, for then the initial and final velocities, $\mathbf{v}_i^0$ and $\mathbf{v}_i$, are uniform over extended periods before and after the interaction. Expressed in terms of the velocity changes, a measurement of the mass ratio, m_i/m_j, of a pair of interacting bodies yields

$$m_i(\mathbf{v}_i - \mathbf{v}_i^0) = -m_j(\mathbf{v}_j - \mathbf{v}_j^0) \tag{2.5}$$

in place of the Law of Reactions (2.3). In the more general case of a system of interacting mass-particles, (2.4) is replaced by

$$\boxed{\sum_i m_i \mathbf{v}_i = \sum_i m_i \mathbf{v}_i^0.} \tag{2.6}$$

Thus the sum of the products of mass and velocity is *conserved* throughout motions of an isolated system of interacting particles.

Quantities that persist unchanged can be regarded as having some kind of "existence" and it becomes appropriate to give them names. The conserved quantity of (2.6) is called the total momentum of the system of particles and such a product as $\mathbf{p}_i = m_i \mathbf{v}_i$ is called a momentum of the mass m_i. Then the result (2.6) represents the conservation of the total momentum of the system.

The Principle that *the total momentum of any isolated system is conserved* forms part of the basic framework on which all physical theory has been constructed.

An exceptionless adherence to the Principle has been borne out by experience in that, for every observed momentum change, it has always turned out possible to identify a compensatory momentum change, in one form or another.* The "isolation" of a system is judged by its conformity to the Principle, inasmuch as additional forms of momentum are looked for whenever the Principle seems to fail. That may make the Principle the result of adopting a certain point of view: the total momentum of an isolated system is conserved because any lack of conservation is taken to be evidence for its nonisolation. It may be regarded as significant that it has so far always been found possible to formulate physical theories on the basis that conserved momenta are always definable. On the other hand, it may merely testify to the mental agility of theorists in rationalizing experience.

The ostensibly fundamental character of the momentum conservation principle gives the Law of Reactions (2.3) the status of a special consequence for pairs of interacting bodies; their reaction relation is obtained as a time-derivative of their momentum conservation (2.5).

2.3 FORCE AND THE EQUATIONS OF MOTION

For a "system" composed of a single, isolated particle, the Principle of Momentum Conservation requires the particle to continue in a state of rest, or of uniform motion with whatever velocity it had at the beginning of its period of isolation. That an unvarying velocity indicates no disturbance of the isolation was already a conclusion from the relativity principle, and now the momentum principle indicates that disturbances should be formulated as sources of *momentum*-change. Some such sources have already been considered—the presence of other particles. Newton generalized the theory of motion by formulating disturbances that need not be explicitly based on other particles but are representable by "force" vectors. A force, $\mathbf{F}_i$, on the ith particle is to be held responsible for changes in its momentum $\mathbf{p}_i = m_i \mathbf{v}_i$ according to

$$\frac{d\mathbf{p}_i}{dt} = \mathbf{F}_i. \tag{2.7}$$

This may be considered an operational definition of force—as measured by the rate of momentum change it produces. The concept has been referred to

* To maintain the point of view, sometimes other types of momentum-possessing entities than point-particles—namely, *fields*—have had to be envisaged, as will eventually be seen.

before as an abstraction from experience with pushing and pulling masses about. When the experimenter's own hand may be the source of momentum changes, then the advantage of foregoing a description of the experimenter himself as a body of particles, and representing his effect by a force vector instead, becomes especially evident.

The Second Law and Phases of Motion

The equations (2.7), one for each particle m_i, constitute Newton's fundamental (Second) Law of Motion, also written as*

$$\boxed{m_i \ddot{\mathbf{r}}_i(t) = \mathbf{F}_i}, \qquad i = 1, 2, 3, \ldots. \tag{2.8}$$

This form emphasizes that predictions of the motion $\mathbf{r}_i(t)$ are to be deduced from these "Equations of Motion" (one vector equation for each m_i) when the forces are *given*.

How the predictive procedure works is quite well illustrated even by the trivial case of a free (isolated) particle, subject to no force. With $\mathbf{F} = 0$, a first integration of $\ddot{\mathbf{r}} = 0$ yields $\dot{\mathbf{r}}$ = an arbitrary constant vector. The integration constant here is obviously determined when the *initial velocity* $\mathbf{v}_0$ is given: $\dot{\mathbf{r}} = \mathbf{v}_0$. A second integration gives $\mathbf{r} = \mathbf{v}_0 t$ + another arbitrary constant vector. This second constant is determined when the *initial position* $\mathbf{r}_0$ is given. The prediction is $\mathbf{r}(t) = \mathbf{v}_0 t + \mathbf{r}_0$, the straight-line motion mentioned in (1.55).

It should be stressed that the Equations of Motion can predict a completely specific path, for a given particle subject to a given force, only if still further information is also given. The equations only specify the rate at which the velocity *changes*; hence a starting velocity must be known before a specific future velocity can be predicted. Then, after the velocity is known for future times, this only specifies the rate at which the position *changes*, and so a starting position must also be known before future positions become specifically predictable. This is a natural cost of generality, of adaptability to many specific situations. The prototype problem of Dynamics may be said to be the derivation of motions $\mathbf{r}_i(t; \mathbf{r}^0, \mathbf{v}^0)$ under a given force and starting from *given initial conditions*: $\mathbf{r}_i(0) = \mathbf{r}_i^0$, $\mathbf{v}_i(0) = \mathbf{v}_i^0$. Giving the conditions in this way is

* Newton's First Law, $\ddot{\mathbf{r}}_i = 0$ if $\mathbf{F}_i = 0$, now appears to be a special case of his Second Law. The separate statement may have been meant to counter very specifically an ancient notion that force is also needed to keep a body moving steadily; as now viewed, such force is added only to cancel out formerly unregarded frictional forces to a zero resultant.

sometimes spoken of as a specification of the initial "state" or "phase" of the motion.

The form taken here by the procedure for predicting motion might have been anticipated as soon as it was decided that a state of uniform motion should be treated as a "natural" one, and that only changes of it should be generated by forces. The predictive procedure can be taken up at any moment, t_1, at which conditions are representable by

$$\mathbf{r}_i(t) = \mathbf{r}_i(t_1) + \mathbf{v}_i(t_1)(t - t_1),$$

with restriction to $t - t_1 \to 0$ for a continuously curved path. The motion is thus treated as a succession of "momentary states" or "phases,"

$$\mathbf{r}_i(t_1),\mathbf{v}_i(t_1) \to \mathbf{r}_i(t_2),\mathbf{v}_i(t_2) \to \cdots, \tag{2.9}$$

or a series of "transitions" between momentary states, induced by the applied force. The treatment of a state of uniform motion as a "natural" one, in this way, is a consequence of special relativistic considerations, and therefore the conclusion that a phase of it is specified by giving $\mathbf{r}_i$, $\mathbf{v}_i$ can be said to be another.

Besides the phases, or instantaneous states, referred to here, "stationary" or "steady" states of motion are sometimes spoken of. The latter terms refer to an entire orbit, like $z(x)$ of (1.58) or $r(\varphi)$ of (1.59), being followed during the motion. As long as some given orbit, specifiable independently of time, is followed, the system is said to remain in a given stationary state.

Force-Functions

The preceding introduced the relation $\mathbf{F} = m\ddot{\mathbf{r}}$ as a definition of force and also as a Law of Motion. One may well ask how a mere definition can also have content from which predictions are to be deduced. The point is that the relation can be used in both ways. First, it is used for *investigating* what forces arise in what circumstances. Then the information so gained provides a basis for *postulating* forces for situations in which motion is to be predicted. The forces embody part of what it is essential to discriminate in circumstances to enable prediction (see Introduction).

Naturally, the *investigations* of force have by now been extensively codified into what may be called "Theories of Force." The theory of gravitation, and of the electromagnetic field, are among the best-developed examples. Thus, in many situations, the *postulation* of forces referred to above can be replaced by well-founded theoretical expectations.

It is customary to regard the detailed theories of force as outside the domain of Mechanics "proper." There is already a broad field to cultivate in just undertaking to find the motions under *given* forces. However, a rigid demarcation of "mechanics proper" from separate "theories of force" cannot, and should not, be maintained. For instance, it may be necessary to draw on facts "belonging" to theories of force in order to judge what solutions of the equations of motion are pertinent to actual physical situations.

Theories of force, and the investigations of forces in general, may permit expressing the results in terms of force *functions*. For example, it might be found that whenever a particle comes to a given position $\mathbf{r}$, then it always suffers acceleration there by a force $\mathbf{F}(\mathbf{r})$. The force may also be found to depend on the particle's velocity $\mathbf{v}$, and perhaps also on the time t after some significant event, yielding a force-function $\mathbf{F}(\mathbf{r}, \mathbf{v}, t)$ of all those variables.

The possibility that the force should be treated as a function of higher derivatives of the position than the velocity is rarely considered, and this may need comment. First, it is clear that the force might be treated as a function of time only; for example, when given as $\mathbf{F}(\mathbf{r}, \mathbf{v}, t)$, it is expressible in terms of the time alone as soon as the positions and velocities are known as functions of the time. Such a representation, by $\mathbf{F}(t)$, is obviously unsatisfactory as part of a predictive procedure. It amounts to choosing for each moment just the accelerations which the motion exhibits at that moment. Much more powerful, from the standpoint of prediction, is a description by forces which grow out of the motion itself, as it progresses, and contingent on the phases that may arise. It is such descriptions that are appropriate for isolated systems, ones conceived as having their behavior determined only by what each, itself, is doing.

It has been seen that knowledge of positions $\mathbf{r}_i$ and velocities $\mathbf{v}_i$ should be sufficient for specifying the momentary state of a particle system, enabling predictions of future phases $\mathbf{r}_i(t)$, $\mathbf{v}_i(t)$, which will be reached under given forces. Thus forces which are to depend only on what the system itself is doing should be functions of the positions and velocities at most:

$$\mathbf{F}(\mathbf{r}_1, \mathbf{v}_1; \mathbf{r}_2, \mathbf{v}_2; \ldots) \equiv \mathbf{F}(\mathbf{r}, \mathbf{v}).$$

This type of force-function should be adequate for isolated systems, independent of anything external to themselves. However, it is sometimes desirable to formulate external influences when those constitute sources of force that are too complex to be included as part of the particle system being given detailed description. Force functions of the type $\mathbf{F}(\mathbf{r}, \mathbf{v})$ may well not be adequate to describe any type of external influence. However, admitting, in addition, an explicit dependence on time should be adequate to meet any situation. Thus, most generally, possibilities of the type $\mathbf{F}(\mathbf{r}, \mathbf{v}, t)$ are considered. Carried out in this way, the admission of the additional, explicit dependence on time becomes equivalent to admitting external influences.

The kinds of motion that various simple force functions are capable of describing will be the chief concern of succeeding chapters. Each function will furnish a "model" found useful in describing some sets of physical circumstances.

Relativistically Invariant Force

Since the relativity principle helped guide its development, the description should be invariant to the Galilean transformation (2.1). These substitutions obviously change* the value of quantities used to describe initial conditions and any momentary state of motion: $\mathbf{r}_i$, $\mathbf{v}_i$ versus $\mathbf{R}_i = \mathbf{r}_i - \mathbf{v}_0 t - \mathbf{r}_0$, $\mathbf{V}_i = \mathbf{v}_i - \mathbf{v}_0$. However, a relation like the momentum conservation principle (2.6),

$$\sum_i m_i(\mathbf{v}_i - \mathbf{v}_i^0) = 0 \leftrightarrow \sum_i m_i(\mathbf{V}_i - \mathbf{V}_i^0) = 0, \tag{2.10}$$

is equally valid with one set of values or the other. Likewise, the Equations of Motion,

$$m_i \dot{\mathbf{v}}_i = \mathbf{F}_i \leftrightarrow m_i \dot{\mathbf{V}}_i = \mathbf{F}_i, \tag{2.11}$$

are the same from one viewpoint or the other, to the extent that exactly the same force values account for a given motion whether this is described in one frame or the other.

The invariance test represented by (2.11) is too superficial when the functional dependence of the forces on the phase variables,

$$\mathbf{F}_i(\mathbf{r}, \mathbf{v}, t) \equiv \mathbf{F}_i(\mathbf{r}_1, \mathbf{v}_1, \mathbf{r}_2, \mathbf{v}_2, \ldots, t), \tag{2.12}$$

is taken into account. That becomes especially evident when some specific example is considered, such as $\mathbf{F}_i = -k\mathbf{v}_i$, interpreted as a "frictional" force proportional to the speed. The Galilean transformation (2.1) substitutes $m_i \dot{\mathbf{V}}_i = -k(\mathbf{V}_i + \mathbf{v}_0)$ for $m_i \dot{\mathbf{v}}_i = -k\mathbf{v}_i$. The value of the force is not changed by the substitution, as indicated in (2.11). However, in the frame in which the mass m_i has the velocity $\mathbf{V}_i$, the force depends on the motion $\mathbf{v}_0$ of the frame as if it were possible to decide that it is the one that is moving, in some

* Such changes are referred to formally as "covariances."

absolute sense contradicting the relativity principle. It is as if the "friction" $\mathbf{F}_i = -k\mathbf{v}_i$ arose from motion of the mass with respect to some "medium," fixed absolutely in space, relative to which a motion of the second frame can be detected. Then, to regain a relativistic invariance, the description must be made more complete by recognizing that a medium capable of making itself evident as through the friction, may also have motion attributable to it. Instead of being treated as fixed in space, velocities $\mathbf{v}_m$ and $\mathbf{V}_m = \mathbf{v}_m - \mathbf{v}_0$ should be assigned to the medium, and additional equations of motion should be introduced to help determine them. Now a properly invariant description, with

$$m_i \dot{\mathbf{v}}_i = -k(\mathbf{v}_i - \mathbf{v}_m) \leftrightarrow m_i \dot{\mathbf{V}}_i = -k(\mathbf{V}_{.} - \mathbf{V}_m),$$

may be constructed, if equations for $\dot{\mathbf{v}}_m$ or $\dot{\mathbf{V}}_m$ having similar invariance are available. From this point of view, the original force function $\mathbf{F}_i = -k\mathbf{v}_i$ is proper to a case in which the medium can be treated as having a uniform velocity everywhere, and the frame of $\mathbf{v}_i$ is the one in which $\mathbf{v}_m = 0$. The transformed expression for the force, $-k(\mathbf{V}_i + \mathbf{v}_0)$, is then to be interpreted as depending on the velocity $\mathbf{V}_m = -\mathbf{v}_0$ of the *medium*, as judged from the second frame, rather than on an absolute velocity $\mathbf{v}_0$ of the frame itself.

The example here shows that a noninvariance of an equation of motion may result from incompleteness of description (as when a description of the medium giving rise to friction was left out). Such incomplete descriptions can have value (as did the not manifestly invariant frictional force for the case of a uniform medium) of the same kind as do force functions in which explicit dependence on time is admitted. In either way, effects external to the system being given detailed description are represented.

On the other hand, physical experience as expressed in the relativity principle demands relativistic invariance of any description purporting to represent a completely formulated, isolated system. The above example indicated how attaining a manifest invariance may enjoin the introduction of further variables, and thus an increase in the completeness of the description —to a point at which the resultant system can be regarded as isolated, uninfluenced by anything further, external to it. All sources of momentum that can have appreciable exchanges with the momenta of interest are searched out, thus satisfying the relativistically invariant conservation principle.

In general, it must be regarded as the business of the particular "theories of force" to construct properly relativistic force functions. Meanwhile, it is clear that force functions in which the phase variables occur only as *relative* positions, $\mathbf{r}_i - \mathbf{r}_j$, and *relative* velocities, $\mathbf{v}_i - \mathbf{v}_j$, will have proper invariance to the Galilean transformation (2.1). A sole dependence on relative positions and velocities is to be expected of a description applicable to an isolated system; if an explicit dependence on time is admitted, or if a dependence on some "absolute" position or velocity, relative to something undescribed, is allowed, this destroys the Galilean invariance and the "isolation."

2.4 ENERGY AND WORK

The first integration of an equation of motion,

$$m\dot{\mathbf{r}}(t) = \int_0^t \mathbf{F}\, dt + m\mathbf{v}_0,$$

which permits immediate introduction of an initial condition on the velocity, requires knowing the force as a function of time. When the force depends on the phases **r**, **v** to be generated in the motion, a less direct integration is more appropriate. An attempt at integration with the help of scalar multiplication by the velocity **v** as an integrating factor,

$$\mathbf{F} \cdot \mathbf{v} = m\mathbf{v} \cdot \dot{\mathbf{v}} = d(\tfrac{1}{2}mv^2)/dt, \tag{2.13}$$

using $\mathbf{v} \cdot \mathbf{v} \equiv v^2$, gives rise to the concept of the *kinetic energy*,

$$T = \tfrac{1}{2}mv^2, \tag{2.14}$$

of the particle. The quantity $\mathbf{F} \cdot \dot{\mathbf{r}}\, dt = \mathbf{F} \cdot d\mathbf{r}$ is called the element of *work*,

$$dW = \mathbf{F} \cdot d\mathbf{r}, \tag{2.15}$$

done by the force, on the particle, during a displacement of it by $d\mathbf{r}$. Since

$$dW = dT, \tag{2.16}$$

it is said that work on the particle goes into increasing its kinetic energy. The rate at which the work is supplied,

$$\frac{dW}{dt} \equiv \mathbf{F} \cdot \mathbf{v} = \frac{dT}{dt}, \tag{2.17}$$

is called the *power* being expended by the force.

The element of work, $\mathbf{F} \cdot d\mathbf{r}$, may be negative, as when the momentum of a particle carries it into a direction against that of the force. Then $dW = dT < 0$, and it is said that the particle works against the force at the cost of a diminution in its kinetic energy.

The kinetic energy of the particle helps characterize a phase of its motion **r**, **v**, and is a scalar in reflection of the scalar work which had been done on or by it, just as its vector momentum reflects the directed forces which had been applied to it.

Potentials and Energy Conservation

For further progress toward a first integral of the motion, consider the work done by a force up to some time t, from an initial moment t_0, while the particle moves from the initial position $\mathbf{r}(t_0) \equiv \mathbf{r}_0$ to $\mathbf{r}(t)$. The work done changes the kinetic energy by

$$T - T_0 = \int_{\mathbf{r}_0}^{\mathbf{r}} \mathbf{F} \cdot \mathbf{dr}. \tag{2.18}$$

This result depends only on the final state **r**, **v** and the initial conditions $\mathbf{r}_0$, $\mathbf{v}_0$, and not on the course of the motion between, *if* $\mathbf{F} \cdot \mathbf{dr}$ is a "perfect differential." This means that a function $-V(\mathbf{r})$, of the integration variable **r** alone, can be found such that

$$\mathbf{F} \cdot \mathbf{dr} = -dV(\mathbf{r}) \tag{2.19}$$

and

$$T - T_0 = -V(\mathbf{r}) + V(\mathbf{r}_0). \tag{2.20}$$

Thus $T + V(\mathbf{r})$, at an arbitrary point of the motion, is equal to $T_0 + V_0$, with $V_0 \equiv V(\mathbf{r}_0)$, the initial value. The quantity is *conserved* during the motion, and is called the *total energy*

$$\boxed{E = T + V}, \tag{2.21}$$

consisting of kinetic energy plus a *potential energy*, $V(\mathbf{r})$. Finally, E is a first integration constant, to be evaluated from the initial conditions as $E = T_0 + V_0$, and this may replace the specification of v_0 in giving the initial state.

Putting $-dV$ for the element of work $dW = \mathbf{F} \cdot \mathbf{dr}$ in (2.16) yields

$$d(T + V) \equiv dE = 0, \tag{2.22}$$

representing a constancy of the total energy during every phase of the motion. As the particle moves, its kinetic energy changes because of work done on or by it, but when a potential function $V(\mathbf{r})$ is definable for the force as in (2.19), then the system of particle plus force can be attributed a potential energy which varies in compensation, so that the initial total $E = T_0 + V_0$ is conserved.

The finding here of a way in which energy conservation may be formulated has great importance. Experience seems to indicate that a general *Principle of Total Energy Conservation* has as exceptionless a validity as does the momentum conservation (§2.2). For every energy change, suitably defined, it has always been found possible to identify a compensatory energy change, in one form or another.* As in the case of the Momentum Principle, any apparent failure of energy conservation is regarded as arising from a disturbance in the "isolation" of the system under examination; "external" sources or sinks of energy are immediately suspected and searched out.

Naturally, some sources of energy change are to be found in the presence of other particles. Forces arising from the other particles may be derivable from potentials depending on positions of all the particles present, $V(\mathbf{r}_1, \mathbf{r}_2, \ldots)$. The general investigation of this will be left to Chapter 6.

Gradient Derivatives

With the occurrence of a differential like $dV(\mathbf{r})$ of (2.19), the vector calculus (§1.2) must be extended to the taking of derivatives of scalar functions having vectors as the independent variables.

The problem is reduced to one of ordinary calculus when $\mathbf{r}$ is resolved into components x, y, z on some particular cartesian frame, $\mathbf{i}, \mathbf{j}, \mathbf{k}$. Then V is determined when x, y, z are given, and so can be treated as function of three scalar variables, $V(x, y, z)$. Upon a displacement

$$\mathbf{dr} = \mathbf{i}\,dx + \mathbf{j}\,dy + \mathbf{k}\,dz, \tag{2.23}$$

the function changes by†

$$dV(x, y, z) = \frac{\partial V}{\partial x}\,dx + \frac{\partial V}{\partial y}\,dy + \frac{\partial V}{\partial z}\,dz. \tag{2.24}$$

The three partial differentials are said to be the cartesian components of the *gradient vector*,

$$\nabla V = \mathbf{i}\frac{\partial V}{\partial x} + \mathbf{j}\frac{\partial V}{\partial y} + \mathbf{k}\frac{\partial V}{\partial z}, \tag{2.25}$$

with which the differential (2.24) may be written as the scalar product

$$\nabla V \cdot \mathbf{dr} = dV(\mathbf{r}) \equiv V(\mathbf{r} + \mathbf{dr}) - V(\mathbf{r}). \tag{2.26}$$

---

* The historically most significant finding of such a kind was perhaps the experimental confirmation of the so-called "equivalence" of heat and mechanical energy.

† $V(x, y, z)$ changes whenever any one of the variables does. The change of V due to dx alone takes place at the rate $\partial V/\partial x$, and so on.

Numbers presumed to be independent of the reference frame (scalars) emerge and this is sufficient to show that ∇V may be treated as a vector.

The symbol ∇ is sometimes written as "**grad**." It is called the gradient *operator* and a cartesian resolution of it may be written

$$\mathbf{grad} \equiv \nabla = \mathbf{i}\frac{\partial}{\partial x} + \mathbf{j}\frac{\partial}{\partial y} + \mathbf{k}\frac{\partial}{\partial z}. \tag{2.27}$$

Of course, this gains meaning only after operation on some function of $\mathbf{r}$.

The definitive meaning of the gradient vector is implicit in (2.26); it represents rate of change with position in such a way that its projection on a displacement $\mathbf{dr}$ in a given direction gives the rate of change with displacement in that direction. The projection is naturally largest, for a given magnitude $|\nabla V|$, when the direction of $\mathbf{dr}$ is chosen parallel to that of the gradient vector; hence the latter is said to have the "direction of steepest change."

The fact that it is $\nabla V \cdot \mathbf{dr} = dV(\mathbf{r})$ which gives meaning to the gradient must be kept in mind when finding other representations of it. When cylindrical variables are used, $V(\mathbf{r}) \equiv V(r, \varphi, z)$ and

$$dV = \frac{\partial V}{\partial r}dr + \frac{\partial V}{\partial \varphi}d\varphi + \frac{\partial V}{\partial z}dz.$$

However,

$$\mathbf{dr} = \mathbf{e}_r\,dr + \mathbf{e}_\varphi\,r\,d\varphi + \mathbf{k}\,dz, \tag{2.28}$$

and hence

$$\nabla = \mathbf{e}_r\frac{\partial}{\partial r} + \mathbf{e}_\varphi\frac{\partial}{r\,\partial \varphi} + \mathbf{k}\frac{\partial}{\partial z}. \tag{2.29}$$

Similarly, in the spherical representation for which $\mathbf{dr}$ has the form (1.31),

$$\nabla = \mathbf{e}_r\frac{\partial}{\partial r} + \mathbf{e}_\vartheta\frac{\partial}{r\,\partial \vartheta} + \mathbf{e}_\varphi\frac{\partial}{r\sin\vartheta\,\partial \varphi}. \tag{2.30}$$

Every component must have the dimension of an inverse length.

Line Integrals

Integrations over a vector variable like the work integral $\int \mathbf{F}\cdot\mathbf{dr}$ of (2.18) are not in general sufficiently defined when only the endpoints $(\mathbf{r}_0, \mathbf{r})$ of the integration are given. A path for the variable to follow between the end points must be chosen. The integral is by definition a limiting sum of contributions from a succession of line-elements $\mathbf{dr}$ which may form some curve C, as in Figure 2.1, and then the result

$$\int_C \mathbf{F}\cdot\mathbf{dr} = W(C), \tag{2.31}$$

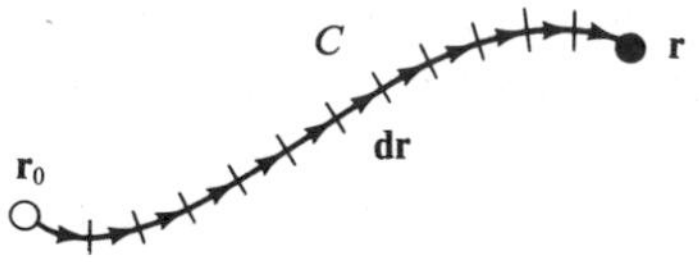

Figure 2.1

called the " line-integral " over C, is a function of the entire curve, of the history of the particle between $\mathbf{r}_0$ and $\mathbf{r}$. An exception occurs when $\mathbf{F}$ is a gradient vector, $-\nabla V$. Then $W = V(\mathbf{r}_0) - V(\mathbf{r})$ is determined by the end points alone and is independent of the path between, as has already been pointed out in (2.20).

A corollary of the last result is that the integral over a curve forming a complete loop ($\mathbf{r} \to \mathbf{r}_0$) vanishes. Loop integrals are symbolized as in

$$\oint \mathbf{F} \cdot d\mathbf{r} = -\oint \nabla V \cdot d\mathbf{r} = -\oint dV(\mathbf{r}) = 0. \tag{2.32}$$

Thus work done by the force is exactly canceled by work against it, in the description of a loop, when the force direction varies with position in the way represented by the gradient of a scalar function.

Force Fields

The foregoing discussion yields a characterization of a " conservative force," one which permits the definition of a conserved energy E, as in (2.21). To conserve energy in the latter way, the force must be of a particular type that is derivable from some scalar potential,

$$\boxed{\mathbf{F} = -\nabla V(\mathbf{r})}\,. \tag{2.33}$$

Such a force can depend on position in space only. Whenever a vector $\mathbf{F}(\mathbf{r})$ is assigned to each point $\mathbf{r}$ of a region in space, it is said that the region contains a " vector field." An example is provided by the uniform gravitational field, $\mathbf{F} = m\mathbf{g}$, in any limited region just above the earth's surface. At whatever point of this field a " test mass " m is placed, it makes the field detectable through the acceleration $\mathbf{g} = \mathbf{F}/m$ that it exhibits.

Whenever a scalar number $V(\mathbf{r})$ is assigned to each point of a region, a " scalar field " is said to be present there. Clearly a scalar potential field may characterize a force field quite as well as the vector field itself, insofar as the latter can be derived from it via $\mathbf{F} = -\nabla V$. If it is $\mathbf{F}(\mathbf{r})$ that is given, the potential is obtainable as

$$V(\mathbf{r}) = -\int_{\mathbf{r}_0}^{\mathbf{r}} \mathbf{F} \cdot d\mathbf{r} + \text{arbitrary constant.} \tag{2.34}$$

The arbitrary additive constant here displays the fact that the potential field is not uniquely determined when **F** is given; after all, V need only be such as to give a derivative (2.33) equal to $(-)\mathbf{F}$. When some particular value (relative to the initial point $\mathbf{r}_0$ of the integration) is given to the additive constant, it is said that a particular potential "gauge" has been chosen. When the constant of (2.34) is put equal to zero and some particular initial point $\mathbf{r}_0$ is chosen for the integration, the potential $V(\mathbf{r})$ is gauged by the work done by or against the force, *starting from* $\mathbf{r}_0$. Then $V(\mathbf{r}_0) \equiv V_0 = 0$ in (2.20). For the case of $\mathbf{F} = m\mathbf{g}$,

$$V(\mathbf{r}) = -m\mathbf{g} \cdot \int_{\mathbf{r}_0}^{\mathbf{r}} d\mathbf{r} = -m\mathbf{g} \cdot (\mathbf{r} - \mathbf{r}_0), \tag{2.35}$$

or $V = mg\,(z - z_0)$ if the z-axis is chosen to have the direction of $-\mathbf{g}$. Notice that when V is a function of z only, then $\nabla_x V = \nabla_y V = 0$ and $F_z = -dV/dz = F$ is the entire force.

The arbitrariness in the gauge of the potential means that the energy $E = T + V$ does not have a definite value until a zero point for it has been chosen. Of course, such conditional measures are familiar; for example, **r** is not definite until a frame origin is chosen!

Whenever $V(\mathbf{r}) \equiv V(x, y, z) =$ a constant, c, can be solved for $z = f(x, y, c)$, a *surface* is defined. In general, the definition of a surface is already implicit in $V(\mathbf{r}) = c$. Because the potential is the same at all points of such a surface, it is called an "equipotential." A continuous number of equipotential surfaces may be definable by varying the choice of the parameter c continuously. If $d\mathbf{r}_c$ is a displacement lying in an equipotential surface, then

$$dV = V(\mathbf{r} + d\mathbf{r}_c) - V(\mathbf{r}) = \nabla V \cdot d\mathbf{r}_c = 0. \tag{2.36}$$

Thus, the force $F(\mathbf{r}) = -\nabla V$ at any point of the field is perpendicular to the equipotential surface passing through the point. If $d\mathbf{r} \to d\mathbf{r}_n$ is a displacement *normal* to an equipotential, $F = -dV/|d\mathbf{r}_n|$; that is, force and work are generated by ascending or descending from one equipotential to another.

More specific examples of potentials will be examined in the succeeding chapters. Here some remarks will be made about their roles in general physical theory. The remarks will acquire substantial significance only after much further development of the theory and are inserted only as points to watch out for in the development.

Like the force functions introduced in §2.4, so potentials from which the forces are derivable may be used to represent the action of "sources" that are too complex to include as part of the system being given detailed treatment.

For example, the motion of a particle attached to a spring is treated with the help of a simple force function proportional to the displacement of the spring, instead of making an attempt to follow the motions of particles forming the spring. In this case, the potential from which that force is derivable may be considered energy "stored" in the spring. Such uses of the potential are sometimes called "phenomenological" but that is only a matter of degree.

Other examples of the potential, such as the field $V(\mathbf{r})$ about a point charge, may be regarded as rudimentary steps toward a type of concept which has at least as wide an importance in physical theory as does the concept of mass-particle. Referred to here are "entities," sometimes called *physical* fields, which, although spread out over space instead of being localized like a particle, can yet be attributed energy and momentum as is the particle. After this concept is fully developed, it will be possible, for example, to speak of a system consisting of a particle and a field which are "isolated together." Then a conserved energy will have a profounder meaning, as a characteristic of a complete isolated system, also having a conserved momentum. It requires deeper inquiry into "theories of force" before the part of the momentum possessed by the field can be adequately described. An example of such a description will be discussed in Chapter 15.

2.5 INERTIAL FORCES AND INERTIAL FRAMES

The foregoing discussed forms in which force functions might be given by some "theories of force," but no particular theory of force, as descriptive of some specific circumstances of motion, has been presented (discounting the sketchy example of a frictional force, used for illustrating the strictures of special-relativistic invariance, in §2.3, and the reference to gravitational force in (2.35)). Here, a particular source of forces will be given, one arising from purely kinematical circumstances and, for this reason, the consequent so-called "inertial forces" are frequently regarded as somehow "fictitious." This attitude may merely be a remnant of a near-anthropomorphism that has led to conceiving of forces as some kind of substantial "causal agencies." A newer attitude regards them all as helping describe circumstances, none any more real or fictitious than any other. The newer attitude has motivated attempts, like Einstein's, to unify all sources of force as manifestations of a "general relativity." The attempts have only been partially successful, as might be expected of physical theories aimed at complete comprehensiveness, and thus undergoing never-ending development. Moreover, the mathematical equipment introduced here so far is grossly inadequate for treating the attacks in question. Any considerations at this point must therefore be circumspect, yet will have value in helping clarify the roles of "inertial forces" and "inertial frames" in Classical Mechanics.

Forces in Relatively Accelerated Frames

A "general" relativity is immediately suggested by the opening clause in the statement of the special relativity principle in §2.1: "only relative motion has meaning." Reference frames in arbitrary relative motion, like those of §1.4, ought to be considered, and not only the uniform motions on which special relativity is based. If a force $\mathbf{F} = m\mathbf{a}$ is needed to account for a given motion as described in one frame, then, according to (1.53), a force

$$\mathscr{F} = m\mathbf{A} = \mathbf{F} - m\dot{\mathbf{v}}_0 - m\dot{\boldsymbol{\Omega}} \times \mathbf{R} - m\boldsymbol{\Omega} \times (\boldsymbol{\Omega} \times \mathbf{R}) - 2m\boldsymbol{\Omega} \times \mathbf{V} \qquad (2.37)$$

will have to be postulated to account for the same motion as "seen" from another frame, which in the first frame is ascribed the translational velocity $\mathbf{v}_0(t)$, and the rotation $\boldsymbol{\Omega}(t)$.

The formal need for a different force is already evident for the simple case of a uniform motion, say $x = vt$ along the x-axis of the "fixed" frame. Consider the description of the same motion with respect to a frame having a uniform rotation, $\boldsymbol{\Omega}$, relative to the fixed frame. Let respective axes of the two frames coincide at $t = 0$ and choose the rotation axis $\hat{\boldsymbol{\Omega}}$ for both the z-axis and Z-direction. Then $X = vt \cos \Omega t$, $Y = -vt \sin \Omega t$, and the motion appears to follow a spiral orbit, $R = -v\Phi/\Omega$, as viewed from the rotating frame. For the spiral to follow from the equations of motion, a force with components $\mathscr{F}_X = m\ddot{X} = -m\Omega^2 X - 2mv\Omega \sin \Omega t$, $\mathscr{F}_Y = -m\Omega^2 Y - 2mv\Omega \times \cos \Omega t$, or $\mathscr{F}_R = -m\Omega^2 R$, $\mathscr{F}_\Phi = -2mv\Omega$, will be needed. The radial and transversal components of this force are just (§1.4) the "centrifugal" and "Coriolis" terms of (2.37); of course, $\mathbf{F} = 0$, $\mathbf{v}_0 = 0$, and $\dot{\boldsymbol{\Omega}} = 0$ in this example.

All the terms of the force (2.37) that arise from the acceleration of the frame are proportional to the mass that the force is to accelerate. This accounts for the name "inertial forces" for the terms. The centrifugal force, $-m\boldsymbol{\Omega} \times (\boldsymbol{\Omega} \times \mathbf{R})$, and the Coriolis force, $-2m\boldsymbol{\Omega} \times \mathbf{V}$, are the only ones that arise from uniform rotation. Centrifugal forces are familiar to anyone who has sat in a speeding car while it followed a turn in a road. Anyone who has tried crossing the aisle of a train while it was rounding a curve in its track has experienced Coriolis force, and found it best to move slowly because of it. The other inertial forces may be the "jerks" one feels when in a nonuniformly moving vehicle.

The detectability of inertial forces implies the existence of a favored set of reference frames in the universe: any which may be so chosen that all inertial forces disappear from them. They are called "inertial frames" (sic!)

and it is usually supposed that they are at rest, or in uniform motion, with respect to the distant ("fixed") stars. This conclusion is motivated by such observations as those on the so-called "Foucault pendulum," which consists of a heavy ball suspended by a long wire from a point fixed to the earth's surface. When it is set swinging as frictionlessly as possible, it is found that the plane of the oscillations retains a fixed orientation with respect to the stars, but rotates with respect to the earth during its motion.

The existence of any special set of inertial frames seems inconsistent with a "general" relativity—that is, the equivalence of all frames, even of pairs in nonuniform relative motion. It may seem to imply that if a single body were alone in space, it could have rotation despite the absence of any other object relative to which its orientation could be judged; the rotation would be evident at least from the occurrence of centrifugal force. From this Mach argued that the existence of at least the distant stars is as essential to the detectability of centrifugal force as to the observability of orientation. It should be considered idle to speak of objects "alone in space," and to expect to form theories about all imaginable universes from observations on this one. Einstein has substantiated such considerations. He showed that a "general relativity" becomes tenable in a broader context than as discussed above, when "non-Euclidean" frames, defining curved nets of reference surfaces and not merely planes, are also admitted; the "space curvatures" are related to the inertial forces, and the distribution of matter in space and time is given an essential role in determining both. The mathematical languages introduced here so far are grossly inadequate for describing these matters more precisely right now.

"This Earthly Sphere"

A reference frame fixed to the earth's surface is of obvious importance to observers on earth. Through reference to the fixed stars, the earth is known to have a daily rotation about its polar axis, $\Omega = \pi/12$ per hour, as well as a yearly rotation about the sun. The diurnal rotation is 365 times as rapid as the annual one. Moreover, the displacements of objects moving near the earth's surface are much larger fractions of the earth's radius than of its distance from the sun. As a consequence, the inertial effects of the annual motion are less considerable than the already small effects of the diurnal one, and so only the daily round will be given attention here.

Resolved on axes **I**, **J**, **K** fixed to the earth's surface as indicated in Figure 2.2,

$$\mathbf{\Omega} = \Omega(\mathbf{I} \cos \lambda + \mathbf{K} \sin \lambda), \tag{2.38}$$

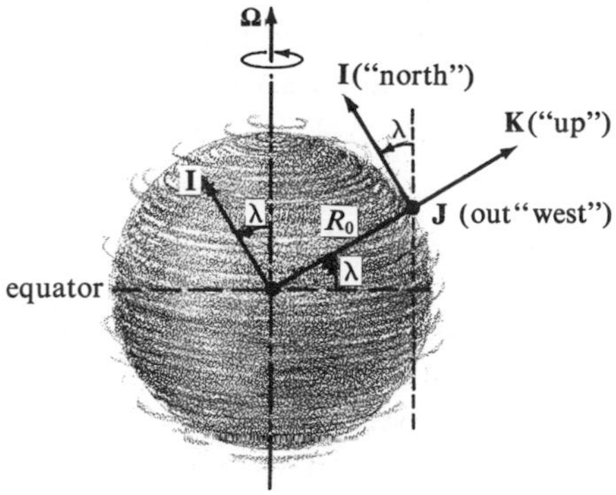

Figure 2.2

where λ is the latitude angle. The results (1.53) for the acceleration can be used most explicitly, and without affecting the conclusions, if the origin of the rotating frame is shifted to the earth's center, so that $\mathbf{r}_0 = \mathbf{v}_0 = \dot{\mathbf{v}}_0 = 0$. An object resting on the earth's surface will have the position vector $\mathbf{R} = \mathbf{K}R_0$ and its "absolute" velocity (relative to the earth's center) will be $\mathbf{v} = \mathbf{\Omega} \times \mathbf{R} = -\mathbf{J}\Omega R_0 \cos\lambda$, eastward and roughly 1000 m.p.h. in magnitude at the equator. The "absolute" acceleration $\mathbf{a}$ corresponding to any observed acceleration $\mathbf{A}$ is

$$\mathbf{a} = \mathbf{A} + \mathbf{\Omega} \times [-\mathbf{J}\Omega R_0 \cos\lambda] + 2\mathbf{\Omega} \times \mathbf{V}, \tag{2.39}$$

where $\hat{\Omega} \times \mathbf{J} = \mathbf{K}\cos\lambda - \mathbf{I}\sin\lambda$. A velocity $\mathbf{V}$ can be allowed without admitting an appreciable change of $\mathbf{R}$ in units of the earth's radius.

Suppose the object is at rest ($\mathbf{V} = 0$) and subject only to the absolute gravitational acceleration $\mathbf{a} = \mathbf{g} = -\mathbf{K}g$. The observed acceleration will be

$$\mathbf{A} \equiv \mathbf{g}' - -\mathbf{K}(g - \Omega^2 R_0 \cos^2\lambda) - \mathbf{I}\Omega^2 R_0 \sin\lambda\cos\lambda. \tag{2.40}$$

There is nominally a reduction in magnitude of the gravitational effect by the centrifugal force, amounting to about 0.3% at the equator. There is also a deflection of any plumb line, from pointing directly at the earth's center toward the equator; the deflection is greatest at 45° latitude and is of the order of a tenth of a degree, too small to be readily identifiable.

Cyclones in the northern hemisphere are supposed to develop their counterclockwise rotations (as seen from above) around their low-pressure centers because of Coriolis force, $-2m\mathbf{\Omega} \times \mathbf{V}$, accelerating the inward flow of the air masses. The effect is somewhat obscured by viscosity and other perturbations, but its existence is attested by comparison to cyclones in the southern hemisphere. There, the angular velocity is directed into, rather than out of, the ground, and the cyclones are found to develop clockwise flow. The Foucault pendulum also exhibits Coriolis acceleration (Exercise 4.2).

The detectability of inertial forces in frames fixed to the earth, arising from its rotation with respect to the stars, definitely indicates that such frames are not among the inertial ones.

Gravitation as an Inertial Force

That the inertial effects of the earth's motion appear in (2.40) as corrections to the gravitational acceleration has significance. The force of gravitation, $m\mathbf{g}$, has in common with inertial forces the characteristic of proportionality to the accelerated mass. This suggests the possibility of a viewpoint from which gravitation may appear as another type of inertial force. The idea was developed in Einstein's General Theory of Relativity.

Consider a moving frame that is falling freely toward the surface of the earth; an elevator that has broken away from its cables is commonly used for illustrating this. The elevator (and any object resting in it) has the gravitational acceleration $\dot{\mathbf{v}}_0 = \mathbf{g}$ relative to the earth. An observer within the falling elevator will find that objects he releases do not "fall" to the floor; he finds no force of gravitation in his frame. Of course, observers on the earth attribute his finding to the fact that the falling observer and his objects are all equally accelerated by the forces of gravity. However, from a General Relativistic point of view, the elevator forms as valid a reference frame as does the earth. From that frame, the earth appears to be undergoing acceleration $-\mathbf{g}$ toward the elevator; the force of gravity observed on earth becomes merely an inertial force arising from that relative acceleration.

This example is meant only to suggest how gravitation may become comparable to an inertial force. A complete investigation would have to take into account the fact that, as a whole, the earth's gravitation is not unidirectional and not uniform; moreover, finite propagation times for the forces must be allowed for, as Einstein discovered.

Isolated Systems and Invariances

Throughout this chapter of definitions and introductions of principles, there has been frequent allusion to "isolated systems."* Already in §2.2 the isolation of a pair of masses had to be stipulated, to ensure a definition of mass-measure which would give results invariant to circumstances in which the masses may be found. There and in §2.4, it was pointed out that the identification of momenta and of energies is subject to their conservation in isolated systems. In Chapter 4 still another conserved quantity, angular

* Some physicists prefer to call them "closed" systems.

momentum, will be found to have a similar provenance. In §2.3 discussions of force functions implied that the most complete descriptions, lending the predictive procedures their greatest power, are ones that depend only on what the system described itself is doing, as to be expected if it is isolated.

In the last section of §2.3, it was stressed that formulations purporting to describe isolated systems must have relativistic invariance. As the Galilean transformation (2.1) was defined, it includes a test for a simpler type of invariance as well: to changes of reference frames that are merely static translations, frames that are relatively at rest so that $\mathbf{v}_0 = 0$, $\mathbf{V} = \mathbf{v}$, and $\mathbf{R} = \mathbf{r} - \mathbf{r}_0$. Invariance to such space displacements is automatic for descriptions in terms of position vectors if it is understood that the vectors may be given any origin; it is made "manifest" if the description contains only relative position vectors $\mathbf{r}_i - \mathbf{r}_j$, since these are invariant if only the vectors have a common origin. The simple property in question here may be regarded as an expression of the "homogeneity of space" for an isolated system, independent of anything external to itself.

Space ought also to be isotropic for an isolated system; that is, a description of such a system should be invariant to the orientation of the frame. This again is automatic for a description in terms of vectors when it is understood that only their relative orientations have significance. However, making this invariance manifest leads to implications that will be made explicit for the first time in Chapter 4.

Finally, the description of an isolated system ought to be invariant to the particular moment chosen for $t = 0$. This accounts for the fact that explicit dependence on time is excluded from force functions that help to describe motions internal to an isolated system. It was stressed in §2.3 that admission of explicit dependence on time is used to represent *external* effects on a non-isolated system.

In sum, a description of an isolated system is characterized by invariance to space translations and rotations, and to translations in time. It must also be consistent with conservation principles, for momentum, angular momentum, and energy. The two sets of characterizations are actually equivalent, as will be shown after a sufficiently general basis for considering this is prepared (Chapter 7).

The discussion so far is predicated on the possibility of isolating *several* systems in the universe. In that case, each one ought to serve equally well as a reference base for laws of motion common to all of them. However, it was suggested in the preceding subsections that such a "general relativistic" point of view cannot be attained until even distant stars are included to complete an isolated system, and hence that only one genuinely isolated system, consisting of the entire universe, can really be defined.

The advantage of treating various, limited systems as isolated is nevertheless not to be denied. Such subsystems of the universe can be so treated insofar as inertial forces arising from accelerated motions relative to the stars can be ignored for the purposes of the description, or if the subsystem is adequately treated by assuming it to have at most a uniform motion relative to the fixed stars. For most purposes, the rotating, orbiting earth provides a "sufficiently inertial" frame. Indeed, for some purposes, even a single atom may be treated as an "island universe," isolated in space.

To avoid considering the usually undetectable corrections by the cosmic inertial forces, only a "special" relativity is enforced. In the Classical Mechanics, an invariance to Galilean transformations, as regards relatively moving frames, is almost always adequate.

EXERCISES

2.1. The Galilean transformation is sometimes regarded as a relic of the horse-and-buggy era! Use it to find the distribution of "absolute velocities" (relative to a road) of the points on the vertical diameter of a buggy wheel that is rolling along at a uniform speed v_0. Clods of mud are released from various points of the rim; from which quadrant comes the mud which flies highest, if the radius is a and $v_0 > (ga)^{1/2}$? At what position relative to the hub of the wheel does a rider's elbow protrude, if the highest-flying mud just manages to reach it? [At a height $(v_0^4 + (ga)^2)/2gv_0^2$ directly above the hub.]

2.2. A unit mass is sent into head-on collision with an unknown mass at rest. The two masses are then observed to recoil from each other with equal and opposite velocities, each just half the initial speed. What is the unknown mass on the presumption that momentum was conserved? Is the result consistent with kinetic energy conservation ("elasticity")? What would have been observed if the role of the two masses had been interchanged, and the second sent toward the first one at rest? Show that the same result is obtainable through an appropriate Galilean transformation of the original circumstances.

2.3. A rocket is projected straight up and explodes into three equally massive fragments just as it reaches the top of its flight. One of the fragments is observed to come straight down in a time t_1, while the other two land at a time t_2, after the burst. Find the height $h(t_1, t_2)$ at which the fragmentation occurred and the distances apart of the landing spots. [$h = \frac{1}{2}gt_1t_2(t_1 + 2t_2)/(2t_1 + t_2)$.] How much energy had been released as kinetic energy per unit mass?

2.4. A particle m is sent in directly toward a central point $x = 0$ with an initial (at $x = \infty$) speed v_0. It is observed that, as the particle approaches $x = 0$, it slows down gradually, and eventually turns back at $x = x_0 > 0$. Its motion during the times $t < t_0$ (where t_0 is the moment of turnabout) is found to be fitted by the formula

$$x - x_0 = s \ln \cosh \frac{v_0(t_0 - t)}{s},$$

with a certain value for the length s. Show that the observed motion can be understood as derivable from a potential

$$V(x) = Be^{-2|x|/s},$$

where B is predicted to be the least energy the particle would need to be given in order to pass through $x = 0$ without getting turned back. [$\cosh \zeta \equiv \frac{1}{2}(e^{\zeta} + e^{-\zeta})$.]

2.5. Show that

$$\nabla f(|\mathbf{r} - \mathbf{r}_0|) = f'(|\mathbf{r} - \mathbf{r}_0|) \frac{\mathbf{r} - \mathbf{r}_0}{|\mathbf{r} - \mathbf{r}_0|},$$

where f' is the derivative of f with respect to its argument.

2.6. A particle is subject to the "dipole force":

$$\mathbf{F} = \frac{D}{r^3} [(2 \cos \vartheta)\mathbf{e}_r + (\sin \vartheta)\mathbf{e}_\vartheta.]$$

By explicit evaluation of the work integral, find the work done by this force while the particle is displaced along a circular arc of radius a, centered on $r = 0$, between $\vartheta = 0$ and $\vartheta = \frac{1}{4}\pi$. Perform a similar evaluation for a second path between the same end points, but first proceeding radially inward to $r = \varepsilon < a$, then along an arc of radius $r = \varepsilon$ before going radially outward to the end point. You should conclude that the force is derivable from a potential and find this. Check that it yields all the above elements of work correctly.

2.7. Make a sketch indicating the equipotential surfaces of the dipole force field (Exercise 2.6). Also indicate lines having the directions of the potential gradients, and interpret the significance of the signs of the radial gradients just above and below the origin, in terms of the attractions or repulsion of charges. This will tell you why it is called a "dipole" force.

2.8. Draw a diagram which shows explicitly how the spiral discussed on page 49 is generated.

2.9. The findings in connection with (2.40) indicate that $\dot{Z} \approx -gt$, with g constant, is a very good approximation to an object's "vertical" speed at a time t after it begins to fall to the surface of the earth from a distance not too far out. Suppose the object starts from "absolute rest" (with respect to the "fixed stars") at a point a distance h above the earth's surface at a latitude λ in the northern hemisphere. Show that the main

Coriolis effect of the earth's diurnal rotation is to make the object land approximately a distance

$$|Y| = \left(\frac{2h}{3}\right)\left(\frac{2h}{g}\right)^{1/2} \Omega \cos \lambda$$

to the east of the point of the earth's surface directly under the starting point. [With only the Coriolis effect to be taken into account, $\ddot{\mathbf{R}} \approx \mathbf{g} - 2\mathbf{\Omega} \times \mathbf{V}$ with $\mathbf{g} \approx -g\mathbf{K}$ and $\mathbf{V} \approx -\mathbf{K}gt$, in the terms used in Figure 2.2.]

2.10. Argue that the Coriolis effect of the earth's diurnal rotation on any motion, with $\mathbf{V} = \mathbf{I}V_X + \mathbf{J}V_Y$, confined to a plane tangent to the earth's surface in a northern latitude, is to make the path deviate toward the right as judged from any point on it. (Begin by arguing that only the part $\mathbf{\Omega}' = \mathbf{K}\Omega \sin \lambda$ of the earth's rotation contributes to the Coriolis effect in the tangent plane.)

CHAPTER 3

LINEAR MOTIONS

Analytic descriptions of motion can be derived from the Law of Motion when the forces are given as functions—of position, velocity and/or time. The kinds of motions various simple force functions are capable of describing will be investigated next. The examples will all be chosen for their importance as prototypes and idealizations of far more complicated physical situations. In this chapter, attention will be restricted to particles subject only to forces which accelerate them along the line in which the motion starts, so that the particles never leave the straight line. It is then adequate to describe a particle's position by a variable distance $x(t)$ along the line.

It will be helpful to summarize the findings in the preceding chapters that pertain here, in simplified versions that are now sufficient. The motions $x(t)$ are to be derived from *equations of motion*,

$$m\ddot{x} = F, \tag{3.1}$$

for various force functions $F(x, \dot{x}, t)$ to be given. A general solution of one such second-order differential equation will contain two arbitrary constants: $x(t, c_1, c_2)$. The arbitrariness may be removed, and a specific motion obtained, if *initial conditions* $x(0, c_1, c_2) = x_0$ and $\dot{x}(0, c_1, c_2) = v_0$ are given. Such relations are just sufficient to determine c_1 and c_2.

There is always interest in examining the rate of work, or power, $dW/dt = F\,dx/dt$, being expended by the force at any phase of the motion. When positive, it goes into increasing the kinetic energy, $T = \frac{1}{2}mv^2 = \frac{1}{2}m\dot{x}^2$, at the rate

$$\frac{dT}{dt} = Fv. \tag{3.2}$$

In those cases in which the force is derivable from some potential, $V(x)$, according to

$$F(x) = -\frac{dV}{dx}, \tag{3.3}$$

the total energy defined by

$$E = T + V \tag{3.4}$$

is constant throughout the motion. The amount of energy in the motion depends on the start, x_0, v_0, given it:

$$E = \tfrac{1}{2}mv_0^2 + V(x_0). \tag{3.5}$$

The number E may sometimes be given, in place of one of the pieces of information x_0 or v_0, when specifying the starting conditions.

3.1 THE LINEAR HARMONIC OSCILLATOR

After the case of a constant force, already treated in §1.5, perhaps the simplest is the so-called "Hooke's Law" force. This is proportional to displacement x from an "equilibrium position" ($x = 0$, where $F = 0$) and directed toward it. If ω_0^2 is the force strength per unit mass and unit displacement,

$$m\ddot{x} = F = -m\omega_0^2 x. \tag{3.6}$$

The sign here is such that the acceleration tends to bring m back to the equilibrium point $x = 0$, for either positive or negative displacements x from it.

The importance of Hooke's Law stems from the fact that, for small displacements x from equilibrium, under any force function of position permitting it,

$$F(x) \approx F(0) + x\left(\frac{\partial F}{\partial x}\right)_0 + \tfrac{1}{2}x^2\left(\frac{\partial^2 F}{\partial x^2}\right)_0 + \cdots, \tag{3.7}$$

and $F(0) = 0$ by definition of equilibrium. The term $\sim x$ tends to be the only considerable one if x is small enough, and $(\partial F/\partial x)_0 < 0$ is necessary for $x = 0$ to be a "stable" equilibrium point, as will be seen. The next higher-order term is called an "anisotropic perturbation," having the same sign on either side of equilibrium.

The Energy

A first integral of (3.6) is obtained after multiplying it with the velocity $\dot{x}$. This was the procedure which was seen, in (2.13) to (2.21), to yield an energy:

$$\tfrac{1}{2}m\dot{x}^2 = -\tfrac{1}{2}m\omega_0^2 x^2 + E, \tag{3.8}$$

E being formally a first arbitrary constant of integration. It equals a conserved total energy, since

$$V(x) = \tfrac{1}{2}m\omega_0^2 x^2 \tag{3.9}$$

is just a potential energy from which the force can be derived, via $F = -dV/dx$ (3.3).

The Motion

To complete the integration of the equation of motion, it is convenient to define a length $a = (2E/m\omega_0^2)^{1/2}$ as the arbitrary integration constant in place of E;

$$E = \tfrac{1}{2}m\omega_0^2 a^2 \tag{3.10}$$

is plainly just the potential energy $V(a)$ at $x = a$, and hence the latter is a position at which no kinetic energy is left, and the velocity is reduced to zero.

The energy relation (3.8) yields for the velocity at any instant

$$\dot{x} \equiv \frac{dx}{dt} = \pm\omega_0(a^2 - x^2)^{1/2}, \tag{3.11}$$

which is easily integrated after a "separation of variables":

$$\pm\int \frac{dx}{(a^2 - x^2)^{1/2}} = \omega_0 t + \varphi_0 \text{ (the second constant of integration)}$$

$$= \pm\sin^{-1}\left(\frac{x}{a}\right).$$

Thus the motion is described by

$$x(t) = a \sin(\omega_0 t + \varphi_0), \tag{3.12}$$

after the alternative sign is dropped because it merely corresponds to increasing φ_0 by π, and φ_0 is an arbitrary constant anyway. Another way of writing the argument of the sine function, its "phase" $\omega_0 t + \varphi_0$, is as $\omega(t - t_0)$; then $t_0 \equiv -\varphi_0/\omega_0$ is a moment at which $x = 0$ and $\dot{x} = +\omega_0 a$, all the energy being in the kinetic form.

The motion is an oscillation of *amplitude a* (that is, $-a \leq x \leq a$). It is repeated in *periods* $T = 2\pi/\omega_0$, with frequency $\nu \equiv 1/T$. The quantity $\omega_0 = 2\pi\nu$ is sometimes called the "angular" or "circular" frequency. The constant φ_0 may be referred to as the "initial phase" of the oscillation.

The form (3.12) for the solution is most apt when the amplitude a or the energy E (3.10) is given initially, together with the initial phase, φ_0. If the initial state is given as x_0, v_0 instead,

$$a = [x_0^2 + (v_0/\omega_0)^2]^{1/2} \qquad \text{and} \qquad \varphi_0 = \tan^{-1}(\omega_0 x_0/v_0), \tag{3.13}$$

as follows from (3.12) evaluated at $t = 0$: $x(0) = a \sin \varphi_0 = x_0$ and $\dot{x}(0) = \omega_0 a \cos \varphi_0 = v_0$.

If the term of the general force expression (3.7) proportional to x is still the only considerable one but $(dF/dx)_0 > 0$, corresponding to reversing the sign of the force (3.6), then the solutions would be arbitrary linear combinations of the exponentials

$$x \sim \exp\colon \pm(dF/m\,dx)_0^{1/2} t. \tag{3.14}$$

These give infinite displacements for the remote future or past, and describe an "instability" of the system, in which the force does *not* tend to send the particle back to an equilibrium point from any other.

Energy Diagrams

The motion in a given potential field $V(x)$ can be understood quite completely from a diagram in which the energies involved, E, $T = \frac{1}{2}m\dot{x}^2$ and $V(x)$, are plotted against the position x. The case of an arbitrary $V(x)$ having an equilibrium point taken as $x = 0$, is plotted in Figure 3.1. Near the equilibrium point,

$$V(x) \approx V(0) + xV'(0) + \tfrac{1}{2}x^2 V''(0) + \tfrac{1}{6}x^3 V'''(0) + \cdots, \tag{3.15}$$

with $V'(0) = -F(0) = 0$ by definition of equilibrium, and $V(0) = 0$ in a suitable gauge. If the quadratic term is the only considerable one—that is, the

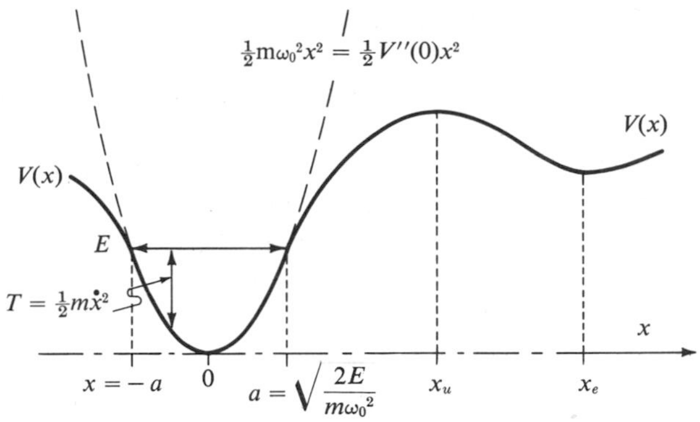

Figure 3.1

amplitude of a motion started in the neighborhood of $x = 0$ is small enough —then the potential $V(x)$ is effectively representable by the indicated parabola with $\omega_0^2 = V''(0)/m$. The complete potential indicated has a second equilibrium point at $x = x_e$, and a center of instability $x = x_u$. The remainder of the discussion will be focused on the motions near $x = 0$, for which $\omega_0^2 = -(dF/m\,dx)_0 > 0$.

If the motion is started with low enough energy E, as indicated by the horizontal line, then the simple harmonic oscillation (3.12), found as a solution of the equation of motion (3.6), may furnish an adequate description of the motion. The amount by which the potential energy $V(x)$ at x is less than the total energy E is just the kinetic energy, also indicated. The kinetic energy disappears at $x = \pm a = \pm(2E/m\omega_0^2)^{1/2}$, and in this way the diagram yields the amplitude of the oscillations. It is essential for the stable repetition of the motion that $V''(0) = -(dF/dx)_0 > 0$; a negative curvature of the potential, as at $x = x_u$, yields an instability as in (3.14). The positive curvature $V''(0)$ leads to slopes $F(\pm a) = -(dV/dx)_{\pm a}$ of such signs that the motion is turned back at $x = \pm a$; hence these are called "turning points" of the motion.

Oscillations Displaced by a Constant Force

The results for the simple harmonic oscillator are easy to modify to the case in which it is simultaneously subjected to a constant force, as is a particle suspended from a spring in the gravitational field:

$$m\ddot{z} = F = -m\omega_0^2 z - mg = -m\omega_0^2(z + g/\omega_0^2), \tag{3.16}$$

when $z = 0$ is the equilibrium point without gravity. The oscillation is then shifted to a new equilibrium point at $z = -g/\omega_0^2$ and

$$z(t) = -g/\omega_0^2 + a \sin(\omega_0 t + \varphi_0) \tag{3.17}$$

describes the motion. An energy diagram with

$$V(z) = \tfrac{1}{2}m\omega_0^2 z^2 + mgz = \tfrac{1}{2}m\omega_0^2(z + g/\omega_0^2)^2 - mg^2/2\omega_0^2$$

may help clarify the situation. The example here has provided an instance of a significant result: that "internal motion," like the oscillation relative to the resultant equilibrium point, is unaffected by uniform gravitation (§6.3).

3.2 A DAMPING FORCE

The simplest force depending on velocity is one proportional to it (see page 40). This is supposed to be characteristic of some types of friction (or air resistance), which always acts to slow down ("damp") the motion. If the constant γ is used to measure the force strength per unit momentum,

$$m\dot{v} = F = -\gamma m v \qquad (3.18)$$

is the equation of motion. Notice that $\gamma > 0$ gives deceleration regardless of the sign of the velocity v. "Separation of variables" yields

$$\int_{v_0}^{v} \frac{dv}{v} = \ln\left(\frac{v}{v_0}\right) = -\gamma t \qquad \text{or} \qquad v = v_0 e^{-\gamma t}, \qquad (3.19)$$

where v_0, the first integration constant, is determined by the initial speed. The result (3.19) is easy to integrate again, to obtain $x(t)$, but the exponential "decay" of the speed exhibited by (3.19) has the principal interest.

A force dependent on the velocity cannot be derived from a scalar potential function of position only, and this inexpressibility of a conserved energy corresponds to the fact that energy lost to friction is irretrievable* for the motion. The kinetic energy at any moment is the steadily diminishing one

$$T = \tfrac{1}{2}mv^2 = T_0 e^{-2\gamma t}, \qquad (3.20)$$

where $T_0 = \tfrac{1}{2}mv_0^2$ is its initial value. Thus 2γ turns out to be just the fractional rate of energy loss,

$$-\frac{1}{T}\frac{dT}{dt} = 2\gamma. \qquad (3.21)$$

* To restore energy conservation, the "medium" responsible for the friction would also have to be described, and the energy transferred to it formulated (perhaps as a "thermal motion"). See page 44.

After each time interval $\Delta t = 1/2\gamma$, the energy is reduced to a fraction $1/e$ and this characterizes the definition of a "mean life," $1/2\gamma$, for the energy here. Given even a large loss rate 2γ, the damping force $\sim v$ still requires an infinite time to reduce the energy to zero; this is obviously because the resistance weakens as the energy is diminished.

When gravitation, $-mg$, is superposed, as for a falling body subject to air resistance, the solution becomes

$$v = (\text{constant})e^{-\gamma t} - g/\gamma. \tag{3.22}$$

Thus the speed of fall reaches a limit, g/γ, not surpassed even in an infinite time of fall. The arbitrary (constant) may be evaluated as $(v_0 + g/\gamma)$, and then the solution (3.22) properly reduces to $v = v_0 - gt$ as $\gamma \to 0$.

3.3 COMPLEX NUMBERS

For the treatment of oscillations, which will be continued in this chapter, it will be convenient to employ the mathematical vocabulary of the complex numbers. These are numbers of the form $\zeta = \xi + i\eta$, where ξ and η are real and i is the "imaginary unit," having the properties $i^2 = -1$, $i^3 = -i$, $i^4 = 1$, $i^5 = i$, ...

The Complex Plane

The real numbers, including the integers, rational fractions, irrationals, and transcendentals, can be put into correspondence with the points of a continuous line. To have a point in one-to-one correspondence with every pair of real numbers ξ, η, which can specify a complex number $\zeta = \xi + i\eta$, a whole plane of points is needed. Such a "complex plane" may be defined as suggested by Figure 3.2. The "real part," $\xi \equiv \text{Re}(\zeta)$, and the "imaginary part," $\eta \equiv \text{Im}(\zeta)$, may serve as cartesian coordinates of the point in correspondence to ζ. A "polar representation" ("Argand diagram") may be adopted instead,

$$\zeta = |\zeta| (\cos \varphi + i \sin \varphi), \tag{3.23}$$

in which $|\zeta| = (\xi^2 + \eta^2)^{1/2}$ is called the "absolute value" or the "magnitude" of ζ, while $\varphi = \tan^{-1}(\eta/\xi)$ is its "phase." Sometimes $|\zeta|$ is called the "modulus" and is denoted $\text{mod}(\zeta)$, while φ is referred to as the "argument" of ζ and is denoted $\arg(\zeta)$.

The point of the complex plane corresponding to $\zeta = \xi + i\eta$ may also

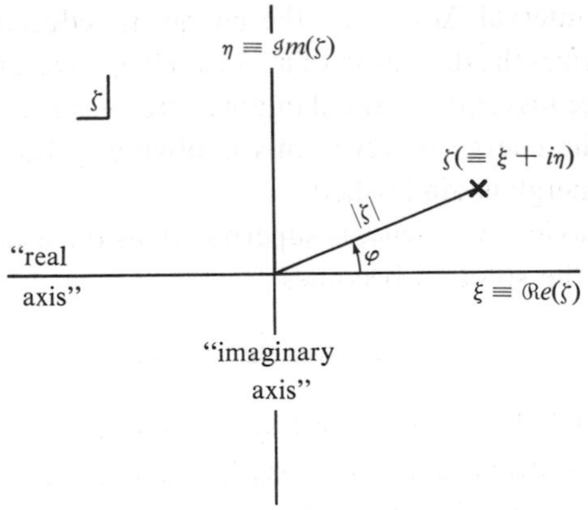

Figure 3.2

be located with the help of a real vector $\boldsymbol{\zeta} = \mathbf{i}\xi + \mathbf{j}\eta$; that is, a correspondence $\zeta \leftrightarrow \boldsymbol{\zeta}$ is adopted. Then the sum of two complex numbers,

$$\zeta_1 + \zeta_2 = (\xi_1 + \xi_2) + i(\eta_1 + \eta_2), \tag{3.24}$$

corresponds to the vector resultant $\boldsymbol{\zeta}_1 + \boldsymbol{\zeta}_2$. However, the product

$$\zeta_1\zeta_2 = \xi_1\xi_2 - \eta_1\eta_2 + i(\xi_1\eta_2 + \xi_2\eta_1) \tag{3.25}$$

does not have direct correspondence with either the scalar product $\boldsymbol{\zeta}_1 \cdot \boldsymbol{\zeta}_2 = \xi_1\xi_2 + \eta_1\eta_2$, or the vector product $|\boldsymbol{\zeta}_1 \times \boldsymbol{\zeta}_2| = |\xi_1\eta_2 - \xi_2\eta_1|$.

Just as vector algebra does not require resolutions into cartesian components, so the complex number algebra may be carried on without breaking up the numbers into their real and imaginary parts. In this connection, the *complex conjugate*

$$\zeta^* = \xi - i\eta \tag{3.26}$$

of $\zeta = \xi + i\eta$ may be defined; ζ^* is just the "mirror image" of ζ as reflected in the real axis. Then explicit reference to the real and imaginary parts of ζ may be eliminated through the substitutions

$$Re(\zeta) \equiv \xi = \tfrac{1}{2}(\zeta + \zeta^*) \qquad \text{and} \qquad \mathrm{Im}(\zeta) \equiv \eta = \frac{1}{2i}(\zeta - \zeta^*). \tag{3.27}$$

Thus ζ and ζ^* may be used as independent parameters covering the complex ζ-plane in place of ξ and η. Giving them determines the magnitude and phase of ζ via

$$|\zeta| = (\zeta^*\zeta)^{1/2} \qquad \text{and} \qquad \tan\varphi = \frac{\zeta - \zeta^*}{i(\zeta + \zeta^*)}. \tag{3.28}$$

Multiplying the complex conjugate of a number into a second complex number yields

$$\zeta_1^*\zeta_2 = (\xi_1\xi_2 + \eta_1\eta_2) + i(\xi_1\eta_2 - \xi_2\eta_1) \\ = \boldsymbol{\zeta}_1 \cdot \boldsymbol{\zeta}_2 + i(\boldsymbol{\zeta}_1 \times \boldsymbol{\zeta}_2)_3, \tag{3.29}$$

with the indicated relation to the real scalar product, and the "third" (and only) vector product component, of the vectors $\boldsymbol{\zeta}_1$ and $\boldsymbol{\zeta}_2$.

Rotations of Complex Numbers

Consider the complex number $i\zeta$ that results from the "operation" of multiplying ζ by i:

$$i\zeta = i(\xi + i\eta) = -\eta + i\xi.$$

This plainly has the same magnitude as ζ but has a phase rotated by 90° from that of ζ (Figure 3.3). Repetition of the operation gives $i^2\zeta = -\zeta$, rotated by 180° from ζ. In general, i^n is an operator which rotates complex numbers through the angle $n\pi/2$.

Next, an operator is sought which will rotate ζ through an arbitrary angle, $\varphi' - \varphi$. First consider an infinitesimal rotation by $d\varphi$, in which ζ is taken to $\zeta + d\zeta = \xi + d\xi + i(\eta + d\eta)$. For this to be merely a rotation, $|\zeta + d\zeta| = |\zeta|$ is necessary, and Figure 3.3 helps make clear that this will be so if $d\xi = -(|\zeta|\, d\varphi) \sin\varphi$ and $d\eta = (|\zeta|\, d\varphi)\cos\varphi$ are chosen. Thus,

$$d\zeta = d\xi + i d\eta = |\zeta|\, d\varphi(-\sin\varphi + i\cos\varphi) = i\zeta\, d\varphi,$$

having the direction of $i\zeta$ perpendicular to ζ. Straightforward integration (a process of addition, already defined) of $d\zeta/\zeta = i\, d\varphi$ gives

$$\ln \zeta \Big|_{\zeta}^{\zeta'} = i(\varphi' - \varphi) = \ln\frac{\zeta'}{\zeta}$$

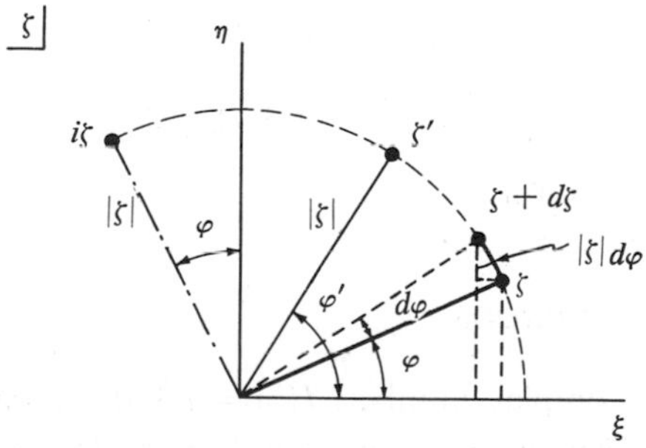

Figure 3.3

and

$$\zeta' = e^{i(\varphi' - \varphi)}\zeta, \tag{3.30}$$

the operation sought.

As a real number, $|\zeta|$ has the direction of the positive real axis. To attain the direction of $\zeta = |\zeta|(\cos\varphi + i\sin\varphi)$ it must be rotated through the angle φ; hence

$$\zeta = |\zeta|\, e^{i\varphi} \tag{3.31}$$

presents a compact expression of any complex number in terms of its magnitude and phase. Implicit here is the famous De Moivre theorem,

$$e^{i\varphi} = \cos\varphi + i\sin\varphi, \tag{3.32}$$

which yields

$$\cos\varphi = \frac{e^{i\varphi} + e^{-i\varphi}}{2} \quad \text{and} \quad \sin\varphi = \frac{e^{i\varphi} - e^{-i\varphi}}{2i}. \tag{3.33}$$

The representation (3.31) also makes the expression of products, powers, and roots of complex numbers simple. Thus

$$\zeta_1\zeta_2 = |\zeta_1|\,|\zeta_2|\, e^{i(\varphi_1 + \varphi_2)},$$

having implicit in it the trigonometric relations

$$\begin{aligned} \cos(\varphi_1 + \varphi_2) &= \cos\varphi_1 \cos\varphi_2 - \sin\varphi_1 \sin\varphi_2, \\ \sin(\varphi_1 + \varphi_2) &= \sin\varphi_1 \cos\varphi_2 + \cos\varphi_1 \sin\varphi_2. \end{aligned} \tag{3.34}$$

Similarly,

$$\zeta^n = |\zeta|^n\, e^{in\varphi} = |\zeta|^n(\cos n\varphi + i\sin n\varphi), \tag{3.35}$$

with the special cases

$$\begin{aligned} i^n &= e^{in\pi/2} = (-)^{n/2}, \\ \sqrt{i} &= e^{i\pi/4} = (1 + i)/\sqrt{2}, \end{aligned} \tag{3.36}$$

which are particularly useful.

Complex Functions

Assigning a complex number, $f(\zeta) = u + iv$, to each point ζ of the complex plane defines a complex function of the complex variable ζ. Naturally, real functions may also be defined ($v \equiv 0$), as well as complex functions of a real variable (for example, $f(\xi) = u(\xi) + iv(\xi)$ "on the real axis" of the ζ-plane). The power ζ^n of (3.35) is an example of a complex function of the complex variable ζ, and of the real variables $|\zeta|$ and φ.

Any $f(\zeta)$ can also be considered a function of the real variables ξ, η when $\zeta = \xi + i\eta$. However, ξ and η occur in expressions $f(\zeta)$ only in the combination $\xi + i\eta$. An arbitrary function of ξ, η, such as $F(\xi, \eta) = F(\frac{1}{2}(\zeta + \zeta^*), (1/2i)(\zeta - \zeta^*))$, is generally a function of both ζ and ζ^*. The real and imaginary parts u, v of a function $f(\zeta)$ will obviously be real functions of the real variables ξ, η:

$$f(\zeta) = u(\xi, \eta) + iv(\xi, \eta).$$

It is then clear that expressions in terms of ζ as the only complex quantity have the property

$$f(\zeta^*) = u - iv = f^*(\zeta). \tag{3.37}$$

More general expressions of a complex function in terms of real quantities and i can be transformed into their complex conjugates simply by reversing the sign of i everywhere it occurs.

The equation of two complex expressions implies that their real and imaginary parts are separately equal. Equal complex numbers correspond to the same point of a complex plane, with individually equal real and imaginary "cartesian components." An example is provided by the identification of the real and imaginary parts of such a function as $f(\zeta) \equiv \zeta^{-1}$:

$$u + iv \equiv \frac{1}{\zeta} = \frac{\zeta^*}{|\zeta|^2} = \frac{\xi - i\eta}{\xi^2 + \eta^2}.$$

The real functions

$$u(\xi, \eta) = \xi/(\xi^2 + \eta^2) \qquad \text{and} \qquad v(\xi, \eta) = -\eta/(\xi^2 + \eta^2)$$

can now be read off. (See Exercise 3.13 for more examples.)

3.4 FORCED OSCILLATIONS

Consider the effect of a force that is explicitly dependent on the time. As discussed in §2.3, such a force may be interpreted as an interference with the motion of the particle by an external, otherwise undescribed, agency. An important case is the periodic impulsion provided by the force $F = F_0 \sin \omega t$. Quite arbitrarily time-dependent functions can be represented as superpositions (sums) of harmonics like $\sin \omega t$, with various frequencies ω, as will eventually be learned (Chapter 12).

The force $F = F_0 \sin \omega t$ will here be applied to a linear oscillator which would, without this interference, perform oscillations with its "natural" frequency, ω_0. It is thus assumed that

$$m\ddot{x} = -m\omega_0^2 x + F_0 \sin \omega t. \tag{3.38}$$

Since $e^{i\omega t} = \cos \omega t + i \sin \omega t$, this equation is just the imaginary part of

$$m\ddot{\zeta} = -m\omega_0^2 \zeta + F_0 e^{i\omega t}, \tag{3.39}$$

if $\zeta = \xi + ix$. It proves somewhat more convenient to deal with the complex exponentials, in place of the trigonometric functions, as the problem is elaborated in succeeding sections.

A standard way to approach solution of "linear" differential equations (those containing the dependent variable only in terms linear in it or its derivatives) is by trying forms like $\zeta = A$ exp: αt. Since $\dot{\zeta} = \alpha\zeta$ and $\ddot{\zeta} = \alpha^2\zeta$ in this trial form, (3.39) will be satisfied by it if

$$m(\alpha^2 + \omega_0^2)Ae^{\alpha t} = F_0 e^{i\omega t}.$$

This makes it obvious that the equation will be satisfied at any time with the choices

$$\alpha = i\omega \qquad \text{and} \qquad A = F_0/m(\omega_0^2 - \omega^2)$$

in the trial form. If the imaginary part of $\zeta = A$ exp: $i\omega t$ is called η, it becomes clear that

$$\eta(t) = [F_0/m(\omega_0^2 - \omega^2)] \sin \omega t \tag{3.40}$$

is a possible solution, $x = \eta(t)$, of the equation of motion (3.38). It is easy to check this by direct substitution into (3.38). Plainly, the real part of $\zeta(t)$ gives a motion, $\xi(t)$, which would arise from applying $F_0 \cos \omega t$, in place of $F_0 \sin \omega t$.

The solution (3.40) is a special one, directly proportional to the imposed force. There are no arbitrary integration constants in it, and so it can represent the displacement $x(t)$ only if the motion were started with $x(0) = 0$ and $\dot{x}(0) = \omega F_0/m(\omega_0^2 - \omega^2)$. To be adaptable to arbitrary starting conditions, x_0, v_0, a general solution is needed.

The general solution of (3.38) can be obtained by adding to the special solution $\eta(t)$ of the "inhomogeneous" equation (containing the "inhomogeneity" $F_0 \sin \omega t$), a general solution of the "homogeneous" equation

$$m\ddot{X} = -m\omega_0^2 X, \tag{3.41}$$

found to be

$$X = a \sin(\omega_0 t + \varphi)$$

in §3.1. The sum $x = X + \eta$ not only satisfies the full equation (3.38) but has the requisite two arbitrary constants a, φ which enable its adaptation to any starting conditions.

The result,

$$x = X + \eta = a \sin(\omega_0 t + \varphi) + [F_0/m(\omega_0^2 - \omega^2)] \sin \omega t, \tag{3.42}$$

shows that the applied force merely superposes a motion of its own frequency on the natural oscillation. It also shows that if the forced oscillation is applied with exactly the natural frequency, $\omega = \omega_0$, then the oscillations get out of hand ($x \to \infty$). This is the phenomenon called "resonance" of the applied force with the natural oscillations. Actually the wide oscillations and high speeds that arise as $\omega \to \omega_0$ may bring into play frictional forces which might otherwise have been negligible. The effect of such damping on oscillations will be considered next.

3.5 DAMPED OSCILLATIONS

To be investigated first is the effect of a damping force, $-\gamma m v$, on *natural* oscillations:

$$\ddot{x} = -\omega_0^2 x - \gamma \dot{x}. \tag{3.43}$$

If solutions of the form $x = A$ exp: αt are sought,

$$\alpha^2 = -\omega_0^2 - \gamma\alpha$$

follows as the condition on α. The equation of motion (3.43) is satisfied with *any* amplitude A, if only

$$\alpha = -\tfrac{1}{2}\gamma \pm \sqrt{\tfrac{1}{4}\gamma^2 - \omega_0^2}\,.$$

Two solutions have thus been found:

$$A \text{ exp: } -\tfrac{1}{2}\gamma + \sqrt{\tfrac{1}{4}\gamma^2 - \omega_0^2}$$

and

$$A' \text{ exp: } -\tfrac{1}{2}\gamma - \sqrt{\tfrac{1}{4}\gamma^2 - \omega_0^2}\,.$$

The sum of these is also a solution (as is readily verified; it is generally true that sums of solutions of "linear" homogeneous equations are also solutions). Now at hand is the *general* solution,

$$x(t) = e^{-(1/2)\gamma t}(Ae^{+\sqrt{(1/4)\gamma^2-\omega_0^2}t} + A'e^{-\sqrt{(1/4)\gamma^2-\omega_0^2}t}), \tag{3.44}$$

having the requisite two arbitrary constants, A and A'. Since $\frac{1}{2}\gamma > |\sqrt{\frac{1}{4}\gamma^2 - \omega_0^2}|$, the displacement $x(t)$ decreases toward $x(\infty) = 0$, and the motion is damped. When $\frac{1}{2}\gamma > \omega_0$, all quantities in (3.44) are real and the motion is without oscillation; large enough friction will damp down any displacement of an oscillator without permitting it to carry out an oscillation.

For cases of smaller damping, $\frac{1}{2}\gamma < \omega_0$, it is convenient to define the real frequency,

$$\omega_\gamma \equiv \sqrt{\omega_0^2 - \tfrac{1}{4}\gamma^2}.$$

Then (3.44) may be rewritten as

$$x(t) = e^{-(1/2)\gamma t}(Ae^{i\omega_\gamma t} + A^*e^{-i\omega_\gamma t}),$$

with $A' = A^*$ for a real x (compare (3.27)). The solution is still general if A is complex, for a complex parameter is equivalent to two real ones. The solution may also be presented as

$$x(t) = e^{-(1/2)\gamma t}\, a \sin(\omega_\gamma t + \varphi), \tag{3.45}$$

with $ae^{i\varphi} \equiv 2iA$; here, a and φ replace A and A^* as the two arbitrary constants of integration. Thus the smaller damping permits oscillations, but drags their frequency down to $\omega_\gamma = \sqrt{\omega_0^2 - \frac{1}{4}\gamma^2}$, lower than the undamped frequency. The effective amplitude of the oscillations, $a \exp\colon -\frac{1}{2}\gamma t$, decreases steadily.

Cases of $\frac{1}{2}\gamma \gtreqless \omega_0$ are compared against the undamped oscillation in Figure 3.4; each corresponds to a start from $x(0) = 0$ with the same impulse, mv_0. The comparisons acquire interest from the fact that many measurement devices (for example, galvanometers) make use of the response of some damped oscillator to an impulse to be measured. The instrument has the greatest sensitivity when the first maximum in the response $x(t)$ is as large as possible; the first maximum has the magnitude $x_1(t_1) = (v_0/\omega_0) \exp\colon -\frac{1}{2}\gamma t_1$, with t_1 the time at which it occurs. This makes it advantageous to have as small a damping as possible, yet it is also desirable to eliminate oscillations in order that the instrument be ready for a new measurement as soon as possible. Consequently, it is the so-called "critical damping", with $\frac{1}{2}\gamma = \omega_0$, that is usually introduced.

In the case of the critical damping, the solutions (3.44) and (3.45) degenerate to

$$(A + A') \exp\colon -\omega_0 t \equiv (a \sin \varphi) \exp\colon -\omega_0 t,$$

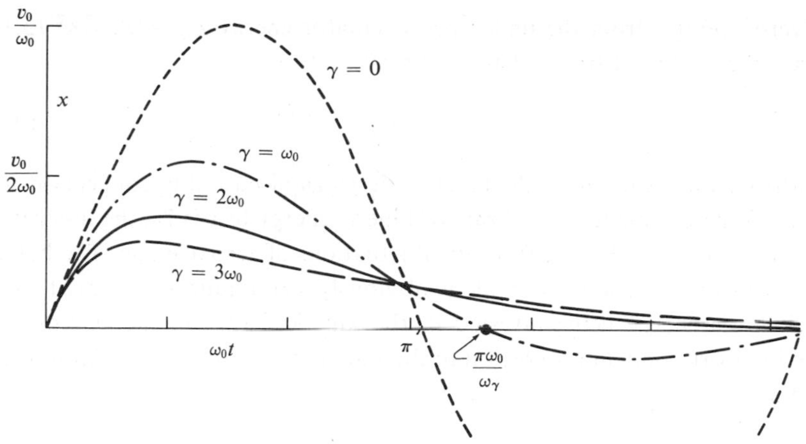

Figure 3.4

and so retain effectively just one arbitrary constant, $A + A' \equiv a \sin \varphi$; they are no longer general enough for adaptability to arbitrary starting conditions. As $\frac{1}{2}\gamma \to \omega_0$, and hence $\omega_\gamma \to 0$ is *approached* in (3.45),

$$x(t) = e^{-(1/2)\gamma t}[(a \cos \varphi) \sin \omega_\gamma t + (a \sin \varphi) \cos \omega_\gamma t]$$
$$\to e^{-\omega_0 t}[(a \cos \varphi)\omega_\gamma t + (a \sin \varphi)],$$

showing that a form linear in t, multiplied by the exponential, obeys the equation of motion in the limit of $\frac{1}{2}\gamma \to \omega_0$. It is easy to check that such a form, which may be written

$$x(t) = e^{-\omega_0 t}[(v_0 + \omega_0 x_0)t + x_0], \tag{3.46}$$

is a general solution of the equation of motion (3.43) with $\frac{1}{2}\gamma = \omega_0$, if v_0 and x_0 are arbitrary (see also Exercise 3.16).

The rate of energy loss in work against the friction is $(m\gamma\dot{x})\dot{x} = 2\gamma(\frac{1}{2}m\dot{x}^2)$, with 2γ being the fractional rate of kinetic energy loss just as in (3.21). Since the damped oscillator's equation of motion leads to

$$-\frac{dE}{dt} \equiv \dot{x}(m\gamma\dot{x}) = -m\dot{x}(\ddot{x} + \omega_0^2 x), \tag{3.47}$$

it is appropriate to call $E(t) \equiv \frac{1}{2}m(\dot{x}^2 + \omega_0^2 x^2)$ the instantaneous (unconserved) energy of the oscillator. For oscillations of the form (3.45), this can be calculated to be

$$E(t) = \tfrac{1}{2}m\omega_0^2 a^2 e^{-\gamma t}[1 - (\gamma/2\omega_0) \sin 2(\omega_\gamma t + \varphi')], \tag{3.48}$$

with $\tan 2(\varphi' - \varphi) = \gamma/2\omega_\gamma$, decaying exponentially, aside from fluctuations of zero average, from the undamped oscillator energy $E_0 \equiv \frac{1}{2}m\omega_0^2 a^2$ (3.10). The "mean life" of the oscillation can be said to be

$$\tau = 1/\gamma, \tag{3.49}$$

just twice the mean life of the kinetic energy as discussed in connection with (3.21). Whereas the fractional rate of kinetic energy loss is 2γ, the loss rate of $E(t)$ is $-dE/dt = \gamma E$ aside from the fluctuations of zero average. The halving of the average fractional loss rate is obviously a consequence of only half the oscillator's energy being invested in the kinetic form, on the average; the friction works only in proportion to the kinetic energy of the oscillator at any instant.

3.6 FORCED OSCILLATIONS OF A DAMPED OSCILLATOR

The periodic force, $F_0 \sin \omega t$, will now be applied to the damped oscillator. As in (3.42), the motion may be analyzed into a sum, $x = X + \eta$, of natural and forced motions, with

$$\begin{aligned} \ddot{X} &= -\omega_0^2 X - \gamma \dot{X}, \\ \ddot{\eta} &= -\omega_0^2 \eta - \gamma \dot{\eta} + (F_0/m) \sin \omega t, \end{aligned} \tag{3.50}$$

being the homogeneous and inhomogeneous equations, respectively. The part $X(t)$ is to contain all the arbitrary integration constants, and so is determined by the "accidents" of the start given to the motion. It has one of the forms found in the preceding section (whether (3.44), (3.45), or (3.46) depends on $\frac{1}{2}\gamma \gtrless \omega_0$), and all of the forms tend to vanish as time goes on. Thus the effects of any particular starting conditions are only *transient*. In contrast to this, the forced part $\eta(t)$ will persist as long as the force is applied.

The same procedure that led to (3.40) will here yield

$$x(t) = \eta(t) = \frac{F_0}{m} \frac{\sin(\omega t - \delta)}{[(\omega_0^2 - \omega^2)^2 + \gamma^2\omega^2]^{1/2}},$$

with (3.51)

$$\delta = \tan^{-1}[\gamma\omega/(\omega_0^2 - \omega^2)],$$

as the motion settled on after the transients become negligible. With friction present ($\gamma \neq 0$), this eventually established response is out of phase with the exciting force, $F_0 \sin \omega t$, as represented by $\delta \neq 0$. A range of phase shifts

$0 \le \delta < \pi$ can represent the entire range of responses obtained for various γ and various applied frequencies, as is shown by the following considerations.

When $\omega \ll \omega_0$, the formula yields

$$x(t) \to (F_0 \sin \omega t)/(m\omega_0^2),$$

as is representable by $\delta \to 0$. This result demonstrates that a very gradually applied force gives the motion time to adjust itself to following its variation exactly (in phase). Moreover, no appreciable friction is called into play during the slow motion, as the absence of γ from the result indicates.

When $\omega \gg \omega_0$, as is representable by a $\delta \to \pi$,

$$x(t) \to -(F_0 \sin \omega t)/m\omega^2[1 + (\gamma/\omega)^2]^{1/2}.$$

Thus, when the impulses are applied with great rapidity, they must keep catching the particle in an opposing phase of motion, if the response is to be an eventually steady one.

Finally, when the applied impulses just match the natural oscillations in frequency ($\omega = \omega_0$), the response no longer tends to infinity as in (3.40), because of the damping now present. Instead, the finite oscillation

$$x(t) = -(F_0 \cos \omega t)/m\gamma\omega_0$$

is maintained, representable by $\delta = \frac{1}{2}\pi$. Here the force is supplying its maximum impulses, $\pm F_0$, as the particle passes through its equilibrium point ($x = 0$), and none at the moments of greatest displacement, so that the oscillations do not get out of hand.

There is a continuous loss of power to the friction, being $(m\gamma\dot{x})\dot{x} = 2\gamma(\frac{1}{2}m\dot{x}^2)$ as in (3.21) at any instant. The motion persists because the loss is exactly offset by the power supply, $(F_0 \sin \omega t)\dot{x}$, in the average over each period $T = 2\pi/\omega$. The instantaneous power supply is proportional to

$$\sin \omega t \cos(\omega t - \delta) = \sin \omega t \cos \omega t(\cos \delta) + \sin^2 \omega t(\sin \delta),$$

and though the first term has a zero average, the remainder averages to

$$\sin \delta \int_0^T (dt/T) \sin^2 \omega t = \tfrac{1}{2} \sin \delta.$$

This shows that the energy transfer will proceed at a maximum rate when $\delta = \pi/2$—that is, when $\omega \to \omega_0$, a condition which may be taken to define "resonance." Such a definition would coincide with the condition for a maximum in the average *kinetic* energy of the oscillations:

$$\int_0^T \frac{dt}{T} (\tfrac{1}{2}m\dot{x}^2) = \frac{\omega^2 F_0^2/4m}{(\omega_0^2 - \omega^2)^2 + \gamma^2\omega^2}.$$

Half the peak value here, $F_0^2/4m\gamma^2$, is attained when ω is "off resonance" by $|\omega - \omega_0| \approx \frac{1}{2}\gamma$, in the limit of small loss rates ($2\gamma \ll \omega_0$). How this result can be obtained will be illustrated in connection with a second definition of resonance, below. The energy transfer rate ($\sim \sin \delta$) is not then reduced to half its peak value, but only by a factor of $\sin(\pi/4) = 2^{-1/2}$.

The definition of "resonance" described in the preceding paragraph does not quite coincide with the condition for a peak value in the amplitude, $a(\omega)$, of the response (3.51). A redefinition in accord with the latter condition gains significance when the interest is in applying the force $F_0 \sin \omega t$ only until the motion (3.51) is established, as a means of starting a natural oscillation of frequency ω_0 with as large an amplitude as possible. If the power supply is cut off just as the displacement reaches a maximum $a(\omega)$, then an initial state will have been established for the natural oscillations, having the energy (3.10):

$$E_0(\omega) = \tfrac{1}{2}m\omega_0^2 a^2(\omega) = \frac{\omega_0^2 F_0^2/2m}{(\omega_0^2 - \omega^2)^2 + \gamma^2\omega^2}$$

$$= \frac{\omega_0^2 F_0^2/2m}{(\omega_r^2 - \omega^2)^2 + \gamma^2\omega_\gamma^2}, \tag{3.53}$$

if $\omega_\gamma^2 = \omega_0^2 - \frac{1}{4}\gamma^2$ as in the damped oscillation (3.45), and

$$\omega_r^2 = \omega_0^2 - \tfrac{1}{2}\gamma^2. \tag{3.54}$$

This gives the redefinition $\omega = \omega_r$ for the resonance frequency.

The effectiveness of having used various frequencies ω to excite the natural oscillator energy, $E_0(\omega)$, is exhibited in Figure 3.5. There is a fall-off on both sides of the peak value, $E_0(\omega_r) = \omega_0^2 F_0^2/2m\gamma^2\omega_\gamma^2$. The "resonance width" is defined as the difference between the two frequencies $\omega_\pm \gtrless \omega_r$ at which half the peak value is attained, and so $\omega_\pm^2 - \omega_r^2 = \pm\gamma\omega_\gamma$. It is easy to find from this that

$$\omega_\pm - \omega_r = \pm\gamma\omega_\gamma/(\omega_\pm + \omega_r) \approx \pm\tfrac{1}{2}\gamma \tag{3.55}$$

in the limit of slow loss rates, $\gamma \ll \omega_0$, when $\omega_r \approx \omega_\gamma \approx \omega_0$ and $\omega_\pm^2 \approx \omega_0(\omega_0 \pm \gamma)$. Thus, the breadth of the resonance in the response to an excitation frequency ω is $\Delta\omega = \omega_+ - \omega_- \approx \gamma$.

The excitation energy $E_0 = \frac{1}{2}m\omega_0^2 a^2$ will not be conserved during ensuing natural oscillations if the fractional loss rate, γ, continues to operate. According to (3.49) it will decay exponentially with the mean life $\tau = 1/\gamma$. The significant result found in this section is that the resonance width equals the inverse of the mean life:

$$\boxed{\gamma = 1/\tau}. \tag{3.56}$$

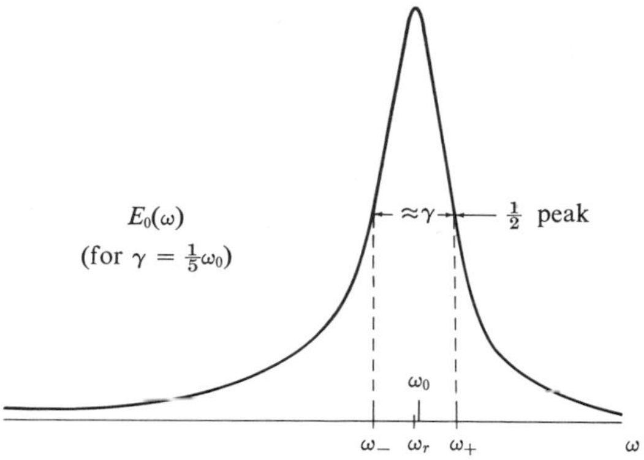

Figure 3.5

This is a highly general characteristic of physical states of motion. There is always an inverse relationship between the mean life of a state and the breadth of its resonance with exciting agencies. Narrow resonances characterize long-lived states of motion.

3.7 COUPLED OSCILLATORS

It will be interesting to see the effect of reactions back on an agency which excites motion of a system like the oscillator. For this purpose, a description of the "agency" must be included. In the example here, this will be taken to be a second particle-oscillator, moving colinearly with the first, as indicated in Figure 3.6. The fixed distance l is taken as the separation of the two equilibrium points, while x_1 and x_2 are the displacements from equilibrium of the individual masses m_1 and m_2. If the only forces were ones, $-m_1\omega_1^2 x_1$ and $-m_2\omega_2^2 x_2$, attaching each mass to its equilibrium point, then the oscillators would move independently,

$$x_{1,2} = \left[\frac{2E_{1,2}}{m_{1,2}\,\omega_{1,2}^2}\right]^{1/2} \sin(\omega_{1,2}t + \varphi_{1,2}), \tag{3.57}$$

each conserving its energy, E_1 and E_2, to itself. To have energy interchanges, some kind of "coupling" between the two oscillations must be introduced. This will be taken to be a force $-\kappa(x_1 - x_2)$ on m_1, together with its reaction $+\kappa(x_1 - x_2)$ on m_2, which tends to maintain the equilibrium separation; that is, all forces vanish when $x_1 = x_2 = 0$ and $l + x_2 - x_1 = l$.

The equations of motion for the coupled-oscillator system are

$$\begin{aligned} m_1\ddot{x}_1 &= -m_1\omega_1^2 x_1 - \kappa(x_1 - x_2) = F_1, \\ m_2\ddot{x}_2 &= -m_2\omega_2^2 x_2 - \kappa(x_2 - x_1) = F_2. \end{aligned} \tag{3.58}$$

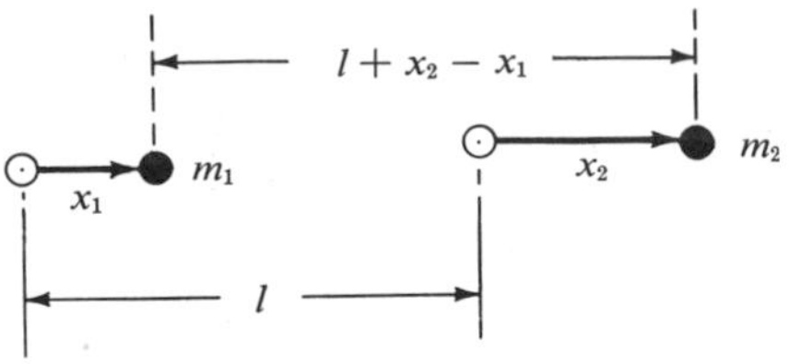

Figure 3.6

The forces $F_{1,2}$ on each particle are derivable from a potential via $F_{1,2} = -\partial V/\partial x_{1,2}$, where

$$V(x_1, x_2) = \tfrac{1}{2}m_1\omega_1^2 x_1^2 + \tfrac{1}{2}m_2\,\omega_2^2\,x_2^2 + \tfrac{1}{2}\kappa(x_1 - x_2)^2. \tag{3.59}$$

The last term is the "coupling energy," which may be thought of as residing in a joining "spring," the mutual bond between the two oscillators.

The Normal Frequencies of Identical Coupled Oscillators

The solution of the coupled equations (3.58) in the general case leads to rather lengthy expressions. Most of the phenomena that arise are more transparently exhibited by the special case of identical oscillators: $m_1 = m_2 = m$ and $\omega_1 = \omega_2 = \omega_0$. Then the sum of the coupled equations yields a simple oscillator equation for the "mean displacement," $\bar{x} \equiv \tfrac{1}{2}(x_1 + x_2)$, independent of the coupling $\sim\kappa$:

$$\ddot{\bar{x}} = -\omega_0^2\,\bar{x}. \tag{3.60}$$

The difference of the coupled equations likewise yields a simple oscillator equation, this time for the relative displacement $x \equiv x_1 - x_2$:

$$\dot{x} = -(\omega_0^2 + 2\kappa/m)x. \tag{3.61}$$

Thus, the general solution of the identical oscillator problem is obtainable from

$$\begin{aligned} \bar{x} &= \tfrac{1}{2}(x_1 + x_2) = a\sin(\omega_0\,t + \alpha), \\ x &= x_1 - x_2 = b\sin(\omega t + \beta), \end{aligned} \tag{3.62}$$

where

$$\omega = \omega_0[1 + 2\kappa/m\omega_0^2]^{1/2} \tag{3.63}$$

and a, α, b, β are the requisite arbitrary integration constants, adaptable to any start given the motion. Each of the two simple harmonic motions (3.62) conserves an energy to itself,

$$\bar{E} = \tfrac{1}{2}(2m)(\dot{\bar{x}}^2 + \omega_0^2 \bar{x}^2) \qquad \text{and} \qquad \varepsilon = \tfrac{1}{2}\left(\frac{m}{2}\right)(\dot{x}^2 + \omega^2 x^2), \qquad (3.64)$$

respectively. The mass coefficients adopted for these conserved quantities are such that the sum of the two kinetic energies is just $\frac{1}{2}m(\dot{x}_1^2 + \dot{x}_2^2)$, and the sum of the potential energies equals (3.59). Notice that $\bar{E}$ and ε are conserved separately even when the oscillators are uncoupled ($\kappa = 0$ and $\omega = \omega_0$).

The individual motions following from (3.62),

$$x_{1,2}(t) = \bar{x} \pm \tfrac{1}{2}x = a \sin(\omega_0 t + \alpha) \pm \tfrac{1}{2}b \sin(\omega t + \beta), \qquad (3.65)$$

are superpositions of the two simple harmonic oscillations. The simply harmonic variables $\bar{x}$ and x are sometimes called the "normal coordinates" of the system, and ω_0 and ω are called the "normal frequencies." The solutions $x_{1,2}(t)$ for the more general case of $m_1 \neq m_2$ and $\omega_1 \neq \omega_2$ are also superpositions of two simple harmonic oscillations and the corresponding normal coordinates are again linear combinations of x_1 and x_2 (see Chapter 11).

The Motions of Identical Coupled Oscillators

Suppose first that the identical oscillators are started in phase with each other (that is, with $x_1 = x_2$ and $\dot{x}_1 = \dot{x}_2$ at $t = 0$). Then the amplitude $b = 0$ in (3.65) and the motion continues with $x_1 = x_2$ at all times. Its frequency is just the natural one ω_0, as if the oscillators were uncoupled, because the particles stay just their equilibrium distance (l) apart throughout their motion, as indicated in Figure 3.7(a). If the oscillators are started "in opposition," with $x_2 = -x_1$ and $\dot{x}_2 = -\dot{x}_1$ at $t = 0$, then $a = 0$ in (3.65) and a simply harmonic oscillation of frequency ω ensues, indicated in Figure 3.7(b). The frequency ω is higher than the natural frequency because the "spring" of the mutual coupling is now added to the forces.

The transfer of energy from one oscillator to the other may be studied by considering a start in which $x_1(0) = x_2(0) = 0$, and also $\dot{x}_2(0) = 0$, with an initial impulse $m\dot{x}_1(0) = mv_0$ given only to the first oscillator. Now $\alpha = \beta = 0$ in (3.65), and $\omega_0 a = \frac{1}{2}\omega b = \frac{1}{2}v_0$, so that

$$\begin{aligned} x_1 &= \frac{v_0}{2}\left(\frac{\sin \omega_0 t}{\omega_0} + \frac{\sin \omega t}{\omega}\right), \\ x_2 &= \frac{v_0}{2}\left(\frac{\sin \omega_0 t}{\omega_0} - \frac{\sin \omega t}{\omega}\right). \end{aligned} \qquad (3.66)$$

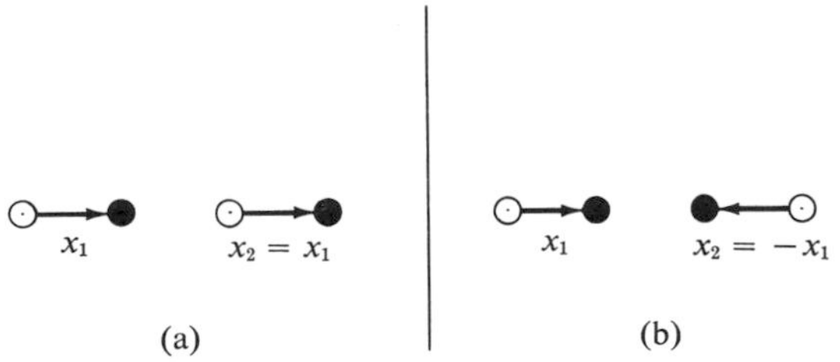

Figure 3.7

These are plotted in Figure 3.8 for the "weak coupling limit"; that is, $\kappa \ll \frac{1}{2}m\omega_0^2$ so that $\omega = \omega_0[1 + 2\kappa/m\omega_0^2]^{1/2} \approx \omega_0$. The difference of ω from ω_0, $\delta\omega \equiv \omega - \omega_0 \approx \kappa/m\omega_0$, may now be neglected in the *amplitudes* of (3.66) but not in the phases, since $(\delta\omega)t$ can become of order π in a finite time, $\pi/\delta\omega$. When $\omega = \omega_0 + \delta\omega$ is put into the phases of (3.66),

$$\begin{aligned} x_{1,2} &\approx (v_0/2\omega_0)[\sin \omega_0 t(1 \pm \cos \delta\omega t) \pm \cos \omega_0 t \sin \delta\omega t] \\ &= c_{1,2}(t) \sin [\omega_0 t + \gamma_{1,2}(t)], \end{aligned} \tag{3.67}$$

where

$$c_{1,2}^2 = (v_0^2/2\omega_0^2)(1 \pm \cos \delta\omega t), \tag{3.68}$$

and the phases $\gamma_{1,2}$ also vary slowly, with the period $2\pi/\delta\omega \gg \pi/\omega_0$. The motions (3.67) can be viewed as oscillations of frequency ω_0 which have amplitudes modulated with the frequency $\delta\omega \approx \kappa/m\omega_0$. Only the motion $x_1(t)$ is plotted in detail in Figure 3.8, but "envelopes" proportional to $\pm|c_{1,2}|$ are plotted for both x_1 and x_2. Plainly, the energy is passed back and forth between the two oscillators, *in toto* even for a very small coupling $\sim\kappa$; a smaller coupling merely protracts the time, $\pi/\delta\omega$, in which a total energy exchange takes place. In the approximation of (3.67), the energy given the second oscillator has the expression

$$E_2 \equiv \tfrac{1}{2}m(\dot{x}_2^2 + \omega_0^2 x_2^2) \approx \tfrac{1}{4}m v_0^2(1 - \cos \delta\omega t), \tag{3.69}$$

having the periodicity claimed.

For very strong coupling (that is, κ large enough so that $\omega \gg \omega_0$), the oscillations (3.66) approach the equality

$$x_1 = x_2 = (v_0/2\omega_0) \sin \omega_0 t. \tag{3.70}$$

This fits exactly the initial condition $\dot{x}_1(0) = \dot{x}_2(0) = \frac{1}{2}v_0$—that is, the initial impulse mv_0 being supplied in equal shares to both masses, as when they are rigidly bound together. It is of course unsurprising that a large enough coupling causes the two masses to move together, as if they were attached to the ends of a massless rigid rod.

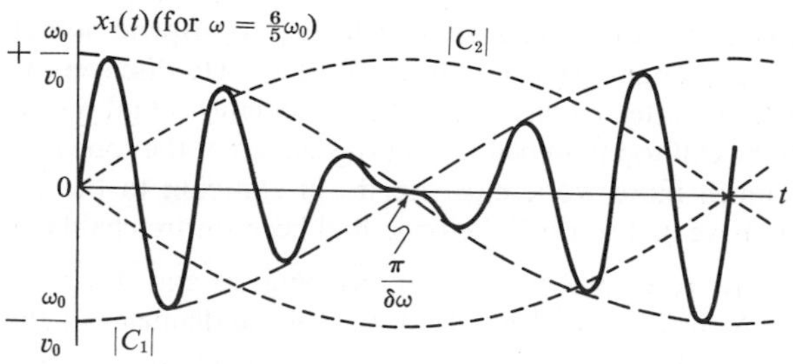

Figure 3.8

A Coupled Oscillator as an "External Agency"

When the coupling force $-\kappa(x_1 - x_2)$ is expressed as an explicit function of time with the use of the solution (3.62), then the equation of motion for the first oscillator becomes

$$m\ddot{x}_1 = -m\omega_0^2 x_1 - \kappa b \sin(\omega t + \beta). \tag{3.71}$$

This amounts to treating the second oscillator as an external agency which applies a forced oscillation of frequency ω to the first particle. The solution (3.42) should apply, with $F_0 = -\kappa b$ and $\beta = 0$. In the case of (3.66), $b = v_0/\omega$. Then (3.42) gives

$$x_1 = a \sin(\omega_0 t + \varphi) + [\kappa v_0/m\omega(\omega^2 - \omega_0^2)] \sin \omega t,$$

which can be seen to agree with (3.66) $[a = v_0/2\omega_0$ and $\omega^2 = \omega_0^2 + 2\kappa/m]$.

Using an identical oscillator as an "external agent" does not permit exhibiting the phenomenon of resonance, since the forcing frequency ω is then necessarily larger than the natural frequency ω_0 of the isolated oscillators. However, an essential phenomenon has been displayed: that an "exciting agency," through its coupling to a system it perturbs, furnishes energy to the system, but, if it persists, the same coupling may deplete that energy again.

EXERCISES

3.1. The "horses" on a merry-go-round oscillate up and down with an amplitude of two feet, four times in each revolution of the platform. What is the greatest rate of revolution which will not unseat the passengers?

3.2. A mountain climber steps off a rock shelf to suspend himself from the end of a rope hanging down a sheer cliff face. He discovers that the rope is elastic, and after he has stopped his oscillations find himself 50% farther from the cliff top than before. On the basis that the rope is perfectly elastic, how much more work must he do in climbing to the cliff top than he would have had to do if the rope had been unstretchable? (25% more.)

3.3. You observe a smooth cylinder bobbing up and down, erectly, in quiet water. You feel confident of estimating the diameter of the cylinder, and have a watch adequate for timing the oscillation period. Is the information so obtained sufficient for an estimate of the weight of the cylinder? Derive a formula you can use.

3.4. Suppose that the force expression (3.7), besides having $(\partial F/\partial x)_0 = -m\omega_0^2 < 0$, has $(\partial^2 F/\partial x^2)_0 = +2m\omega_0^2/s > 0$ in its only other considerable term. With the help of an appropriate energy diagram, find the least energy a particle started from $x_0 \gg s$ would have to be given if it is to penetrate to $x < 0$. If given this energy, at what point will it be turned back? Next suppose that $(\partial^2 F/\partial x^2)_0 = 0$ but that $(\partial^3 F/\partial x^3)_0 = 6m\omega_0^2/s^2 > 0$. The oscillator is given the energy $\frac{1}{2}m\omega_0^2 a^2$ with $a \ll s$. Show that the consequent amplitude is smaller than a, by the fraction $3a^2/4s^2$.

3.5. A charge $+e$ is borne by a point-mass m which moves on the line joining two equal, repelling point-charges, $+Ze$ each, fixed at $x = \pm a$. Show that m may execute small simple harmonic oscillations near $x = 0$, with the circular frequency $\omega_0 = 2(Ze^2/ma^3)^{1/2}$. Draw the energy diagram as a function of $-a < x < a$, and find the energy the particle must have to oscillate with the amplitude $\frac{1}{2}a$. What is then the velocity of its passage through $x = 0$? $[(4Ze^2/3ma)^{1/2}.]$ How does this compare with the approximate speed calculated for the simple harmonic oscillation with the same amplitude?

3.6. When a unit impulse, $mv_0 = 1$, is delivered at $t = 0$ to a linear oscillator in equilibrium, then the function

$$x_1(t \leq 0) = 0, \qquad x_1(t \geq 0) = \frac{\sin \omega_0 t}{m\omega_0}$$

describes the motion, according to (3.12) and (3.13). Suppose, instead, that an arbitrarily time-dependent force, $F(t)$, is applied. Show that

$$x(t) = a \sin(\omega_0 t + \varphi) + \int_{-\infty}^{t} x_1(t - t_1)F(t_1)\, dt_1$$

is a general solution of the appropriate equation of motion. It must be remembered that $x_1(t - t_1) = 0$ for $t \leq t_1$, according to the above definition of this so-called "Green's function" of the oscillator. The element $F(t_1)\, dt_1$ represents the number of units of impulse delivered in the time dt_1, and $x_1 = [\sin \omega_0(t - t_1)]/m\omega_0$ the contribution of each such unit to $x(t > t_1)$; this conforms to a "principle of causality" in that the impulse delivered at t_1 has effects only at $t > t_1$. (The reader may want to check that the procedure here will yield (3.42) correctly.)

3.7. Using the result of Exercise 3.6, find the motion which will ensue from imposing the force

$$F(t) = F_0(1 - e^{-t/\tau})$$

on the oscillator, starting from equilibrium at $t = 0$. This corresponds to displacing the oscillations by a constant force, as in (3.17), but only *eventually*, after $t \gg \tau$. Show that as a result of the gradual ($\tau \neq 0$) introduction of the force, the energy it eventually gives the oscillator is less by the factor $1/(1 + \omega_0^2 \tau^2)$ than it would have been if $\tau = 0$. Notice that a slow enough imposition of the force ($\omega_0 \tau \to \infty$) results in a vanishingly small imparting of energy (the process is then called "adiabatic," resulting in a negligible "perturbation" of any initial oscillation energy).

3.8. Other examples in which a "principle of causality" can be said to be invoked are ones dealing with observations delayed by the finite propagation speed of sound, as in the following elementary illustration. The depth d of a well is to be deduced from the time T it takes a stone to drop down to the water *and* the sound of the splash to return, knowing g and the speed of sound, c. Developing an explicit formula for d will involve choosing the root of a quadratic in conformity with the "causality" implied by the above interpretation of T. Show that if the stone is regarded as staying well below the speed of sound during its fall, then the formula is properly approximated by

$$d \approx \tfrac{1}{2}gT^2[1 - gT/c + \cdots],$$

using the expansion

$$\sqrt{1 + \varepsilon} \approx 1 + \tfrac{1}{2}\varepsilon - \tfrac{1}{8}\varepsilon^2 + \tfrac{1}{16}\varepsilon^3 + \cdots,$$

correct for $\varepsilon \ll 1$.

3.9. Find the limiting speed of fall in air if the resistance is proportional to the kinetic energy of the particle, rather than to its momentum [say it is given by βmv^2 instead of γmv, as in (3.18)]. Check that the rate of work against the friction equals the loss rate of gravitational potential energy, during the fall with the limiting speed.

3.10. Suppose that the resistance to a boat's progress through water is proportional to its speed. If a steady "push," P per unit mass, maintains the boat at a speed v, in what distance $s(v, P)$ will it come to rest after the power is cut off? How fast would it still be going at that distance if the resistance were proportional to the square of the speed, instead? [$s = v^2/P$, $\dot{x} = v/e$.]

3.11. A plane flying through still air at speed v_0 jettisons an object as it approaches a position over the edge of a river. Suppose the object meets air resistance of force γ per unit of momentum component relative to the air, and assume that the plane is high enough for the object to reach near-terminal velocity while still in the air. How far from the river edge was the drop made if the object just clears the edge? (v_0/γ).

3.12. In reference to the situation mentioned in §2.1, an object is dropped from the top of a bare vertical mast, on a vessel having speed v_0 relative to still air. Under the same kind of air resistance as in Exercise 3.11, where will the object land relative to the foot of the mast, if the time of fall is just $t = 1/\gamma$? How high is the mast if this is the time of fall? $(g/\gamma^2 e)$.

3.13. Find the real and imaginary parts of the functions (a) $1/(1 + i\zeta)$, (b) $e^{i\zeta}$, (c) $\cos\zeta$, (d) $\ln\zeta$[$\text{Re}\{\ln\zeta\} = \ln|\zeta|$, $\text{Im}\{\ln\zeta\} = \arg\zeta \equiv \varphi$ in the last case!]

3.14. Find $f(\zeta)$, *not* $f(\zeta, i)$, such that

$$\text{Re}\{f\} = (1 + \xi)^2 - \eta^2.$$

3.15. Show that $|\sin\zeta| > 1$ is possible, and find all the solutions of $\sin\zeta = 0$.

3.16. Whenever trial of a given form for the solution of an equation like (3.43) yields only a special, but not a general, solution as in the case of critical damping, then a new form is tried, consisting of a product, like $x = u(t)e^{-\omega_0 t}$, of the special solution (here, exp: $-\omega_0 t$) with a new function, $u(t)$, to be found. Obtain the differential equation which $u(t)$ must satisfy, in the case of critical damping, and show how it leads to the result (3.46).

3.17. Plot the phase shift, $0 < \delta < \pi$, of (3.51) as a function of $0 < \omega < \infty$, for the cases of $\gamma = \frac{1}{2}\omega_0$, $\frac{1}{5}\omega_0$ and 0, noting its behavior in the passage through resonance. Draw in a straight line, through the point $\omega = \omega_0$, $\delta = \frac{1}{2}\pi$, and with the slope $(d\delta/d\omega)_{\omega_0}$. For each of the three cases, find the breadth, $\Delta\omega$, of the span in which the *straight line* varies from $\delta = 0$ to $\delta = \pi$. (This may serve as a new measure of the resonance breadth.)

3.18. It was shown in §3.6 that a force $F_0 \sin\omega t$ works on the damped oscillator at the maximum average rate when ω just equals the natural frequency ω_0. Show that it works at a rate slower by the factor $(\omega_r/\omega_\gamma)^2$ at the "resonance" frequency $\omega = \omega_r$.

3.19. Check the statements made in the text just after (3.64).

3.20. Plot a diagram like that in Figure 3.8, but for the case $\omega = 2\omega_0$.

CHAPTER 4

ORBITAL MOTIONS OF A PARTICLE

Motion is not necessarily confined to a straight line even under forces of constant direction, unless the particle is started precisely along the force direction. An example of curvilinear motion under a unidirectional force of gravity was given in §1.5. That example also showed that the geometrical *orbit* of a motion may have greater interest than its specification in time, $\mathbf{r}(t)$.

4.1 THE ISOTROPIC OSCILLATOR

Suppose that the simple oscillator force already investigated in §3.1 is *isotropic* (that is, independent of the direction of the displacement $\mathbf{r}$ from the equilibrium point at $\mathbf{r} = 0$). Then

$$m\ddot{\mathbf{r}} = \mathbf{F} = -m\omega_0^2\,\mathbf{r}. \qquad (4.1)$$

This may be referred to a cartesian frame and to a zero of time chosen so as to make the representation of the ensuing motion as simple as possible.

However the motion is started, it will stay in a plane determined by the starting velocity $\mathbf{v}_0$ and position vector $\mathbf{r}_0$. That is because the initial force coincides in direction with $-\mathbf{r}_0$ and so no acceleration component can develop to take the particle out of the initial plane. The general problem reduces to a planar one and the x, y axes may be chosen to lie in the plane of motion. It will simplify matters further to choose the x-axis to lie along the direction of $\mathbf{r}_0$, so that $x_0 = r_0$ and $y_0 = 0$.

The Component Motions

However the x and y axes are oriented,

$$\begin{aligned} m\ddot{x} &= F_x = -m\omega_0^2 x, \\ m\ddot{y} &= F_y = -m\omega_0^2 y \end{aligned} \tag{4.2}$$

are the corresponding components of (4.1). The motion is thus separated into two orthogonal motions, as in the constant force problem of (1.57). The x and y "degrees of freedom" are said to be "uncoupled" because there is no passing back and forth of energy between them, in the sense that

$$E_x = \tfrac{1}{2}m(\dot{x}^2 + \omega_0^2 x^2) \qquad \text{and} \qquad E_y = \tfrac{1}{2}m(\dot{y}^2 + \omega_0^2 y^2) \tag{4.3}$$

are each separately conserved.

Each degree of freedom undergoes a simple harmonic variation such as is described by (3.12). Each velocity component $\dot{x}$, $\dot{y}$ vanishes periodically and so there will be no real loss of generality in choosing $t = 0$ to be a moment at which $\dot{x} = 0$. Then the starting conditions $x(0) = x_0$, $y(0) = 0$, $\dot{x}(0) = 0$, $\dot{y}(0) = v_0$ lead to

$$\begin{aligned} x(t) &= r_0 \cos \omega_0 t, \\ y(t) &= (v_0/\omega_0) \sin \omega_0 t, \end{aligned} \tag{4.4}$$

as is easily verified by evaluation at $t = 0$; that the equations of motion are satisfied may be reconfirmed by substitution into (4.2).

The Orbit

Eliminating the time parameter between the equations (4.4) yields

$$\frac{x^2}{r_0^2} + \frac{y^2}{(v_0/\omega_0)^2} = 1, \tag{4.5}$$

which describes an elliptical orbit, as does (1.60). The particle passes repeatedly through the phase $x = r_0$, $y = 0$, $\dot{x} = 0$, $\dot{y} = v_0$, which had been taken as the initial one, and this occurs at the end of one of the principal radii. A quarter of a revolution later, the radial component of the velocity has again disappeared, the phase now being $x = 0$, $y = v_0/\omega_0$, $\dot{x} = -\omega_0 r_0$, $\dot{y} = 0$. In polar coordinates centered on the equilibrium point at $x = y = 0$ (rather than on a focus, as in (1.59)):

$$r = r_0 \Big/ \left[1 - \left(1 - \frac{\omega_0^2 r_0^2}{v_0^2}\right) \sin^2 \varphi\right]^{1/2}. \tag{4.6}$$

Either of the forms (4.5) or (4.6) shows that the orbit is circular if the motion is started with $v_0 = \omega_0 r_0$.

The description in cartesian coordinates is just as easy to find in the case of the anisotropic oscillator, which is represented by replacing ω_0^2 in (4.2) with different force constants, say ω_x^2 and ω_y^2, in each of the two component equations of motion. The resultant orbits form so-called "Lissajou figures." These are closed loops only when the ratio ω_x/ω_y is rational. They become ellipses or straight linear segments when $\omega_x = \omega_y = \omega_0$, as above (see Exercises 4.3 and 4.4).

4.2 CENTRAL FORCES AND ANGULAR MOMENTUM CONSERVATION

The isotropic oscillator force, $\mathbf{F} = -m\omega_0^2 \mathbf{r}$, is an example of a *central* force—that is, a spherically symmetric force field directed to or from a point-center: $\mathbf{F} = \hat{\mathbf{r}} F(r)$. It is noteworthy that any radially directed force must also be spherically symmetric, $F(\mathbf{r}) \equiv F(r)$, if it is to be derivable from a potential. For then, $\mathbf{F} = -\nabla V = -\hat{\mathbf{r}}\, \partial V/\partial r$ and the transverse components $-\partial V/r\, \partial\vartheta$ and $-\partial V/r \sin\vartheta\, \partial\varphi$ of the gradient vector (2.25) must each vanish; this means that the potential field V is independent of the orientation angles ϑ, φ and $F = -dV/dr$ depends only on the distance r from the force center. In the case of the isotropic oscillator

$$F = -\frac{dV}{dr} = -m\omega_0^2 r \qquad \text{and} \qquad V = \tfrac{1}{2} m\omega_0^2 r^2, \tag{4.7}$$

if the potential gauge with $V(r = 0) = 0$ is chosen.

Torque and Angular Momentum

The equation of motion under a radially directed force,

$$\dot{\mathbf{p}} = \mathbf{F} = \hat{\mathbf{r}} F, \tag{4.8}$$

has a simple "first integral" apart from the energy integral arising when the force is derivable from a potential. The new integral can be found by using a *cross-product* of the radius vector $\mathbf{r}$ into the force as an "integrating factor":

$$\mathbf{r} \times \mathbf{F} = \mathbf{r} \times \dot{\mathbf{p}}. \tag{4.9}$$

The quantity on the left here is called the "moment of the force" or "torque" about the point $\mathbf{r} = 0$. Its magnitude is Fb, if b is the length of the perpendicular dropped from the "momental center" ($\mathbf{r} = 0$) to the line of the force as it acts on the particle; the length b is frequently called the "lever arm" of the moment. Figure 4.1 shows a force which does exert torque about $\mathbf{r} = 0$; clearly, a radially directed force has no lever arm about this as a momental center and its torque therefore vanishes.

The right side of (4.9) is just the time-derivative of the vector

$$\mathbf{L} \equiv \mathbf{r} \times \mathbf{p} \equiv \mathbf{r} \times m\mathbf{v}, \tag{4.10}$$

for $\dot{\mathbf{L}} = \dot{\mathbf{r}} \times \mathbf{p} + \mathbf{r} \times \dot{\mathbf{p}} = \mathbf{r} \times \dot{\mathbf{p}}$, since $\dot{\mathbf{r}} \times m\mathbf{v} \equiv \mathbf{v} \times m\mathbf{v} \equiv 0$. The "moment of the momentum" $\mathbf{L}$ here is most frequently called the "angular momentum" of the mass m about the momental center, but sometimes "orbital momentum" instead. In general, (4.9) yields

$$\dot{\mathbf{L}} = \mathbf{r} \times \mathbf{F}, \tag{4.11}$$

showing that angular momenta are changed by torques as directly as linear momenta are changed by forces (4.8). Plainly, the angular momentum $\mathbf{L}$, which is perpendicular to the radius vector $\mathbf{r}$, provides some measure of rotational motion relative to the momental center; a component of velocity transverse to $\mathbf{r}$ must exist if $\mathbf{L}$ is to exist.

In the special case of a radially directed force, the torque in the equation of angular motion (4.11) vanishes and $\dot{\mathbf{L}} = 0$. Thus the *angular momentum is conserved* under radially directed forces. Of course this refers to angular momentum about the particular point to or from which the forces are directed. Whatever starting phase $\mathbf{r}_0$, $\mathbf{v}_0$ is given the motion,

$$\mathbf{L} = m\mathbf{r} \times \mathbf{v} = m\mathbf{r}_0 \times \mathbf{v}_0 \tag{4.12}$$

retains its initial value as $\mathbf{r}$ and $\mathbf{v}$ individually change in the course of the motion. In the example of the isotropic oscillator of §4.1, $L = L_z = xp_y - yp_x = mr_0 v_0$ at every phase of the motion. This can be verified by substituting the solution (4.4) into $L_z = xp_y - yp_x$.

Since the angular momentum $\mathbf{L}$ is constant only if its direction, as well as its magnitude, is fixed, the motion under the radial force is always

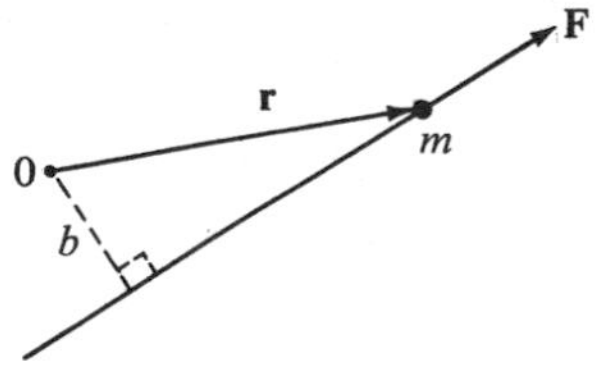

Figure 4.1

planar. It takes place in the plane determined by $\mathbf{r}$ and $\mathbf{p}$, which is always normal to the constant direction of $\mathbf{L} = \mathbf{r} \times \mathbf{p} = m\mathbf{r}_0 \times \mathbf{v}_0$.

The finding of descriptions in which an angular momentum is conserved has great importance. Experience seems to indicate that a general *Principle of Total Angular Momentum Conservation* has as exceptionless a validity as do linear momentum and energy conservations (§2.2, §2.4). It is another earmark of an isolated system (§2.5).

The Polar Representation

It is exceptional for a radial force, even a spherically symmetric central one $\mathbf{F} = \hat{\mathbf{r}}F(r)$, to allow separation of the motion into two independent cartesian components, as was the case in the isotropic oscillator problem of (4.2). For instance, $m\ddot{x} = F_x(r) = F(r)x/r$ depends on y through $r = (x^2 + y^2)^{1/2}$, except when $F \sim r$ (proportional to r) as for the oscillator. In general, the motions $x(t)$ and $y(t)$ are "coupled" to each other, the energy being passed back and forth between the two degrees of freedom, in contrast to (4.3).

On the other hand, for a central force the radial motion can be obtained independently of the angular displacement. This can be found by using the polar resolution $ma_r = F$, $ma_\varphi = 0$ of the equation of motion (4.8). According to the polar representation of the acceleration (1.38),

$$\begin{aligned} m(\ddot{r} - r\dot{\varphi}^2) &= F(r), \\ m(r\ddot{\varphi} + 2\dot{r}\dot{\varphi}) &= 0. \end{aligned} \tag{4.13}$$

The second of these equations is independent of the force and can be integrated without specifying it. After multiplication by the radius r, which is essentially the step taken from (4.8) to (4.9), it becomes

$$m(r^2\ddot{\varphi} + 2r\dot{r}\dot{\varphi}) = \left(\frac{d}{dt}\right)mr^2\dot{\varphi} = 0. \tag{4.14}$$

The constant quantity here is just the magnitude of the angular momentum

$$L = mr^2\dot{\varphi} = mv_\varphi r, \tag{4.15}$$

expected to be conserved. The identification here can be verified by noting that the definition of L (4.10) yields $L = L_z = rp_\varphi = mv_\varphi r$.

The radial equation of motion in (4.13) now becomes

$$m\ddot{r} = F(r) + mv_\varphi^2/r = F + L^2/mr^3. \tag{4.16}$$

As to be expected, radial accelerations $\ddot{r}$ arise not only from the applied force F but also from centrifugal force, $mr\dot{\varphi}^2 = mv_\varphi^2/r = L^2/mr^3$, which will exist if the motion is started with some angular momentum $\mathbf{L} = \mathbf{r}_0 \times m\mathbf{v}_0 \neq 0$. With L staying constant during the motion, the centrifugal force behaves like a central repulsion, inversely proportional to the cube of the distance, and the motion $r(t)$ may be obtained independently of the angle φ, as already asserted.

The "Centrifugal Potential"

The centrifugal force, L^2/mr^3, that exists for a given conserved angular momentum L is derivable from a spherically symmetric "potential" $V_c(r)$ according to

$$L^2/mr^3 = -dV_c/dr \rightarrow V_c(r) = L^2/2mr^2. \tag{4.17}$$

In the gauge $V_c(\infty) = 0$ adopted here, this "centrifugal potential" is just the kinetic energy invested in rotational motion

$$T_\varphi = \tfrac{1}{2}mv_\varphi^2 = \tfrac{1}{2}mr^2\dot{\varphi}^2 = L^2/2mr^2 = V_c, \tag{4.18}$$

since $L = mr^2\dot{\varphi}$. It behaves like a "potential" energy for the radial motion because less rotational kinetic energy is needed to maintain the constant angular momentum $L = mv_\varphi r$ when the lever arm r happens to increase during the motion, but the rotational velocity v_φ must be speeded up during phases of the motion in which the particle approaches close to the center.

When the central force $F = -dV/dr$ is derivable from a potential $V(r)$, the conservation of the total energy

$$\begin{aligned} E &= \tfrac{1}{2}m(v_r^2 + v_\varphi^2) + V \\ &= \tfrac{1}{2}m\dot{r}^2 + L^2/2mr^2 + V(r) \end{aligned} \tag{4.19}$$

ensues. This can be verified directly from the equation of motion (4.13) by adding the results of multiplying the first by $v_r = \dot{r}$ and the second by $v_\varphi = r\dot{\varphi}$, essentially the step in the approach toward the energy integral used in (2.13). As the total E is conserved during motion, there is a passing back and forth of energy between the radial kinetic energy $\tfrac{1}{2}m\dot{r}^2$ and energy otherwise invested: in rotation, as well as in the "genuine" potential field, $V(r)$.

The energy equation (4.19) and the angular momentum equation (4.15) amount to first integrals of the two equations of motion (4.13), with E and L serving as the first two integration constants. They are themselves sufficient for deriving the complete description of the motion: $r(t)$, $\varphi(t)$. In principle, $r(t;\, r_0,\, E,\, L)$ may be found by integrating (4.19), and the substitution of the result into (4.15) should yield $\varphi(t;\, r_0,\, \varphi_0,\, E,\, L)$. However, the orbit $r(\varphi)$ may be obtained directly, though eliminating the parametrization by the time with the use of

$$\dot{r}(\varphi) = \frac{dr}{d\varphi}\,\dot{\varphi} = \frac{L}{mr^2}\frac{dr}{d\varphi}. \tag{4.20}$$

Now the energy equation (4.19) gives

$$\frac{dr}{d\varphi} = \pm\frac{mr^2}{L}\left\{\frac{2}{m}\,[E - V(r)] - \frac{L^2}{m^2r^2}\right\}^{1/2}, \tag{4.21}$$

which may be integrated by "separation of variables." For example, this yields the orbit (4.6) quite directly for the case $V = \frac{1}{2}m\omega_0^2 r^2$ and the initial conditions $\varphi = 0$, $r = r_0$, $L = mv_0 r_0$, $E = \frac{1}{2}m(v_0^2 + \omega_0^2 r_0^2)$ (see Exercise 4.5).

4.3 INVERSE-SQUARE LAW OF FORCE: BOUND STATES

The Kepler orbits (1.59) can now be understood on the basis of the Newtonian dynamics. The kinematical laws found by Kepler can be used to determine the type of force that can account for them.

Kepler's (second) "Law of Areas" (1.61) is just an expression of the Conservation of Angular Momentum, about the sun as momental center. The sectorial velocity is just $\dot{A} = L/2m$, according to the polar representation of the angular momentum L in (4.15). The implication is that the force is directed toward the sun, $\mathbf{F} = \hat{\mathbf{r}}F$, since it exerts no torque capable of changing the angular momentum of the planet.

The force function magnitude F may be obtained most directly by substituting the formal description (1.59) of Kepler's first law into the radial equation of motion (4.16). For this, the radius $r(\varphi) = b^2/[a - c\cos\varphi]$ must first be differentiated to yield the radial velocity, which is

$$\dot{r} = -\frac{r^2}{b^2}(c\sin\varphi)\dot{\varphi} = -\frac{L}{mb^2}\,c\sin\varphi, \tag{4.22}$$

after the introduction of the constant $L = mr^2\dot{\varphi}$. Another differentiation then leads to

$$m\ddot{r} = -\frac{L}{b^2}(c\cos\varphi)\dot{\varphi} = \frac{L^2}{mr^3} - \left(\frac{aL^2}{mb^2}\right)\frac{1}{r^2}, \tag{4.23}$$

after use of the orbit expression $r(\varphi)$ to eliminate the phase angle φ. The first of the two final terms is just the centrifugal force in the equation of radial motion (4.16). The second implies that there must also exist a central force of attraction,

$$F(r) = -\frac{K}{r^2}, \tag{4.24}$$

having some force-strength constant K which helps determine the orbit dimensions according to

$$b^2/a = L^2/mK, \tag{4.25}$$

if the motion is started with a given orbital momentum L.

The conclusion about the force also follows from a similar treatment of the energy relation (4.19), on the presumption that the force is a central one derivable from some potential $V(r)$. The outcome (Exercise 4.6) is

$$V(r) = -\frac{K}{r} \rightarrow F = -\frac{dV}{dr} = -\frac{K}{r^2}, \tag{4.26}$$

in a potential gauge having $V(\infty) = 0$. In this gauge, the expression

$$E = -L^2/2mb^2 = -K/2a \tag{4.27}$$

emerges for the total energy. Its negative value is merely a result of the choice of gauge for the potential; the negative values $V = -K/r$ assigned to the potential energy exceed the positive kinetic energy in magnitude at every point of the orbit.

Newton's Law of Gravity

More information about the force constant K emerges from Kepler's third law (1.64). Using the relation (4.25) to eliminate the sectorial velocity $\dot{A} = L/2m$ from the expression (1.63) for the period yields

$$T = 2\pi(m/K)^{1/2}a^{3/2}. \tag{4.28}$$

The finding that the proportionality constant between T and $a^{3/2}$ is the same for all the planets implies that K is independent of any accidents in the way the planetary motions got started, as the occurence of a, b and L in (4.25) might have suggested, but is simply proportional to the planet's mass: $K \sim m$.

At about this juncture, Newton formed the hypothesis which provided a striking example of how expectations of simple symmetry lead scientists to make their most comprehensive predictions. He supposed that the attraction of the sun and a given planet, to each other, is basically of the same type as that responsible for the gravitation of falling objects to the earth and for holding the moon in an orbit around the earth. He expected the force constant to be proportional not only to the mass of the one partner in their mutual attraction but also to the mass of the other, simply because the two partners ought to enter the relationship in a symmetrical way. Then Kepler's value for K/m is the same for all the planets because in each case the same sun is a partner. In general, for any pair of masses m, M,

$$K = GmM \qquad \text{and} \qquad \mathbf{F} = -\frac{\hat{\mathbf{r}}GmM}{r^2}, \tag{4.29}$$

where G is some "universal" constant depending numerically on the units in which the masses are measured. The gravitational acceleration $\mathbf{g}$ found at the earth's surface appears constant only because it is tested over heights of fall which are small compared to the earth's radius, R. Thus $g = GM_{\text{earth}}/R^2$ is expected,* and

$$\mathbf{F}_{\text{earth}} = -\hat{\mathbf{r}}mg\left(\frac{R}{r}\right)^2 \tag{4.30}$$

is the force of gravitational attraction to the earth at any distance $r > R$. The first test of this was to see whether the period consequently predicted for the moon's orbit around the earth,

$$T = 2\pi(K_{\text{earth}}/m_{\text{moon}})^{-1/2}a^{3/2} = 2\pi(a/R)(a/g)^{1/2}, \tag{4.31}$$

comes out to be one lunar month. After a sufficiently accurate number for the earth's radius R became available, late in Newton's life, the agreement found was striking indeed (there is a small correction to be discussed in §5.1). Meanwhile, Kepler's value for $K_{\text{sun}}/m \equiv GM_{\text{sun}} = (gR^2)M_{\text{sun}}/M_{\text{earth}}$ offers a means for "weighing the sun" in units of earth mass.

The diminution of gravity with distance above the earth, represented in (4.30), causes deviations in the parabolic orbits (1.58) of missiles flung from

---

* This presumes that the earth's mass can for this purpose be treated as concentrated at a point, the earth's center. Proof that this is so will be postponed until "potential theory" is discussed in connection with electrostatics (however, see Exercise 6.6).

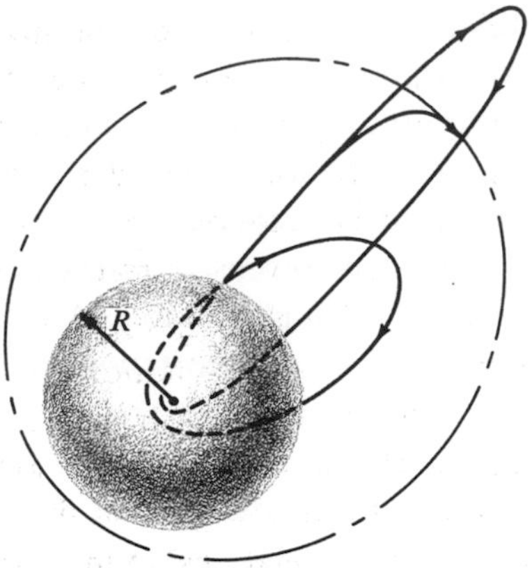

Figure 4.2

the earth. The orbits become more nearly sectors of ellipses, with one focus at the earth's center, as indicated in Figure 4.2. A missile flung from the earth returns to it because of the periodicity of elliptical orbits. In order to make an artificial satellite enter a stable orbit that will not return it to the earth, its velocity must be reoriented, as by rocket motors, after the initial launch.

The Energy Diagram

An inverse-square law of force, $F = -K/r^2$, was also found by Coulomb for the attractions between unlike electric charges and for repulsions of like charges. In the case of an electron with charge $-e$ in the field of an atomic nucleus with charge $+Ze$, $K \equiv Ze^2$. Thus, the study of the inverse-square law of force has more general interest than just for the gravitational problem. It corresponds to a uniform "flux of impulse" (momentum change $F\,\Delta t$) into all directions, since $4\pi r^2 F\,\Delta t = -4\pi K\,\Delta t$ is the same on spheres of any radii r.

How general a solution is presented by the elliptical orbit $r = b^2/[a - c\cos\varphi]$ of (1.59) is best understood with the help of an energy diagram modeled on the example of Figure 3.1. The energies plotted are terms of the expression (4.19), with $V = -K/r (K > 0)$. For Figure 4.3, it is supposed that the motion was started at some finite initial radius r_i, with some angular velocity $\dot{\varphi}_i$, so that $L = mr_i^2\dot{\varphi}_i \neq 0$. The corresponding centrifugal potential $L^2/2mr^2$, which is just the rotational kinetic energy $T_\varphi = \frac{1}{2}mr^2\dot{\varphi}^2$ of (4.18), is represented by the dashed curve. The curve formed by adding it to the "genuine" potential $V = -K/r$ represents an effective potential energy for the radial motion. The result has a minimum at a radius r_m,

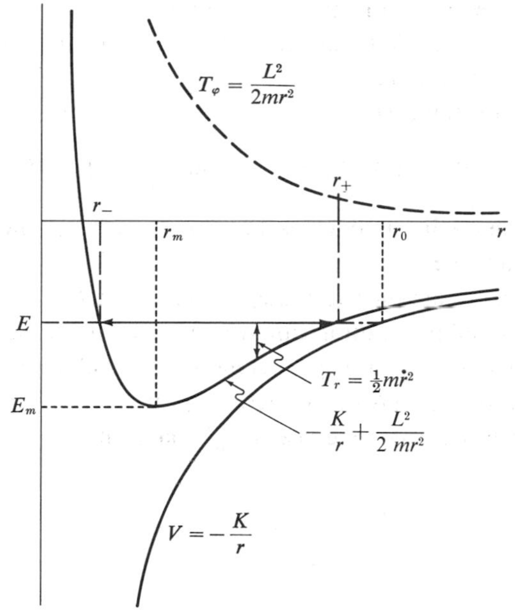

Figure 4.3

where the gradient derivatives representing the centrifugal force and the attraction just balance,

$$L^2/mr_m^3 = K/r_m^2 \rightarrow r_m = L^2/mK. \tag{4.32}$$

The centrifugal repulsion wins out over the attraction as $r \rightarrow 0$.

With the given rotational momentum L, the total energy must exceed the value of the effective potential at its minimum,

$$E \geqslant -K/r_m + L^2/2mr_m^2 = -mK^2/2L^2 \equiv E_m. \tag{4.33}$$

For the diagram, it is supposed that the starting kinetic energy, $T_i = \frac{1}{2}m(\dot{r}_i^2 + r_i^2\dot{\varphi}_i^2)$, was low enough so that the total energy $E = T_i - K/r_i$ is negative. The excess of E over the effective potential curve represents the radial kinetic energy, $T_r = \frac{1}{2}m\dot{r}^2$, during various phases of the motion. This vanishes at certain maximum and minimum radii, $r_\pm$, found by equating E to the effective potential:

$$\begin{aligned} E &= -K/r_\pm + L^2/2mr_\pm^2, \\ \rightarrow r_\pm &= K/2|E| \pm [K^2/4E^2 - L^2/2m|E|]^{1/2}. \end{aligned} \tag{4.34}$$

The radius of the motion must oscillate between these turning points and this is so only for negative energies; the case of a positive total energy is left to the next section. Because the particle cannot escape to $r = \infty$ with the negative energy, the particle is said to be "bound" or moving in a "bound stationary state." If it were begun with only a radial velocity ($L = mr_i^2\dot{\varphi}_i = 0$), then $\varphi = \varphi_i$ is a constant, with

$$r_+(L = 0) = +K/|E| \equiv r_0 \qquad \text{and} \qquad r_-(L = 0) = 0. \tag{4.35}$$

The motion is then a linear one, oscillating between r_0 and the origin (if no new forces intervene).

The energy diagram has made it plain that the radial oscillations characteristic of the elliptic motion can be expected only for negative energies. That the most general orbit of negative energy is the ellipse can be found by integrating (4.21) for $r(\varphi)$. [A like integration will be explicitly demonstrated in (4.49) and (4.50). Also see Exercise 4.8.] When the result is written in the form

$$r(\varphi) = b^2/[a - c \cos \varphi],$$

with $b^2 \equiv a^2 - c^2$, then

$$r_+ = r(0) = a + c \qquad \text{and} \qquad r_- = r(\pi) = a - c. \tag{4.36}$$

Comparison with (4.34) shows that the parameters a, b, c, are determined by the starting conditions E (<0) and L according to

$$a = K/2|E| \qquad \text{and} \qquad b^2 = L^2/2m|E|, \tag{4.37}$$

already seen in (4.27). The greater the angular momentum, for a given energy, the more nearly circular ($b \to a$) the ellipse tends to be. It does become circular for $L = K(m/2|E|)^{1/2}$.

The time-dependence can be measured by the phase angle $\varphi(t)$, made with the major radius, as determined from integrating $dt = (m/L)r^2(\varphi)\, d\varphi$, a fact already utilized in (1.62). The result for $t(\varphi)$ is unfortunately a transcendental function, and $\varphi(t)$ is then easier to obtain, for specific cases, by numerical evaluation. On the other hand, the time for a full revolution has the simple expression

$$T = \pi(K/|E|)(m/2|E|)^{1/2}, \tag{4.38}$$

which is equivalent to (1.63) and (4.28), and is obtainable in the same way.

When the given energy is $E = E_m$ of (4.33), the minimum possible for a given angular momentum, there is no radial kinetic energy and the motion is confined to a constant radius $r \equiv r_m = r_+ = r_- = a = b$, according to the energy diagram. The orbit is a circle of radius $r = r_m$. The centrifugal and attractive forces balance, as in (4.32), at every point of the orbit and that at once leads to

$$E \equiv E_m = -K/r_m + L^2/2mr_m^2 = -K/2a. \tag{4.39}$$

The final expression has already been seen to be generally valid, in (4.37). In the circular orbit, the kinetic energy is always just half the potential energy, K/a, each being separately conserved. The angular momentum is $L = mva = (2m|E_m|)^{1/2}a = K(m/2|E_m|)^{1/2}$, as already noted above.

For energies exceeding the minimum E_m by a small enough amount, the radial oscillations about the circle $r = r_m$ are small, and the effective potential may be approximated as

$$V + T_\varphi \approx E_m + \tfrac{1}{2}(r - r_m)^2(-2E_m)^3/K^2 + \cdots, \tag{4.40}$$

with the linear term $\sim(r - r_m)$ dropping out because the forces balance to zero at $r = r_m$. The result simulates a simple harmonic oscillator potential, as in (3.15) with $V(r_m) = E_m$ and $V''(r_m) > 0$; the circular frequency of the oscillation is

$$\omega_0 = 2[-2E_m^3/mK^2]^{1/2}. \tag{4.41}$$

in consistency with (4.38) via $T = 2\pi/\omega_0$. In the approximation here,

$$\dot{\varphi} \approx L/mr_m^2 = mK^2/L^3 = \omega_0 \quad \text{and} \quad \varphi \approx \omega_0 t.$$

Then, since $c = \frac{1}{2}(r_+ - r_-)$ is small, the ellipse $r = b^2/[a - c\cos\varphi]$ is approximated by

$$r \approx \frac{b^2}{a}\left(1 + \frac{c}{a}\cos\varphi\right) = r_m\left[1 + \left(1 - \frac{E}{E_m}\right)^{1/2}\cos\omega_0 t\right], \tag{4.42}$$

as is to be expected of a simple harmonic motion of energy $(E - E_m) > 0$, relative to an appropriately gauged potential.

4.4 INVERSE-SQUARE LAW OF FORCE: UNBOUND STATES

The energy diagram of Figure 4.3 shows how, as energy is added until it becomes positive, the outer radius of the motion becomes unlimited: $r_+ \to \infty$. The outer "potential barrier" is surmounted and the particle escapes to infinity. Starting from any finite radius, the particle must be given a certain minimum "escape velocity," corresponding to the kinetic energy difference between $E = 0$ and the potential energy. For a mass on the surface of the earth ($r = R$), the escape velocity is computed from

$$E = \tfrac{1}{2}mv^2 - K/R = 0 \rightarrow v = (2gR)^{1/2}, \tag{4.43}$$

since $K = mgR^2$ according to (4.30); the value here is about $v \approx 24{,}000$ m.p.h. The work needed may be diminished slightly from $\frac{1}{2}mv^2 = mgR$ because the earth's diurnal spin can be used to provide help from centrifugal force. Escape to the moon is further helped by the lowering of the potential barrıer by the moon's attraction.

When the particle has enough energy to escape to $r = \infty$, where $V = -K/r \rightarrow 0$, it is sometimes said to be in a "free" state of motion, a nomenclature more generally reserved for uniform motions, with $V = 0$ everywhere. "Free" or "unbound" states are the only kind possible when the force is repulsive, having $K < 0$ so that $V = +|K|/r$ and $F = +|K|/r^2$. Examples of such repulsions are provided by encounters of positively charged atomic nuclei; when a pair has the nuclear charges $+ze$ and $+Ze$, then $K = -zZe^2$ and $V = +zZe^2/r$.

Energy Diagram and Initial Conditions

It will be most economical to treat the unbound states under the attractive and repulsive forces together, using $V = -K/r$ with K positive or negative. The energy diagram for the case of attraction, in Figure 4.3, is supplemented by one for the case of repulsion in Figure 4.4.

For a given total energy $E > 0$ and angular momentum $L \neq 0$ there is a "distance of closest approach," where the radial velocity disappears. This is determined by the equation of E to the effective potential in (4.34),

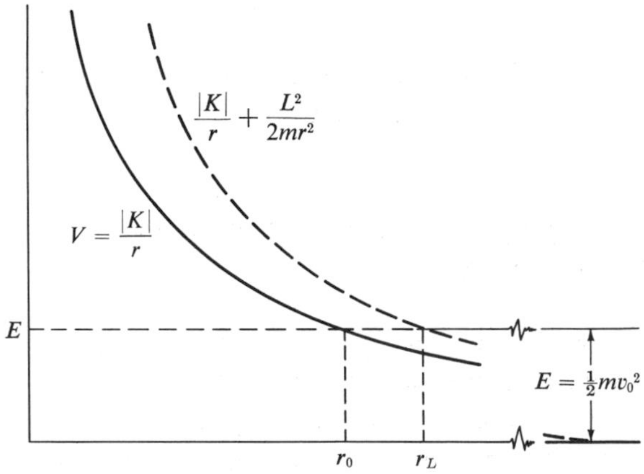

Figure 4.4

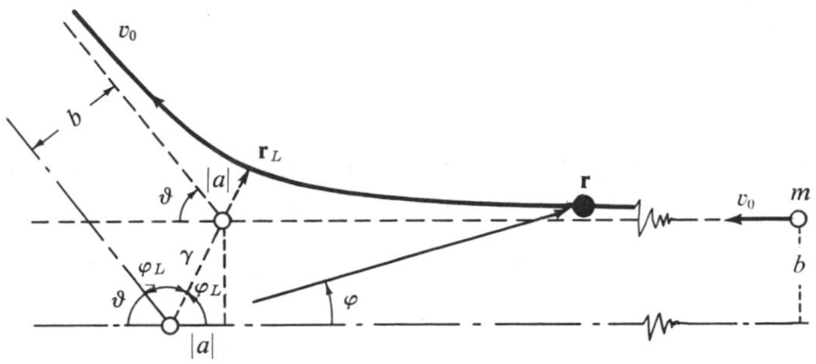

Figure 4.5. Orbit of a particle under a repulsive central force.

which now affords only one positive root

$$r_L = (K^2/4E^2 + L^2/2mE)^{1/2} - K/2E, \tag{4.44}$$

being the greater in the case of the repulsion ($K < 0$), for a given E and L. For "head-on" approach, when $L = 0$ and the velocity is completely radial, $r_0 = |K|/E$ (compare (4.35)) in the case of repulsion, and $r_0 \to 0$ for attraction.

In a phase of the motion at which $r \to \infty$, the force disappears and the particle moves uniformly with a velocity $v_0 = (2E/m)^{1/2}$. This velocity is not necessarily radial but may miss being directed "head-on" by a lever arm b, in which case $L = mv_0 b$. The energy diagrams make clear that the entire motion may consist of a phase of approach to the radius r_L, followed by a phase of retreat to $r = \infty$. The most complete orbits will be obtained when the particle is started from infinity with an inward velocity component:

$$\begin{aligned} &r(0) \to \infty, \qquad \varphi(0) = \sin^{-1}[b/r(0)] \to 0, \\ &\dot{r}(0) = -v_0 \cos \varphi(0) = -v_0, \qquad \dot{\varphi}(0) = v_0 \sin \varphi(0)/r(0) = 0, \end{aligned} \tag{4.45}$$

as reference to Figure 4.5 helps make clear. After the encounter, the particle retreats to $r(\infty) = \infty$ and energy conservation demands that it attain the velocity $v_0 = (2E/m)^{1/2}$ again, while angular momentum conservation requires that the lever arm resume the value $b = L/mv_0$. The retreat will be along some new direction, deflected by some angle ϑ from the initial one, with ϑ to be found from an asymptote of the orbit. The corresponding diagram for the attractive case, with the same initial conditions, differs from Figure 4.5 in that the distance of closest approach is smaller, $r_L < b$, and the final deflection ϑ has the opposite sense.

Orbits and Deflections

The orbit $r(\varphi)$ is found by integrating (4.21) with $V = -K/r$. The expression (4.21) may be written more simply by introducing, besides

$b = L/mv_0$, the "length"

$$a = K/2E = K/mv_0^2, \tag{4.46}$$

which is respectively taken to be positive and negative for attraction and repulsion with $E > 0$ (compare (4.37)), and the length

$$\gamma = +(a^2 + b^2)^{1/2} \tag{4.47}$$

versus $c = +(a^2 - b^2)^{1/2}$ of (1.59). In terms of those lengths, the distance of closest approach (4.44) may be written

$$r_L = \gamma - a = \gamma \mp |a|, \tag{4.48}$$

the last for attraction and repulsion, respectively.

The orbit derivative (4.21) may now be written

$$\begin{aligned} \frac{dr}{r\,d\varphi} &= \frac{\pm [r^2 + 2ar - b^2]^{1/2}}{b} \\ &= \pm \left(\frac{r}{b}\right) \left[\left(\frac{\gamma}{b}\right)^2 - \left(\frac{b}{r} - \frac{a}{b}\right)^2\right]^{1/2}. \end{aligned} \tag{4.49}$$

It vanishes at $r = r_L$, which was found as a root of the bracket in the first place, and hence the orbit is normal to the radius of closest approach, as expected. The integration becomes very simple after the substitution $u = b/r - a/b$, having $du = -b\,dr/r^2$:

$$\int du/[(\gamma/b)^2 - u^2]^{1/2} = -\cos^{-1}(bu/\gamma) = \mp(\varphi - \varphi_L),$$

with φ_L the integration constant. The final result is

$$r(\varphi) = b^2/[a + \gamma \cos(\varphi - \varphi_L)], \tag{4.50}$$

as against (1.59) for the bound state orbits. The reason for denoting the integration constant φ_L is now clear; it is the polar angle corresponding to r_L, the distance of closest approach, since $r(\varphi_L) = b^2/(a + \gamma) = \gamma - a = r_L$.

The orbit (4.50) is obviously symmetrical around the angle $\varphi = \varphi_L$. It has asymptotes ($r \to \infty$) at angles φ for which $\cos(\varphi - \varphi_L) = -a/\gamma$. The adopted initial conditions require that the integration constant φ_L be so chosen that one asymptote lies at $\varphi = 0$; hence a $\varphi_L < \pi$ is determined by

$$\cos \varphi_L = -a/\gamma. \tag{4.51}$$

In the repulsive case ($a < 0$) this cosine is positive and $\varphi_L < \frac{1}{2}\pi$, and in the attractive case $\frac{1}{2}\pi < \varphi_L < \pi$. A second asymptote will occur, at an angle to be called $\varphi = \pi \pm \vartheta$, in the cases of attraction and repulsion, respectively:

$$\cos(\pi \pm \vartheta - \varphi_L) = \mp |a|/\gamma = \cos \varphi_L .$$

Hence

$$\vartheta = |\pi - 2\varphi_L| < \pi \tag{4.52}$$

is the deflection angle, as expected from the symmetry about $\varphi = \varphi_L$. Now $|\cos \varphi_L| = |a|/\gamma = \sin \frac{1}{2}\vartheta = |a|/(a^2 + b^2)^{1/2}$, and so

$$\tan \tfrac{1}{2}\vartheta = \frac{|a|}{b} = \frac{|K|}{mv_0^2\, b}. \tag{4.53}$$

This shows that the deflection approaches the backward direction, $\vartheta \to \pi$, as the approach nears being "head-on" ($b \to 0$). It also shows that the deflection increases with the force strength $|K|$, and that the more energetic the particle the less it is deflected.

In view of the properties already pointed out for it, it is not surprising that the orbit defined by (4.50) is a branch of a *hyperbola*. The polar form may be put into a possibly more familiar cartesian one by adopting an x-axis lying along $\varphi = \varphi_L$ such that $x = r\cos(\varphi - \varphi_L) - \gamma$ and $y = r \sin(\varphi - \varphi_L)$. Then (4.50) leads to

$$\frac{x^2}{a^2} - \frac{y^2}{b^2} = 1, \tag{4.54}$$

with the repulsive branch at $x > 0$ and an attractive branch at $x < 0$. (See Exercise 1.18.)

EXERCISES

4.1. Investigate the effect of a uniform field on an orbital motion by considering the effect of uniform gravity on the isotropic oscillator mass of §4.1.

(a) Find the orbit if the particle is started from the (oscillator-) force *center*. with an initial horizontal speed v_0.

(b) Give arguments that, if the particle is given a more general start, its orbit will be planar only if the initial velocity lies in a vertical plane through the force center.

(c) Show that the projection of a *completely* arbitrary orbit on a vertical plane will be an ellipse centered on a point below the force center. [$Ax^2 + Bxy + Cy^2 + Dx + Ey + F = 0$ describes an ellipse if $B^2 - 4AC < 0$.]

4.2. A long pendulum (length L) when constrained to some "vertical" plane fixed to the earth, performs small oscillations which may be treated as straight-line displacements,

$$X = a \sin[(g/L)^{1/2} t],$$

if $a \ll L$. The frequency $\omega = (g/L)^{1/2}$ should be familiar from elementary physics. Now suppose the constraint to the vertical plane is removed, so that the pendulum becomes one of the "Foucault" type (see page 50), free to change its plane of oscillation. Show that the Coriolis effect of the earth's diurnal rotation (§2.5) is equivalent to the earth's rotating "under" the pendulum, with its plane having a fixed orientation relative to the stars. Reference to Exercises 2.9 and 2.10 may help. Any motions out of the plane tangent to the earth's surface are obviously to be neglected.

4.3. A convenient way to graph any "Lissajou orbit" for a plane nonisotropic oscillation, like (see page 85)

$$x = a \sin(\omega_x t + \alpha), \qquad y = b \sin(\omega_y t + \beta),$$

is to use the fact that x is a projection, on a line parallel to a diameter of a circle having radius a, of a point moving on the circle with angular velocity ω_x, and y is a similar projection from a circular motion of radius b and angular velocity ω_y. Use such constructions to plot orbits for several frequency ratios ω_x/ω_y.

4.4. Show that the trajectory of the motion

$$x = a \sin \omega t, \qquad y = a \sin(\omega + \Delta\omega)t,$$

with $\Delta\omega \ll \omega$, may be described by

$$\left(\frac{x'}{\cos \frac{1}{2} \Delta\omega t}\right)^2 + \left(\frac{y'}{\sin \frac{1}{2} \Delta\omega t}\right)^2 = 2a^2,$$

where $x' = 2^{-1/2}(x + y)$ and $y' = -2^{-1/2}(x - y)$. Make a sketch which shows how this motion tranverses near-elliptic paths, with continuously changing major and minor radii, the major axis switching between the two orientations making 45° angles with x- and y-axes.

4.5. Derive the orbit (4.6) in the way suggested immediately below (4.21).

4.6. Derive the result (4.26) in the way suggested just above it.

4.7. The integrations relating time to the phase of motions in an inverse-square force field become simple to treat adequately for the following.

(a) Show that if the particle starts from rest at a distance r_0 from the force center, it will "fall" into the center in a time proportional to $r_0^{3/2}$.

(b) If, instead, the particle has, at r_0, the inward speed it would have gained if it had started from rest "at infinity," then it will traverse the remaining distance (r_0) in a time shorter than in (a) by just the factor $4/3\pi$. Show this.

4.8. Derive the elliptical ($E < 0$) orbits in a inverse-square force field by the same method that led to the hyperbolic orbits (4.50).

4.9. Suppose a missile is launched from the earth's surface with a velocity $v_0(<(gR)^{1/2})$ large enough for it to be essential to take into account the variation of gravity with height. Paying no attention to the earth's rotation and to air resistance, show that the missile will attain maximum range for a given launch speed if its initial angle of elevation (with the plane tangent to the earth at the launch site) is

$$\alpha = \tan^{-1}[1 - v_0^2/gR]^{1/2}.$$

Compare this with the result for the parabolic path (1.58). Moreover, show that the elliptical orbit which yields the maximum range has its second focus half-way along the chord through the earth which joins the launch and landing sites. Further, show that the angle of elevation just halves the angle between the chord and the direction of gravity, as in the case of parabolic orbits (Exercise 1.16). Finally, show that with $v_0 > (gR)^{1/2}$, it is possible for the missile launcher to shoot himself in the backside!

4.10. To see how important the earth's rotation can be, consider an artificial satellite which seems fixed in the sky, as viewed from any point on earth. Find the orbit radius, and compare the speed in orbit with the velocity the satellite had before it was launched, while at rest on the surface of the rotating earth. [The interested reader may now want to investigate various idealized problems quite obviously associated with the launching, both from equatorial sites and from other latitudes. The results of Exercise 4.9 may be used for help with "ballistic" launchings; rocket motors that supply (relatively) instantaneous impulses may be assumed to help in "adjustments to orbit," after some point of the desired orbit is reached.]

4.11. Find the "Kepler orbit in momentum space," as in a diagram having p_x, p_y as orthogonal axes. Show that the "orbit" is a circle centered on a point with $|p| = Lc/b^2$, and having radius La/b^2. [This curious result was brought to the author's attention in a lecture by Dr. M. Gutzwiller.]

4.12. A "mu-mesic" helium atom consists (essentially) of a point-nucleus of charge $+2e$, a "muonic cloud" of total charge $-e$, and an electron of charge $-e$. As a consequence, the electron may be treated as moving in the field of potential energy

$$V = -\left(\frac{e^2}{r}\right)[1 + e^{-\alpha r}(1 + \tfrac{1}{2}\alpha r)],$$

where $(2\alpha)^{-1}$ is the so-called "muonic Bohr radius." Check that this expression conforms to what might be expected for $r \to 0$ and for $r \to \infty$. Find the total energy the electron must have, in terms of e, α and a, to move in a stable circular orbit of radius a. ($E = -(e^2/2a)[1 + e^{-\alpha a} \times (1 - \tfrac{1}{2}\alpha^2 a^2)]$.)

CHAPTER 5

INTERACTIONS OF TWO PARTICLES

A system of two interacting particles is the simplest one that may be formulated as an *isolated* system without restricting it to uniform motions. To be considered isolated, the system must have a conserved energy, momentum and angular momentum, and a description that is invariant to translations in time and space and to reorientations in space (see the final subsection of Chapter 2).

The only means for describing energy conservation in nonuniform motions which have been made available so far require that the forces be derivable from a scalar potential, $V(\mathbf{r}_1, \mathbf{r}_2)$. For this reason, attention will here be restricted to forces $\mathbf{F}_{1,2}$, on each of the particles, which are of the types

$$\mathbf{F}_1 = -\nabla_1 V(\mathbf{r}_1, \mathbf{r}_2) \qquad \text{and} \qquad \mathbf{F}_2 = -\nabla_2 V(\mathbf{r}_1, \mathbf{r}_2), \tag{5.1}$$

with $\nabla_{1,2}$ denoting gradients in the dependences on the respective position-vectors, $\mathbf{r}_{1,2}$. A significant step here is the exclusion of an explicit dependence on time. Admitting an explicit time-dependence would plainly prevent the constancy of the total energy and, as already pointed out in §2.3, is used to represent external effects which destroy isolation. The relation suggested here between invariance to time shifts and energy conservation will eventually be seen to be a highly general one (§7.5).

To be invariant to the position chosen for the frame origin, the potential can depend only on the *relative* position $\mathbf{r} = \mathbf{r}_1 - \mathbf{r}_2$ of the two particles: $V(\mathbf{r}_1 - \mathbf{r}_2) \equiv V(\mathbf{r})$. This leads to

$$\mathbf{F}_1(\mathbf{r}) = -\nabla V(\mathbf{r}) = +\nabla_2 V = -\mathbf{F}_2(\mathbf{r}), \tag{5.2}$$

where ∇ denotes a gradient derivative with respect to the separation vector, $\mathbf{r}$. Thus the force is restricted to a *mutual* one, $\mathbf{F}(\mathbf{r}) = \mathbf{F}_1 = -\mathbf{F}_2$. The force on the second particle is a reaction from its force on the first, so that $m_2 \dot{\mathbf{v}}_2 = -m_1 \dot{\mathbf{v}}_1$. The total momentum $\bar{\mathbf{p}} = m_1 \mathbf{v}_1 + m_2 \mathbf{v}_2$ is conserved as a consequence. This relation between invariance to location in space and momentum conservation is also a basic one. Note that, with only relative displacements involved, the description is automatically invariant to the Galilean transformations (page 41).

Finally, for invariance to orientation, the potential can depend at most on the magnitude r, of the separation, and not on its orientation, $\hat{\mathbf{r}}$. This makes

$$\mathbf{F}(\mathbf{r}) = -\hat{\mathbf{r}}\, dV/dr = \hat{\mathbf{r}} F(r) \tag{5.3}$$

a "central" force, and permits angular momentum conservation, as will be checked explicitly in the next section. In a space which must be isotropic for an isolated system, there is no other direction that can be singled out for a mutual force, between just two point-particles, than the direction from one particle to the other.

It may still be objected that the intrusion of an energy-possessing entity like the potential field $V(r)$ destroys the isolation of the two particles by themselves. However, a potential energy depending only on the distance apart of the two particles may be regarded as possessed by them as a system. The treatment retains consistency with the concept of isolation because the instantaneous reaction of each particle to the other, implicit in the equality $\mathbf{F}_2 = -\mathbf{F}_1$, allows momentum lost by either particle to reappear immediately as momentum of the other (this makes the treatment essentially what is sometimes called an "impulse approximation").

5.1 TWO-PARTICLE EQUATIONS OF MOTION

The introductory paragraphs of this chapter supplied reasons for an interest in the equations of motion,

$$\begin{aligned} m_1 \ddot{\mathbf{r}}_1 &= \mathbf{F}(\mathbf{r}) = \hat{\mathbf{r}} F(r), \\ m_2 \ddot{\mathbf{r}}_2 &= -\mathbf{F}(r), \end{aligned} \tag{5.4}$$

with $\mathbf{r} \equiv \mathbf{r}_1 - \mathbf{r}_2$.

The Motion as a Body

The sum of the equations (5.4) is independent of the mutual force and merely reiterates the Law of Reactions (2.3):

$$m_1\ddot{\mathbf{r}}_1 + m_2\ddot{\mathbf{r}}_2 = 0.$$

As already seen in (2.5), a first integration yields the conservation of the total momentum:

$$m_1\dot{\mathbf{r}}_1 + m_2\dot{\mathbf{r}}_2 \equiv \bar{\mathbf{p}}, \quad \text{constant.} \tag{5.5}$$

The whole system, "as a body," has the mass $M = m_1 + m_2$ and the velocity average

$$\bar{\mathbf{v}} = \bar{\mathbf{p}}/M = (m_1\dot{\mathbf{r}}_1 + m_2\dot{\mathbf{r}}_2)/M \tag{5.6}$$

may be defined for it. A corresponding "average position," $\bar{\mathbf{r}}$, may now be defined so that $\dot{\bar{\mathbf{r}}} = \bar{\mathbf{v}}$; the position

$$\bar{\mathbf{r}} \equiv (m_1\mathbf{r}_1 + m_2\mathbf{r}_2)/M \tag{5.7}$$

is called the *center-of-mass* (CM) or *centroid* of the two-particle system (see Exercise 1.4). It is an average in which the individual particle positions are weighted according to the fraction of the whole mass that is associated with each; if $m_2 = 2m_1$, then m_2 has two-thirds of the whole mass, and the CM is then at a position two-thirds of the way from m_1, on the line from m_1 to m_2.

The equation governing the motion of the whole mass M is

$$\boxed{\dot{\bar{\mathbf{p}}} \equiv M\ddot{\bar{\mathbf{r}}} = 0}, \tag{5.8}$$

with the trivial solution, $\bar{\mathbf{v}} =$ constant. It describes a uniform motion which can be "transformed away" merely by referring it to a uniformly moving frame, as befits an isolated system; that is, it conforms to Galilean invariance. The internal forces do not disturb this motion as a body any more than anybody can "lift himself by his bootstraps."

The "Internal Motion"

Consider now the motion (variation), $\mathbf{r}(t) \equiv \mathbf{r}_1 - \mathbf{r}_2$, of the relative position-vector joining the two particles. It follows from the equations of motion (5.4) that

$$\ddot{\mathbf{r}} \equiv \ddot{\mathbf{r}}_1 - \ddot{\mathbf{r}}_2 = \left(\frac{1}{m_1} + \frac{1}{m_2}\right)\mathbf{F}. \tag{5.9}$$

The coefficient of the force here has the dimensions of an inverse mass called the "reduced mass" of the two particles:

$$\mu = \left(\frac{1}{m_1} + \frac{1}{m_2}\right)^{-1} = \frac{m_1 m_2}{M}. \tag{5.10}$$

Now,

$$\boxed{\mu \ddot{\mathbf{r}} = \mathbf{F}}. \tag{5.11}$$

With this, the two equations of motion (5.4) have been replaced by a new pair (5.8) and (5.11), corresponding to the variable substitutions $\mathbf{r}_1, \mathbf{r}_2 \to \bar{\mathbf{r}}, \mathbf{r}$. The substitutions are advantageous because the motion "as a body," $\bar{\mathbf{r}}(t)$, was shown to be a trivial uniform one, and the relative motion $\mathbf{r}(t)$ is no more complicated than that of a single particle in a fixed-center force field. These motions are uncoupled while the individual particle motions $\mathbf{r}_{1,2}(t)$ exchange energy.

The equation for the "internal motion" (5.11) will clearly, for a given $\mathbf{F}$, have the kind of solutions already found in Chapter 4. It is only necessary to substitute in them the reduced mass μ of the two particles in place of the single-particle mass, m, and to reinterpret the significance of the position vector $\mathbf{r}$ appropriately. The reinterpretation is very simple: $\mathbf{r}(t)$ describes the trajectory of either one (to be called m_1) of the two masses as "seen" from the second mass (m_2) as origin.

Writing the expression for the reduced mass as a fraction of m_1 or m_2,

$$\mu = \left(\frac{m_2}{m_1 + m_2}\right) m_1 = \left(\frac{m_1}{m_1 + m_2}\right) m_2,$$

makes it plain that it is smaller than either m_1 or m_2. The reduction is greatest (μ is the least fraction of either mass) for equal masses, when $\mu = \frac{1}{2}m_1 = \frac{1}{2}m_2$. On the other hand, if one of the masses, say m_2, is so large that adding the other to it makes no appreciable correction, the reduced mass becomes indistinguishable from the lighter one: $\mu \approx m_1$. It becomes clear that the fixed-center force cases of Chapter 4 correspond to isolated, two-particle systems in which the mass of one of the particles is so heavy as to be practically immovable: $|\dot{\mathbf{v}}_2| = (m_1/m_2)|\dot{\mathbf{v}}_1| \ll |\dot{\mathbf{v}}_1|$.

The latter is the situation in the motion of most of the planets around the sun. The result (4.28) for the "annual" periods should properly be amended to*

$$T = 2\pi(\mu/K)^{1/2} a^{3/2} = 2\pi[(m_1 + m_2)G]^{-1/2} a^{3/2},$$

* Notice that the mutual of gravitational force attraction may now be written $F = -G\mu M/r^2$, where μ is the reduced mass and $M = m_1 + m_2$ is the total mass.

but even for the heaviest planet, Jupiter, m_1 is only about 5% of the sun's mass, m_2, and hence, to a fair approximation, $\mu/K \approx m_1/K$ as assumed in (4.28). The earth is about 80 times as heavy as its moon, and hence the reduced mass correction diminishes the prediction (4.31) for the monthly period by about 0.6%.

Another simple example of the reinterpretations needed in going from fixed-force problems to relative motions under a mutual force is concerned with harmonic oscillations. Suppose two masses m_1, m_2 are connected by a relatively weightless spring, which provides a "Hooke's Law" mutual force. Suppose also that the force strength is measured by fixing m_2 and finding the frequency ω_1 with which m_1 then oscillates. Then, according to $m_1 \ddot{x}_1 = -m_1\omega_1^2 x_1$ (3.6), the force strength is $m_1 \omega_1^2$ per unit of displacement. Next, consider the oscillation of m_2 with m_1 fixed; since $m_2 \ddot{x}_2 = -m_1\omega_1^2 x_2$ for the same spring, m_2 will oscillate with the frequency $\omega_2 = (m_1\omega_1^2/m_2)^{1/2} = (m_1/m_2)^{1/2}\omega_1$. Finally, let both masses oscillate freely under the influence of the same connecting spring. Their separation, $x = x_1 - x_2$, will then oscillate according to $\mu\ddot{x} = -m_1\omega_1^2 x$; the frequency of this will be $\omega = (m_1/\mu)^{1/2}\omega_1 = (M/m_2)^{1/2}\omega_1$, greater than either ω_1 or ω_2.

The Internal Dynamical Variables

The various dynamical quantities used in describing motion must be suitably interpreted when they refer to the relative motion. For example,

$$\mathbf{v} \equiv \dot{\mathbf{r}} = \dot{\mathbf{r}}_1 - \dot{\mathbf{r}}_2 = \mathbf{v}_1 - \mathbf{v}_2 \tag{5.12}$$

is the *relative* velocity of the two particles, independent of any external frame of reference with respect to which the individual velocities $\mathbf{v}_1$ and $\mathbf{v}_2$ may be defined. The "internal momentum" $\mathbf{P}$ which is generated directly by the mutual force in accordance with $\dot{\mathbf{P}} = \mathbf{F}$ is plainly

$$\mathbf{P} = \mu\mathbf{v} = \frac{m_1 m_2}{M}(\mathbf{v}_1 - \mathbf{v}_2) = \frac{m_2}{M}\mathbf{p}_1 - \frac{m_1}{M}\mathbf{p}_2, \tag{5.13a}$$

as follows from the equation of internal motion (5.11). No comparable significance can generally be given to $\mathbf{p}_1 - \mathbf{p}_2$, although

$$\mathbf{P} = \tfrac{1}{2}(\mathbf{p}_1 - \mathbf{p}_2), \qquad \text{for} \quad m_1 = m_2, \tag{5.13b}$$

is the relative, "internal," momentum of a pair of *equal* masses.

Any "relative torque" or "couple," $\mathbf{r} \times \mathbf{F}$, existing between the two particles would generate angular momentum according to

$$\mathbf{r} \times \mathbf{F} = \mathbf{r} \times \mu\ddot{\mathbf{r}} = d(\mathbf{r} \times \mathbf{P})/dt. \tag{5.14}$$

Thus an "internal" angular momentum is defined by

$$\mathbf{l} \equiv \mathbf{r} \times \mu\mathbf{v} \equiv \mathbf{r} \times \mathbf{P}. \tag{5.15}$$

It is actually conserved, $\dot{\mathbf{l}} = 0$, when only the mutual force directed from one particle to the other exists, since this provides no couple ($\mathbf{r} \times \mathbf{F} = 0$). The conservation law fits in with the assumption of isolation and can be traced to the orientational invariance requirement which led to the adoption of the "central" character (5.3) for the mutual force.

The power expended on the relative displacements by the mutual force is

$$\mathbf{F} \cdot \mathbf{v} = d(\tfrac{1}{2}\mu v^2)/dt \tag{5.16}$$

according to the equation of internal motion. Thus,

$$\tau = \tfrac{1}{2}\mu v^2 \tag{5.17}$$

is definable as the kinetic energy of the internal motion. With a mutual force derivable from a potential according to $\mathbf{F} = -\nabla V(r) = -\hat{\mathbf{r}}V'(r)$, the total internal energy

$$\varepsilon = \tau + V \tag{5.18}$$

is conserved, a result obtainable in the same way as the conserved energy of (2.21).

The relation of the internal dynamical variables $\mathbf{P}$, $\mathbf{l}$, and τ to ones which can be attributed to each of the particles separately will be explored in §5.4.

5.2 CLASSICAL COLLISION THEORY IN THE "LABORATORY FRAME"

The types of two-particle interactions given more extended attention here are ones occurring in circumstances described as "collisions," or "scattering." Processes of this kind have a unique importance in that they form the essentials of physical observation. An object is seen when "particles of light" fall on it and are reflected ("scattered") to the eye. In modern laboratories also, massive particles are used for such purposes as in electron microscopes. Atoms are probed by shooting beams of mass-particles onto them and observing the number and the directional distribution of those that are scattered. Finally, collision processes also play an essential role in the statistical mechanics of gases.

Initial and Final States

Under so-called "laboratory" collision conditions, an incident particle of momentum $m_1\mathbf{v}_0$ is sent toward a "target" particle of mass m_2 at rest. What will happen depends on the distance b (to be called the "impact parameter") by which the velocity $\mathbf{v}_0$ misses being aimed at "head-on" collision with the force-center on the target particle. As a result of the collision, the incident particle will have been deflected through some angle $\vartheta(b)$ and will go off along the deflected direction with some velocity $\mathbf{v}_1$. The target particle will recoil at some angle $i(b)$ to the direction of incidence, and gain some velocity $\mathbf{v}_2$. Both initially and finally the particles are far apart, outside the effective "range" of their mutual interaction. The initial and final states are characterized by uniform velocities $\mathbf{v}_1^0 = \mathbf{v}_0$, $\mathbf{v}_2^0 = 0$ and $\mathbf{v}_1$, $\mathbf{v}_2$. The transition from one to the other takes place during an intermediate period of impact in which the deflection $\vartheta(b)$ and the momentum transfer $m_2\mathbf{v}_2$ into the direction $i(b)$ are generated by the mutual force of interaction. For a pair of point-particles forming an isolated system, the force must be a central one, as argued above, and then the whole process takes place in a plane determined by the vector $\mathbf{v}_0$ and the initial position of the target particle.

The Conservation Conditions

With the initial and final states as defined here, the conservation conditions expected to characterize the isolated two-particle system are easy to apply. The potential energy is supposed negligible outside the force range; hence

$$\tfrac{1}{2}m_1 v_0^2 = \tfrac{1}{2}m_1 v_1^2 + \tfrac{1}{2}m_2 v_2^2 \tag{5.19}$$

expresses the conservation of the energy. The recoil energy,

$$\Delta E = \tfrac{1}{2}m_2 v_2^2 = \tfrac{1}{2}m_1({v_0}^2 - v_1^2), \tag{5.20}$$

is also called the "energy transfer" to the target particle, resulting from the collision. With the kinetic energies conserved in this way, the collision is called "elastic."

The conservation of the momentum is most succinctly expressed in a diagram (Figure 5.2). For analytic treatment, various resolutions of the momentum vectors in the diagram may be employed. When they are resolved on the direction of incidence and normally to it,

$$\begin{aligned} m_1 v_0 &= m_1 v_1 \cos\vartheta + m_2 v_2 \cos i, \\ 0 &= m_1 v_1 \sin\vartheta - m_2 v_2 \sin i. \end{aligned} \tag{5.21}$$

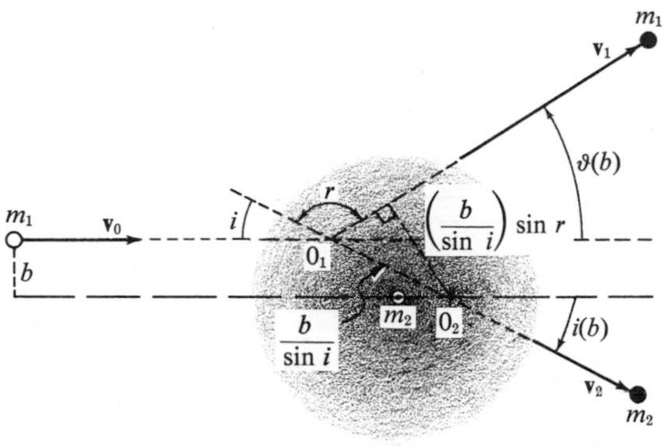

Figure 5.1

These, together with the energy conservation condition (5.19), provide three equations for determining v_1, v_2 and the deflection ϑ, for a given initial velocity v_0 and recoil angle $i(b)$. Actually given is not i but the collision parameter b; how the two are related depends on the specific nature of the collision force, and hence the equation of motion must still be solved: to give the particle orbits having "asymptotes" along $\mathbf{v}_0$ and $\mathbf{v}_1$ for m_1, and along $\mathbf{v}_2$ for m_2.

Still other conservation conditions, on the *angular* momenta as evaluated about various momental centers, may seem available to help determine the results independently of the particular force responsible for the interaction. An initial angular momentum

$$L = m_1 v_0 b, \tag{5.22}$$

about the initial position of the target particle, is introduced into the system by the incident particle. To express the final angular momenta about that momental center requires knowledge of the lever arms of $m_1\mathbf{v}_1$ and $m_2\mathbf{v}_2$ relative to the initial position of m_2; those are not forthcoming until after the equation of motion is solved. However, the initial angular momentum is the same, $L = m_1 v_0 b$, about a momental center lying at the intersection of the recoil direction with the line of $b = 0$ (the momental center marked O_2 in Figure 5.1). Then the recoil momentum $m_2\mathbf{v}_2$ makes no contribution toward the conservation of L thus taken. To find the lever arm of the deflected momentum, $m_1\mathbf{v}_1$, consider still another momental center (marked O_1 in Figure 5.1) defined by the intersection of the recoil direction with the line of the initial velocity $\mathbf{v}_0$. No angular momentum about this center is introduced initially, the recoil momentum again contributes nothing, and so the projection back of the final velocity $\mathbf{v}_1$ must also intersect the same momental center (O_1). Then the geometry (see Figure 5.1) makes it clear that the desired

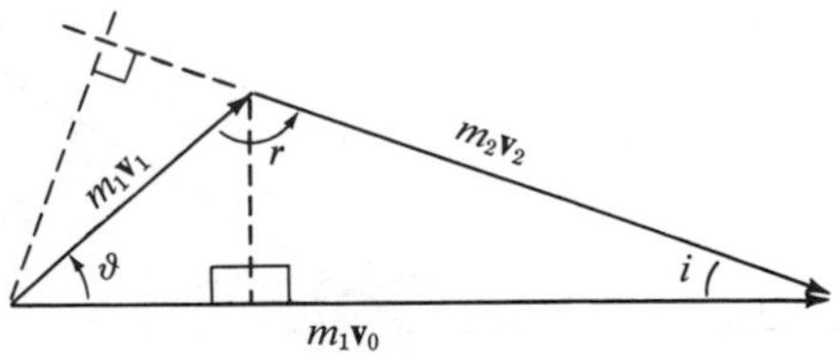

Figure 5.2

lever arm of $m_1\mathbf{v}_1$ is just $b \sin r/\sin i$, where $r = \pi - i - \vartheta$ is the "reflection angle" shown both in Figure 5.1 and in the momentum-conservation diagram, Figure 5.2. Thus

$$L = m_1 v_0 b = m_1 v_1 b \sin r/\sin i. \tag{5.23}$$

Unfortunately, this adds nothing to the determination of the results not already represented in the linear momentum conservation. The component of the latter normal to the recoil momentum (see the dotted lines in Figure 5.2) is just

$$m_1 v_0 \sin i = m_1 v_1 \sin r,$$

and is also implicit in the analytical expression (5.21). That other choices of momental center can therefore give no more help* will be clarified in Chapter 6.

Head-on Collisions

The case of head-on ($b = 0$) collisions, in which the recoil can only be forward ($i = 0$), is particularly simple because it is essentially a one-dimensional problem. For it, the momentum conservation conditions simplify to

$$m_1 v_0 = m_1 v_1 + m_2 v_2, \tag{5.24}$$

if it is agreed that a negative scattered velocity v_1 is to stand for a backward deflection (only positive values of v_0 and v_2 are still to be admitted, as for all the velocity magnitudes in the more general conditions (5.21)). Together with

* Some insight into why angular momentum conservation provides no help additional to that provided by linear momentum conservation can perhaps be gained as follows. An arbitrary interaction described by a force $\mathbf{F}(\mathbf{r}_1 - \mathbf{r}_2)$ may be regarded as a distribution of points at which momentum exchanges of various magnitudes take place. Suppose that at a definite point, $\mathbf{r}_1 = \mathbf{r}_2 = \mathbf{r}_0$, there are momentum changes $\mathbf{p}_1, \mathbf{p}_2 \rightarrow \mathbf{p}_1 + \Delta\mathbf{p}_1, \mathbf{p}_2 + \Delta\mathbf{p}_2$. Momentum conservations requires $\Delta\mathbf{p}_2 = -\Delta\mathbf{p}_1$. The concomitant changes of angular momentum are $\Delta\mathbf{L}_1 = \mathbf{r}_0 \times \Delta\mathbf{p}_1$ and $\Delta\mathbf{L}_2 = \mathbf{r}_0 \times \Delta\mathbf{p}_2$. Thus $\Delta(\mathbf{L}_1 + \mathbf{L}_2) = 0$ is equivalent to the momentum conservation.

the energy conservation condition (5.19), this leads to

$$v_1 = \frac{m_1 - m_2}{m_1 + m_2} v_0 \qquad \text{and} \qquad v_2 = \frac{2m_1}{m_1 + m_2} v_0 . \tag{5.25}$$

The collision results are thus completely determined, independently of the force responsible (it must only be there, to avoid the trivial solution, $v_1 = v_0$, $v_2 = 0$, describing a "miss").

The results (5.25) show that m_1 rebounds backward if it is lighter than the target mass m_2; with a target so heavy that m_1 is negligible by comparison, it will rebound without energy loss, having $v_1 \to -v_0$. On the other hand, when it is the target mass that is negligible, then m_1 goes on practically undisturbed, with $v_1 \to +v_0$, and the light mass m_2 chases ahead with $v_2 \to 2v_0$. For the intermediate case of equal masses, the incident particle stops dead, $v_1 = 0$, and the target particle takes up all the momentum and energy: $v_2 = v_0$.

An amusing example of a series of head-on collisions is provided by suspending two balls as pendulums that collide head-on when both pass through their vertical positions. Suppose the motion is started with one mass, M, at rest in its vertical position and by allowing the second mass, m, to swing into it with velocity v_0. After this initial collision, M will swing away with a velocity $V_1 = 2mv_0/(m + M)$ and m rebounds (if small enough) with velocity $v_1 = (M - m)v_0/(M + m)$. Both pendulums will subsequently return to the vertical after periods which are precisely the same if $V_1 = v_1$ (that is, if $M = 3m$ had been chosen). With that choice, a second collision takes place at the vertical configuration, both masses entering it with equal and opposite velocities of magnitude $\frac{1}{2}v_0$. The result of the second collision is to reestablish the first starting conditions again—that is, with m having v_0 and $M = 3m$ at rest. The process will therefore continue repeating itself, as long as energy and momentum are conserved.

Encounters of Equal Masses

When the considerations are not restricted to head-on collisions, they are about the simplest for the case of equally massive collision partners. The masses then drop out of the conservation conditions and the energy conservation relation (5.19) is simplified to $v_0^2 = v_1^2 + v_2^2$. This shows that the velocity vectors form a right triangle and the momentum conservation diagram of Figure 5.2 becomes the right triangle of Figure 5.3. The striking consequence here is that every collision of equal masses results in their coming away at right angles to each other, a fact much relied upon in the game of billiards.

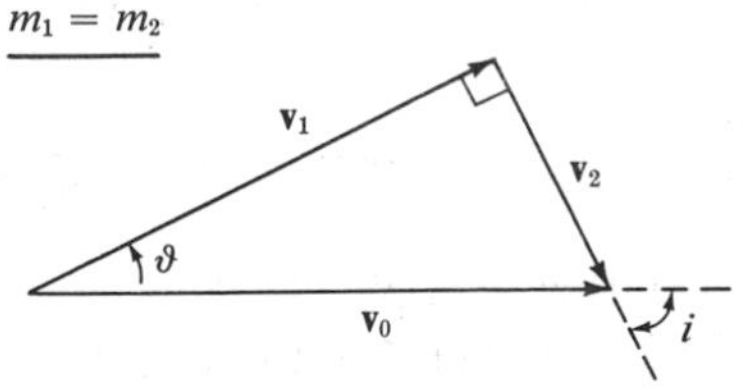

Figure 5.3

With recoil and deflection angles complementary ($i + \vartheta = \frac{1}{2}\pi$), the deflection is restricted to the range $0 \leq \vartheta \leq \frac{1}{2}\pi$. The resultant velocities, for $m_1 = m_2$, are

$$v_1 = v_0 \cos\vartheta = v_0 \sin i,$$
$$v_2 = v_0 \sin\vartheta = v_0 \cos i. \tag{5.26}$$

The energy transferred to the target particle (5.20) becomes

$$\Delta E = E_0 \sin^2\vartheta = E_0 \cos^2 i. \tag{5.27}$$

It is complete ($\Delta E \to E_0 \equiv \frac{1}{2}m_1 v_0^2$), for the maximum deflections $\vartheta \to \frac{1}{2}\pi$, and these occur in head-on collisions ($i \to 0$ and $v_1 \to 0$).

General Conclusions from the Conservation Relations

Although the conservation conditions, by themselves, determine the kinematics of a collision completely only when it is head-on,* they still lead to noteworthy conclusions, independent of the interactions involved, about more general collisions.

The momentum-conservation relations (5.21) may be solved for the recoil momentum, $m_2 v_2$, by first eliminating the deflection angle:

$$(m_1 v_0 - m_2 v_2 \cos i)^2 + (m_2 v_2 \sin i)^2 = (m_1 v_1)^2.$$

Next, v_1 is eliminated by using the energy conservation condition (5.19):

$$m_1^2 v_0^2 - 2m_1 m_2 v_0 v_2 \cos i + m_2^2 v_2^2 = m_1^2 v_0^2 - m_1 m_2 v_2^2.$$

Then

$$m_2 v_2 = 2\mu v_0 \cos i (>0 \text{ for } 0 \leq i \leq \tfrac{1}{2}\pi), \tag{5.28}$$

where $\mu = m_1 m_2/(m_1 + m_2)$ is just the reduced mass introduced in (5.10).

*Actually, whenever the recoil direction $i(b)$ is given, whether this be $i(0) = 0$ or not.

This result bears out the expectation that the recoil angle should be confined to forward directions, $i < \frac{1}{2}\pi$. The corresponding recoil energy, transferred to the target particle, is

$$\Delta E = \tfrac{1}{2}m_2 v_2^2 = 4(\mu/M)E_0 \cos^2 i, \tag{5.29}$$

where $M = m_1 + m_2$ is the total mass of the system and $E_0 = \frac{1}{2}m_1 v_0^2$ is the energy brought into it by the incident particle.

The findings so far yield significant conclusions about collisions against a very heavy target mass, $m_2 \gg m_1$. Such a mass absorbs the finite momentum $m_2 v_2 = 2m_1 v_0 \cos i$, but without moving appreciably, since $v_2 = 2(m_1/m_2)v_0 \cos i \to 0$ as $m_2/m_1 \to \infty$. It absorbs no appreciable energy, since this is $(m_2 v_2)^2/2m_2 \to 0$ for finite $m_2 v_2$, and as a consequence the incident particle can be treated as conserving its energy to itself, even though its momentum is not conserved by itself. The momentum condition can then be ignored, leaving the heavy mass to absorb any amount necessary without taking any appreciable energy. This can be regarded as the situation in each of the fixed force-center problems of the preceding chapter.

The fractional energy transfer $\Delta E/E_0$ is proportional to a mass ratio $\mu/M = m_1 m_2/(m_1 + m_2)^2$ which is symmetrical in the two particles, and this has the consequence that the maximal fraction is transferred between equal masses: $\Delta E/E_0 = \cos^2 i$ (5.27). It is for this reason that passage through hydrogenous substances is a most efficient way to slow down neutrons; the mass of an hydrogen atom about equals that of a neutron.

The conservation conditions also establish a relation between the deflection and recoil angles ϑ and i:

$$\tan \vartheta = \sin 2i/[(m_1/m_2) - \cos 2i]. \tag{5.30}$$

The deflection ϑ here is plotted in Figure 5.4, as a function of the recoil angle for various values of the mass ratio m_1/m_2. Perhaps most striking is the discontinuity from an unlimited deflection ($\vartheta_{max} \to \pi$), when the incident mass m_1 is at all smaller than the target mass m_2, to a $\vartheta_{max} = \frac{1}{2}\pi$ for equal masses. Of course, when the target mass excess is only slight, the deflections $\vartheta > \frac{1}{2}\pi$ are very unenergetic ($i \approx 0$ and the energy transfer (5.29) is nearly as complete as it can be for $m_1 = m_2$). When the incident mass is the heavier ($m_1 > m_2$), the maximum possible deflection is still more restricted, by

$$\vartheta < \vartheta_{max} = \sin^{-1}(m_2/m_1), \tag{5.31}$$

with $\vartheta = \vartheta_{max}(<\frac{1}{2}\pi)$ occurring for $2i = \frac{1}{2}\pi - \vartheta_{max}$ (determining the line of loci indicated in the figure). In this range, the recoil angles for a given deflection are double-valued: $i \lessgtr \frac{1}{2}(\frac{1}{2}\pi - \vartheta_{max})$. Although the deflection angle is then

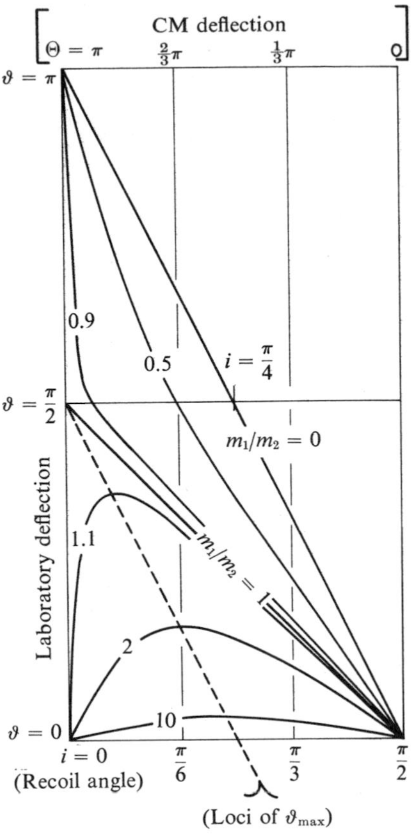

Figure 5.4

the same for two different recoil angles, the deflected velocity v_1 is not the same, since the unequal recoil angles plainly lead to different recoil velocities (5.28).

The deflected velocity v_1 is obtained as a single-valued function of recoil angle i when v_2 and ϑ are eliminated among the conservation conditions (5.19) and (5.21):

$$v_1 = v_0[1 - 4(\mu/M)\cos^2 i]^{1/2}. \tag{5.32a}$$

This properly reduces to the head-on collision result (5.25) for $i = 0$, and to the equal-mass result (5.26) for $M = 4\mu$. When the recoil angle is eliminated in favor of the deflection ϑ,

$$v_1 = (v_0/M)[m_1 \cos\vartheta \pm (m_2^2 - m_1^2 \sin^2\vartheta)^{1/2}]. \tag{5.32b}$$

The positive one of the alternative signs here must be used for $m_2 > m_1$

in order that v_1 remain properly positive, since the absolute value of the radical is then greater than $m_1 \cos \vartheta$. However, both signs are needed to yield the two positive values of v_1 that are possible for each deflection $\vartheta < \vartheta_{\max} \equiv \sin^{-1}(m_2/m_1)$ when $m_2 < m_1$; this is the double-valuedness to which attention has already been called in the preceding paragraph.

5.3 SCATTERING CROSS SECTIONS

Results that depend on the kind of force responsible for a collision interaction, and are thus revelatory of its character, are obtained when the frequency and angular distribution of many scatterings, with a variety of impact parameters b, are studied. This is a standard way of probing intra- and interatomic forces when techniques are not available for holding the particles in a variety of relative positions (as done with gross charged bodies by Coulomb) and measuring the forces directly. The scattering experiments are performed as indicated in Figure 5.5. A unit incident beam of monoenergetic particles, consisting of one particle per unit cross-sectional area per unit time, is sent into a target consisting of a material sheet containing the atoms to be probed. The numbers of beam particles deflected by various angles are counted, per atom in the path of the beam. The measurements can be discussed as "per atom" because even a solid material is mostly "empty," on the atomic scale, and so is easy to make thin enough for multiple collisions of a given incident particle to be reduced to negligibility. The count of scattered particles per atom and per unit beam,

$$\sigma = \frac{\text{Number of particles scattered per atom per second}}{\text{Number of beam particles per cm}^2\text{ per sec}}, \tag{5.33a}$$

is called the total scattering "cross-section" of an atom and obviously has the dimensions of an area. When only the particles scattered by an angle within the solid angle element $d\Omega = \sin \vartheta \, d\vartheta \, d\varphi$ are counted, the result is called the "differential cross-section":

$$\frac{d\sigma}{d\Omega}. \tag{5.33b}$$

The limits on the angles are $0 < \varphi < 2\pi$, $0 < \vartheta < \pi$; hence $+1 > \cos \vartheta > -1$ and $\oint d\Omega = 4\pi$ is the complete, spherical solid angle. Clearly, a cross section σ or $d\sigma/d\Omega$ may be thought of as a transverse area that intercepts as many particles from the unit beam as undergo the processes to be represented by the cross section.

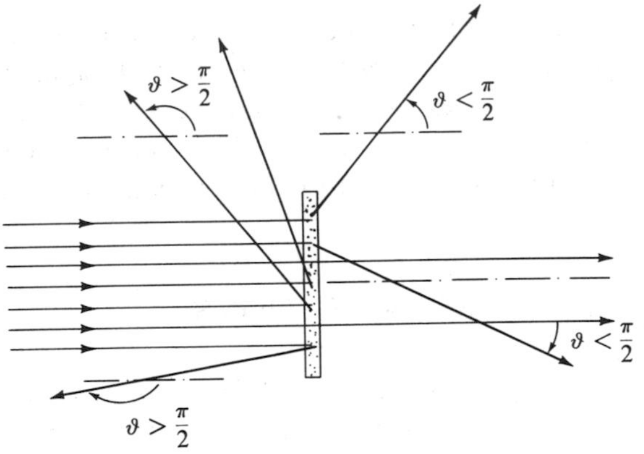

Figure 5.5

Theoretical expectations for cross sections can be derived from knowledge of "deflection functions", $\vartheta(b)$. There is to be expected a correspondence between a given impact parameter b and a deflection angle ϑ, determined by the particular force of interaction. All the incident particles having a b within a ring of breadth db must be deflected by some $\vartheta(b)$ in a range $d\vartheta = (d\vartheta/db)\,db$. A unit beam sends $2\pi b\,db$ particles per second through that ring, and therefore $d\sigma = 2\pi b\,db = \pi(db^2/d\vartheta)\,d\vartheta$ when the particles scattered into the range $d\vartheta$ are counted. If only those within a solid angle element $d\Omega = d\varphi\,d(\cos\vartheta)$ are counted,

$$d\sigma/d\Omega = b\,db/d(\cos\vartheta). \tag{5.34}$$

Of course, the total cross section follows from integration:

$$\sigma = \oint [b\,db/d(\cos\vartheta)]\,d\Omega = 2\pi \int b\,db. \tag{5.35}$$

The last form is to include all elements db that yield processes to be counted in σ.

For gauging the significance of any specific result for the differential cross section $d\sigma/d\Omega$, it is valuable to be familiar with the expression for the case in which the scattering is isotropic (the same into all directions). Then through every unit area of any very large sphere, radius $r \to \infty$, centered around the intersection of beam and target, must pass the same number of deflected particles: $\sigma/4\pi r^2$ per second per atom. An area element on the sphere, $r^2 \sin\vartheta\,d\vartheta\,d\varphi \equiv r^2\,d\Omega$, has $\sigma\,d\Omega/4\pi$ particles passing through it. Thus, for isotropy,

$$d\sigma/d\Omega = \sigma/4\pi, \tag{5.36}$$

an expression independent of ϑ. On the other hand, through the ring area on the sphere subtended by $d\vartheta$, $(\sigma/4\pi r^2)2\pi r^2 \sin\vartheta\, d\vartheta$ particles pass; hence

$$\frac{d\sigma}{d\vartheta} = \tfrac{1}{2}\sigma \sin\vartheta \qquad \text{and} \qquad \frac{d\sigma}{d(\cos\vartheta)} = \tfrac{1}{2}\sigma.$$

All these expressions are used as representatives of isotropic scattering.

The Rutherford Formula

An immediate example of a deflection function is provided by (4.53), which, for example, applies to scattering of charges ze borne by masses m against a fixed point-charge, Ze. In such a case

$$b^2(\vartheta) = \frac{(K/mv_0^2)^2}{\tan^2 \frac{1}{2}\vartheta} = \left(\frac{K}{2E_0}\right)^2 \frac{1 + \cos\vartheta}{1 - \cos\vartheta} \tag{5.37}$$

with $K = zZe^2$, and then

$$\frac{d\sigma}{d\Omega} = \frac{db^2}{2\, d(\cos\vartheta)} = \frac{(zZe^2/2mv_0^2)^2}{\sin^4 \frac{1}{2}\vartheta}. \tag{5.38}$$

This is the famous formula used by Rutherford to interpret his probings of atoms by α-particles ($z = 2$). He found that this angular distribution, characteristic of a point-charge coulomb field, was followed quite accurately even at near-backward ($\vartheta \to \pi$) angles. Such backward deflections could have arisen only in nearly head-on collisions in which the α-particles approach to the distance $r_0 = 2Ze^2/E_0$ (4.44) from the force center. Rutherford used α-particles of sufficient energy to make $r_0 \lesssim 10^{-12}$ cm, about 1/10,000*th* the radius known to be occupied by an entire atom, inclusive of its neutralizing cloud of electrons. Thus arose the picture of an atom as consisting of a tiny nucleus in which most of the mass and all the positive charge of the atom are concentrated, surrounded by a much less concentrated cloud of the negatively charged electrons.

The electrons are known to be light enough not to deflect the 8000 times as heavy α-particles appreciably. The atomic nucleus itself is usually much heavier than the α-particles, and treating it as fixed is a good approximation; however, the nuclear recoil is not difficult to take into account and this will be done in §5.4.

The point-charge angular distribution (5.38) is strongly peaked toward forward angles, and becomes divergently large as $\vartheta \to 0$. In consequence, the integrated cross section becomes infinite. This must be regarded as a result of overidealization, inherent in treating the inverse-square law of force as extending uninterrupted to infinite distances. Taken into account are divergently numerous near-zero deflections arising from incidences with very large impact parameters, b. All the appreciable deflections actually come from impact parameters well within the negative charge cloud. Moreover, the practically unobservable near-zero deflections are reduced further by the intervention of negative charge between positive nucleus and passing alpha particle; this screening is practically complete for impact parameters of the order 10^{-8} cm, measuring the radius of the entire neutral atom.

Hard-Sphere Collisions

A case for which the deflection function $\vartheta(b)$ is reasonably simple to calculate in the laboratory frame (Figure 5.6) is the elastic collision of impenetrable and smooth spheres. This model was much used to represent atoms (or molecules) in the early attempts at a kinetic theory of gases. The colliding spheres may have arbitrary radii a_1, a_2 and, if impenetrable, interact only during an instant of contact when their centers are a distance $a = a_1 + a_2$ apart. The problem directly reduces to the incidence of a *point*-mass m_1 on a "sphere of contact," of radius a; hence Figure 5.6 is drawn for that case. It is clear from the diagram that the impact parameter b determines an "angle of incidence,"

$$i = \sin^{-1}(b/a), \tag{5.39}$$

between the incidence direction and the normal to the sphere at the point of contact.

The collision is supposed to be elastic, which means that the kinetic energy conservation relation (5.19) applies. It is also supposed to be "smooth," which means that the component of the incident momentum tangential to the surface of the sphere, at the point of contact, slips by unchanged: $m_1 v_0 \sin i = m_1 v_1 \sin r$. This is all of one component of the momentum conservation condition, since in the smooth collision the target sphere can recoil only in the direction of the normal, at an angle i equal to the angle of incidence (the momental center O_2 of Figure 5.1 now coincides with the initial position of the target center). The same conserved momentum component is equivalent, after multiplication by b, to the conserved angular momentum $L = mv_0 b$ of (5.23), now definitely applying to the initial position of the target mass as momental center. The force in play, although "impulsive" (instantaneous), is nevertheless clearly a central one, and the angular momentum conservation should be expected. In sum, the one supplementation of the conservation conditions needed, to characterize the particular force in play, is provided when the recoil angle $i(b)$ is identified with the angle of incidence (5.39).

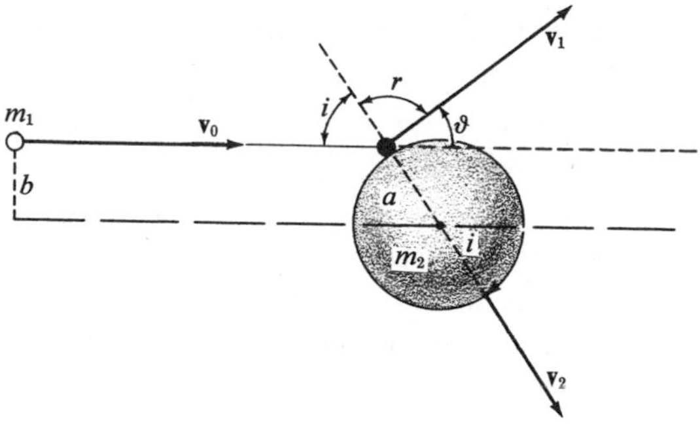

Figure 5.6

Adhering to the resolution of the momentum conservation condition on the directions tangential and normal to the sphere at the point of contact (the broken lines of Figure 5.2),

$$
\begin{aligned}
m_1 v_0 \sin i &= m_1 v_1 \sin r, \\
m_1 v_0 \cos i &= m_2 v_2 - m_1 v_1 \cos r,
\end{aligned}
\tag{5.40}
$$

leads, with the help of the consequence (5.28) for the recoil momentum, most directly to the expression

$$
\tan r = \frac{m_1 v_0 \sin i}{m_2 v_2 - m_1 v_0 \cos i} = (\tan i)\frac{m_2 + m_1}{m_2 - m_1} \tag{5.41}
$$

for the "angle of reflection," r. This shows that the angle of reflection equals the angle of incidence only if the sphere is practically immovable ($m_2 \gg m_1$), like a wall ($\equiv a \to \infty$), Otherwise, the angle of reflection is larger, becoming $r = \pi/2$ for $m_1 = m_2$, as in Figure 5.3. For an incident mass exceeding the target mass ($m_1 > m_2$) the reflection angle varies with the recoil angle only within the limits $r = \pi/2$ to $r = \pi$ until, for $m_1 \to \infty$, $r = \pi - i$ and the deflection $\vartheta = \pi - r - i \to 0$.

The most economical expression for the deflection function $\vartheta(b)$ still has the general form (5.30), except that now the connection between recoil angle i and impact parameter b is known, so that the substitutions

$$
\sin 2i = 2(b/a)[1 - (b/a)^2]^{1/2}, \qquad \cos 2i = 1 - 2(b/a)^2 \tag{5.42}
$$

can be made to present $\vartheta(b)$ explicitly. It can be readily checked that the result properly gives $\vartheta = \pi$ and 0 for head-on ($b = 0$) collisions with $m_1 < m_2$ and $m_1 > m_2$, respectively; moreover, $\vartheta \to 0$, regardless of mass, for "grazing" collisions with $b \to a$.

In principle, the deflection function $\vartheta(b)$, as given by (5.30) and (5.42), should now be used in (5.34) to yield a theoretical cross section. However, the present calculations, in the laboratory frame, are unnecessarily complicated for the general case ($m_1 \gtrless m_2$), as will be seen in the succeeding sections. Only the cases $m_1 \ll m_2$ and $m_1 = m_2$ are simple in the laboratory frame, and only these will be presented here.

$m_1 \ll m_2$. This is the case of the practically "fixed" force-center when the angle of reflection is equal to the angle of incidence and the deflection function is

$$\vartheta \equiv \pi - r - i = \pi - 2i = \pi - 2\sin^{-1}(b/a). \tag{5.43}$$

The inverse function $b^2(\vartheta) = a^2 \sin^2 i = \frac{1}{2}a^2(1 + \cos\vartheta)$ is more directly used in the differential cross-section expression (5.34), $d\sigma/d\Omega = db^2/2d(\cos\vartheta)$, and the result is

$$d\sigma/d\Omega = \tfrac{1}{4}a^2 = \sigma/4\pi. \tag{5.44}$$

Thus the scattering from a fixed sphere is isotropic, and the total cross section σ is just 4π times $\frac{1}{4}a^2$: $\sigma = \pi a^2$. Of course, this is the total cross section to be expected of the sphere, fixed or otherwise, since πa^2 is the cross-sectional area presented to the incident beam. Notice that $\sigma = 2\pi\int_0^a b\,db = \pi a^2$, according to (5.35).

$m_1 = m_2$. In this case, it is known that the deflection and recoil angles are always complementary: $\vartheta = \frac{1}{2}\pi - i$ and

$$\vartheta(b) = \tfrac{1}{2}\pi - \sin^{-1}(b/a), \tag{5.45}$$

just half the result (5.43) for the fixed sphere. Now, $b = a\sin i = a\cos\vartheta$ and

$$d\sigma/d\Omega = a^2\cos\vartheta, \tag{5.46}$$

where $0 \le \vartheta \le \frac{1}{2}\pi$. The scattering is into the forward directions only, the angular distribution pattern (intensity plotted radially as a function of deflection angles) being a sphere passing through the scattering center. The total cross section is

$$\sigma = 2\pi\int_0^{\pi/2} d\vartheta \sin\vartheta \cdot a^2\cos\vartheta = \pi a^2 \tag{5.47}$$

again, as expected.

Another quantity of interest in this case is a cross section that counts processes in which the scattered particle emerges with an energy E within a range dE about

$$E = \tfrac{1}{2}mv_1^2 = E_0 - \tfrac{1}{2}mv_2^2 = E_0\cos^2\vartheta. \tag{5.48}$$

The expression here follows from $\Delta E = \frac{1}{2}mv_2^2$ as given in (5.27). Now

$$\frac{d\sigma}{dE} = \frac{d\sigma}{d(\cos\vartheta)}\frac{d(\cos\vartheta)}{dE} = 2\pi a^2 \cos\vartheta \frac{d(\cos\vartheta)}{dE} = \frac{\sigma}{E_0}. \tag{5.49}$$

This signifies that the scattered energies are uniformly distributed in the range $0 \le E \le E_0$. The simplicity of this result helps make the problem of treating the energy losses of neutrons passing through hydrogen (see page 113) particularly easy.

5.4 DESCRIPTION RELATIVE TO THE MASS-CENTER

No advantage was taken in the treatment of collisions, so far, of the findings in §5.1. Considerable simplification should be expected from separating off the mass motion, $\bar{\mathbf{r}}(t)$, since this motion is uniform (5.8) and the internal motion (5.11) is like that of a single particle in a fixed-center force field. Indeed, the hard-sphere collision problem is practically the only one which is tolerably tractable without such a separation, and problems in which the target is very heavy are simple only because the mass-center (5.7) then coincides with the heavy mass. The reader will therefore be asked to imagine himself sitting at the center of mass, moving with it, and watching the positions $\mathbf{R}_1(t)$ and $\mathbf{R}_2(t)$ of the two particles from that vantage point. It takes a mere Galilean transformation to change to the uniformly moving CM reference frame, in which the mass-center has the steady position $\bar{\mathbf{R}} = 0$.

The Position-Vector Transformations

The individual particle positions relative to the mass-center are obviously defined by

$$\mathbf{R}_1 = \mathbf{r}_1 - \bar{\mathbf{r}} \quad \text{and} \quad \mathbf{R}_2 = \mathbf{r}_2 - \bar{\mathbf{r}}, \tag{5.50}$$

where $\bar{\mathbf{r}} = (m_1\mathbf{r}_1 + m_2\mathbf{r}_2)/M$ (5.7). Naturally,

$$\bar{\mathbf{R}} = (m_1\mathbf{R}_1 + m_2\mathbf{R}_2)/M = 0,$$

and hence $\mathbf{R}_1$ and $\mathbf{R}_2$ are oppositely directed, with the constant ratio

$$\mathbf{R}_1/\mathbf{R}_2 = -m_2/m_1.$$

The transformation relations for $\mathbf{r}_1, \mathbf{r}_2 \to \mathbf{r}, \bar{\mathbf{r}}$ follow from the definition of $\bar{\mathbf{r}}$ and of the relative position-vector, $\mathbf{r} \equiv \mathbf{r}_1 - \mathbf{r}_2$:

$$\begin{aligned} \mathbf{r}_1 &= \bar{\mathbf{r}} + \mathbf{R}_1 = \bar{\mathbf{r}} + (m_2/M)\mathbf{r}, \\ \mathbf{r}_2 &= \bar{\mathbf{r}} + \mathbf{R}_2 = \bar{\mathbf{r}} - (m_1/M)\mathbf{r}. \end{aligned} \tag{5.51}$$

Thus $\mathbf{R}_1$ and $\mathbf{R}_2$ are each a constant fraction of the interparticle vector $\mathbf{r}$, such that $\mathbf{R}_1 - \mathbf{R}_2 = \mathbf{r} = \mathbf{r}_1 - \mathbf{r}_2$. The equations of motion,

$$m_1\ddot{\mathbf{R}}_1 = \mu\ddot{\mathbf{r}} = \mathbf{F} \qquad \text{and} \qquad m_2\ddot{\mathbf{R}}_2 = -\mu\ddot{\mathbf{r}} = -\mathbf{F}, \tag{5.52}$$

follow directly from the equation of internal motion (5.11). These are also to be expected from the Galilean transformation, which leaves (5.4) invariant.

Velocities and Momentum

The time derivatives of the positions (5.51) yield the velocities

$$\begin{aligned} \mathbf{v}_1 &= \bar{\mathbf{v}} + \mathbf{V}_1 = \bar{\mathbf{v}} + (m_2/M)\mathbf{v}, \\ \mathbf{v}_2 &= \bar{\mathbf{v}} + \mathbf{V}_2 = \bar{\mathbf{v}} - (m_1/M)\mathbf{v}, \end{aligned} \tag{5.53}$$

where $\mathbf{V}_1 = \dot{\mathbf{R}}_1$ and $\mathbf{V}_2 = \dot{\mathbf{R}}_2$ are the individual particle velocities relative to the mass-center. They have the invariant difference $\mathbf{V}_1 - \mathbf{V}_2 = \mathbf{v} = \mathbf{v}_1 - \mathbf{v}_2$.

The individual momenta relative to the mass-center are

$$\begin{aligned} \mathbf{P}_1 &= m_1\mathbf{V}_1 = \mu\mathbf{v} = \mathbf{P}, \\ \mathbf{P}_2 &= m_2\mathbf{V}_2 = -\mu\mathbf{v} = -\mathbf{P}, \end{aligned} \tag{5.54}$$

where $\mathbf{P} = \mu\mathbf{v}$ is the "internal" momentum (5.13). This is now seen to equal a momentum of *either* particle relative to the mass-center. The individual momenta must be equal and opposite, $\mathbf{P}_2 = -\mathbf{P}_1$, since the "body" motion of the mass-center itself was defined to carry the total, net momentum of the system as a whole. The equations of motion (5.52) have already displayed the fact that $\dot{\mathbf{P}}_1 = -\dot{\mathbf{P}}_2$.

The Angular Momenta

The individual angular momenta about the mass-center, $\mathbf{L}_1 = \mathbf{R}_1 \times \mathbf{P}_1$ and $\mathbf{L}_2 = \mathbf{R}_2 \times \mathbf{P}_2$, are easily seen to add up to

$$\mathbf{L}_1 + \mathbf{L}_2 = \mathbf{r} \times \mu\mathbf{v} \equiv \mathbf{l}, \tag{5.55}$$

the total "internal angular momentum" of (5.15). Individually,

$$\mathbf{L}_1 = (m_2/M)\mathbf{l}, \qquad \mathbf{L}_2 = (m_1/M)\mathbf{l}, \tag{5.56}$$

each having the same sense, and *each* conserved as is $\mathbf{l}$, under a "mutual" force.

If the total angular momentum about a point fixed in space is considered, $\mathbf{L} = \mathbf{r}_1 \times \mathbf{p}_1 + \mathbf{r}_2 \times \mathbf{p}_2$, it will be found that

$$\mathbf{L} = \mathbf{l} + \bar{\mathbf{r}} \times \bar{\mathbf{p}}. \tag{5.57}$$

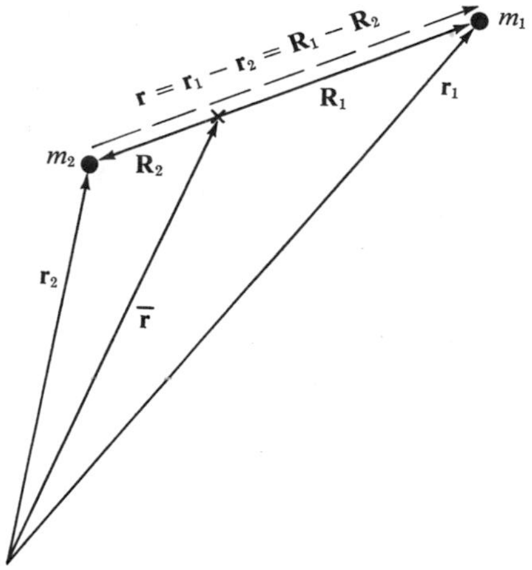

Figure 5.7

That is, it differs from **l** by the angular momentum (if any) of the centroidal motion, $\bar{\mathbf{L}} = \bar{\mathbf{r}} \times \bar{\mathbf{p}}$.

The Energy

If the body motion, $\bar{\mathbf{v}}$, is separated off by using $\mathbf{v}_{1,2} = \bar{\mathbf{v}} + \mathbf{V}_{1,2}$ (5.53) in the kinetic energy,

$$T = \tfrac{1}{2}m_1 v_1^2 + \tfrac{1}{2}m_2 v_2^2 = \tfrac{1}{2}M\bar{v}^2 + (\tfrac{1}{2}m_1 V_1^2 + \tfrac{1}{2}m_2 V_2^2), \tag{5.58}$$

it separates into a kinetic energy $\bar{T} = \frac{1}{2}M\bar{\mathbf{v}}^2$ of the body motion, plus the kinetic energy of the particles relative to the center of mass. There is no coupling of the two motions because cross-terms, $m_1 \mathbf{V}_1 \cdot \bar{\mathbf{v}} + m_2 \mathbf{V}_2 \cdot \bar{\mathbf{v}} = (\mathbf{P}_1 + \mathbf{P}_2) \cdot \bar{\mathbf{v}}$, vanish on account of the zero total momentum relative to a mass-center. It is easy to show that the kinetic energy relative to the mass-center is exactly the same as

$$\tau = \tfrac{1}{2}\mu v^2 = \tfrac{1}{2}m_1 V_1^2 + \tfrac{1}{2}m_2 V_2^2, \tag{5.59}$$

the "internal kinetic energy" introduced in (5.17). The individual shares are

$$\tau_1 = \tfrac{1}{2}m_1 V_1^2 = (m_2/M)\tau, \qquad \tau_2 = \tfrac{1}{2}m_2 V_2^2 = (m_1/M)\tau, \tag{5.60}$$

in inverse ratio to the masses as to be expected for the equal, though oppositely directed, momenta of the particles.

For the uniformly moving isolated systems, the centroidal kinetic energy $\bar{T} = \frac{1}{2}M\bar{v}^2$ is conserved by itself. Only the internal kinetic energy τ

is available for exchange with the mutual potential energy, V, as in the conserved $\varepsilon = \tau + V$ of (5.18).

More specific descriptions of motions relative to the mass-center will be taken up in the next section, for cases of unbound motion. There is less profit in considering such descriptions of bound motions, beyond the points already made about "internal motions" in §5.1.

5.5 CLASSICAL COLLISION THEORY IN THE CM FRAME

Through reference to a frame in which the mass-center of the collision partners is at rest, motions independent of (and hence effects extraneous to) the interaction are eliminated. There remains just the relative motion which reflects directly the characteristics of the mutual force of interaction, through $\mu\ddot{\mathbf{r}} = \mathbf{F}$.

Transformations between Laboratory and CM Frames

In the laboratory frame of §5.2, a collision is initiated by sending a particle of momentum $m_1\dot{\mathbf{r}}_1 = m_1\mathbf{v}_0$ toward a target particle at rest ($m_2\dot{\mathbf{r}}_2 = 0$), with some collision parameter b. The centroid is thereby given the initial velocity

$$\bar{\mathbf{v}} = (m_1\dot{\mathbf{r}}_1 + m_2\dot{\mathbf{r}}_2)/M = (m_1/M)\mathbf{v}_0, \tag{5.61}$$

and, with only a mutual force coming into play, this velocity persists throughout the process. The centroidal kinetic energy, $\bar{T} \equiv \frac{1}{2}M\bar{\mathbf{v}}^2 = (m_1/M)E_0$, is separately conserved, so that only the internal kinetic energy part,

$$\tau = \tfrac{1}{2}\mu v_0^2 = (m_2/M)E_0, \tag{5.62}$$

of the laboratory energy $E_0 = \frac{1}{2}m_1 v_0^2$ is available for work against the internal force, F. Since the internal energy is also conserved, the magnitude of the relative velocity of the two particles is the same before and after sojourn within the force range: $v_0 = |\dot{\mathbf{r}}_1 - \dot{\mathbf{r}}_2| = |\mathbf{v}_1 - \mathbf{v}_2| \equiv v$.

Reference to the CM frame is introduced when all the laboratory velocities have $-\bar{\mathbf{v}}$ added to them. Then, as viewed in the CM frame, both the incident and the target particles are initially in motion, approaching each other as indicated in the velocity diagrams of Figure 5.8. The incident particle has $\mathbf{V}_1^0 = \mathbf{v}_0 - \bar{\mathbf{v}} = (m_2/M)\mathbf{v}_0$ and the target moves with $\mathbf{V}_2^0 = -\bar{\mathbf{v}} = -(m_1/M)\mathbf{v}_0$. After collision, both particles fly apart in a deflected direction,

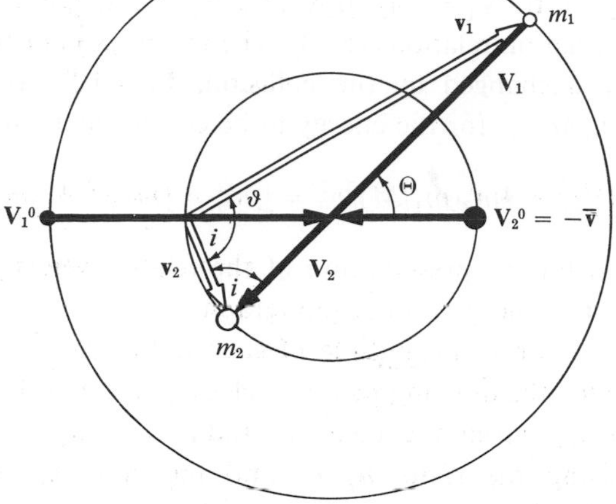

Figure 5.8(a). Case $m_1 < m_2$, with m_2 initially at rest in the laboratory frame.

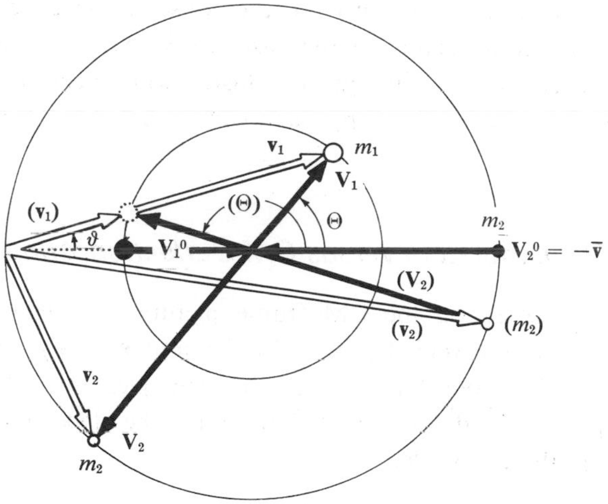

Figure 5.8(b). Cases $m_1 > m_2$, with equal deflections, ϑ. The relative velocity v_0 is the same as in (a), but this time it is the lighter mass that is initially at rest in the laboratory frame.

with velocities $\mathbf{V}_1$ and $\mathbf{V}_2$. These are again in opposite directions, conserving the zero initial total momentum:

$$m_1(\mathbf{v}_0 - \bar{\mathbf{v}}) + m_2(-\bar{\mathbf{v}}) = 0 = m_1\mathbf{V}_1 + m_2\mathbf{V}_2, \tag{5.63}$$

so that $\mathbf{V}_2 = -(m_1/m_2)\mathbf{V}_1$. Each is a fraction of the relative velocity $\mathbf{v} = \mathbf{v}_1 - \mathbf{v}_2 = \mathbf{V}_1 - \mathbf{V}_2$, with $\mathbf{V}_1 = (m_2/M)\mathbf{v}$ and $\mathbf{V}_2 = -(m_1/M)\mathbf{v}$, as already seen in the transformation relations (5.53). The magnitudes of the individual velocities must be unchanged by the collision, $V_1 = V_1^0 = (m_2/M)v_0$ and $V_2 = V_2^0 = \bar{v} = (m_1/M)v_0$, for the energy to be conserved as well:

$$\tfrac{1}{2}m_1(m_2/M)^2v_0^2 + \tfrac{1}{2}m_2(m_1/M)^2v_0^2 = \tfrac{1}{2}\mu v_0^2 = \tfrac{1}{2}m_1V_1^2 + \tfrac{1}{2}m_2V_2^2. \quad (5.64)$$

This is also needed for the preservation of the relative velocity magnitude, $v = v_0$, pointed out in the preceding paragraph.

The preservation in magnitude of each individual particle velocity, V_1 and V_2, prompted the drawing of the circles (spheres) in the diagrams of Figure 5.8. These are given the constant radii $V_1 = (m_2/M)v_0$ and $V_2 = (m_1/M)v_0 = \bar{v}$, having the ratio m_2/m_1 and the sum v_0. Each of the oppositely directed velocity vectors $\mathbf{V}_1$ and $\mathbf{V}_2$, resulting from a given collision, must extend to some point on its respective sphere. They will have been deflected by some angle Θ, with the initial direction, as seen in the CM frame. The CM deflection Θ may fall anywhere in the range $0 < \Theta < \pi$; no restriction on it has arisen from the conservation conditions, as on the laboratory frame deflection in the cases $m_1 \geq m_2$. The particular CM deflection Θ that will occur in a given collision will reflect directly the effect of the mutual interaction, in an approach with a particular impact parameter b: $\Theta(b)$.

The Laboratory versus CM Deflection Angles

Superposed on the CM frame pictures of Figure 5.8 are vector diagrams giving the velocities $\mathbf{v}_1 = \mathbf{V}_1 + \bar{\mathbf{v}}$ and $\mathbf{v}_2 = \mathbf{V}_2 + \bar{\mathbf{v}}$, which will be observed in the laboratory frame. The triangles formed by $\mathbf{v}_1 = \mathbf{V}_1 + \bar{\mathbf{v}}$ show how the CM deflection Θ determines the deflection ϑ that will be seen in the laboratory. Plainly

$$\tan\vartheta = \frac{V_1 \sin\Theta}{\bar{v} + V_1\cos\Theta} = \frac{m_2\sin\Theta}{m_1 + m_2\cos\Theta} \quad (5.65)$$

in each case. The triangles for $\mathbf{v}_2 = \mathbf{V}_2 + \bar{\mathbf{v}}$ have isosceles forms because $V_2 = \bar{v}$. They show that the recoil angle that will be seen in the laboratory is

$$i = \tfrac{1}{2}(\pi - \Theta). \quad (5.66)$$

When this is used to eliminate Θ from the laboratory deflection expression (5.65), the relation (5.30) already found in the laboratory frame emerges.

The last result shows that Figure 5.4 can be adapted to exhibiting the relation between the laboratory and CM deflections, ϑ and Θ, simply by substituting a linear scale $\pi > \Theta > 0$ for $0 < i < \frac{1}{2}\pi$, as indicated on the top of the diagram in Figure 5.4. The diagram then clearly indicates how $\vartheta \to 0$ independently of Θ for a very heavy incident mass ($m_1/m_2 \to \infty$). It also shows how the restrictions $\vartheta < \vartheta_{max} = \sin^{-1}(m_2/m_1)$ on the laboratory deflections arise for $m_1 > m_2$, despite the lack of restriction on the CM deflection: $0 < \Theta < \pi$. The restrictions are results of the choice merely of frame, and emphasize again that it is within the CM frame that more significant effects arise. Reference to Figure 5.8(*b*), which applies to $m_1 > m_2$, now makes it transparent how the double-valuedness of the recoil angles for a given laboratory deflection (pointed out in connection with Figure 5.4) develops.

Velocity and Energy Transformations

The CM deflection $\Theta(b)$ that occurs not only determines the laboratory deflection as in (5.65), but also the deflected velocity

$$\begin{aligned} v_1 &= (\bar{v}^2 + V_1^2 + 2\bar{\mathbf{v}} \cdot \mathbf{V}_1)^{1/2}, \\ &= \frac{v_0(m_1^2 + m_2^2 + 2m_1 m_2 \cos\Theta)^{1/2}}{M}, \end{aligned} \tag{5.67}$$

derivable from the triangle for $\mathbf{v}_1 = \bar{\mathbf{v}} + \mathbf{V}_1$. The triangle for $\mathbf{v}_2 = \bar{\mathbf{v}} + \mathbf{V}_2$ yields the recoil velocity as

$$v_2 = 2\bar{v}\cos i = 2(m_1/M)v_0 \sin\tfrac{1}{2}\Theta. \tag{5.68}$$

These expressions coincide with (5.32*a*) and (5.28), respectively, when put in terms of the recoil angle $i = \frac{1}{2}(\pi - \Theta)$ as given by (5.66). Notice that the deflected velocity v_1 is a single-valued function of the CM deflection Θ even when $m_1 > m_2$, unlike its connection (5.32*b*) to the laboratory deflection ϑ for such cases. Figure 5.8(*b*) shows the two different CM deflections and the two corresponding deflection velocities that can occur with the same laboratory deflection angle ϑ. The two corresponding recoil velocities and directions are also indicated.

The energy transfer that follows from (5.68) is

$$\Delta E = \tfrac{1}{2}m_2 v_2^2 = 4(\mu/M)E_0 \sin^2\tfrac{1}{2}\Theta, \tag{5.69}$$

checking with (5.29) since $\Theta = \pi - 2i$. A perspicuous way of expressing this energy loss by m_1, despite the constancy of its velocity magnitude $V_1^0 = V_1 = (m_2/M)v_0$ in the CM frame, is provided by

$$\Delta E = \tfrac{1}{2}m_1(v_0^2 - v_1^2) = \tfrac{1}{2}m_1(|\bar{\mathbf{v}} + \mathbf{V}_1^0|^2 - |\bar{\mathbf{v}} + \mathbf{V}_1|^2), \tag{5.70}$$

where $\mathbf{V}_1 = (m_2/M)\mathbf{v}$ is different only in direction from $\mathbf{V}_1^0 = (m_2/M)\mathbf{v}_0$. The expression readily reduces to

$$\Delta E = \bar{\mathbf{v}} \cdot \mu(\mathbf{v}_0 - \mathbf{v}) = \bar{\mathbf{v}} \cdot \Delta\mathbf{P}, \tag{5.71}$$

where $\Delta\mathbf{P}$ is the merely deflectional change, from $\mu\mathbf{v}_0$ to $\mu\mathbf{v}$, in the internal momentum $\mathbf{P}$ of (5.54). The energy transfer is proportional to just the projection $\mu v_0(1 - \cos\Theta) = 2\mu v_0 \sin^2 \frac{1}{2}\Theta$ of $\Delta\mathbf{P}$ onto the mass velocity $\bar{\mathbf{v}}$. The result reduces to (5.69) when the target is initially at rest in the laboratory, so that $\bar{\mathbf{v}} = (m_1/M)\mathbf{v}_0$, but is valid also for a moving target, as the form of (5.70) should make clear.

Laboratory versus CM Cross Sections

A final transformation between the laboratory and CM frames should be explored—that connecting the differential cross sections as defined in the two frames. The CM cross section defined by

$$d\sigma/d\Omega = db^2/2d(\cos\Theta), \tag{5.72}$$

in accordance with (5.34) but in terms of the CM deflection Θ, is the significant one for theoretical consideration. The relative motion of the collision partners reflects most directly the character of their interaction, and the CM deflection function $\Theta(b)$ is a direct result of solving the theoretical problem ($\mu\ddot{\mathbf{r}} = \mathbf{F}$). For example, the CM angular distribution may be merely isotropic —that is, $d\sigma/d\Omega = \sigma/4\pi$ (5.36)—yet quite a complicated angular distribution may be observed under the laboratory conditions (the target initially at rest).

The differential cross section observed in the laboratory will now be denoted $d\sigma/d\omega$, where $d\omega = d(\cos\vartheta)\,d\varphi$ versus $d\Omega = d(\cos\Theta)\,d\varphi$, above. Experimenters customarily transform the $d\sigma/d\omega$ they observe to the CM frame, before submitting their results to theoretical consideration. How they do this follows from the relation (5.65) between the CM and laboratory deflections. First, the counts found at the angle ϑ in the laboratory must be ascribed to an angle Θ in the CM frame that follows from (5.65). Second, it must be recognized that the function $d\sigma(\Theta)/d\omega$ which results in that way is not yet equal to $d\sigma(\Theta)/d\Omega$ because the solid angle elements in which the counts are to be made are not equal in the two frames. Thus

$$\frac{d\sigma/d\Omega}{d\sigma/d\omega} = \frac{d\omega}{d\Omega} = \frac{d(\cos\vartheta)}{d(\cos\Theta)} = \frac{m_2^2(m_2 + m_1\cos\Theta)}{(m_1^2 + m_2^2 + 2m_1 m_2 \cos\Theta)^{3/2}} \tag{5.73}$$

follows from (5.65).

When the target is much heavier than the incident mass ($m_1 \ll m_2$), so that there is no discernible difference between laboratory and CM frames, then the transformation factor (5.73) properly reduces to unity. When the masses are equal, the transformation is particularly simple, for then the laboratory deflection is just half the CM deflection, $\vartheta = \frac{1}{2}\Theta$, as follows from (5.65) or Figure 5.4. Thus, for $m_1 = m_2$,

$$\frac{d\sigma}{d\Omega} = \frac{d\sigma}{d\omega}\frac{d[\cos\frac{1}{2}\Theta]}{d\cos\Theta} = \frac{d\sigma/d\omega}{4\cos\frac{1}{2}\Theta}. \tag{5.74}$$

This means that an isotropic cross section in the CM frame will be observed as one proportional to $4\cos\frac{1}{2}\Theta = 4\cos\vartheta$ in the laboratory (compare (5.44) and (5.46)!). In cases $m_1 > m_2$, the two possible CM deflections corresponding to each observed laboratory deflection ϑ must be distinguished; that is relatively easy for the experimenter, since the observed velocity magnitudes are different in the two cases.

Hard-Sphere Scattering

In the CM frame, the collision diagram of Figure 5.6 is replaced by the more symmetrical one of Figure 5.9. The reflection angle (of $\mathbf{V}_1$), as seen in the CM frame, must equal the angle of incidence i fundamentally because the equation for the relative motion, $\mu\ddot{\mathbf{r}} = \mathbf{F}$, is like that for a particle in a fixed force field. A consequence is that the CM deflection is given by

$$\Theta = \pi - 2i, \tag{5.75}$$

just like the laboratory deflection $\vartheta = \pi - 2i$ (5.43) against an infinite target mass. Quite consistently, the laboratory recoil angle (of $\mathbf{v}_2$) equals the incidence angle, as in the lower part of Figure 5.9.

Unlike the laboratory deflection $\vartheta(b)$ in §5.2, the CM deflection function $\Theta(b)$ is independent of the mass ratio, being given by

$$b = a\sin i = a\cos\tfrac{1}{2}\Theta. \tag{5.76}$$

The consequent CM cross section is an isotropic one,

$$d\sigma/d\Omega = \tfrac{1}{4}a^2 = \sigma/4\pi, \tag{5.77}$$

just like the laboratory frame cross section (5.44) in the case of a fixed target.

The energy distribution found in (5.49) for equally massive collision partners is now easy to generalize to an arbitrary mass-ratio. From the deflected velocity (5.67) it follows that

$$E = \tfrac{1}{2}m_1 v_1^2 = E_0(m_1^2 + m_2^2 + 2m_1 m_2\cos\Theta)/M^2, \tag{5.78a}$$

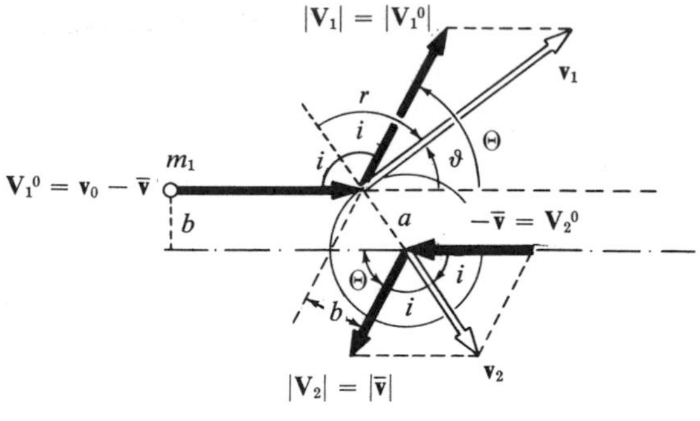

Figure 5.9

having the limits

$$[(m_1 - m_2)/(m_1 + m_2)]^2 E_0 \leqslant E \leqslant E_0 \equiv \tfrac{1}{2} m_1 v_0^2, \tag{5.78b}$$

attained at $\Theta = \pi$ and $\Theta = 0$, respectively. Then

$$\frac{d\sigma}{dE} = \frac{d\sigma}{d(\cos\Theta)} \frac{d(\cos\Theta)}{dE} = \frac{M}{4\mu} \frac{\sigma}{E_0}. \tag{5.79}$$

Thus the distribution is also uniform in the general case, although the energies in it have a finite lower limit when $m_1 \neq m_2$. It is easy to check that $\int dE(d\sigma/dE) = \sigma$, properly.

Coulomb (Rutherford) Scattering

The coulombic scattering of charges by charges is an important process because of the prevalence of charged particles in matter. The equivalence of the problem $\mu\ddot{\mathbf{r}} = -\hat{\mathbf{r}}K/r^2$ to the fixed-center problem of §4.4 now provides indispensable help. Only some reinterpretations are needed to convert the orbit diagram of Figure 4.5 into a collision diagram in the CM frame. It is only necessary to remember that $\mathbf{R}_1 = (m_2/M)\mathbf{r} = -(m_2/m_1)\mathbf{R}_2$. Figure 5.10 presents the case of repulsion; in the corresponding case of attraction, the hyperbolas lead to opposite deflections and cross each other. Adding the CM velocity $\bar{\mathbf{v}}$ to all the velocities in the CM diagram would convert it into a laboratory frame picture in which the velocities of m_2 have only positive components along the incidence direction (of $\mathbf{v}_0$ and $\mathbf{V}_1^0$), with end results corresponding to Figure 5.1.

The deflection function (4.53) of the $m_2 = \infty$ case is converted into an expression for the CM deflection through the replacements $\vartheta \to \Theta$, $m \to \mu$, and $\frac{1}{2}mv_0^2 \to \frac{1}{2}\mu v_0^2$:

$$\tan \tfrac{1}{2}\Theta = \frac{|K|}{\mu v_0^2 b}. \tag{5.80}$$

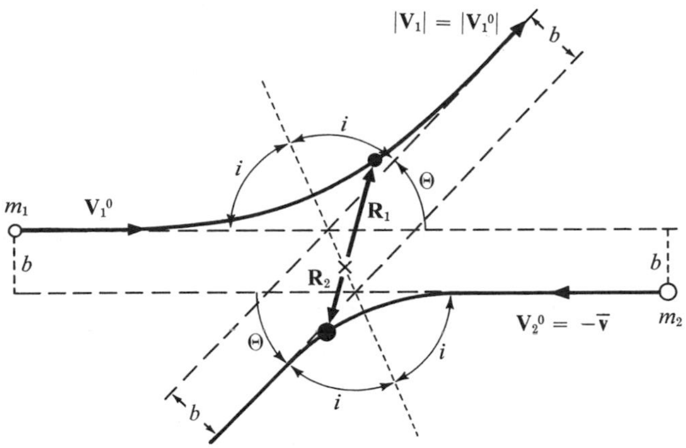

Figure 5.10

This gives a $b^2(\Theta)$ exactly like (5.37) of the fixed-center case except that $m \to \mu$ and $\vartheta \to \Theta$; hence the CM cross section is like (5.38):

$$d\sigma/d\Omega = (K/2\mu v_0^2)^2/\sin^4 \tfrac{1}{2}\Theta. \tag{5.81}$$

If this were converted to a laboratory cross section, $d\sigma(\vartheta)/d\omega$, through the use of the $\vartheta(\Theta)$ relation (5.65) and the solid-angle ratio (5.73), the result would contain the recoil effects that were neglected in (5.38) and mentioned in connection with it. The predominance of small deflections shown by (5.81) is intensified when the target particles are the light electrons, for then the laboratory deflection is restricted to $\vartheta < \sin^{-1}(m_2/m_1)$ even for $\Theta \to \pi$. This accounts for the straightness of tracks made by heavier charged particles passing through cloud chambers.

The CM cross section is as easy to convert into an "energy-loss cross section" as it was in the hard-sphere case (5.79). Since the scattered energy (5.78*a*) may also be written (compare (5.32a) and (5.67))

$$E = E_0\left[1 - 4\left(\frac{\mu}{M}\right)\sin^2 \tfrac{1}{2}\Theta\right], \tag{5.82}$$

it follows that

$$\frac{d\sigma}{dE} = \frac{m_1}{m_2}\frac{\pi}{E_0}\left(\frac{K}{E_0 - E}\right)^2, \tag{5.83}$$

with the limits on the energy E again given by (5.78*b*). This result indicates the prevalence of small energy losses, $\Delta E = E_0 - E$, which is to be expected from the predominance of small angle collisions.

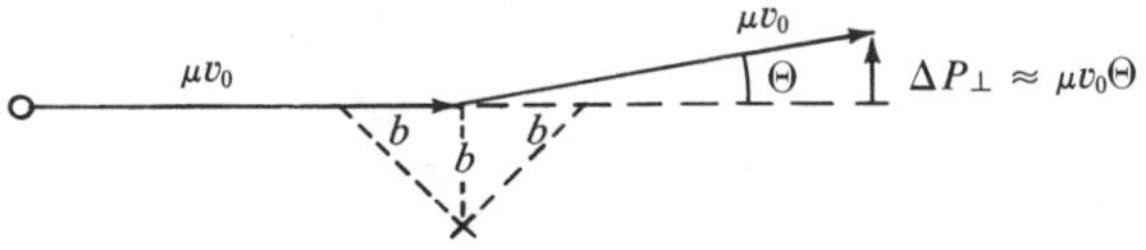

Figure 5.11

The dominance of small deflections and energy losses makes useful a perspicuous approximate treatment of the scattering, not requiring the derivation of the hyperbolic orbit. The situation is pictured in Figure 5.11. It is supposed that the transverse momentum change $\Delta P_\perp \approx \mu v_0 \Theta$ is generated by the force $|K|/b^2$ acting over some "collision time" interval, Δt, so that

$$\mu v_0 \Theta \approx |K| \Delta t / b^2. \tag{5.84}$$

A choice $\Delta t \approx 2b/v_0$ seems reasonable (see Exercise 5.9), and the result then agrees with the exact one (5.81) for $\Theta \to 0$:

$$d\sigma/d\Omega \approx |d(b^2)/d(\Theta^2)| = (2K/\mu v_0^2)^2/\Theta^4. \tag{5.85}$$

With the energy loss $\Delta E = E_0 - E \approx \bar{v}\, \Delta P_{\|}$ (5.71) and $\Delta P_{\|} \approx \mu v_0(1 - \cos\Theta)$ $\approx \frac{1}{2}\mu v_0 \Theta^2$, the energy-loss cross section (5.83) is reproduced exactly.

EXERCISES

5.1. A mass-particle m_1 moving along a line with speed $u_1(>0)$, collides head-on with a particle m_2 which is not at rest in the observer's frame, but moving along the same line with a velocity $u_2 < u_1$ (and $u_2 \gtrless 0$ possible).

(a) Directly from momentum and kinetic energy conservation, as applied in the observer's frame, find an expression $v_1(u_1, u_2)$ for the resultant velocity of m_1, and then show that the expression for $v_2(u_1, u_2)$ is the same except for an interchange of all the numerical subscripts 1, 2.

(b) Show that the same expressions could have been obtained from (5.25) merely by "boosting" all the velocities there by u_2 (that is, by performing an appropriate Galilean transformation).

(c) Show that both the expressions in (a), and (5.25), can be obtained from transformations like (5.53), with appropriate choices of centroidal and relative velocities.

(d) Show that the expressions found in (a) yield an energy transfer $\Delta E = \frac{1}{2}m_2(v_2^2 - u_2^2)$ in agreement with a direct calculation from (5.71).

5.2. Investigate the momentum transfer from one particle to another, of the same mass, as mediated by elastic collisions with a third, *lighter*, particle in the following circumstances. The three particles are ranged in a straight line at some distances apart, with the light one in the middle, so that, when the first is given an initial impulse along the line, only head-on collisions ensue. Find the fraction of the initial momentum that is finally transferred to the last particle if the mediating particle is just one-third as heavy as either of the other two [15/16]. (It is most profitable to develop formulas for an arbitrary mass-ratio, first. The reader may want to construct other problems suggested by this one. For example, the third particle may be replaced by an elastic wall, and the final momentum, with which the first particle is "reflected," found.)

5.3. Suppose that a particle of momentum $m_0 v_0$ has a "head-on" collision with a linear oscillator having its mass-center at rest, and consisting of two equal masses, m each, coupled only to each other and undergoing the motions

$$x_{1,2}(t) = \mp [\tfrac{1}{2}l + a \sin(\omega t + \alpha)]$$

before the collision at $t = 0$. "Head-on" here means that $m_0 v_0$ is incident on the first oscillator mass (at x_1) in line with the vibrations. The length l is obviously the equilibrium separation of the oscillating masses, and the initial oscillation energy is $\varepsilon_0 = m\omega^2 a^2$. The results may be studied as a function of the phase in which the oscillator is "caught" by the collision, as measured by $\varphi \equiv \dot{x}_1(0)/v_0$, and as a function of the mass-ratio $\rho = m_0/m$.

(a) Find the kinetic energy imparted to the oscillator's mass center: $\bar{E}(\rho, \varphi, E_0)$ where $E_0 = \frac{1}{2} m_0 v_0^2$. Show that this is the same as the energy imparted by $m_0 v_0$ to a *free* particle of mass m, if the oscillator is caught in the phase $\varphi = -0.414$.

(b) Show that the oscillatory motion has the excitation energy

$$\varepsilon - \varepsilon_0 = \frac{2\rho}{(\rho + 1)^2} (1 - \varphi) \left[1 + \left(1 + \frac{2}{\rho} \right) \varphi \right] E_0$$

imparted to it. From this, find the phase $\varphi(\rho)$ in which the oscillator must be "caught" to receive maximum excitation, and those in which it will suffer de-excitation. Why is $\varphi > 1$ meaningless for this problem?

5.4. According to "Newton's Rule" for the *in*elastic impact of solid spheres, the component of their relative velocity along their line of centers is reduced by the collision, to some fraction e of its initial value. The constant e is to be characteristic of the materials involved, having the value one for perfectly elastic solids, and approaching zero for puttylike substances. It is called the "coefficient of restitution" because at least the *head-on* collisions are pictured as taking place in two distinct stages: in the first, the spheres compress each other until they travel together with one velocity, all their relative momentum μv_0 (5.13) thus being absorbed; in the second stage, there is re-expansion, with a "restitution" of relative momentum $e\mu v_0$. Investigate the following points, specifically for collisions in which a sphere of mass m_1 is incident, with velocity v_0, on a sphere m_2 at rest.

(a) The proper expression for the energy loss due to the inelasticity has $\frac{1}{2}\mu v_0^2$ as a maximum.

(b) In the relation (5.41), between the angles of incidence and reflection, the denominator $m_2 - m_1$ is to be replaced by $em_2 - m_1$ when $e \neq 1$.

(c) The cross section for collisions in which the total kinetic energy is reduced to a value E in the range dE represents a uniform energy distribution between the limits

$$E_{min} = \frac{m_1 + e^2 m_2}{m_1 + m_2} E_0$$

and $E_0 = \frac{1}{2} m_1 v_0^2$.

5.5. A particle of momentum $m_0 \mathbf{v}_0$ collides head-on with one of the atoms of a homopolar diatomic molecule, consisting of two equal masses, m each, connected by a rigid linear bond. Suppose that the collision is elastic, that the initial motion of the molecule is negligible (as compared to v_0) and that the bond line initially makes some angle i with the incidence direction (of $m\mathbf{v}_0$).

(a) Give arguments for assuming that the second atom of the molecule can get an impulse only along the direction of the bond line.

(b) Find an expression for the final velocity of the incident particle that reduces to plausible values for $i = 0$ and for $i = \frac{1}{2}\pi$.

(c) Find the direction and magnitude of the velocity imparted to the molecule's center of mass.

(d) Show that the ratio of the rotational to the translational energy imparted to the molecule is independent of the mass, and cannot exceed unity.

5.6. The scattering of a neutron by an atomic nucleus is sometimes treated as a passage through a fixed spherical volume, having some "nuclear radius" R, in which the particle acquires a mean, negative potential energy $V = -D$ (and then D is called the "depth" of a "spherical potential well"). Taken literally, this corresponds to the neutron's receiving an inward radial impulse whenever it crosses the nuclear "surface" (and any "nucleons" with a total energy less than zero are "bound" within the nuclear radius!).

(a) Deduce and sketch the "orbits" that will be followed by neutrons which are incident with exactly the kinetic energy $\frac{1}{2}mv_0^2 = +D$ and with impact parameters $b = 0$, $R/\sqrt{2}$, and R.

(b) Show that the maximum deflection of a neutron incident with the kinetic energy $\frac{1}{2}mv_0^2 = fD$ is given by

$$\vartheta_{max} = \cos^{-1}[(f-1)/(f+1)].$$

(c) Suppose now that the incident neutrons are replaced by protons of charge $+e$, and that the nucleus has the charge $+Ze$. Find the total cross section for protons of a given energy $\frac{1}{2}mv_0^2$, to be deflected by the forces inside R (the "specifically nuclear forces"). What is the least energy (the "barrier height") the protons must have, to come within the reach of those forces at all?

5.7. A neutron of kinetic energy E_0 is incident on a nucleus just seven times as massive, and is absorbed by it with a total "capture" cross section, σ_0. After a time in which the motion is undisturbed, the "compound nucleus" that has been formed undergoes a "fission process" in which it separates into two equal fragments, "α-particles," each just four times as massive as a neutron. Suppose that the relative kinetic energy with which the α-particles spring apart just equals the energy that had been absorbed (not actually quite correct when so-called "mass-defects" are taken into account). On the basis that the directions of the springing apart in the fission process have relative probabilities that are isotropic in the CM frame, what is the differential cross section for observing an α-particle emerge at a laboratory angle ϑ to the incident neutron beam?

$$\left(\frac{d\sigma}{d\omega}=\frac{\sigma_0}{2\pi}\frac{[(6+\cos^2\vartheta)^{1/2}+\cos\vartheta]^2}{[7(6+\cos^2\vartheta)]^{1/2}}\right).$$

Also show that the α-particle energies are distributed uniformly within a certain range.

5.8. Investigate the classical Coulombic scattering of protons by protons ($z=Z=1$). Show that the angular distribution in which protons will be observed to emerge may be represented by

$$d\sigma/d\omega=(e^2/E_0)^2[\sin^{-4}\vartheta+\cos^{-4}\vartheta]\cos\vartheta,$$

where ϑ is the angle with the incident proton direction as observed in the laboratory (target protons at rest). Check this conclusion by comparing it with the scattering of indistinguishable hard spheres (for which an easily interpretable, finite total cross section is obtainable).

5.9. Show that the "reasonable assumption," $\Delta t \approx 2b/v_0$, used for (5.84), can be avoided by integrating the impulses imparted to a stationary center during the entire passage of the charge along the straight line path; that is,

$$\Delta P_\perp \approx \int_{-\infty}^{\infty}\frac{|K|}{r^2}\frac{b}{r}\,dt,$$

in which $dt = dx/v_0$.

CHAPTER 6

MANY-PARTICLE SYSTEMS

Consider what can now be said about a system composed of more than just two particles. There are as many equations of motion as there are particles, having masses $m_1, m_2, \ldots, m_i, \ldots$ and subject to forces $\mathbf{F}_1, \mathbf{F}_2, \ldots, \mathbf{F}_i, \ldots$. More conclusions can be made explicit if the possibility is considered that, among the forces making up the resultant $\mathbf{F}_i$ on the ith particle, there are ones of the type (5.3) found for isolated pairs of particles: mutual ones, $\mathbf{F}_{ij} = \hat{\mathbf{r}}_{ij} F(r_{ij}) = -\mathbf{F}_{ji}$. This is done by introducing a resultant $\mathbf{f}_i$ to stand for the difference between $\mathbf{F}_i$ and the resultant of all the mutual forces on the ith particle. Accordingly, the forms

$$m_i \ddot{\mathbf{r}}_i = \sum_j \mathbf{F}_{ij} + \mathbf{f}_i \tag{6.1}$$

will be assumed by the equations of motion. Notice that the reaction relationship $\mathbf{F}_{ij} = -\mathbf{F}_{ji}$ implies $\mathbf{F}_{ii} = 0$, and so the summation

$$\sum_j \mathbf{F}_{ij} \equiv \mathbf{F}_{i1} + \mathbf{F}_{i2} + \cdots + \mathbf{F}_{ij} + \cdots \tag{6.2}$$

does not include a force of the particle on itself.

6.1 TRANSLATION AS A BODY

As for the particle-pairs in §5.1, a center-of-mass, $\bar{\mathbf{r}}(t)$, may be defined for any several-particle system in such a way that its motion with velocity $\dot{\bar{\mathbf{r}}} \equiv \bar{\mathbf{v}}$, when attributed to the entire mass $\sum_i m_i \equiv M$, forms the total momentum:

$$\bar{\mathbf{p}} = \sum_i m_i \dot{\mathbf{r}}_i \equiv M\dot{\bar{\mathbf{r}}}. \tag{6.3}$$

Accordingly, the definition

$$\bar{\mathbf{r}} = \sum_i m_i \mathbf{r}_i / M \tag{6.4}$$

is adopted. Any variation of this "average" position with time, $\bar{\mathbf{r}}(t)$, may be considered a translation of the swarm of particles as a whole, a motion "as a body." Motions of the individual particles with respect to it, as given by $\mathbf{R}_i(t) = \mathbf{r}_i(t) - \bar{\mathbf{r}}(t)$, may be thought of as "internal motions" within the body of particles. Formally, this could even be done for particles that pay no attention to each other but, naturally, it is most advantageous to isolate for such treatment a set with mutual interactions of some kind.

An equation of motion for the mass-center is obtained by adding together all the individual equations of motion (6.1), to obtain

$$\sum_i m_i \ddot{\mathbf{r}}_i = M\ddot{\bar{\mathbf{r}}} = \sum_i \left(\sum_j \mathbf{F}_{ij}\right) + \sum_i \mathbf{f}_i .$$

Here the "double summation," which is also frequently written $\sum_{i,j} \mathbf{F}_{ij}$, stands for

$$\begin{aligned}\sum_{i,j} \mathbf{F}_{ij} &\equiv \sum_i (\mathbf{F}_{i1} + \mathbf{F}_{i2} + \mathbf{F}_{i3} + \cdots) \\ &\equiv 0 + \mathbf{F}_{12} + \mathbf{F}_{13} + \cdots \\ &\quad + \mathbf{F}_{21} + 0 + \mathbf{F}_{23} + \cdots \\ &\quad + \mathbf{F}_{31} + \mathbf{F}_{32} + 0 + \cdots \\ &\quad + \cdots .\end{aligned}$$

For every $\mathbf{F}_{ij}$ in this vector sum there occurs an $\mathbf{F}_{ji} = -\mathbf{F}_{ij}$, and hence the resultant of all the mutual forces of particle-pairs vanishes. It follows that

$$\boxed{M\ddot{\bar{\mathbf{r}}} = \mathbf{f},} \tag{6.5}$$

where $\mathbf{f} = \sum_i \mathbf{f}_i$ is the resultant of all the forces excluding the mutual ones of pairs.

It can be taken as a matter of observation that no body, however constructed, can ever start motion-as-a-body of itself, "lift itself by its own bootstraps." Any acceleration $\ddot{\bar{\mathbf{r}}}$ of the body as a whole is attributed to external forces, disturbing its isolation; hence the resultant $\sum_i \mathbf{F}_i = \mathbf{f}$ can contain uncanceled only *external* forces. This would follow naturally from $\mathbf{f} = \sum_i \mathbf{f}_i$ if each $\mathbf{f}_i$ were only an external force and the pair-forces, $\mathbf{F}_{ij} = -\mathbf{F}_{ji}$, were the only internal ones. It has never been found really essential, in any analyses into point-particles, to consider the existence of any internal forces among them except the mutual, two-body ones; attempts at using formulations with internal forces requiring the presence of three or more particles have rarely proved helpful. Accordingly, it will hereafter be assumed that each of the "residue" forces $\mathbf{f}_i$ is solely an external one, absent whenever the system is isolated.

The reference here to the observational experience that no body starts motion by itself is of course no more than a restatement of Newton's First Law, except that now the notion of "body" is extended to any system of particles. If that Law is understood to apply only to the individual particles, then Newton's Third Law (of Reactions), also plays a role in embodying the experience. Finally, equation (6.5) for the center-of-mass motion may be considered a restatement of Newton's Second Law, inclusive of bodies analyzable into several particles.

The resemblance of the body equation of motion, $M\ddot{\bar{\mathbf{r}}} = \mathbf{f}$, to a single-particle equation provides partial justification for restricting attention to a discrete number of points when treating the motion of a body having extension. The points treated may each be considered a mass-center of some neighborhood of points ignored, and the mass associated with a point treated to be the total mass of its whole neighborhood. There are restrictions on the validity of such treatments. For the equation of motion (6.5) to be useful, the net force $\mathbf{f}$ must be insensitive to any internal motions within the neighborhood.* A force function of individual neighborhood variables, such as $\mathbf{f}(\mathbf{r}_1, \mathbf{r}_2, \mathbf{r}_3, \ldots)$, must be either adequately approximated by $\mathbf{f}(\bar{\mathbf{r}}, \bar{\mathbf{r}}, \bar{\mathbf{r}}, \ldots)$, or a replacement of it by some average force must be sufficient for the kind of predictions to be made. Otherwise, more individual points must be discriminated and treated.

6.2 ROTATION AS A BODY

The motions may also be discussed as rotations about some chosen center. The choice of center may be made independently of the motions to be analyzed, and treated as a momental center "fixed in space." A choice defined

* This is so in ideally rigid bodies (Chapter 10), which have no internal motions.

by the motions themselves, the possibly moving mass-center, will be considered immediately afterward.

Rotation about a Point Fixed in Space

It is convenient to make whatever momental center is chosen the origin for the position-vectors. Then $\mathbf{L}_i \equiv \mathbf{r}_i \times \mathbf{p}_i = \mathbf{r}_i \times m_i \dot{\mathbf{r}}_i$ is the angular momentum of the *i*th particle about the chosen center. As for (4.11), $\dot{\mathbf{L}}_i = \mathbf{r}_i \times m_i \ddot{\mathbf{r}}_i$ and the equations of motion (6.1) yield

$$\dot{\mathbf{L}}_i = \sum_j \mathbf{r}_i \times \mathbf{F}_{ij} + \mathbf{r}_i \times \mathbf{f}_i, \tag{6.6}$$

with the position-vector "cross-multiplied" into every term of the force, yielding various torques about the chosen momental center. When all such equations, one for each particle, are added together, the result gives the time-rate of change of the total angular momentum $\mathbf{L} = \sum_i \mathbf{L}_i$:

$$\dot{\mathbf{L}} = \sum_{i,j} \mathbf{r}_i \times \mathbf{F}_{ij} + \sum_i \mathbf{r}_i \times \mathbf{f}_i .$$

The last summation here, to be symbolized by $\mathbf{N} \equiv \sum_i \mathbf{r}_i \times \mathbf{f}_i$, is the resultant of all the external torques on the particles of the system. The double-summation of torques arising from the internal forces consists of pairs of terms like

$$\mathbf{r}_i \times \mathbf{F}_{ij} + \mathbf{r}_j \times \mathbf{F}_{ji} = (\mathbf{r}_i - \mathbf{r}_j) \times \mathbf{F}_{ij} \equiv \mathbf{r}_{ij} \times \mathbf{F}_{ij},$$

where $\mathbf{r}_{ij} = \mathbf{r}_i - \mathbf{r}_j = -\mathbf{r}_{ji}$ is an interparticle position vector. It can now be insisted that at least the resultant, $\frac{1}{2}\sum_{i,j} \mathbf{r}_{ij} \times \mathbf{F}_{ij}$, must vanish because no body can start moving by itself, from the effect of internal torques alone. Actually, each individual "couple," $\mathbf{r}_{ij} \times \mathbf{F}_{ij}$, vanishes if the internal forces are restricted to the mutual ones, $\mathbf{F}_{ij} = \hat{\mathbf{r}}_{ij} F(r_{ij})$, directed along the interparticle position-vector, $\mathbf{r}_{ij}$. The consequence is that only a nonvanishing resultant *external* torque $\mathbf{N}$ can change the angular momentum, and

$$\dot{\mathbf{L}} = \mathbf{N} \tag{6.7}$$

is an equation of motion for the rotation as a body.

The motions measured by angular momenta are called rotations because it is essential for $\mathbf{L}_i = \mathbf{r}_i \times \mathbf{p}_i$ to exist that the linear momentum $\mathbf{p}_i$ has a component transverse to the radial vector $\mathbf{r}_i$. However, it is fairly obvious, and will be confirmed shortly, that the translational motion $\bar{\mathbf{p}} = M\dot{\bar{\mathbf{r}}}$ already described in (6.5), contributes the angular momentum $\bar{\mathbf{L}} = \bar{\mathbf{r}} \times \bar{\mathbf{p}}$ to the motion relative to a point fixed in space as described by (6.7). To get rotational motion independent of the translation already treated, a momental center moving so that the total momentum relative to it disappears ($\bar{\mathbf{P}} = 0$) should be chosen. This is done next.

Rotations about the Mass-Center

The individual momenta relative to the mass-center are $\mathbf{P}_i = m_i \dot{\mathbf{R}}_i = m_i(\dot{\mathbf{r}}_i - \dot{\bar{\mathbf{r}}})$ and their total is

$$\bar{\mathbf{P}} = \sum_i \mathbf{P}_i = \sum m_i \dot{\mathbf{r}}_i - M\dot{\bar{\mathbf{r}}} = 0, \tag{6.8}$$

because of $\sum m_i \mathbf{r}_i = M\bar{\mathbf{r}}$(6.4). The definition of the center of mass was of course designed to produce just this result. The total angular momentum about the mass-center is

$$\begin{aligned} \mathbf{l} &= \sum_i \mathbf{R}_i \times \mathbf{P}_i = \sum_i m_i(\mathbf{r}_i - \bar{\mathbf{r}}) \times (\mathbf{v}_i - \bar{\mathbf{v}}) \\ &= \sum_i \mathbf{r}_i \times m_i \mathbf{v}_i - \left(\sum_i m_i \mathbf{r}_i\right) \times \bar{\mathbf{v}} - \bar{\mathbf{r}} \times \left(\sum_i m_i \mathbf{v}_i\right) + \left(\sum_i m_i\right) \bar{\mathbf{r}} \times \bar{\mathbf{v}}. \end{aligned} \tag{6.9}$$

Each of the last three terms represents either a subtraction or addition of the same moment, $\bar{\mathbf{L}} = \bar{\mathbf{r}} \times M\bar{\mathbf{v}}$, of the total momentum about the same momental center as that of $\mathbf{L} = \sum_i \mathbf{r}_i \times \mathbf{p}_i$, considered in the preceding subsection. The end result is that the angular momentum about a point fixed in space is

$$\mathbf{L} = \mathbf{l} + \bar{\mathbf{L}}, \tag{6.10}$$

having a moment $\bar{\mathbf{L}} = \bar{\mathbf{r}} \times \bar{\mathbf{p}}$ of the translational center-of-mass motion $\bar{\mathbf{p}}$ added to the "intrinsic" angular momentum $\mathbf{l}$ about the mass-center.

Now $\dot{\bar{\mathbf{L}}} = \bar{\mathbf{r}} \times \dot{\bar{\mathbf{P}}} = \bar{\mathbf{r}} \times \mathbf{f}$ follows from equation (6.5) for the mass-center position. This, together with the mass-rotation equation (6.7), yields

$$\dot{\mathbf{l}} = \dot{\mathbf{L}} - \dot{\bar{\mathbf{L}}} = \sum_i (\mathbf{r}_i - \bar{\mathbf{r}}) \times \mathbf{f}_i .$$

The right side is

$$\mathbf{n} \equiv \sum_i \mathbf{R}_i \times \mathbf{f}_i , \tag{6.11}$$

the resultant of torques about the mass-center arising from just the external forces. Internal torques

$$\sum_{i,j} \mathbf{R}_i \times \mathbf{F}_{ij} = \tfrac{1}{2} \sum_{i,j} (\mathbf{R}_i - \mathbf{R}_j) \times \mathbf{F}_{ij} = \tfrac{1}{2} \sum_{i,j} \mathbf{r}_{ij} \times \mathbf{F}_{ij}$$

vanish as before, since $\mathbf{R}_i - \mathbf{R}_j \equiv \mathbf{r}_i - \mathbf{r}_j$. Now the equation

$$\boxed{\dot{\mathbf{l}} = \mathbf{n}} \tag{6.12}$$

describes rotational motion-as-a-body about the mass-center, additional to and independent of the translation-as-a-body described by $\dot{\mathbf{P}} = \mathbf{f}$ (6.5).

The Center of Gravity

A common case of external influence on bodies near the earth's surface is the uniform gravitational acceleration providing $\mathbf{f}_i = m_i \mathbf{g}$. The resultant of these forces is $\mathbf{f} = (\sum_i m_i)\mathbf{g} = M\mathbf{g}$, and so the equation (6.5) for the motion of the mass-center reduces to

$$\ddot{\bar{\mathbf{r}}} = \mathbf{g}. \tag{6.13}$$

The mass-center "falls" exactly like a single particle. The total external torque relative to a momental center fixed in space is

$$\mathbf{N} = \sum_i \mathbf{r}_i \times m_i \mathbf{g} = \left(\sum_i m_i \mathbf{r}_i\right) \times \mathbf{g} = M\bar{\mathbf{r}} \times \mathbf{g}. \tag{6.14}$$

Thus the total force of gravity, $\mathbf{f} = M\mathbf{g}$, exerts a torque on the body of particles as if their mass were concentrated at their mass-center. The torque about the mass-center is

$$\mathbf{n} = \sum_i \mathbf{R}_i \times m_i \mathbf{g} = \left(\sum_i m_i \mathbf{R}_i\right) \times \mathbf{g} = 0, \tag{6.15}$$

since it has the position $\bar{\mathbf{R}} = (\sum_i m_i \mathbf{R}_i/M) = 0$ relative to itself. No rotation ($\dot{\mathbf{l}} = \mathbf{n} = 0$) about the mass-center can be generated by uniform gravity. Any freely falling body will only retain whatever internal angular momentum it had at the beginning of its fall.

The result (6.15) may be expressed by saying that the mass-center is a "point of balance" in the gravitational field. This, together with the behavior represented by (6.14), accounts for the fact that the center of mass is frequently referred to as the center of gravity.

6.3 INTERNAL MOTIONS AND ENERGIES

The two equations for the motions of the particles as a body, $\dot{\mathbf{P}} = \mathbf{f}$ and $\dot{\mathbf{l}} = \mathbf{n}$, give no details about the individual particle motions. In principle, these should be obtained from the equations for $\mathbf{R}_i = \mathbf{r}_i - \bar{\mathbf{r}}$ following from (6.1) and (6.5):

$$m_i \ddot{\mathbf{R}}_i = m_i(\ddot{\mathbf{r}}_i - \ddot{\bar{\mathbf{r}}}) = \sum_j \mathbf{F}_{ij} + (\mathbf{f}_i - m_i \mathbf{f}/M). \tag{6.16}$$

Thus the "internal" motions, relative to the mass-center, depend solely on internal forces not only when external forces are absent, but also when the external force on each particle is just the fraction m_i/M of the resultant external force on the system. The latter is rather a special case, being characteristic of uniform gravitational accelerations, $\mathbf{f}_i = m_i \mathbf{g}$. A conclusion is that *uniform* gravity does not disturb any of the internal motions of a system, just as it does not disturb the internal angular momentum (6.15). (See Exercise 6.3 for an example of what may happen in a *non*uniform field.)

Even when dealing with an isolated system, for which $\mathbf{f}_i = 0$, there are still as many equations (6.16) to solve as there are particles, and that is rarely feasible without approximation of some kind, except in the two-particle case. The latter reduces to a one-particle problem as seen in the preceding chapter, and the corresponding simplification for several particles, arising from the condition

$$\sum_i m_i \mathbf{R}_i = \sum_i m_i(\mathbf{r}_i - \bar{\mathbf{r}}) = 0, \tag{6.17}$$

is only to reduce an N-particle problem to an $(N-1)$-particle problem. The condition (6.17) allows evaluation of an Nth position vector only after $(N-1)$ others are known.

Some general conclusions can still be adduced about the energies.

Kinetic Energies

The total kinetic energy $T = \sum_i \frac{1}{2} m_i v_i^2$ may be analyzed by putting $\mathbf{v}_i = \bar{\mathbf{v}} + \mathbf{V}_i$ for the individual velocities:

$$\begin{aligned} T &= \sum_i \tfrac{1}{2} m_i(\bar{v}^2 + 2\bar{\mathbf{v}} \cdot \mathbf{V}_i + V_i^2) \\ &= \tfrac{1}{2}\Big(\sum_i m_i\Big)\bar{v}^2 + \bar{\mathbf{v}} \cdot \sum_i m_i \mathbf{V}_i + \sum_i \tfrac{1}{2} m_i V_i^2. \end{aligned}$$

The first term is $\bar{T} = \frac{1}{2} M \bar{v}^2$, the kinetic energy of the total mass M by virtue of center-of-mass translation, $\bar{\mathbf{v}}$. The middle term vanishes because $\sum_i \mathbf{P}_i = 0$ (6.8). The final term,

$$\tau = \sum_i \tfrac{1}{2} m_i V_i^2, \tag{6.18}$$

is the "internal" kinetic energy of all the individual particles in motion relative to their center-of-mass. The total kinetic energy is then

$$T = \bar{T} + \tau, \tag{6.19}$$

in a frame in which the mass-center moves with velocity $\bar{v} = (2\bar{T}/M)^{1/2}$.

Kinetic energy is generated by power expended by the forces,

$$\frac{dT}{dt} = \sum_i m_i \mathbf{v}_i \cdot \dot{\mathbf{v}}_i = \sum_{i,j} \mathbf{F}_{ij} \cdot \mathbf{v}_i + \sum_i \mathbf{f}_i \cdot \mathbf{v}_i,$$

according to the equations of motion (6.1). During displacements $\mathbf{dr}_i = \mathbf{v}_i\, dt$,

$$dT = \sum_{i,j} \mathbf{F}_{ij} \cdot \mathbf{dr}_i + \sum_i \mathbf{f}_i \cdot \mathbf{dr}_i. \tag{6.20}$$

The work by the internal forces can be written as a sum of pairs of terms arising from "interparticle displacements" $\mathbf{dr}_{ij} = d(\mathbf{r}_i - \mathbf{r}_j) \equiv d(\mathbf{R}_i - \mathbf{R}_j)$:

$$\sum_{i,j} \mathbf{F}_{ij} \cdot \mathbf{dr}_i = \tfrac{1}{2} \sum_{i,j} \mathbf{F}_{ij} \cdot \mathbf{dr}_{ij}, \tag{6.21}$$

since $\mathbf{F}_{ji} = -\mathbf{F}_{ij}$.

Potential Energies

When the external forces $\mathbf{f}_i$ are absent, the system is isolated and its energy should be conserved. This can be represented by internal forces $\sum_j \mathbf{F}_{ij}(r_{ij}) = \mathbf{F}_i(r_{12}, r_{13}, r_{14}, \ldots)$, which are derivable from a scalar potential $U(r_{12}, r_{13}, r_{23}, r_{14}, \ldots)$ according to

$$\mathbf{F}_i = -\nabla_i U = -\sum_j \left(\frac{\partial U}{\partial r_{ij}}\right) \nabla_i r_{ij}.$$

Now (see Exercise 2.5)

$$\nabla_i r_{ij} \equiv \nabla_i |\mathbf{r}_i - \mathbf{r}_j| = (\mathbf{r}_i - \mathbf{r}_j)/|r_i - r_j| = \hat{\mathbf{r}}_{ij}, \tag{6.22}$$

the unit vector, as can be found for each cartesian component of the gradient applied to $|\mathbf{r}_i - \mathbf{r}_j| = [(x_i - x_j)^2 + (y_i - y_j)^2 + (z_i - z_j)^2]^{1/2}$, for example. Thus, the mutual force in one pair may be identified with

$$\mathbf{F}_{ij} = -\cdot_{ij} \frac{\partial}{\partial r_{ij}} U(r_{12}, r_{13}, r_{23}, r_{14}, \ldots) = -\mathbf{F}_{ji}, \tag{6.23}$$

since $\mathbf{r}_{ji} = -\mathbf{r}_{ij}$. This is more general than it need be to represent the mutual forces of the character, $F_{ij}(r_{ij})$, assumed from the beginning. A more special form $U = \frac{1}{2}\sum_{i,j} U_{ij}(r_{ij})$, a sum of "pair-potentials," would do for the latter.

Since $r_{ij} = r_{ji}$ and $dr_{ij} = \hat{\mathbf{r}}_{ij} \cdot \mathbf{dr}_{ij}$,

$$\begin{aligned} dU(r_{12}, r_{13}, r_{23}, r_{14}, \ldots) &= \tfrac{1}{2} \sum_{i,j} (\partial U/\partial r_{ij})\, dr_{ij} \\ &= \tfrac{1}{2} \sum_{i,j} (\partial U/\partial r_{ij}) \hat{\mathbf{r}}_{ij} \cdot \mathbf{dr}_{ij} \\ &= -\tfrac{1}{2} \sum_{ij} \mathbf{F}_{ij} \cdot \mathbf{dr}_{ij}, \end{aligned} \tag{6.24}$$

which is just the negative of the element of work (6.21) by the internal forces. Then (6.20) yields $d(T + U) = 0$ in the absence of external forces ($\mathbf{f}_i = 0$). In the latter case, the CM kinetic energy $\bar{T} = \frac{1}{2}M\bar{v}^2$ is conserved by itself as shown by $M\dot{\mathbf{V}} = \mathbf{f} = 0$ (6.5), hence $d(\tau + U) = 0$ and the "internal energy"

$$\varepsilon = \tau + U \tag{6.25}$$

is conserved by itself for the isolated several-particle system.

A conserved total energy emerges in the presence of external forces when these are derivable from potentials $V_i(\mathbf{r}_i)$ such that $\mathbf{f}_i = -\nabla_i V_i(\mathbf{r}_i)$. Then the work element $\mathbf{f}_i \cdot d\mathbf{r}_i$ of (6.20) is just $-\nabla_i V_i \cdot d\mathbf{r}_i = -dV_i(\mathbf{r}_i)$ and the expression yields $d(T + U + \sum_i V_i) = 0$ or

$$E = \bar{T} + \sum_i V_i(\mathbf{r}_i) + \varepsilon \tag{6.26}$$

as a conserved total energy. With external forces present, the internal energy ε is in general no longer conserved by itself. It still is in the special cases in which the external forces disappear from the equations of internal motion (6.16), for the conservation of $\varepsilon = \tau + U$ can then be proved directly from those.

The Virial Theorem

A final generalization about systems of many particles will be reviewed because of its usefulness in statistical mechanics. A simple instance of it has already been encountered in connection with (4.39): the kinetic energy in a circular orbit, of a particle moving in an "inverse-square" field of force, is just half the potential energy in magnitude. A corresponding relation also holds in elliptic orbits, but between *averages* of the energies over the orbit. These are manifestations of a considerably more general finding known as the "virial theorem of Clausius."

The result is most readily developed from a consideration of how the quantity defined by

$$S(t) \equiv \sum_i \mathbf{p}_i \cdot \mathbf{r}_i \tag{6.27}$$

varies with time during motions $\mathbf{r}_i(t)$, $\mathbf{v}_i(t) = \mathbf{p}_i/m_i$ of any system of particles. (The quantity S will be met again in Chapter 8, as a "generator" of motions and as a "phase-function" in "configuration waves.")

It follows from the equations of motion, $\dot{\mathbf{p}}_i = \mathbf{F}_i$, that

$$\dot{S} = \sum_i \mathbf{F}_i \cdot \mathbf{r}_i + \sum_i m_i v_i^2.$$

The average of this over any time span T is denoted

$$\langle \dot{S} \rangle \equiv \int_{t_1}^{t_1+T} \frac{dt}{T}\frac{dS}{dt} = \frac{S(t_1 + T) - S(t_1)}{T}, \tag{6.28}$$

and can be held to vanish as $T \to \infty$ if the motion is confined within some bounds on the positions $\mathbf{r}_i$ (as by vessel walls in the situations of interest in statistical physics). It also vanishes if T is a period in any motion that repeats itself cyclically, as in orbital motions. As a consequence, it can be asserted that

$$\left\langle \sum_i \mathbf{F}_i \cdot \mathbf{r}_i \right\rangle = -2\left\langle \sum_i \tfrac{1}{2} m_i v_i^2 \right\rangle \tag{6.29}$$

must hold between time-averages modeled on the integral in (6.28). The quantity on the left is what is called the virial of Clausius.

For the motion of a single particle in a central force-field derivable from a potential $V(r)$, (6.29) leads to an average kinetic energy

$$\langle \tfrac{1}{2} m v^2 \rangle = \tfrac{1}{2} \langle r dV/dr \rangle. \tag{6.30}$$

For any power-law potential, $V = Kr^n$; the average of the kinetic energy will be equal to $\frac{1}{2}n\langle V \rangle$. Large inverse powers ($n < 0$) occur among the molecules of interest in statistical physics. For an isotropic oscillator potential ($n = +2$) the average energies are equal, whereas in an "inverse-square" force-field ($n = -1$), the average kinetic energy is just half the average potential energy in absolute magnitude. Notice that v must describe attraction ($K < 0$ when $n < 0$) to represent the conditions presumed.

Describing an Isolated Several-Particle System

To be summarized finally is how an isolated system is to be described in the terms used here. All external forces $\mathbf{f}_i$ are absent and then the expected conservation principles follow automatically. The mass-center motion governed by $\dot{\mathbf{P}} = \mathbf{f} = 0$ becomes uniform, and no torques exist to change the angular momentum, $\dot{\mathbf{l}} = \mathbf{n} = 0$, about the mass-center; the total linear and angular momenta are conserved:

$$\bar{\mathbf{p}} = \text{constant}, \qquad \mathbf{l} = \text{constant}. \tag{6.31a}$$

Moreover, the internal energy,

$$\varepsilon = \tau + U = \text{constant}, \tag{6.31b}$$

is conserved as already noted in (6.25). The internal motions are governed only by internal forces according to

$$m_i \ddot{\mathbf{R}}_i = -\textstyle\sum_j \hat{\mathbf{r}}_{ij}\, \partial U/\partial r_{ij}, \tag{6.32}$$

and this is about all that can be said quite generally yet simply.

Further Developments of Mechanics

With the considerations of this chapter, a survey of the basic features of the Newtonian formulation may be regarded as completed. What has been surveyed may be thought of as a Mechanics of Particles, since attention was concentrated on the motions of point-positions, $\mathbf{r}_i(t)$, with associated masses, m_i. The many-particle problem can become so complex that whole disciplines have been developed to deal with various aspects of it. Among them is the theory of Rigid Bodies (Chapter 10), Elasticity Theory (of deformable bodies), Fluid Dynamics, Kinetic Theory (of gases), and Statistical Mechanics.

All these further developments may be treated as extensions of the Newtonian formulation. The distinctive feature of the latter may be regarded to be the expression of deviations from uniform motion in terms of forces. Alternative expressions will be surveyed in the next two chapters. However, descriptions of circumstances in terms of forces remain the most important because they are usually found to be more direct abstractions from experience than are the alternatives.

EXERCISES

6.1. A neutral atom consists of a point-nucleus, mass m_0 and charge Ze, and of Z point-electrons, each having mass m and charge $-e$.

(a) Construct an equation of motion for each particle.

(b) From the equations of motion, show that the total angular momentum is conserved about any momental center, fixed or in uniform motion.

6.2. A "body" of particles, of mass $M = \sum m_i$, extends over distances that are small enough compared to its distance from the earth's center that the approximation

$$r_i^n = |\bar{\mathbf{r}} + \mathbf{R}_i|^n = (\bar{r}^2 + 2\bar{\mathbf{r}} \cdot \mathbf{R}_i + R_i^2)^{n/2}$$
$$\approx \bar{r}^n[1 + n(\hat{\mathbf{r}} \cdot \mathbf{R}_i)/\bar{r}],$$

for any power of the particle-distances from the earth's center, is adequate. Here $\bar{\mathbf{r}} \equiv \bar{r}\hat{\mathbf{r}}$ is the position of the body's centroid relative to the earth's center, and $\mathbf{R}_i$ is a particle position relative to the centroid.

(a) Show that, to this approximation, the body falls toward the earth from large distances as if all its mass were concentrated at the centroid.

(b) Show that there is a net gravitational torque about the mass-center which may be expressed as

$$\mathbf{n} = -3(gR_e^2/\bar{r}^3)\hat{\mathbf{r}} \times \sum m_i \mathbf{R}_i(\mathbf{R}_i \cdot \hat{\mathbf{r}}),$$

where R_e is the earth's radius occurring in (4.29). (Some notion as to whether this torque can induce appreciable "tumbling" in an artificial satellite may be gained from the following exercise.)

6.3. Suppose that the "body" of the preceding exercise consists of two equal point-masses, m each, separated by a weightless rigid rod of length s, and that its mass-center describes a circular orbit of radius $\bar{r}$, around the earth.

(a) Evaluate for it the torque of the preceding exercise, in terms of an angle φ made by the rod with the radius vector $\bar{\mathbf{r}}$.

(b) From the equation of motion for the internal angular momentum, $\dot{l} = \frac{1}{2}ms^2\,\ddot{\varphi} = n$, show that the "internal energy" as defined by

$$\varepsilon = \tfrac{1}{4}ms^2\left[\dot{\varphi}^2 + \frac{3gR_e^2}{\bar{r}^3}\sin^2\varphi\right]$$

is conserved.

(c) Sketch an energy diagram for the relevant range of $\varphi(-\frac{1}{2}\pi \leq \varphi \leq \frac{1}{2}\pi)$, and argue from it that the orientations $\varphi = \pm\frac{1}{2}\pi$ are unstable, but that $\varphi = 0$ yields a stable equilibrium.

(d) Find the period of small oscillations about $\varphi = 0$, in ratio to the period in orbit. [$3^{-1/2}$. This makes the oscillation period greater than 49 minutes.]

6.4. A number n of force-centers (such as atoms in a crystal) are distributed over the fixed points $\mathbf{r}_1, \mathbf{r}_2, \mathbf{r}_3, \ldots, \mathbf{r}_n$. A mass-particle m, when at a point $\mathbf{r}$, is subject to the force $-m\sum_s \omega_s^2(\mathbf{r} - \mathbf{r}_s)$.

(a) Find the expression for the equilibrium position, $\bar{\mathbf{r}}$, of the particle amid the given force-centers.

(b) If there were only the one force-center at $\mathbf{r}_1$, then the particle would follow an elliptical orbit with period $2\pi/\omega_1$ (§4.1). Find the orbit and the period when all n forces are in operation.

6.5. Suppose that the discrete force-centers of the preceding exercise are replaced by a *continuum* of them, confined within a spherical volume $V = \frac{4}{3}\pi R^3$, with each volume element $dV(\mathbf{r}_s)$ contributing the potential energy

$$\tfrac{1}{2}m\omega^2(\mathbf{r} - \mathbf{r}_s)^2\, dV/V,$$

for $0 < r_s < R$, to a mass-particle m at $\mathbf{r}$. Show that the resultant force on the particle is independent of the size of the sphere.

6.6. Apply the procedure suggested by the preceding exercise to showing that the *gravitational* potential energy of a mass-particle m at any point *inside* the earth is

$$V(r \leq R) = (mg/2R)(r^2 - 3R^2),$$

in a gauge that matches up with $V(r \geq R) = -mgR^2/r$, of (4.26) and (4.30), at $r = R$.

Check by explicit integration that, *outside* a spherically symmetric earth, the attraction is indeed equivalent to that by a *point*-mass at its center, as assumed in Chapter 4.

6.7. Apply the results of the preceding exercise to the following.

(a) Suppose a hole were dug straight through the center of the earth to the antipodes, and a mass m were dropped into it. On the basis of a frictionless passage, how soon would it return?

(b) Suppose a subway direct from Washington to Moscow were constructed, consisting of a straight tube through the earth, and that the trains used power only to cancel frictional effects. What would be the time for one trip? (About 42 minutes.)

6.8. Show that the relation between the kinetic and potential energies in the circular orbits of Exercise 4.12 is consistent with the virial theorem.

CHAPTER 7

DESCRIPTIONS IN CONFIGURATION SPACES

Schemes alternative to the Newtonian formulation have been developed: d'Alembertian, Lagrangian, Variational, Hamiltonian, and others. Each can be made to describe successfully the same body of experience and serves as another representation of the same basic theory: the Classical Mechanics. That is demonstrated by the fact that each scheme is so constructed that the Newtonian formulation may be derived from it. The chief value of such generalizations is that they offer more ways to attack the problems of describing motion. A greater variety of viewpoints is made available and this helps clarify just what it is essential to describe about the circumstances of any motion.

7.1 THE d'ALEMBERTIAN FORMULATION

The viewpoint associated with d'Alembert grew from the study of conditions for the static equilibrium of a system. In the resulting formulation, motion is regarded as a "stationary" state of *dynamic equilibrium* between forces like $\mathbf{F}_i$ applied to m_i and its "kinetic reaction," $-m_i\ddot{\mathbf{r}}_i$.

Constraints

To see how the formulation arises, the conditions for the *static* equilibrium of a system of particles will be examined. These turn out to be almost trivially obvious; the problems of statics become really substantial in engineering applications, when the equilibrium of extended rigid and tensile bodies is to be represented in terms of some systems of forces. Some of that substance is introduced into problems of the equilibrium of particles when "geometrical constraints" are considered. These may not be regarded as of fundamental interest, but they have played a great enough role in the development of Mechanics for at least an acquaintance with them to be desirable. Moreover, the generality of the Newtonian formulation, from a mathematical point of view, may be questioned until constraints are allowed for.

A particle's motion may be restricted to the top of a table or to some other "surface of constraint." This type of condition may be formulated geometrically, by giving a relation $G(x, y, z) = 0$ that must exist among its coordinates if the particle is to stay on the surface. More generally, several such conditions, $G_s(\mathbf{r}_1, \mathbf{r}_2, \ldots, t) = 0$, each relating several particles, may exist. An explicit dependence on time may be used, for instance, to describe a surface of constraint having some *given* motion. There may also be conditions not expressible in terms of functions like the G's, so-called "nonholonomic" constraints. For example, there may be a stage at which the particle moving on a table top falls off the edge!

The restriction of a particle to a table top may be formalized by $G \equiv z - h = 0$, where h is the height of the table. An example of a restriction relating two particles is provided by a "dumbbell" model (of a rigid diatomic molecule, for example). This may be described by $G \equiv |\mathbf{r}_1 - \mathbf{r}_2| - l = 0$ where l is the length of a joining rod. For a molecule that is not rigid, but vibrating with a *given* frequency ω_0 for instance, $l = a + b \sin \omega_0 t$. A simple example of a nonholonomic system is provided by a *rolling* disc. An angular displacement $d\vartheta$ of the disc is related to the displacement ds along a trajectory by $ds = a\, d\vartheta$, where a is the radius of the disc.

From the Newtonian point of view, constraints such as those described here are obeyed because certain "forces of constraint" compel obedience. A purely Newtonian formulation can be retained by first somehow deriving the forces from the nature of the constraints. An example familiar in elementary physics is provided by the so-called "Atwood machine." Two masses are suspended, under constant gravity, from the two ends of a flexible, weightless, inextensible string that passes over a smooth, or inertialessly rotating, pulley. The lengths of string on each side of the pulley, z_1 and z_2, are then related as $G \equiv z_1 + z_2 - l = 0$. The Newtonian equations would be written $m_1\ddot{z}_1 = m_1 g - C$ and $m_2\ddot{z}_2 = m_2 g - C$, where C is an initially unknown constraint force "transmitted" by the inextensible string. The condition $G = 0$ requires that $\ddot{z}_2 = -\ddot{z}_1$, and so $C = 2m_1 m_2 g/(m_1 + m_2) \equiv 2\mu g$. As a consequence, $\ddot{z}_1 = -\ddot{z}_2 = (m_1 - m_2)g/(m_1 + m_2)$.

Even in this simple example a series of assumptions is involved in reducing the problem to a matter of finding the single constraint force, C. It may not be regarded as quite obvious that it should be the same on both sides of the pulley while the system is undergoing acceleration. There is an "ad hoc" dipping into theorizing about forces, going beyond the principles of Mechanics "proper," when there is reference to an instantaneous "transmission" of constraint force in a string. Of course, some such considerations are at least implicit in any representation of circumstances by a force; Newton can be said to have derived the gravitational force from constraints of the planetary motions into elliptical orbits on the basis that it is transmitted instantaneously. In any case, a formulation that makes direct use of more immediately given conditions on the coordinates, without requiring preliminary derivations of corresponding constraint forces, clearly has advantages. Provision for such direct use is made in all the formulations introduced hereafter.

Formulation in terms of forces rather than geometrical constraint conditions is sometimes regarded the more "fundamental." A geometrical condition involves a degree of idealization which may easily make it inadequate for sufficient precision of description. An ideal plane may be adequate for representing the effect of a table top in most observations, but more refined observations may, for example, reveal a lack of smoothness. Representation by forces seems more adaptable to such greater ranges of possibility, and then analysis into forces in all situations tends to make the over-all picture more unified. This is not a conclusive argument, however; there may also be overidealization in specifying any definite force function. In the end, it is best to admit every possible viewpoint for whatever value it may have.

The Principle of Virtual Work

Each particle m_i of a system in static equilibrium will remain unaccelerated only if the resultant, $\mathbf{F}_i$, of all the forces on it, vanishes. To make sure that the system will remain static, all the forces $\mathbf{A}_i$, $\mathbf{B}_i$, $\mathbf{C}_i$, ... that may be acting on m_i must be sought out and seen to balance:

$$\mathbf{F}_i = \mathbf{A}_i + \mathbf{B}_i + \mathbf{C}_i + \cdots = 0, \tag{7.1}$$

there being such a condition for each particle, $i = 1, 2, 3, \ldots$.

Another approach will allow dispensing with any constraint forces in the balance. *Imagine* displacing each particle by a small, but otherwise arbitrary, amount: the ith particle by $\boldsymbol{\delta}\mathbf{r}_i$. The symbol $\boldsymbol{\delta}\mathbf{r}_i$, rather than $\mathbf{dr}_i$, is used to emphasize that it refers to an imagined, "virtual," displacement, and not to any actual motion $\mathbf{dr}_i = \mathbf{v}_i\, dt$, not amenable to arbitrary choice in given circumstances.

During the virtual displacement $\boldsymbol{\delta}\mathbf{r}_i$, the forces on the ith particle do "virtual" work: $\mathbf{A}_i \cdot \boldsymbol{\delta}\mathbf{r}_i$, $\mathbf{B}_i \cdot \boldsymbol{\delta}\mathbf{r}_i$, $\mathbf{C}_i \cdot \boldsymbol{\delta}\mathbf{r}_i$, ... Clearly, the virtual work by all the forces on all the particles is

$$\delta W = \sum_i (\mathbf{A}_i + \mathbf{B}_i + \mathbf{C}_i + \cdots) \cdot \boldsymbol{\delta}\mathbf{r}_i \equiv \sum_i \mathbf{F}_i \cdot \boldsymbol{\delta}\mathbf{r}_i = 0, \tag{7.2}$$

vanishing under the equilibrium conditions (7.1). The single statement $\delta W = 0$ follows from the N statements for N particles, $\mathbf{F}_i = 0$, but it may not be quite as obvious that the N latter statements also follow from the single one, $\delta W = 0$. This is so by virtue of the proviso that N displacements $\boldsymbol{\delta}\mathbf{r}_i$ may be *arbitrarily* chosen. Then δW cannot vanish because of cancellations among terms referring to different degrees of freedom, for that would require related choices—of $\boldsymbol{\delta}\mathbf{r}_i$ and $\boldsymbol{\delta}\mathbf{r}_j$, for example. Expression (7.2) can vanish only if the coefficient, $\mathbf{F}_i$, of each arbitrary displacement, $\boldsymbol{\delta}\mathbf{r}_i$, vanishes independently. Thus, a single, encompassing "principle of virtual work," $\delta W = 0$, can be regarded as equivalent to the N equilibrium conditions, $\mathbf{F}_i = 0$

The development here amounts to mere formality except for the opportunity it affords for taking geometrical constraints into account. For the force equilibrium conditions (7.1), initially unknown constraint forces $\mathbf{C}_1, \mathbf{C}_2, \ldots$ would have to be found as a preliminary step. Those can remain unknown for an equilibrium test by the principle of virtual work. It is only necessary that the choices of the virtual displacements not be made completely arbitrarily, but *in consistency* with the constraints: $\boldsymbol{\delta}'\mathbf{r}_1, \boldsymbol{\delta}'\mathbf{r}_2, \ldots$. By this is meant choices such that

$$\delta G_s = \sum_i \boldsymbol{\delta}'\mathbf{r}_i \cdot \boldsymbol{\nabla}_i G_s = 0$$

for every constraint relation $G_s = 0$ that exists; nonholonomic constraint conditions may be in a form like $\sum_i \boldsymbol{\delta}'\mathbf{r}_i \cdot \mathbf{f}_i = 0$ from the beginning (for example, $\delta' s - a\delta'\vartheta = 0$ for the rolling disc mentioned before). If no constraint is violated, no constraint forces $\mathbf{C}_i$ need be called upon to enforce it, and contributions $\sum_i \mathbf{C}_i \cdot \boldsymbol{\delta}'\mathbf{r}_i = 0$ do not exist in the expression for the virtual work:

$$\delta' W = \sum_i \mathbf{F}_i' \cdot \boldsymbol{\delta}'\mathbf{r}_i = 0, \tag{7.3}$$

where the $\mathbf{F}_i'$ are resultants of forces *aside* from any constraints unviolated by the $\boldsymbol{\delta}'\mathbf{r}_i$.

A simple illustration is provided by the equilibrium of two mass-particles connected by a string hanging, under gravity, down the curved sides of a cylinder, without leaving its surface. If the positions of m_1 and m_2 are given by angles ϑ_1 and ϑ_2 from the vertical, then $\delta'\vartheta_1 = -\delta'\vartheta_2$ and the principle of virtual work requires that

$$(m_1 g \sin \vartheta_1)\delta'\vartheta_1 + (m_2 g \sin \vartheta_2)(-\delta'\vartheta_1) = 0;$$

hence $\vartheta_2 = \sin^{-1}[(m_1/m_2) \sin \vartheta_1]$. Once this is known, the constraint forces maintaining a constant length for the string can be deduced from trying arbitrary displacements, $\delta\vartheta_{1,2}$, since then

$$(m_1 g \sin \vartheta_1 - C_1)\delta\vartheta_1 + (m_2 g \sin \vartheta_2 - C_2)\delta\vartheta_2 = 0$$

implies $C_1 = m_1 g \sin \vartheta_1$ and $C_2 = m_2 g \sin \vartheta_2 = m_1 g \sin \vartheta_1 = C_1$. This is a trivial problem, but it illustrates that knowledge of constraint forces may be dispensed with in finding equilibrium configurations, and it also shows how such forces may afterward be found.

d'Alembert's Principle

Writing the Newtonian equations of motion as

$$\mathbf{F}_i - m_i \ddot{\mathbf{r}}_i = 0 \qquad (i = 1, 2, \ldots, N) \tag{7.4}$$

permits regarding them as conditions like (7.1), but for a "dynamic" equilibrium, attained when the accelerations $\ddot{\mathbf{r}}_i$ come into existence.

Again consider the effects of virtual displacements, $\boldsymbol{\delta}\mathbf{r}_i$, imagined carried out at some instant t, from the instantaneous positions $\mathbf{r}_i(t)$ then actually existing. A principle of virtual work

$$\delta W = \sum_i (\mathbf{F}_i - m_i \ddot{\mathbf{r}}_i) \cdot \boldsymbol{\delta}\mathbf{r}_i = 0, \tag{7.5}$$

may be promulgated for the instant, when work by the "kinetic reactions," $-m_i\ddot{\mathbf{r}}_i$, is included. This can be regarded as d'Alembert's formulation of Mechanics. It leads right back to the Newtonian formulation for completely arbitrary virtual displacements, such as are permissible when the $\mathbf{F}_i$ include any constraint forces that exist. It has the advantage over the Newtonian formulation that any constraint forces may be omitted from the resultants $\mathbf{F}_i$, if only the virtual displacements $\boldsymbol{\delta}\mathbf{r}_i$ are restricted to consistency with the constraints. Of course, any geometrical relationships among the position-vectors $\mathbf{r}_i$ also imply relationships among the accelerations, $\ddot{\mathbf{r}}_i$.

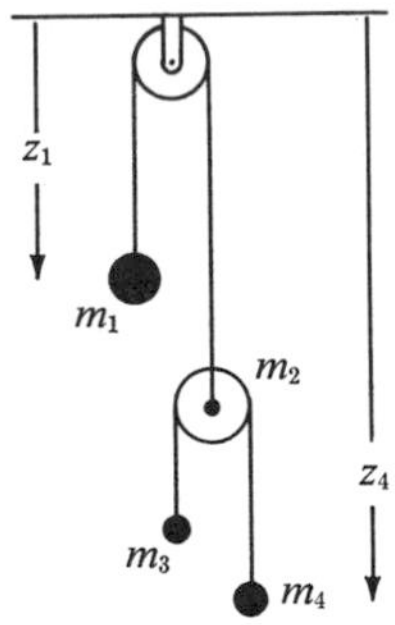

Figure 7.1

The d'Alembertian Principle of Virtual Work (7.5) embodies an approach to investigating a system that is instinctive with monkeys and humans: "jiggling" parts of it to see what happens. Here are rudiments of an important technique known as the "variational method," to be developed further in §7.4.

A simple problem in which the d'Alembertian approach shows to good advantage is provided by the "compound" Atwood machine indicated in Figure 7.1. By properly restricting the displacements in the virtual work $\sum_{i=1}^{4} m_i(g - \ddot{z}_i)\delta z_i = 0$, according to $\delta(z_1 + z_2) = 0$ and $\delta(z_3 + z_4 - 2z_2) = 0$, it is only a matter of algebra to find at once

$$\ddot{z}_1 = -\ddot{z}_2 = \frac{m_1 - (m_2 + 4\mu_{34})}{m_1 + (m_2 + 4\mu_{34})}\, g, \tag{7.6a}$$

where $\mu_{34} \equiv m_3 m_4/(m_3 + m_4)$. This properly reduces to $-g$ for $m_1 = 0$ and to a simple Atwood machine result for $m_3 = m_4$. Similarly,

$$\ddot{z}_3 = \frac{(m_3 - m_4)(m_1 + m_2) - 2m_4(m_1 - m_2) + 4m_3 m_4}{(m_3 + m_4)(m_1 + m_2) + 4m_3 m_4}\, g \tag{7.6b}$$

and $\ddot{z}_4 = -2\ddot{z}_1 - \ddot{z}_3 = \ddot{z}_3(m_3 \leftrightarrow m_4)$. This properly reduces to a simple Atwood machine result for $m_1 = m_2 = \infty$ or for $m_3 = m_4$, besides giving $\ddot{z}_3 = g$ for $m_4 = 0$. Now secure grounds are provided for coming to conclusions about the constraint forces, $C_i = m_i(g - \ddot{z}_i)$: $C_3 = C_4$ but $C_2 \neq C_1$, in general $(C_2 = C_1 - C_3 - C_4)$.

Another instructive problem is provided by a mass at the end of a string, swinging around a cylindrical post and winding up on it, gravity being ignored (Figure 7.2). Only "constraint forces" are involved and these can do no work during displacements consistent with the constraints, according to d'Alembert's principle; hence the particle's kinetic energy must be conserved. This may also be fairly evident from the fact that the constraint force on the particle must be directed along the string and its velocity can only

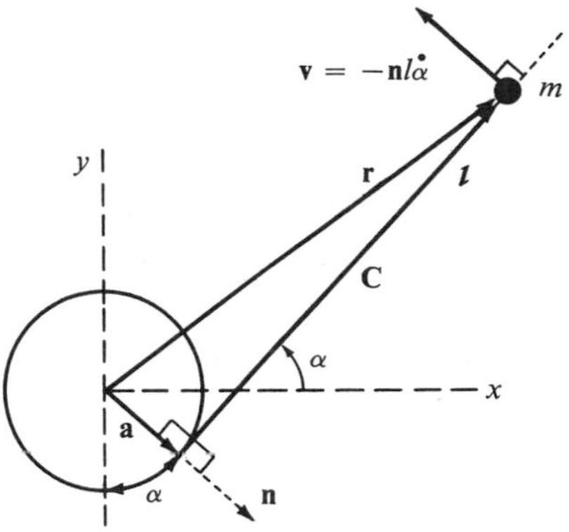

Figure 7.2

be perpendicular to it, and so the force can do no work during the motion. Instead, the angular momentum about the post center is unconserved, being proportional to the length of string left unwound. The constraint condition in this problem may be given as $\delta l = -a\delta\alpha$.

7.2 GENERALIZED COORDINATES

The value of substitutions for the cartesian coordinates was evident when the polar coordinates, r and φ, were used for the central force problems of Chapter 4. In substitutions such as $r = (x^2 + y^2)^{1/2}$ and $\varphi = \tan^{-1}(y/x)$, the new coordinates of each point depend only on the old coordinates of the same point. An example in which the new coordinates depend on more than one position is provided by the center-of-mass and relative coordinates, $\bar{\mathbf{r}} = (m_1\mathbf{r}_1 + m_2\mathbf{r}_2)/M$ and $\mathbf{r} = \mathbf{r}_1 - \mathbf{r}_2$, used for describing two-particle systems in Chapter 5.

Lagrange investigated what can be done in general with substitutions $\mathbf{r}_1, \mathbf{r}_2, \ldots \to q_1, q_2, \ldots,$ where the q's may be any quantities that can be observed to change with the motion of the system. The new "generalized coordinates" need not even be geometrical quantities (they might be deflections of a speedometer needle, for example, but also such things as electric current intensities or "fourier amplitudes," in suitable circumstances). However, to give as adequate a description as do the position-vectors, the latter should be derivable from them through some functional relationships,

$$\mathbf{r}_i = \mathbf{r}_i(q_1, q_2, \ldots, t). \tag{7.7}$$

For example, these may stand for $x = r\cos\varphi$, $y = r\sin\varphi$ in the transformation to polar coordinates. In the case of the string winding up on a post in

Figure 7.2, the motion may be described simply by giving $q \equiv l(t)$, the length of string still unwound at a given phase, or, instead, by giving $q \equiv \alpha(t)$, the angle at the post center subtended by string wound up since the phase $\alpha = 0$. Then the relations (7.7) have cartesian components:

$$x = [l_0 - a\alpha) \cos \alpha + a \sin \alpha] \quad \text{or} \quad [l \cos(l_0 - l)/a + a \sin(l_0 - l)/a],$$
$$y = [l_0 - a\alpha) \sin \alpha - a \cos \alpha] \quad \text{or} \quad [l \sin(l_0 - l)/a - a \cos(l_0 - l)/a].$$

The corresponding expressions for the conserved kinetic energy become simply

$$T = \tfrac{1}{2}ml^2\dot{\alpha}^2 = \tfrac{1}{2}m(l_0 - a\alpha)^2\dot{\alpha}^2 = \tfrac{1}{2}m(l/a)^2\dot{l}^2 = \tfrac{1}{2}mv_0^2.$$

The unconserved angular momentum about the post center is $L = mv_0 \, l(t)$.

An example in which the time enters explicitly into the relationship between a point-position vector and a generalized coordinate is provided by a pendulum with a point of suspension forced to oscillate horizontally with a given frequency (ω), as indicated in Figure 7.3. The motion in the plane can be

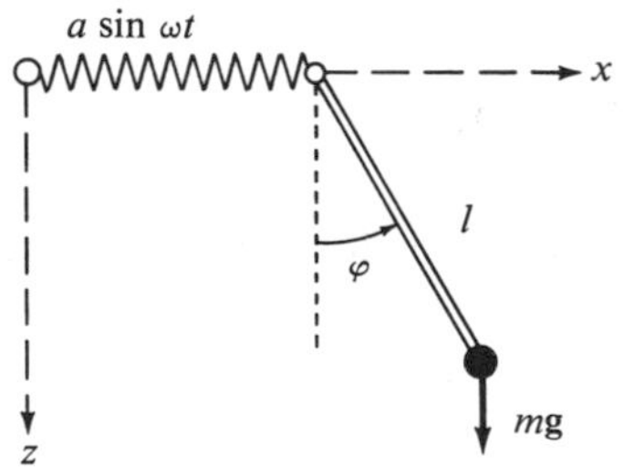

Figure 7.3

described by $q \equiv \varphi(t)$, the angle with the direction of the earth's gravity. Then

$$z = l \cos q,$$
$$x = l \sin q + a \sin \omega t.$$

If the pendulum bob has the only considerable mass, the kinetic energy is

$$T = \tfrac{1}{2}m[l^2\dot{q}^2 + 2al\omega\dot{q} \cos q \cos \omega t + a^2\omega^2 \cos^2 \omega t].$$

The potential energy is simply $V = -mgl \cos q$.

An example in which the potential, rather than kinetic, energy becomes explicitly time-dependent is provided when the pendulum is suspended from the end of a rigid rod that is forced to rotate uniformly in a vertical plane, as indicated in Figure 7.4. The generalized coordinate $q \equiv \varphi(t)$ may

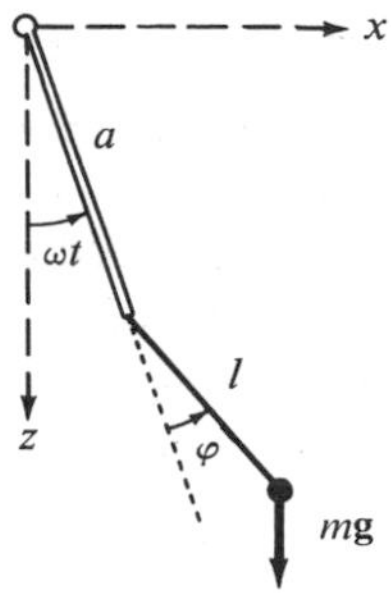

Figure 7.4

now be taken to be the angle made with the rod. Then, in motions confined to the plane of the circling rod,

$$z = a \cos \omega t + l \cos(\omega t + q),$$

$$x = a \sin \omega t + l \sin(\omega t + q).$$

The kinetic energy of the pendulum bob becomes

$$T = \tfrac{1}{2}ml^2(\omega + \dot{q})^2 + mal\omega(\omega + \dot{q}) \cos q + \tfrac{1}{2}m\omega^2 a^2,$$

and

$$V = -mgz = -mg[a \cos \omega t + l \cos(\omega t + q)].$$

The agency maintaining the uniform rotation of the rod in every phase of the motion contributes arbitrarily to the energy, as given by the explicit time-dependence, and provides an "external influence" that prevents conservation of the energy in the system.

Degrees of Freedom

The examples just reviewed call attention to the fact that fewer generalized coordinates than position-vector components are needed to specify the configuration of a system, when constraint relations exist. If there are c such constraints to help determine the $3N$ components of $\mathbf{r}_1, \mathbf{r}_2, \ldots, r_N$, then there need be only $f = 3N - c$ independent values $q_1, \ldots, q_f$ to assign at any phase of motion in order to complete the determination of the position-vectors—that is, only f independent relations

$$q_s = q_s(\mathbf{r}_1, \ldots, \mathbf{r}_N), \qquad s = 1, \ldots, f \tag{7.8}$$

need be implicit in (7.7). The integer f is called the number of the unconstrained "degrees of freedom" of the system. The generalized coordinates $q_1, \ldots, q_f$

are themselves sometimes referred to as the degrees of freedom. Formulation in terms of degrees of freedom may be regarded as an alternative to "analysis into particles" with point-positions $\mathbf{r}_1, \ldots, \mathbf{r}_N$.

Restricting the number of generalized coordinates used to $f = 3N - c$ is the general way in which the formulations hereafter will make use of constraint relations in place of constraint forces. It is of course possible to persist in using $f > 3N - c$ coordinates, but then the equivalent of constraint forces must be given attention, to enforce the constraint relationships remaining among the q's.

All the formulations in generalized coordinates remain valid also for the choices $q_1 \equiv x_1, q_2 \equiv y_1, q_3 \equiv z_1, q_4 \equiv x_2, \ldots$. However, the q's need not form such position-vectors. A problem may even be brought to a conclusion without any configuration in space having been explicitly determined.

Generalized Velocities

Regardless of the interpretation a given "degree of freedom," $q_s(t)$, may have, its rate of change with time, $\dot{q}_s$, will be called a "generalized velocity." Its relation to the ordinary velocities, $\mathbf{v}_i = \dot{\mathbf{r}}_i$, may be found from the transformation relations (7.7):

$$\mathbf{v}_i = \sum_s (\partial \mathbf{r}_i/\partial q_s)\dot{q}_s + \partial \mathbf{r}_i/\partial t. \tag{7.9}$$

The last term, $\partial \mathbf{r}_i/\partial t$, arising from a possible explicit time-dependence of the transformation, is whatever velocity a point $\mathbf{r}_i$ may have even when all the variables q_s are fixed. For example, the q's may be coordinates relative to a frame with a given motion, and then $\partial \mathbf{r}_i/\partial t$ is velocity imparted by the motion of the frame, existing even when the particle is at rest relative to that frame.

The "transformation coefficients" $(\partial \mathbf{r}_i/\partial q_s)$ may be thought of as "generalized direction cosines," since, for example, $\partial x/\partial r = \cos \varphi$ for the polar coordinates. On the other hand, $\partial x/\partial \varphi = -r \sin \varphi$, $\partial x_1/\partial \bar{x} = 1$, and so on, and hence the qualification "generalized." They are plainly functions of only the q's and t at most, and so the velocity, $\mathbf{v}_i(q_1, \ldots, q_f; \dot{q}_1, \ldots, \dot{q}_f; t)$ given by (7.9), is *linear* in the generalized velocities $\dot{q}_s$. It follows that

$$\partial \dot{\mathbf{r}}_i/\partial \dot{q}_\sigma = \partial \mathbf{r}_i/\partial q_\sigma, \tag{7.10}$$

and that this is independent of all the $\dot{q}$'s. (The relation is not mysterious. In the simple case in which $q \equiv x'$ forms a cartesian axis at angle α to the x-axis, $\partial x/\partial x' = \cos \alpha$ and $\partial \dot{x}/\partial \dot{x}' = \cos \alpha$.) It should be emphasized that the partial derivative, $\partial/\partial \dot{q}_\sigma$, used in (7.10) is one taken while holding all $q_1, \ldots, q_f$ constant.

Kinetic Energy

Because the velocity expression (7.9) is linear in the $\dot{q}$'s, the kinetic energy $T = \sum_i \frac{1}{2} m_i v_i^2$ will form an expression quadratic in the generalized velocities. Substitution of (7.9) for each factor of $v_i^2 \equiv \mathbf{v}_i \cdot \mathbf{v}_i$ will yield a kinetic energy expression decomposable into three parts:

$$T = \tau + \sum_s \pi_s \dot{q}_s + \kappa. \tag{7.11}$$

Here τ is to be a generalized "internal" kinetic energy in a "frame" in which $\partial \mathbf{r}_i/\partial t$ of (7.9) vanishes:

$$\tau \equiv \tfrac{1}{2} \sum_i m_i \sum_{s,\sigma} (\partial \mathbf{r}_i/\partial q_s) \cdot (\partial \mathbf{r}_i/\partial q_\sigma) \dot{q}_s \dot{q}_\sigma, \tag{7.12a}$$

in which the factors of $\mathbf{v}_i \cdot \mathbf{v}_i$ are distinguished by using different symbols for their "dummy indices" $s = 1, \ldots, f$ and $\sigma = 1, \ldots, f$.

$$\kappa \equiv \sum_i \tfrac{1}{2} m_i (\partial \mathbf{r}_i/\partial t)^2 = \kappa(q_1, \ldots, q_f; t) \tag{7.12b}$$

is a generalized kinetic energy imparted by "frame motion." Finally, π_s is a "momentum coefficient,"

$$\pi_s \equiv \sum_i m_i (\partial \mathbf{r}_i/\partial q_s) \cdot (\partial \mathbf{r}_i/\partial t) = \pi_s(q_1, \ldots, q_f; t), \tag{7.12c}$$

which helps to express a "dynamic coupling" between the "internal" and "frame" kinetic energies, τ and κ. Of course, $T \to \tau$ with $\pi_s = 0$ and $\kappa = 0$ when there is no explicit time-dependence in the connection (7.7), so that $\partial \mathbf{r}_i/\partial t \equiv 0$.

The important generalized "internal" kinetic energy τ may be expressed as

$$\tau = \tfrac{1}{2} \sum_{s,\sigma} \mu_{s\sigma} \dot{q}_s \dot{q}_\sigma, \tag{7.13}$$

by defining the generalized "mass coefficients":

$$\mu_{s\sigma}(q_1, \ldots, q_f; t) \equiv \sum_i m_i (\partial \mathbf{r}_i/\partial q_s) \cdot (\partial \mathbf{r}_i/\partial q_\sigma) = \mu_{\sigma s}. \tag{7.14}$$

For the transformation to centroidal and relative coordinates,

$$\mathbf{r}_1, \mathbf{r}_2 \to \mathbf{q}_1 \equiv \bar{\mathbf{r}}, \mathbf{q}_2 \equiv \mathbf{r}: \mu_{11} = m_1 + m_2 = M,$$
$$\mu_{22} = m_1(m_2/M)^2 + m_2(m_1/M)^2 = \mu$$

and $\mu_{12} = \mu_{21} = 0$; for the cartesian to polar coordinate transformation,

$$x, y \to q_1 \equiv r, q_2 \equiv \varphi: \mu_{11} = m, \mu_{22} = mr^2$$

and again, $\mu_{12} = \mu_{21} = 0$. Here μ_{11} is an ordinary mass but μ_{22} is a "moment of inertia" and presents an example of a generalized mass-coefficient depending on the coordinates. There are also examples in which the "off-diagonal" coefficients $\mu_{s \neq \sigma}$ do not vanish (resulting, for example, in "products of inertia" for rigid bodies or "dynamic couplings" among oscillations, to be reviewed in Chapters 10 and 11, respectively).

Generalized Momentum

Since the kinetic energy (7.13) is quadratic in the generalized velocities,

$$p_r \equiv \partial\tau/\partial\dot{q}_r = \sum_\sigma \mu_{r\sigma}\dot{q}_\sigma \tag{7.15}$$

is linear in them. This defines a "generalized momentum"; a still more generalized definition, of a "conjugate" momentum, will be introduced later (7.38). For the case $T = \frac{1}{2}M\bar{v}^2 + \frac{1}{2}\mu v^2$, the definition (7.15) yields p's which are components of $\bar{\mathbf{p}} = M\bar{\mathbf{v}}$ and $\mathbf{P} = \mu\mathbf{v}$, respectively [see (5.6) and (5.13)]. In the example of the polar coordinates for a single particle, $T = \frac{1}{2}m(\dot{r}^2 + r^2\dot{\varphi}^2)$ and $p_r = m\dot{r}$; $p_\varphi = mr^2\dot{\varphi}$ is an *angular* momentum (4.15).

The kinetic energy expression (7.13) can be reduced to a single summation by introducing the generalized momenta (7.15) into it:

$$\tau = \tfrac{1}{2}\sum_s p_s\dot{q}_s. \tag{7.16}$$

Expressions like $T = \frac{1}{2}(p_r\dot{r} + p_\varphi\dot{\varphi})$ for the kinetic energy in the polar variables have proved useful (see page 225).

Generalized Force

A "generalized force" may be introduced through the expression for virtual work, (7.2):

$$\delta W = \sum_i \mathbf{F}_i \cdot \boldsymbol{\delta}\mathbf{r}_i. \tag{7.17}$$

Here the $\boldsymbol{\delta}\mathbf{r}_i$ can be viewed as arising from imagined, virtual displacements δq of the q's, tried out *at some moment*. From the transformation (7.7),

$$\boldsymbol{\delta}\mathbf{r}_i = \sum_s (\partial\mathbf{r}_i/\partial q_s)\,\delta q_s, \tag{7.18}$$

much like the motional displacements per unit time (7.9) except that there is no term corresponding to $\partial \mathbf{r}_i/\partial t$, since $\boldsymbol{\delta}\mathbf{r}_i$ refers to displacement at a given moment. Now

$$\delta W = \sum_{i,s} \mathbf{F}_i \cdot (\partial \mathbf{r}_i/\partial q_s)\, \delta q_s = \sum_s K_s\, \delta q_s, \tag{7.19}$$

with

$$K_s = \sum_i \mathbf{F}_i \cdot \partial \mathbf{r}_i/\partial q_s \tag{7.20}$$

having the role of a generalized force responsible for the work per unit of the displacement δq_s. In the example of the cartesian to polar coordinate transformation, $K_r = F_x \cos\varphi + F_y \sin\varphi$ is just the component of the force $\mathbf{F}$ in the radial direction. However,

$$\begin{aligned} K_\varphi &= F_x\, \partial x/\partial\varphi + F_y\, \partial y/\partial\varphi \\ &= -F_x y + F_y x \equiv [\mathbf{r} \times \mathbf{F}]_z \end{aligned} \tag{7.21}$$

is a *torque* rather than a force, as might have been expected from its association with an angular degree of freedom.

When the forces are derivable from some potential $V(\mathbf{r}_1, \ldots, \mathbf{r}_N)$ (that is, $\mathbf{F}_i = -\nabla_i V$), then (7.7) should be used to substitute q's for the $\mathbf{r}$'s in V to produce a function $V(q_1, \ldots, q_f, t)$. Then

$$K_s = -\sum_i \nabla_i V \cdot \partial \mathbf{r}_i/\partial q_s \equiv -\partial V/\partial q_s, \tag{7.22}$$

as might have been expected. In the example of (7.21),

$$K_\varphi = -\partial V/\partial\varphi = -r\nabla_\varphi V = rF_\varphi, \tag{7.23}$$

the expected torque. An example of time-dependence introduced into a potential by a transformation to generalized coordinates was seen in connection with Figure 7.4.

7.3 THE LAGRANGIAN FORMULATION

It remains to be seen how the Newtonian equations of motion are transformed by the substitution of generalized coordinates. That is done most expeditiously by first taking the step to the d'Alembertian formulation (7.5) and calling on the virtual displacement interrelationships (7.18):

$$\sum_{i,s} (\mathbf{F}_i - m_i \ddot{\mathbf{r}}_i) \cdot (\partial \mathbf{r}_i/\partial q_s)\delta q_s = \sum_s \left[K_s - \sum_i m_i \ddot{\mathbf{r}}_i \cdot (\partial \mathbf{r}_i/\partial q_s)\right]\delta q_s = 0. \tag{7.24}$$

In this expression,

$$m_i \ddot{\mathbf{r}}_i \cdot (\partial \mathbf{r}_i/\partial q_s) \equiv (d/dt)[m_i \dot{\mathbf{r}}_i \cdot (\partial \mathbf{r}_i/\partial q_s)] - m_i \dot{\mathbf{r}}_i \cdot (\partial \dot{\mathbf{r}}_i/\partial q_s), \tag{7.25}$$

and the last term is equivalent to

$$-(\partial/\partial q_s)[\tfrac{1}{2}m_i \dot{\mathbf{r}}_i \cdot \dot{\mathbf{r}}_i] = -(\partial/\partial q_s)[\tfrac{1}{2}m_i v_i^2].$$

The term preceding this takes a similar form when $\partial \mathbf{r}_i/\partial q_s$ is replaced by $\partial \dot{\mathbf{r}}_i/\partial \dot{q}_s$, allowed by (7.10). Putting these forms into (7.25) and summing over the particles (i) yields:

$$\sum_i \left[\frac{d}{dt}\frac{\partial}{\partial \dot{q}_s}(\tfrac{1}{2}m_i v_i^2) - \frac{\partial}{\partial q_s}(\tfrac{1}{2}m_i v_i^2)\right] \equiv \frac{d}{dt}\left(\frac{\partial T}{\partial \dot{q}_s}\right) - \frac{\partial T}{\partial q_s}, \tag{7.26}$$

where $T = \sum_i \frac{1}{2}m_i v_i^2$ is the total kinetic energy. With this, d'Alembert's Principle of Virtual Work (7.24) takes the form

$$\sum_s \left[K_s - \frac{d}{dt}\left(\frac{\partial T}{\partial \dot{q}_s}\right) + \frac{\partial T}{\partial q_s}\right]\delta q_s = 0, \tag{7.27}$$

having only q's, and no $\mathbf{r}$'s, explicit in it. The virtual displacements $\delta q_1, \ldots, \delta q_f$ are independently arbitrary if there are no constraint conditions among the q's ($f = 3N - c$ of the preceding section). Even if constraints among them do exist ($f > 3N - c$), arbitrary virtual displacements of all the q's are still allowed if only the generalized forces K_s include the corresponding constraint forces (leaving to these the enforcement of the constraints, as discussed in §7.1). With all the δq_s arbitrary in (7.27), it can vanish only if, for each degree of freedom $s = 1, 2, \ldots, f$,

$$\frac{d}{dt}\left(\frac{\partial T}{\partial \dot{q}_s}\right) - \frac{\partial T}{\partial q_s} = K_s. \tag{7.28}$$

These f equations are the transforms of Newton's equations, as expressed in generalized coordinates.

With the generalized momentum $p_s = \partial T/\partial \dot{q}_s$, as defined in (7.15), the equations of motion may be written

$$\dot{p}_s = K_s + \partial T/\partial q_s, \tag{7.29}$$

where the last term may be thought of as a "generalized inertial force" added to K_s. The name is appropriate in view of the polar-coordinate example $T = \frac{1}{2}m(\dot{r}^2 + r^2\dot{\varphi}^2)$, when

$$\partial T/\partial r = mr\dot{\varphi}^2 = mv_\varphi^2/r \tag{7.30}$$

is just the "inertial" centrifugal force. With $K_r \equiv F_r$ (see the discussion leading to (7.21)), the radial component of (7.29) becomes

$$\dot{p}_r \equiv m\ddot{r} = F_r + mr\dot{\varphi}^2, \tag{7.31}$$

which coincides with the first of the Newtonian equations (4.13). Similar considerations apply to the second.

One property shared by all formulations in generalized coordinates, like the equations of motion (7.28), is a "form-invariance." Whereas the form (7.28) can be used with any choice of coordinates, the original Newtonian equations have forms like $m\ddot{x} = F_x$, $m\ddot{y} = F_y$, ..., in cartesian variables, but, for example, $m\ddot{r} \neq F_r$, as (7.31) shows.

The Lagrangian

In the cases of forces derivable from a potential $V(q_1, \ldots, q_f; t)$, the equations of motion (7.28) can be reduced to conditions on a *single* function which, for those cases, is defined by

$$\begin{aligned} L(q, \dot{q}, t) &\equiv L(q_1, \ldots, q_f; \dot{q}_1, \ldots, \dot{q}_f; t), \\ &= T(q, \dot{q}, t) - V(q, t). \end{aligned} \tag{7.32}$$

Introduced here is a notation in which q and $\dot{q}$ without subscripts stand for the whole sets $q \equiv q_1, \ldots, q_f$ and $\dot{q} \equiv \dot{q}_1, \ldots, \dot{q}_f$. It is easy to see that, when $K_s = -\partial V/\partial q_s$, (7.28) reduces to

$$\boxed{\frac{d}{dt}\left(\frac{\partial L}{\partial \dot{q}_s}\right) - \frac{\partial L}{\partial q_s} = 0 \qquad \text{for} \qquad s = 1, \ldots . f}, \tag{7.33}$$

since $\partial V/\partial \dot{q}_s = 0$.

The Lagrange equations of motion (7.33) may also be used in a wider variety of cases than those for which a scalar potential exists and $L = T - V$. All cases in which the generalized forces are derivable from some "work-function"

$$W(q, \dot{q}, t) = L - T, \tag{7.34}$$

by way of

$$K_s = \left[\frac{\partial}{\partial q_s} - \frac{d}{dt}\left(\frac{\partial}{\partial \dot{q}_s}\right)\right] W(q, \dot{q}, t), \tag{7.35}$$

are likewise encompassed, since for them also the Lagrange equations (7.33) are equivalent to the transforms (7.28) of the Newtonian equations. The cases of forces derivable from a scalar potential are special ones in which $W(q, t) = -V$. The broader class of generalized forces (7.35) includes velocity- and time-dependent ones. As will eventually be seen, the force on a charge moving in a given electromagnetic field, for example, may not be derivable from a scalar potential but yet be derivable from a velocity-dependent work-function (see Exercise 7.12).

It was seen before that the introduction of a potential energy V provides one way to describe an energy-conserving system. The generalization to a "lagrangian," $L(q, \dot{q}, t)$, which is not restricted to $T - V$, will offer more ways to represent energy conservation, as will be seen below. Indeed, it is a simple matter of fact that, for every system which has been successfully treated as isolated, a lagrangian has been found, permitting the Lagrange equations to yield the correct descriptions. Because it seems possible to couch at least all the most complete descriptions in terms of lagrangians, the equations (7.33) have come to be regarded as providing at least as complete a theory of Classical Mechanics as does the Newtonian formulation.

In the Lagrange formulation, it is assumed that it is only necessary to make a suitable choice of coordinates $q \equiv q_1, \ldots, q_f$ and of a *single* lagrangian function, $L(q, \dot{q}, t)$, to describe the circumstances of any given motion. Thus, the postulations of forces, or the several force functions $\mathbf{F}_i(\mathbf{r}, \mathbf{v}, t)$, characteristic of the Newtonian formulation, are now replaced by the postulation of one lagrangian function.

It becomes important to see whether every known description in terms of forces which has been considered satisfactory can be given an equally successful Lagrange description. There are cases, of so-called "dissipative systems," in which a second function besides L is sometimes introduced. The simplest example is provided by the frictional force of §3.2. No time-independent lagrangian expression $L(x, \dot{x})$ can be constructed which will make the Lagrange equation equivalent to the Newtonian one, $m\ddot{x} = -m\gamma x$, despite the fact that the force here is not explicitly time-dependent. A "nonholonomic" variable q, defined by

$$\dot{q} \equiv \dot{x} + \gamma x, \tag{7.36}$$

could be introduced into $L = \frac{1}{2}m\dot{q}^2$; then $\partial L/\partial \dot{q} = m\dot{q}$ and the Lagrange equation correctly yields $m\ddot{q} \equiv m(\ddot{x} + \gamma\dot{x}) = 0$. It is also always possible to use an explicitly time-dependent lagrangian,* such as $L = e^{\gamma t}(m\dot{x}^2/2)$, which leads to $p = e^{\gamma t}(m\dot{x})$, and $\dot{p} = 0$ as the equation of motion. Such devices are sometimes regarded as unsatisfactory, and hence the introduction of the second function alluded to above: Rayleigh's "dissipation function." However, the latter treatment has no very appreciable advantages over a simple reversion to the equations (7.28), and so is given no further attention here.

---

* Suggested to the author by Professor M. Rudermann.

A significant point is that the unmodified Lagrange formulation has always proved adequate for the fundamental, energy-conserving, isolated systems. A single lagrangian function can be held adequate for any classical motion described with a sufficient degree of completeness (as when sources of friction are described in detail).

It must yet be conceded that formulation in terms of forces rather than lagrangians retains an important advantage. Forces are usually more clearly abstractable from experience, and this helps greatly in constructing a description of the circumstances of almost any motion. This is so much the case that most lagrangians have been found only after the force descriptions had already been formulated; the preliminary finding of a potential V from which the forces are derivable, then putting $L = T - V$, is quite typical.

The main advantage of Langrangian formulation stems from the possibility it offers for coming to conclusions entirely in terms of highly generalized coordinates, without the necessity of relating these to some kind of point-position vectors, as in (7.7). This means that attention may be confined to any sort of quantities whose variations are available for observation, as the system under study exhibits its behavior. The emphasis may be on directly observable "degrees of freedom" rather than, for example, on the positions of unobservable particles. A frequently quoted illustration concerns the description of motion inside a "black box" when all that can be observed are the motions of various levers or protuberances, or other manifestations of something going on inside the box. Only enough generalized coordinates are introduced to describe any exterior moving parts, or to measure other manifestations, without presumptions concerning the interior of the box. When a study of the relations between only the observable variables permits predictions about their behavior, the description found can be regarded as the most complete to be warranted by what is observable.*

Examples

The immediately most important examples of Lagrangian formulation are those in which a potential exists and $L = T - V$.

If the potential is expressed as a function of the cartesian coordinates of mass-particles, then the kinetic energy is expressed as $T = \sum \frac{1}{2} m_i(\dot{x}_i^2 + \dot{y}_i^2 + \dot{z}_i^2)$

* It is along such lines that an answer to a persistent question may lie: when an analysis has arrived at a description in terms of some species of "elementary particles," how can assurance be given that these are not further analyzable into "still more elementary" particles? The "elementary particles" should perhaps be treated as "black boxes" with "interiors" of no interest as long as they are not accessible to any kind of observation, and hence without influence on what is being observed.

and $\partial L/\partial \dot{x}_i = m_i \dot{x}_i$, and so on. Each Lagrange equation of motion then takes a form like $m_i \ddot{x}_i = -\partial V/\partial x_i \equiv F_{ix}$ and is no more than a cartesian component of a Newtonian equation for a particle.

The final outcome is similarly trivial for coordinates which are only point-transformations of the cartesian coordinates of particles, such as the the spherical coordinates r, ϑ, φ. Yet already here the Lagrangian formulation is of significant help. It is much easier to arrive at a kinetic energy expression like $T = \frac{1}{2}m(\dot{r}^2 + r^2\dot{\vartheta}^2 + r^2 \sin^2\vartheta\, \dot{\varphi}^2)$ than at the components $a_{r,\vartheta,\varphi}$ (1.39) of the corresponding acceleration. With $L = T - V(r, \vartheta, \varphi)$, $\partial L/\partial \dot{\vartheta} = mr^2\dot{\vartheta}$, for example, and the corresponding Lagrange equation gives

$$mr(r\ddot{\vartheta} + 2\dot{r}\dot{\vartheta}) = -\partial V/\partial \vartheta + mr(r\dot{\varphi}^2 \sin \vartheta \cos \vartheta).$$

This plainly has a_ϑ implicit in it and is one of the simplest ways of deriving it. Similar advantage will be found in the description of rigid body motion in terms of "Euler angles," to be reviewed in Chapter 10.

A simple illustration of how constraints are taken into account is provided by the compound Atwood machine problem of Figure 7.1. Because of the given constraints, only two of the four string lengths, z_1 *or* z_2 and z_3 *or* z_4, need be adopted as generalized coordinates. In terms of z_1 and z_3:

$$L = \tfrac{1}{2}(m_1 + m_2)\dot{z}_1^2 + \tfrac{1}{2}m_3 \dot{z}_3^2 + \tfrac{1}{2}m_4(2\dot{z}_1 + \dot{z}_3)^2$$
$$+(m_1 - m_2)gz_1 + m_3 g z_3 - m_4 g(2z_1 + z_3) + \text{const.}$$

From this, the two Lagrange equations lead to exactly the same results as did the d'Alembertian formulation (page 154).

The problem of Figure 7.2 has $L = T = \frac{1}{2}m(l\dot{l}/a)^2$, as reference to the discussion following (7.7) shows. The Lagrange equations will then be found to give $l\ddot{l} = -\dot{l}^2$, and one integration of this shows that $l\dot{l} = \text{constant} = -av = -av_0$, the expected constancy of the velocity. A final description $l^2(t) = l_0^2 - 2av_0 t$ and $\alpha = (l_0 - l)/a$ is also elementary to obtain.

Generalized Momentum Conservation

The Lagrange equations of motion (7.33) make it obvious that whenever it happens that some one, say q_i, of the adopted coordinates is not needed for expressing the lagrangian, but only $\dot{q}_i$ is (that is,

$$L(q, \dot{q}, t) \equiv L(q_1, \ldots, q_{i-1}, q_{i+1}, \ldots q_f; \dot{q}_1, \ldots, \dot{q}_f; t)$$

so that $\partial L/\partial q_i = 0$), then

$$(d/dt)(\partial L/\partial \dot{q}_i) = 0 \qquad (7.37)$$

and the quantity $\partial L/\partial \dot{q}_i$ is conserved during the motion. The coordinate q_i is then said to be "ignorable."

The quantity defined by

$$\boxed{p_s \equiv \partial L/\partial \dot{q}_s}, \tag{7.38}$$

whether conserved or not, is called the momentum "conjugate" to the degree of freedom q_s. The motive for calling it a "momentum" stems from its identifiability with the generalized momentum $\partial \tau/\partial \dot{q}_s$ of (7.15) in the important special cases when $L = \tau - V(q)$. In the cartesian representation mentioned in the preceding subsection, $p_x \equiv \partial L/\partial \dot{x} = m\dot{x}$ is the ordinary linear momentum, but in the spherical coordinate example $p_\vartheta \equiv \partial L/\partial \dot{\vartheta} = mr^2\dot{\vartheta}$. It is characteristic for the momentum conjugate to an angular degree of freedom to be an angular momentum, in consistency with dimensional requirements.

It becomes appropriate to call the consequence (7.37) of the ignorability of a coordinate a generalized principle of momentum conservation. It is generalized in the sense that it may be expressing either linear or angular momentum conservation, and also the conservation of more arcane conjugate momenta, corresponding to highly generalized degrees of freedom q_i.

Each conclusion that a p_i is a constant in the motion may be itself constitute a more special conservation principle than such as the total linear momentum conservation, $\sum_i \mathbf{p}_i =$ constant, or the total angular momentum conservation $\sum_i \mathbf{L}_i =$ constant. For example, the total momentum of a particle moving under constant gravity, as indicated in Figure 1.11, is not conserved; however, its horizontal cartesian component, $p_x = \partial L/\partial \dot{x} = m\dot{x}$, is conserved. The latter fact is implicit in the absence of x from the lagrangian $L = \frac{1}{2}m(\dot{x}^2 + \dot{z}^2) - mgz$. Thus a special ignorability of a coordinate in a particular system leads to a special conservation property in it.

Plainly the ignorability of one or more of the coordinates in a set $q_1, \ldots, q_f$, and the corresponding generation of a conservation law, depend on the choice of coordinates in the first place. For example, the angular degree of freedom φ is absent from the lagrangian for a central force problem, $L(r, \dot{r}, \dot{\varphi}) = \frac{1}{2}m(\dot{r}^2 + r^2\dot{\varphi}^2) - V(r)$; hence the conservation of the angular momentum $\partial L/\partial \dot{\varphi} = mr^2\dot{\varphi}$ is obvious. However, if the same problem were cast in cartesian coordinates, with $L(x, y, \dot{x}, \dot{y}) = \frac{1}{2}m(\dot{x}^2 + \dot{y}^2) - V[(x^2 + y^2)^{1/2}]$, then neither $p_x = \partial L/\partial \dot{x}$ nor $p_y = \partial L/\partial \dot{y}$ is conserved. Finding the angular momentum conservation principle in these terms would require somehow noticing that the combination $(x\, \partial L/\partial \dot{y} - y\, \partial L/\partial \dot{x})$ is constant during the motion. The moral to be drawn is that a change to appropriate coordinates, ones in which as many as possible become ignorable, should be made in order that conservation properties become *manifest*.

The ignorability of a coordinate expresses an inherent symmetry in the description of a system. Thus the absence of the azimuthal angle φ from $L(r, \dot{r}, \dot{\varphi})$ above follows from the spherical symmetry of the central force problem. It is clear that when arbitrary changes δq_i of a coordinate q_i make no difference to the description by the lagrangian, q_i need not appear in it and the description possesses a symmetry expressable as an "invariance to δq_i." The consequent conservation property testifies to an equivalence between conservation laws and invariances or symmetries, to be explored further in §7.5.

Generalized Energy Conservation

A formal introduction of the energy concepts into the Newtonian description was made when the equations of motion were multiplied by velocity vectors in order to form rates of work by forces (§2.4, §6.3, and others). That development may be paralleled here by first putting each of the Lagrange equations of motion (7.33) into the form $\dot{p}_s - \partial L/\partial q_s = 0$, which follows from the definition (7.38) of the conjugate momentum, then multiplying it with the corresponding generalized velocity $\dot{q}_s$ and summing the results:

$$\sum_s \dot{q}_s\left(\dot{p}_s - \frac{\partial L}{\partial q_s}\right) \equiv \frac{d}{dt}\left(\sum_s p_s \dot{q}_s\right) - \sum_s\left(\ddot{q}_s p_s + \dot{q}_s \frac{\partial L}{\partial q_s}\right) = 0.$$

The last summation can be identified as

$$\sum_s\left(\frac{\partial L}{\partial q_s}\dot{q}_s + \frac{\partial L}{\partial \dot{q}_s}\ddot{q}_s\right) = \frac{dL}{dt} - \frac{\partial L}{\partial t},$$

since $L(q_1, \ldots, q_f; \dot{q}_1, \ldots, \dot{q}_f; t)$ varies with time through its dependence on the variable $q(t)$'s and $\dot{q}(t)$'s, as well as through any explicit dependence on t. As a consequence,

$$\frac{d}{dt}\left(\sum_s p_s \dot{q}_s - L\right) = -\frac{\partial L}{\partial t}. \tag{7.39}$$

Formed here is a quantity,

$$\boxed{H = \sum_s p_s \dot{q}_s - L}, \tag{7.40}$$

which has the property of being conserved, $dH/dt = 0$, whenever the lagrangian description is not explicitly time-dependent, $\partial L/\partial t = 0$. In the case $L(q, \dot{q}) = T - V(q)$, it follows from (7.16) that

$$H = 2T - L = T + V = E, \tag{7.41}$$

the conserved energy as described with the help of a potential. It is therefore appropriate to call H a "generalized energy"; a more usual name is "hamiltonian energy."

The hamiltonian energy may be conserved even when no potential $V(q)$ exists. It is only necessary that $\partial L/\partial t = 0$, an "ignorability" of the time variable. Such representations encompass all the known ways of formulating energy conservation in isolated systems.

The generalized conservation theorems, for p_s of (7.38) and H of (7.40), have in common the fact that the conserved quantity in each case is defined by an invariance of the description. In the case of the generalized energy conservation it is invariance to time shifts (hence to the moment taken as the zero of time). For this reason, H is frequently considered to be "conjugate" to t in something like the same sense as p_s is conjugate to q_s.

7.4 THE VARIATIONAL FORMULATION

The most concise statement of the Laws of Mechanics has the form of a "Variational Principle." This developed out of an old idea, ascribed to Maupertuis, that, out of all distinguishable alternatives, nature will follow a course that is most "economical" as judged by some criterion. Any motion which actually occurs, as compared to alternatives which could be described, presumably requires a least "expenditure," if this be suitably defined. It has indeed proved possible to develop a method of comparing alternatives for a motion and to define "expenditure" in such a way that the Laws of Mechanics can be stated in the desired form.

The Configuration Space

The concept of a "configuration space" has been developed as a convenient means of making well-defined comparisons among various alternatives for a motion.

At any moment t during a motion, a mechanical system under consideration will have definite values for a set of coordinates: $q_1(t), \ldots, q_f(t)$. This set can be taken to be rectangular coordinates for a single point in a "hyperspace" of f dimensions. A set of f mutually perpendicular q_s-axes are to be visualized, and this has less help from normal intuitions than has the construction of a three-dimensional space, but there is no difficulty about manipulating in it mathematically. (Making the effort pleases the ego with the notion that it is evolving into a higher-order animal, through success in dealing with any number of dimensions.)

In the course of the motion, each of the variables $q_1(t), \ldots, q_f(t)$ will vary continuously and their representative point in the configuration space will trace out a continuous path. Clearly, any choice of continuous path will lead to a well-defined set of generalized velocities, $\dot{q}_1(t), \ldots, \dot{q}_f(t)$, at each point of it.

Next, a quantity I is sought which may be appropriately referred to as some kind of "expenditure" during a given motion. It may be conceived as spent at some rate λ during each time-interval dt of motion. The rate λ may be expected to depend on the path chosen and the motion over it, and so should be treated as a function of the momentary phases $q, \dot{q}$. As before, the symbols $q, \dot{q}$ stand for the whole set $q_1, \ldots, q_f; \dot{q}_1, \ldots, \dot{q}_f$. Moreover the expenditure rate may depend explicitly on the time at which it takes place: $\lambda(q, \dot{q}, t)$. Thus, a quantity of the form

$$I = \int_{t_1}^{t_2} \lambda(q, \dot{q}, t)\, dt \tag{7.42}$$

is sought, having different values according to the path $q(t)$ taken between times t_1 and t_2, and having the least value on the actual one of all the paths which might be arbitrarily chosen for evaluating it.

At the times t_1 and t_2, between which the expenditure is to take place, the actual motion will have specific configurations, $q(t_1)$ and $q(t_2)$. Between such specified end points a *unique* result for the actual path of motion is expected, and only after such specifications, or their equivalent, are laid down. For this reason the expenditures which should be compared are ones evaluated on arbitrarily various paths which join the *same* end points $q(t_1)$ and $q(t_2)$.

A question that thus asks "What is the actual motion starting at $q(t_1)$ which arrives at $q(t_2)$?" differs from what was called the "prototype" problem of Mechanics in §2.3. In the latter, $q(t_1)$ and $\dot{q}(t_1)$ are given, rather than $q(t_1)$ and $q(t_2)$; the specifications are equivalent in that, for either, a unique motion is expected to ensue under given forces or a given lagrangian. (Specifying $q(t_1)$ and $q(t_2)$ happens to be better consistent with the Quantum Mechanics, according to which not both $q(t_1)$ and $\dot{q}(t_1)$ *can* be given, and the only questions that may be asked are ones such as "Starting from $q(t_1)$, will the system arrive at some preselected $q(t_2)$?")

The Variational Principle

Paths in a configuration space may be schematized as in Figure 7.5. The point marked $q(t)$ stands for the entire configuration assumed by the system at time t on the actual path, whatever this may be, and $q'(t)$ marks a configuration assigned to the system for the same time on some arbitrarily

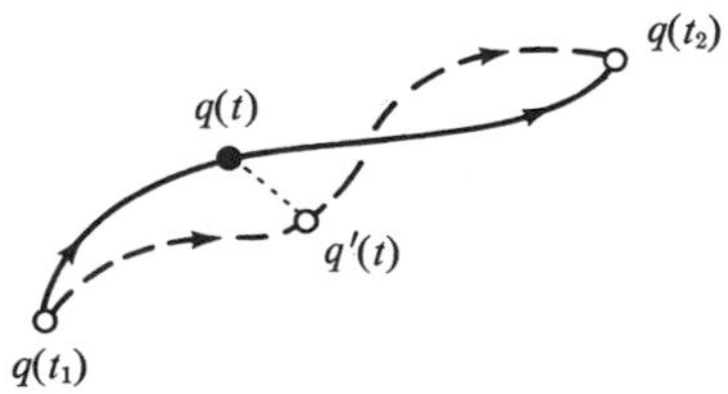

Figure 7.5

chosen alternative path. Only choices with $q'(t_1) = q(t_1)$ and $q'(t_2) = q(t_2)$ are to be considered.

The expenditure I is to be so defined that its value on the actual path is less than it is on any alternative choice of one. Finding the minimum of a function like I, which has a whole path as the independent variable, is done in a manner analogous to finding any "extremum" of an ordinary function. The choice of procedure about to be introduced may not seem unique but that is scarcely material, since mathematical results must in any case be judged relative to the procedures (the logic and the algebra) adopted.

The "extremum" or "stationary" values of an ordinary function are found from comparing its values for "infinitesimally" different choices of the independent variable. In treating the "function of path," I, its value on the path of actual motion, $q(t)$, will be compared against its values on infinitesimally different paths $q'(t) \equiv q(t) + \delta q(t)$, where $\delta q(t)$ is to be treated as vanishingly small but otherwise arbitrary. The generalized velocities will also have small differences, to be denoted $\delta\dot{q}(t) \equiv \dot{q}'(t) - \dot{q}(t)$. It should be remembered that q and $\dot{q}$ each stand for whole sets $q_1, \ldots, q_f$ and $\dot{q}_1, \ldots, \dot{q}_f$; then $\delta q_s(t)$ and $\delta\dot{q}_s(t)$ are called "variations" of $q_s(t)$ and $\dot{q}_s(t)$, respectively, from one path to another. For the paths considered here, $\delta q_s(t_1) = 0$ and $\delta q_s(t_2) = 0$ for every $s = 1, \ldots, f$. Further, just as $\dot{q}_s(t) = dq_s(t)/dt$, so

$$\dot{q}'_s(t) \equiv \dot{q}_s(t) + \delta\dot{q}_s(t) = (d/dt)[q_s(t) + \delta q_s(t)],$$

and therefore

$$\delta\dot{q}_s(t) = (d/dt)\delta q_s(t). \tag{7.43}$$

Of course, generally neither $\delta\dot{q}_s(t_1)$ nor $\delta\dot{q}_s(t_2)$ will be zero.

The difference of expenditure on the paths $q(t)$ and $q'(t) = q(t) + \delta q(t)$ may be evaluated as

$$\begin{aligned} I' - I \equiv \delta I &= \int_{t_1}^{t_2} \lambda(q + \delta q, \dot{q} + \delta\dot{q}, t)\, dt - \int_{t_1}^{t_2} \lambda(q, \dot{q}, t)\, dt \\ &= \int_{t_1}^{t_2} dt \sum_s \left[\frac{\partial\lambda}{\partial q_s}\delta q_s + \frac{\partial\lambda}{\partial\dot{q}_s}\delta\dot{q}_s\right], \end{aligned} \tag{7.44}$$

in the limit $\delta q_s \to 0$ for every s. The square bracket is the obvious differential between $\lambda(q + \delta q, \dot{q} + \delta\dot{q}, t)$ and $\lambda(\dot{q}, \dot{q}, t)$ generated by the variations δq_s and $\delta\dot{q}_s$. There is a difference arising for each $s = 1, \ldots, f$, and hence the summation over s.

Next, the variations $\delta\dot{q}_s$ should be expressed in terms of the δq_s, since it is the latter which are arbitrarily chosen to yield the varied paths. That is done by using (7.43) to write

$$\frac{\partial\lambda}{\partial\dot{q}_s}\,\delta\dot{q}_s = \frac{d}{dt}\left[\frac{\partial\lambda}{\partial\dot{q}_s}\,\delta q_s\right] - \left[\frac{d}{dt}\left(\frac{\partial\lambda}{\partial\dot{q}_s}\right)\right]\delta q_s. \tag{7.45}$$

The time integral of this expression is needed for (7.44). The first of the terms on the right side is easy to integrate and has the result

$$[(\partial\lambda/\partial\dot{q}_s)\,\delta q_s]_{t_1}^{t_2} = 0,$$

because $\delta q_s(t_1) = 0$ and $\delta q_s(t_2) = 0$ for all the paths. For the variation of the expenditure, there is now left

$$\delta I \equiv \delta\int_{t_1}^{t_2}\lambda(q, \dot{q}, t)\,dt = \int_{t_1}^{t_2}dt\sum_s \delta q_s\left[\frac{\partial\lambda}{\partial q_s} - \frac{d}{dt}\left(\frac{\partial\lambda}{\partial\dot{q}_s}\right)\right]. \tag{7.46}$$

The procedure for finding the extremum of I is defined by requiring $\delta I = 0$ for arbitrary variations δq. Whether the outcome for I is a minimum, a maximum, or other "stationary (inflection) point" makes no difference to the advantages to be gained from the formulation.

The requirement $\delta I = 0$ for arbitrary, independent variations of each δq_s in (7.46), and for arbitrary choices of the times, can be achieved only if

$$(d/dt)(\partial\lambda/\partial\dot{q}_s) - \partial\lambda/\partial q_s = 0 \tag{7.47}$$

for each $s = 1, \ldots, f$ and for each moment t. In the general theory of finding extrema of path functions these are known as "Euler conditions." It is the equivalence of these conditions to the succinct statement $\delta I = 0$ that gives the formalism its advantages.

Now recall the Lagrange equations (7.33); those have been found to govern actual motions, and with about as great a generality as could be desired. It has become evident that they are "Euler conditions" equivalent to the "Variational Principle"

$$\boxed{\delta\int_{t_1}^{t_2}L(q, \dot{q}, t)\,dt = 0}\,, \tag{7.48}$$

if the δ-process is defined as having the choices for $\delta q_s(t)$ restricted to forming continuous paths with $\delta q_s(t_1) = \delta q_s(t_2) = 0$, whatever the chosen end times

t_1, t_2 may be. The formulation (7.48) is sometimes called "Hamilton's Principle," to distinguish it from others, usually less satisfactorily complete ones, which have been advanced.

Starting from the Variational Principle as a basis for applications again requires preliminary choices of coordinates and construction of a lagrangian, to be descriptive of the circumstances in which the motion is to take place. Performing the prescribed variations will then develop the various steps leading to the Euler–Lagrange equations. The d'Alembert or Newtonian formulations emerge directly if the coordinates chosen are position-vectors $\mathbf{r}_i$, and a lagrangian of the form $L = \sum_i \frac{1}{2} m_i \dot{r}_i^2 + W(\mathbf{r}, \dot{\mathbf{r}}, t)$ is constructed. In such circumstances, $\delta I = 0$ of (7.46) yields

$$\sum_i \boldsymbol{\delta}\mathbf{r}_i \cdot [\mathbf{F}_i - m_i \ddot{\mathbf{r}}_i] = 0, \tag{7.49a}$$

with

$$\mathbf{F}_i = \partial W/\partial \mathbf{r}_i - (d/dt)\partial W/\partial \dot{\mathbf{r}}_i. \tag{7.49b}$$

It has been found more convenient here to use the notation $\partial/\partial\mathbf{r}_i$ for the gradient vector $\mathbf{\nabla}_i$, so that components like $\partial/\partial\dot{x}_i$, $\partial/\partial\dot{y}_i$, $\partial/\partial\dot{z}_i$ are implied for the similar vector operator $\partial/\partial\dot{\mathbf{r}}_i$. This encompasses all d'Alembertian and Newtonian descriptions in which the forces are derivable from work functions, as in (7.49*b*). These can be held to include all circumstances that are adequately representable by the behavior of discrete point-positions. Meanwhile, the result (7.49*a*) indicates that, by sticking to the correct path, the motion avoids work on displacements straying from it. The minimization of "expenditure" may be regarded as a generalization of a "principle of least work," with work by "kinetic reactions," $-m_i\ddot{\mathbf{r}}_i$, to be included.

The fact that application of the Variational Principle leads merely to application of Lagrangian, Newtonian, or other equations indicates that it offers no *new* advantages in such regards. On the other hand, it also indicates that all the other formulations are special realizations of the variational one; hence the variational formulation offers a highly compact representation of the whole Classical Mechanical theory. This formulation is less committed than any other to special assumptions of the kind involved in adopting a specific set of variables and a specific force function, or lagrangian, and so on. This will be appreciated when "invariance" properties of the Principle are considered next.

Gauge Transformations of the Lagrangian

Even for describing a unique motion $q(t)$, the lagrangian may be chosen in a variety of ways. This possibility of changing the lagrangian may be regarded a generalization of the freedom with which the "gauge" of the potential in $L = T - V$ may be chosen; it is no longer a matter merely of an arbitrary additive constant, because the forces may be more general (7.35) than those derivable from just a position-dependent potential.

Consider the expenditure integrals for two different choices of lagrangian,

$$I = \int_{t_1}^{t_2} L(q, \dot{q}, t)\, dt \quad \text{and} \quad I' = \int_{t_1}^{t_2} L'(q, \dot{q}, t)\, dt, \tag{7.50}$$

both to be evaluated on exactly the same path. The extrema of both, as functions of path, will yield exactly the same results if only the variations are equal, $\delta I = \delta I'$, in variations of the path on which both are evaluated, these variations being restricted to ones with $\delta q(t_1) = 0$ and $\delta q(t_2) = 0$. Then L and L' may differ by

$$L - L' = df(q, t)/dt, \tag{7.51}$$

with arbitrary choices of a differentiable function $f(q, t)$, since for any choice

$$\delta \int_{t_1}^{t_2} dt[df(q, t)/dt] = \delta[f(q(t_2), t_2) - f(q(t_1), t_1)] = 0,$$

vanishing because of $\delta q(t_{1,2}) = 0$.

The result (7.51) may be checked directly, by substituting it into the Lagrange equations (7.33); it will be found that, independently of what $f(q, t)$ may be, L' of (7.51) satisfies the equations if L does. As an example, it will be found that the lagrangian* $L' = \frac{1}{2}m(\dot{x} + i\omega_0 x)^2$ leads to exactly the same simple oscillator equation of motion (3.6) as does the more conventional choice $L = \frac{1}{2}m(\dot{x}^2 - \omega_0^2 x^2)$; here, $f = -\frac{1}{2}im\omega_0 x^2$ has been chosen. The property that a lagrangian may be changed according to (7.51), yet describe the same motion, may be referred to as a "gauge invariance."

A change of lagrangian carries with it a change in the gauging of the momentum conjugate to each degree of freedom, q_s. Since

$$\begin{aligned} df(q, t)/dt &= \sum_s (\partial f/\partial q_s)\dot{q}_s + \partial f/\partial t, \\ p_s' &\equiv \partial L'/\partial \dot{q}_s = p_s - \partial f/\partial q_s, \end{aligned} \tag{7.52}$$

where $p_s \equiv \partial L/\partial \dot{q}_s$. In the simple harmonic oscillator example quoted above, $p' = m(\dot{x} + i\omega_0 x)$ subject to the "force" $\dot{p}' = \partial L'/\partial x = i\omega_0 p'$; in terms of the general solution (3.12), $p' = m\omega_0 a \exp\colon i\omega_0 t$ with a generally complex amplitude, a.

---

* Compare (7.36), with $\gamma = i\omega_0$.

There is a corresponding change in the gauging of the generalized energy

$$H' = \sum_s p'_s \dot{q}_s - L' = H + \partial f/\partial t. \tag{7.53}$$

For the example above, $L' = (p')^2/2m$ and $H' = |p'|^2/2m = \frac{1}{2}m(\dot{x}^2 + \omega_0^2 x^2) = H = T + V$, because $f = -\frac{1}{2}im\omega_0 x^2$ does not depend explicitly on time.

Configuration Spacc Transformations

The transformation of the formalism may be made more complete than just the regauging (7.51) of the way the lagrangian depends on the variables $q(t)$. A concurrent coordinate substitution, from the set q to a new one $Q \equiv Q_1, \ldots, Q_f$, may also be undertaken. After a specific correspondence

$$q_s = q_s(Q_1, \ldots, Q_f; t) \tag{7.54}$$

is adopted, the set $Q(t)$ may be used to describe exactly the same motion as does the correspondent $q(t)$.

A lagrangian $\mathscr{L}(Q, \dot{Q}, t)$, which generates $Q(t)$ via

$$(d/dt)(\partial\mathscr{L}/\partial\dot{Q}_s) - \partial\mathscr{L}/\partial Q_s = 0, \tag{7.55}$$

need not even be proportional to $L(q, \dot{q}, t)$ at the corresponding configuration, as the possibility of the gauge transformation (7.51) makes clear. Not even the expenditures on the corresponding paths $q(t)$ and $Q(t)$,

$$I = \int_{t_1}^{t_2} L(q, \dot{q}, t)\, dt \qquad \text{and} \qquad \mathscr{I} = \int_{t_1}^{t_2} \mathscr{L}(Q, \dot{Q}, t)\, dt, \tag{7.56}$$

need be equal, but may be differently gauged. It is sufficient that the variations be equal, $\delta I = \delta\mathscr{I}$, when the paths $q(t)$ and $Q(t)$ are varied to other corresponding ones.

The consequences of $\delta I = \delta\mathscr{I}$ are clearest when the evaluations of I and $\mathscr{I}$, and of their variations, are carried out in the same one of the two configuration spaces, of q *or* of Q. That may be done, for example, by using the given correspondence (7.54) to write

$$I = \int_{t_1}^{t_2} L(q(Q), \dot{q}(Q, \dot{Q}, t), t)\, dt. \tag{7.57}$$

Clearly the variables q or Q merely have the roles of "dummy" integration variables in the evaluation of the definite integrals, each of which depends only on the motion and a rate $L(t)$ or $\mathscr{L}(t)$, and not on whether the motion is described by $q(t)$ or by $Q(t)$. The use of integration variable substitutions to ease the performance of integrations, without change of the results, is quite familiar. The invariance to the variables used should already have been evident when the form given I of (7.50) was displayed; the integrals reduce to integrations over time alone, as soon as any path $q(t)$ or $Q(t)$ is chosen. One of the significant advantages of the variational formulation is this invariance to choice of variables.

The consequences of $\delta I = \delta \mathscr{I}$ can be argued exactly as was the result (7.51) of $\delta I = \delta I'$; here

$$L(q(Q, t), \dot{q}(Q, \dot{Q}, t), t) - \mathscr{L}(Q, \dot{Q}, t) = df(Q, t)/dt, \tag{7.58}$$

with $f(Q, t)$ again a function that may be chosen arbitrarily. The choice may be carried out by starting with some expression $F(q, Q, t)$ in terms of the symbols q, Q, t and only afterward introducing the substitutions (7.54) to obtain an $f(Q, t) \equiv F(q(Q, t), Q, t)$. Thus, the Q's may still be treated as variables independent of the q's in the expression

$$L(q, \dot{q}, t) - \mathscr{L}(Q, \dot{Q}, t) = dF(q, Q, t)/dt, \tag{7.59}$$

with this then a condition to be satisfied by any proposed variable substitution and gauge change, together. The arbitrariness of F corresponds to the freedom that exists as to the substitutions and the choice of gauge. Making a definite choice for $F(q, Q, t)$ can be regarded as putting conditions on the relationship of the Q's to the q's, as well as $\mathscr{L}$ to L, through (7.59). For that reason F is called a "transformation-generator." Procedures for generating variable transformations as suggested by this will be developed in the next chapter. Meanwhile, some consequences of starting with Q's independent of the q's in (7.59) will be recorded. The right side of (7.59) is linear in the independent $\dot{Q}$'s and $\dot{q}$'s, as the development

$$\frac{dF}{dt}(q, Q, t) = \sum_s \left(\frac{\partial F}{\partial q_s} \dot{q}_s + \frac{\partial F}{\partial Q_s} \dot{Q}_s \right) + \frac{\partial F}{\partial t} \tag{7.60}$$

demonstrates. The derivatives with respect to each $\dot{q}$ and each $\dot{Q}$ on the two sides of (7.59) must match. This results in the $2f$ equations

$$\partial F/\partial q_s = \partial L/\partial \dot{q}_s (\equiv p_s), \tag{7.61}$$

$$-\partial F/\partial Q_s = \partial \mathscr{L}/\partial \dot{Q}_s (\equiv P_s). \tag{7.62}$$

The set (7.62) gives the new generalized momenta which evolve. There will also be a new generalized energy,

$$\mathscr{H} \equiv \sum_s P_s \dot{Q}_s - \mathscr{L}(Q, \dot{Q}, t) = H + \partial F/\partial t, \tag{7.63}$$

differing in value from the old H, at corresponding phases of the motion $q(t), \dot{q}(t) \leftrightarrow Q(t), \dot{Q}(t)$, whenever an explicitly time-dependent generator F is chosen. The use of all these relationships will be demonstrated in the next chapter.

7.5 SPACE-TIME INVARIANCES AND THE FUNDAMENTAL CONSERVATION LAWS

The role of conservation laws as expressions of symmetries characteristic of a description, of invariances to changes of viewpoint or reference frame, have been stressed repeatedly (pages 52 and following, 102 and on, 168, and others). The Variational Principle offers a very general basis on which an expectation of invariance becomes equivalent to an expectation of some kind of conservation law. That becomes manifest when the "expenditures" on two actual paths of motion are compared, as will be done next.

Comparative Expenditures

The expenditures on different *actual* paths that might be followed by a system are now to be compared. The different paths may be the results of different starts given to the motions. On each of the paths to be considered, the extremum condition expressed by the Variational Principle is fulfilled and the consequent Euler-Lagrange equations obeyed.

The expenditure integrals will be evaluated for two neighboring paths, as indicated in Figure 7.6. The motions in the two cases will be represented as $q(t)$ and $q'(t)$, respectively. The expenditure on the path $q(t)$ will be evaluated for the motion between $q(t_1)$ and $q(t_2)$, and that will be compared against the motion between $q'(t_1 + \Delta t_1)$ and $q'(t_2 + \Delta t_2)$ on the second path. The differences between the corresponding end points of the two paths will be respectively denoted

$$\Delta q_s(t_{1,2}) = q_s'(t_{1,2} + \Delta t_{1,2}) - q_s(t_{1,2}) \tag{7.64}$$

for each $s = 1, \ldots, f$. As when comparing actual with "virtual" paths, differences between simultaneous points on the two "actual" paths will be denoted δq_s, so that $q_s'(t) = q_s(t) + \delta q_s(t)$. Then

$$\begin{aligned} \Delta q_s(t_1) &= q_s(t_1 + \Delta t_1) + \delta q_s(t_1 + \Delta t_1) - q_s(t_1) \\ &\approx \delta q_s(t_1) + \dot{q}_s(t_1)\,\Delta t_1, \end{aligned} \tag{7.65}$$

the last being valid in the limit of vanishing differences. A similar relation holds between $\Delta q_s(t_2)$ and $\delta q_s(t_2)$.

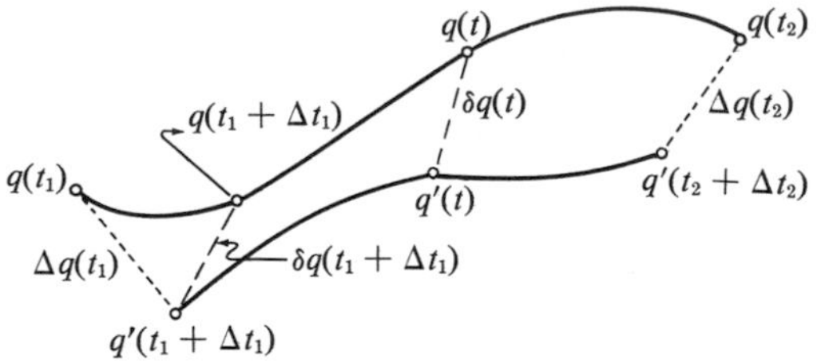

Figure 7.6

The difference between the expenditures on the two paths is

$$\Delta I = \int_{t_1+\Delta t_1}^{t_2+\Delta t_2} L(q + \delta q, \dot{q} + \delta\dot{q}, t)\, dt - \int_{t_1}^{t_2} L(q, \dot{q}, t)\, dt. \tag{7.66}$$

Consider first the integration from t_2 to $t_2 + \Delta t_2$:

$$\int_{t_2}^{t_2+\Delta t_2} L(q + \delta q, \dot{q} + \delta\dot{q}, t)\, dt \approx [L(q, \dot{q}, t)\, \Delta t]_{t_2}.$$

The variations δq and $\delta\dot{q}$ have been neglected here, as of second order relative to the approximation of (7.65). Similarly, the integral from $t_1 + \Delta t_1$ to t_1 will yield

$$-[L(q, \dot{q}, t)\, \Delta t]_{t_1}.$$

There remains only the integration between t_1 and t_2 in the two parts of ΔI (7.66), and this may be evaluated as was δI of (7.44):

$$\int_{t_1}^{t_2} dt \sum_s \left[\frac{\partial L}{\partial q_s}\,\delta q_s + \frac{\partial L}{\partial \dot{q}_s}\,\delta\dot{q}_s\right] = \int_{t_1}^{t_2} dt \sum_s \delta q_s \left[\frac{\partial L}{\partial q_s} - \frac{d}{dt}\left(\frac{\partial L}{\partial \dot{q}_s}\right)\right] + \left[\sum_s \frac{\partial L}{\partial \dot{q}_s}\,\delta q_s\right]_{t_1}^{t_2}.$$

The last term here cannot generally be set equal to zero as for δI of (7.46), since $\delta q_s(t_{1,2}) \neq 0$ is possible for the paths now considered. On the other hand, the remaining integral here vanishes, since the Lagrange equations are obeyed on actual paths of motion.

Collecting the results of the preceding paragraph yields

$$\Delta I = \left[L\, \Delta t + \sum_s (\partial L/\partial\dot{q}_s)\, \delta q_s\right]_{t_1}^{t_2},$$

and (7.65) may be used to replace δq_s by Δq_s in this:

$$\Delta I = \left[\sum_s (\partial L/\partial\dot{q}_s)\, \Delta q_s - \left\{\sum_s (\partial L/\partial\dot{q}_s)\dot{q}_s - L\right\} \Delta t\right]_{t_1}^{t_2}.$$

With the definition (7.38) of the conjugate momentum, and (7.40) of the hamiltonian energy,

$$\Delta I = \Big[\sum_s p_s\,\Delta q_s - H\,\Delta t\Big]_{t_1}^{t_2}. \tag{7.67}$$

Naturally, $\Delta I = 0$ if the compared paths have $\Delta q_s(t_{1,2}) = 0$ and $\Delta t_{1,2} = 0$, since these are conditions for a unique path.

Now there are circumstances in which it should be possible to take the expenditure on two different paths to be the same, to have $\Delta I = 0$ despite $\Delta t \neq 0$ and/or one or more $\Delta q_s \neq 0$. Various such circumstances will be considered next.

Time-Invariance and Energy Conservation

A system should be able to go through exactly the same phases of motion during two different time spans, if it is otherwise given the same start and is undisturbed by arbitrarily intruding external influences, such as could make the two time periods essentially different. This is just the invariance to displacement in time already mentioned as a characteristic to be expected of isolated systems (end of Chapter 2). It should then be possible to have $\Delta I = 0$ when $\Delta q_s = 0$ for every degree of freedom, even though $\Delta t_1 = \Delta t_2 = \Delta t \neq 0$. In such circumstances (7.67) reduces to

$$\Delta I = -[H]_{t_1}^{t_2}\,\Delta t = 0,$$

and this can happen only if

$$H(t_2) = H(t_1), \tag{7.68}$$

that is, if the hamiltonian energy is conserved. Thus displayed is a very general basis on which the expectation of invariance to shifts in time becomes equivalent to an expectation of energy conservation.

The same conclusion has already been reached in connection with the definition of the generalized energy by (7.40). There, the invariance to times was represented by $\partial L/\partial t = 0$, which excludes the explicit appearance of the time from the lagrangian, so that corresponding phases of the motion will have the same forces in effect at the shifted times.

The comparison of paths of motion here makes more graphic the significance of the invariance. It implies that a conserved energy must be definable if experiments are to be reproducible, when due precautions for isolating them from arbitrary external effects are observed. It may even be said that the energy of the entire universe—an isolated system by virtue of there being nothing outside it—must be conserved if physical laws are to be eternal.

The generalization of the energy concept has demonstrated that it is defined as whatever is conserved in consequence of the invariance. This was already implied in §2.4, when energy conservation was discussed as a matter of identifying it suitably.

Momentum Conservation

A given experiment should also be reproducible in different laboratories, when care is taken to isolate it from any external influence. This amounts to the invariance to translation in space already mentioned as a characteristic to be expected of isolated systems. The invariance is equivalent to viewing the same motion from different reference frames which are merely shifted relative to each other, in a space which should be homogeneous for an isolated system. Only the relative disposition of bodies, sources of force, and so on should have significance in such circumstances.

To formulate translation in ordinary space, position-vectors $\mathbf{r}_i$ should be used as the coordinates. Then the system is translated bodily when $\Delta q_s \to \Delta\mathbf{r}_i = \Delta\mathbf{r}$, the same for every position variable. Let $\mathbf{p}_i$ be the momentum conjugate to $\mathbf{r}_i$, with components like $\partial L/\partial \dot{x}_i$. Then (7.67) is written

$$\Delta I = \Big[\sum_i \mathbf{p}_i\Big]_{t_1}^{t_2} \cdot \Delta\mathbf{r}, \tag{7.69}$$

since $\Delta t \equiv 0$ must be chosen to see just the effect of the space-translation. There is invariance of the description if $\Delta I = 0$, achievable only if

$$\Big(\sum_i \mathbf{p}_i\Big)_{t_1} = \Big(\sum_i \mathbf{p}_i\Big)_{t_2}. \tag{7.70}$$

That is, a total, generalized, momentum must be conserved. It may be said that the conserved momentum is defined by the homogeneity of space for the system, whenever this is isolated from disturbances.

Angular Momentum Conservation

It should make no difference to an experiment if the apparatus is merely reoriented, in a space which should be isotropic for a system screened from any external influence.

Let $\hat{\boldsymbol{\omega}}$ be a unit vector pointing out an arbitrarily chosen rotation axis. Then, in a rotation through an angle $\Delta\varphi$ about that axis, any given position vector $\mathbf{r}_i$ suffers the vector displacement

$$\Delta\mathbf{r}_i = \Delta\varphi\hat{\boldsymbol{\omega}} \times \mathbf{r}_i, \tag{7.71}$$

when the origin for the position vectors is chosen to lie on the axis $\hat{\boldsymbol{\omega}}$. In the rotation by $\Delta\varphi\hat{\boldsymbol{\omega}}$ of the entire system, the expression (7.67) becomes

$$\Delta I = \Delta\varphi\Big[\sum_i (\hat{\boldsymbol{\omega}} \times \mathbf{r}_i) \cdot \mathbf{p}_i\Big]_{t_1}^{t_2} = \Delta\varphi\hat{\boldsymbol{\omega}} \cdot \Big[\sum_i \mathbf{r}_i \times \mathbf{p}_i\Big]_{t_1}^{t_2}.$$

The sum here can be identified as the total angular momentum $\mathbf{L} = \sum_i \mathbf{r}_i \times \mathbf{p}_i$ about the chosen origin, and

$$\Delta I = \Delta\varphi\hat{\boldsymbol{\omega}} \cdot [\mathbf{L}]_{t_1}^{t_2} = 0 \tag{7.72}$$

for invariance to the reorientation. Since the rotation center and axis were chosen arbitrarily, the invariance requires the total vector angular momentum about any momental center to be conserved

$$\mathbf{L}(t_1) = \mathbf{L}(t_2). \tag{7.73}$$

It may be said that conserved angular momenta are defined by the isotropy of space for an isolated system.

The Conjugacy of Variables

The preceding subsections exhibited the most basic invariances of description in space-time, the homogeneity in time, space, and direction for isolated systems which require that each possess a conserved energy, momentum, and angular momentum.

More detailed conservation principles can also be understood from the invariance of I (7.67) to shifts of the motion. The conservation of the momentum conjugate to an individual, "ignorable" degree of freedom q_i (7.37) can be seen as a consequence of

$$\Delta I = [p_i]_{t_1}^{t_2} \, \Delta q_i = 0, \tag{7.74}$$

expressing the invariance to shifts Δq_i. The example of this type afforded by the "horizontal" momentum in a uniform "vertical" gravitational field has already been quoted (page 167). The conserved angular momentum about the special point serving as the center of a spherically symmetric force field is another example. In such cases there can be said to be "isolation" of just one degree of freedom of the system.

Perhaps the most important result of considering the possibility of invariance to a shift in a single degree of freedom is the insight this gives into the property of conjugacy, in connection with that degree of freedom by itself. The relation between conjugate variables q_s and p_s is such that isolation of one entails conservation of the other. The conjugate of a variable is something identifiable as persisting with a definite, constant value whenever an experiment is arranged in which the system's behavior is made independent of any variations in that variable. Of course, several "control" experiments may be necessary before the correspondence can be uniquely identified.

EXERCISES

7.1. Two beads, of masses m_1 and m_2, may slide without friction on a circular wire hoop.

(a) First, the hoop is fixed in a vertical plane and the beads are connected by a weightless cord shorter than a diameter. Show how the Principle of Virtual Work leads to an equilibrium configuration in which at least one bead stays above the hoop center, and exerts a gravitational torque about it equal and opposite to that of the second bead.

(b) Suppose instead that the hoop is free to rotate about a pivot point on its rim, and that the cord is used to suspend m_1 from the same pivot, but m_2 is left free to slide to any point of the hoop. Show that the equilibrium tension in the cord will be

$$m_1 Mgl/[(M - m_1)^2 a^2 + m_1 Ml^2]^{1/2},$$

where M is the sum of all the masses, l is the cord length, and a the hoop radius. Check this result against your expectations for a *fixed* hoop, with the cord to m_1 suspended from its topmost point. (Note that the reaction of the hoop is just $m_1 g$ in the latter case.)

7.2. To see one way in which the tension in a string may be modified, suppose that the string used in the simple Atwood machine described on page 151 has a nonnegligible mass, distributed uniformly with a density ρ per unit length.

(a) Compare the tension at the pulley and at m_1, in the part of the string with the extension z_1, on the basis of instantaneous transmission. Also find the tension $T(z)$ at *any* point $0 < z < z_1$.

(b) The machine may now be modified by removing the masses m_1, m_2, leaving only the heavy string. Show how the application of d'Alembert's Principle to the "summation," $\int \rho \, dz$, of the mass-*elements*, by making it unnecessary to consider the tensions at all, leads directly to the equation of motion

$$\ddot{z}_1 = (g/l)(2z_1 - l)$$

for one end of the string. Check that this is consistent with the rate at which the total kinetic energy should increase as the potential energy decreases.

7.3. Suppose that the mass m_2, hung from the simple Atwood machine discussed on page 151, is replaced by a live monkey, and suppose that the monkey climbs the rope with some speed, $v(t)$, *relative to the rope*. Find the accelerations of $m_1 g$ during phases in which the monkey climbs with uniform speed, and also in phases of $\dot{v}(t) \neq 0$. Is it possible for the monkey to *raise* a weight greater than its own, by climbing vigorously enough?

7.4. Suppose that the simple Atwood machine (page 151), instead of being suspended from a fixed pulley, is pulled upward by a constant force F. Given also that the mass of the pulley is m_0, find the accelerations of m_1 and m_2. Show that, *relative to the pulley*, they are the same as for a fixed-pulley Atwood machine in a gravitational field augmented by the acceleration of the pulley. What is the minimum force F needed to insure an upward acceleration? [$F > (4\mu + m_0)g$.]

7.5. Compare the Newtonian, d'Alembertian and Lagrange formulations of the problem suggested by the figure. The mass-particle m slides freely,

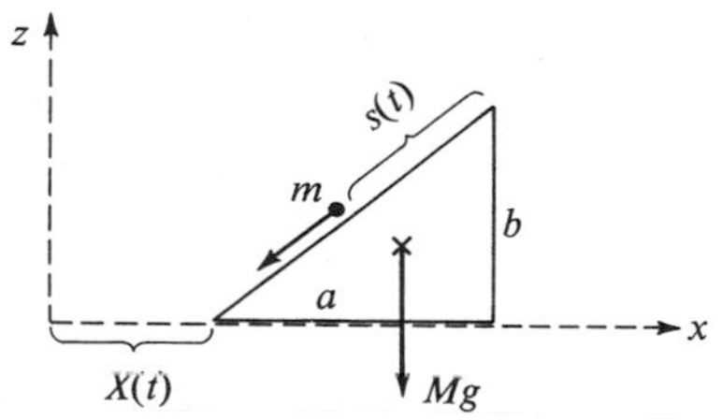

from rest at the top of the slope, on the triangular block of weight Mg. The block can slide freely on the horizontal plane, and is also at rest initially.

(a) From the Newtonian formulation, find the reaction of the block on the particle

$$[R = mg.\ Mac/(Mc^2 + mb^2) \text{ with } c^2 = a^2 + b^2].$$

(b) From the d'Alembertian formulation, find the trajectory in space, $z(x)$, of the particle. For what slope, b/a, will the speed $\dot{X}$ finally imparted to the block have a maximum value?

(c) From a Lagrange formulation, find the acceleration $\ddot{s}$, of the particle relative to the block. Interpret the "ignorability" of the block coordinate $X(t)$.

7.6. A bead of mass m slides on a frictionless, straight wire which is being rotated with a constant angular velocity, ω, about one of its points, and in a horizontal plane. Derive the Lagrange equation of motion for the bead. Will the energy be conserved?
Identify forces acting on the bead and the reaction on the wire, and compare these with the inertial forces discussed in §2.5. What initial speed must be given to the bead if it is to reach the rotation axis from a starting point a distance r_0 away? Show that $r = r_0 e^{-\omega t}$ if the last requirement is only barely met.

7.7. A pendulum consisting of a comparatively weightless rod of length l, and a bob of weight mg, has a second, similar pendulum suspended from its bob. The suspensions are such that only motions in a vertical plane are possible.

(a) Derive equations of motion for $\ddot{\varphi}_1$ and $\ddot{\varphi}_2$, where φ_1 is the angle with the direction of gravity made by the upper rod, while φ_2 is the angle between the directions of the lower and upper rods. (To appreciate the power of the Lagrangian approach, the reader should also derive the same equations from the d'Alembertian and the Newtonian formulations. Both can be done!)

(b) Show that while the oscillations stay small enough ($\varphi_{1,2} \ll 1$ and $l\dot{\varphi}_{1,2}$ much less than the speed, $(2gl)^{1/2}$, acquired by free fall through a distance l)* the equations can be reduced to

$$\ddot{\varphi}_1 \approx -(g/l)(\varphi_1 - \varphi_2), \quad \ddot{\varphi}_2 \approx -(g/l)(3\varphi_2 - \varphi_1).$$

From these find the "normal frequencies" (see §3.7), and sketch configurations assumed during motions with each of the two frequencies.

7.8. Investigate the "juggling" of a heavy ball on top of a comparatively weightless rod. Show that the juggler can be a robot capable only of imparting simply harmonic horizontal displacements, of arbitrary amplitude and frequency. The problem then amounts to treating small oscillations* of the system in Figure 7.3, turned upside down. Show that a disturbance which gives the rod an initial angular velocity $\dot{\varphi}_0$, from the upright position $\varphi_0 = 0$, can be made the start of a stable oscillation which remains small, if a frequency $\omega \gg \dot{\varphi}_0$ is used, and an amplitude $a = -(l\dot{\varphi}_0/\omega)(1 + g/\omega^2 l)$. (A similar treatment may also be applied to the juggling of two heavy-headed canes, one atop the other, as an upside down version of the preceding exercise, and also to problems† of the stability of orbits in high-energy accelerators!)

7.9. In the expression for the kinetic energy in the system of Figure 7.4, identify the individual quantities μ, τ, π, and κ making up the general expression (7.11). Calculate the generalized force according to (7.20) and check its derivability from the potential. Finally, construct the equation of motion (7.28).

7.10. Show that the equation of motion for the system of Figure 7.4 reduces to

$$l\ddot{\varphi} \approx -(\omega^2 a + g \cos \omega t)\varphi - g \sin \omega t$$

during phases of the motion in which‡ $\varphi \ll 1$. Since it is plain that a *low* angular velocity should merely lead to motion relative to the circling suspension point which differs little from ordinary pendular motion about the direction of gravity, investigate what happens for *very large* angular velocities, $\omega \gg (g/a)^{1/2}$. Show that the pendulum *may* then have small

---

* The approximations $\sin \varphi_{1,2} \approx \varphi_{1,2}$ and $\cos \varphi_{1,2} \approx 1$ are to be made.

† A point brought to the author's attention in a lecture by Dr. D. Judd.

‡ See the footnote to Ex. 7.7.

oscillations about the *direction of the rod*, but that forced oscillations, having the frequency of the rotation, are superposed. Moreover, if $l \approx a$, a resonance arises which prevents any deflections from remaining small (compare §3.4).

7.11. A special conservation principle for unconstrained motion of a particle in a uniform gravitational field was pointed out on pages 167 and 181. Find a similarly special conserved quantity in the motion of a mass-particle constrained to a surface that is axially symmetric about the direction of gravity.

(a) Suppose the surface is specified by $r = f(z)$, in cylindrical coordinates. Derive the equations of motion, in terms of f and its derivatives.

(b) For the case of the cone $r = z > 0$ (when $g_z = -g$), describe the kinds of motion that can ensue from a start at some level z_0, with only a *horizontal* component of initial velocity, v_0.

(c) Also for the case (b), find the critical v_0 which will lead to rotation at a steady level z_0, and the maximum amount by which the particle will descend if v_0 has only half the critical value. [$\frac{1}{2}z_0$].

7.12. The force on a particle bearing the charge q through a uniform magnetic field $\mathbf{B}$ is derivable from the work-function [see (7.49*b*)]:

$$W(\mathbf{r}, \mathbf{v}) = \boldsymbol{\mu} \cdot \mathbf{B}, \qquad \boldsymbol{\mu} \equiv q\mathbf{L}/2mc,$$

where $\mathbf{L} \equiv \mathbf{r} \times m\mathbf{v}$ is the particle's orbital angular momentum. (Then $\boldsymbol{\mu}$ is called the "orbital magnetic moment.")

(a) Show that the force rotates the particle's velocity vector according to

$$\dot{\mathbf{v}} = -\boldsymbol{\omega} \times \mathbf{v},$$

where* $\boldsymbol{\omega} = q\mathbf{B}/mc$. Discuss the work and the kinetic energy in these circumstances.

(b) Adopt an orthogonal frame of unit vectors $\hat{\boldsymbol{\omega}}$, $\mathbf{e}_1$ and $\mathbf{e}_2 \equiv \hat{\boldsymbol{\omega}} \times \mathbf{e}_1$, so defined that $\hat{\boldsymbol{\omega}}$ is parallel to $\mathbf{B}$ and the *initial* velocity may be represented as $\mathbf{v}_0 = \hat{\boldsymbol{\omega}} v_\parallel + \mathbf{e}_1 v_\perp$. Show that the velocity at later times may then be represented as

$$\mathbf{v}(t) = \mathbf{v}_\parallel + v_\perp \operatorname{Re}\{(\mathbf{e}_1 - i\mathbf{e}_2)e^{-i\omega t}\}.$$

(c) Show that the trajectory is a circular helix with a radius, $r = v_\perp/\omega$, which could also have been predicted from a simple equation of the magnetic to a centrifugal force.

7.13. Suppose that the magnetic field treated in the preceding exercise is now applied to a point-charge moving in a potential field $V(r, z)$, which is axially symmetric about $\mathbf{B}$, taken parallel to the z-axis of a cylindrical coordinate system.

---

* Sometimes called the "cyclotron frequency" because it can represent the revolution rate of a particle circulating in the magnetic field of a high-energy accelerator.

(a) Show that despite the ignorability of φ in the lagrangian, the angular momentum $L_z = mr^2\dot{\varphi}$ is *not* conserved by itself. How do you reconcile this with the general theorem (7.37)?

(b) Change to a frame which rotates about the z-axis with an angular velocity $\dot{\Phi} - \dot{\varphi} = \omega_L$, where $\omega_L \equiv qB/2mc$. Then show that, in fields weak enough so that $\omega_L^2 \ll \dot{\Phi}^2$, the motion relative to the rotating frame is the same as it would have been in the original frame with **B** absent. (This demonstrates that the effect of a weak field is only to rotate the moving system "bodily," with the so-called "Larmor precession frequency", ω_L.)

(c) In a device known as a "cylindrical magnetron," the axially symmetric potential is simply $V = -|q\mathscr{E}|r$, obtained by applying a voltage, $=\mathscr{E}a$, between a filament along the axis and a cylindrical plate at $r = a$. Electrons with $q = -e$ are emitted by the filament, with negligible initial energies. Find an expression for the magnetic field needed to prevent the electrons from reaching the plate. (The field here *cannot* be considered weak, as in part (b), since $\dot{\varphi}^2 = \omega_L^2$.)

7.14. Suppose that the coordinates $q \equiv q_1, q_2, q_3, \ldots, q_n$ are constrained by $G_k(q, t) = 0$, $k = 1, 2, \ldots, c$, and that a lagrangian $L(q, \dot{q}\ t)$ which contains the effects of all the forces except the constraints is constructed. Show that the Variational Principle

$$\delta \int_{t_1}^{t_2} dt[L(q, \dot{q}\ t) + \sum_{k=1}^{c} \lambda_k G_k(q, t)] = 0$$

will lead to automatic consistency with the constraint relations if the unknown parameters, $\lambda_1, \ldots, \lambda_c$, which have been introduced here, are given arbitrary variations in defining the varied paths (the parameters λ_k are then called "Lagrange multipliers"). By comparing the resultant Euler conditions with (7.28), show that the expression

$$-\lambda_k\, \partial G_k/\partial q_s$$

acquires the role of a generalized constraint force which compels the motion of q_s to obey the constraint $G_k = 0$. Notice that a sufficient number $(n + c)$ of relations are available to determine the unknown λ_k's, as well as the motions.

7.15. Apply the results of the preceding exercise to the four coordinates $z_{1,2,3,4}$ of the compound Atwood machine of Figure 7.1. Show that the constraint forces C_i pointed out in connection with Figure 7.1 can be related to Lagrange multipliers as $C_1 = \lambda_1$, $C_2 = \lambda_1 - 2\lambda_2$, and $C_3 = C_4 = \lambda_2$.

7.16. A much-used approximation in solving the many-body problem posed by the equations (6.32), for motions relative to the mass-center, is the replacement of U by some average potential energy $\bar{U}(\mathbf{R}_i)$ for each particle; that is, the Langrangian is approximated as in

$$L = \sum_i \tfrac{1}{2}m_i \dot{\mathbf{R}}_i^2 - U \approx \sum_i [\tfrac{1}{2}m_i \dot{\mathbf{R}}_i^2 - \bar{U}(\mathbf{R}_i)].$$

The approximation clearly yields independent motions by the individual particles, and, in general, these may lead to a spurious motion of the mass-center, $\bar{\mathbf{R}} = \sum_i m_i \mathbf{R}_i$, "relative to itself," as a result of the approximation. This may be remedied by treating the requirement $\sum_i m_i \mathbf{R}_i = 0$ as a constraint on the motion.

Use the method of Lagrange multipliers (Exercise 7.14) to show that spurious mass-center motion can be avoided by adding, to the force $-\nabla_i \bar{U}(\mathbf{R}_i)$ on each particle, a constraint force $-(m_i/M)\sum_j \nabla_j \bar{U}(\mathbf{R}_j)$. (This can be negligible if m_i is a negligible enough fraction of the total mass M of the system.)

Also check that no such constraint force is needed in the exact equation (6.32), or whenever the total resultant of the internal forces vanishes.

7.17. The classic "brachistochrone" problem can be posed in a form entirely analogous to that of the Variational Principle (7.48). This is the problem of finding the curve $z(x)$ along which a particle should slide, under uniform gravity, in order to move from one point to another in the least time possible (say from rest at $x = z = 0$ to x_1, z_1 in the diagram).

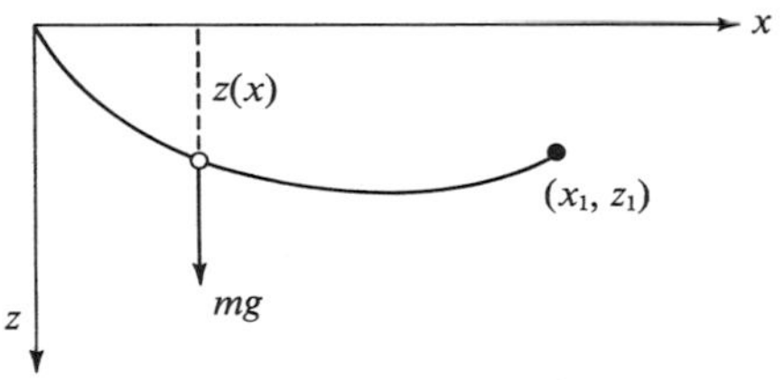

(a) Show that an appropriate reformulation of the Variational Principle is the replacement of t by x, of $q(t)$ by $z(x)$, and of $L(q, \dot{q}, t)$ by

$$L(z(x), z'(x)) = \{[1 + (z')^2]/2gz\}^{1/2},$$

where $z'(x) \equiv dz/dx$.

(b) Show that the analogue of the Hamiltonian, $H = z'(\partial L/\partial z') - L$, will remain constant along the curve.

(c) Show that, in terms of a parameter θ introduced through the substitution $z = (1 - \cos\theta)/4gH^2$, the brachistochrone curve is described by

$$x = (\theta - \sin\theta)/4gH^2.$$

(d) Show that this type of curve (called a "cycloid") is also followed by any one point on the rim of a rolling disk, with θ an angle subtended at the center of the disk.

(e) What value must be chosen for the constant H^2 in the case of $x_1 = 2\pi a$ and $z_1 = 0$? [$H^2 = 1/4ga$.]

CHAPTER 8

DESCRIPTIONS IN PHASE SPACES

Any motion of a system may be regarded as a series of transitions among momentary phases $q(t)$, $\dot{q}(t)$. A distinct one may be started with any initial phase, $q(0)$, $\dot{q}(0)$, and this can be chosen with greater freedom than is exercised by choosing just a starting point, $q(0)$, in the f-dimensional configuration space of $q_1, \ldots, q_f$. A set $\dot{q}(0) \equiv \dot{q}_1(0), \ldots, \dot{q}_f(0)$ is to be chosen as well; hence the choice of initial phase has rather a one-to-one correspondence to points of a $2f$-dimensional domain. It is useful, particularly in extensions to Statistical and Quantum Mechanics, to give explicit recognition to this, by "plotting" points in a $2f$-dimensional "phase space," to represent the system at various moments. Moreover, Hamilton found a way to formulate the Classical Mechanics which treats not only the q's but also their conjugate momenta $p \equiv p_1, \ldots, p_f$ as independent variables, to be determined by equations of motion on an equal basis with the q's. It is quite apparent that giving $q(t)$, $p(t)$ serves to specify a momentary phase of the motion just as well as does giving $q(t)$, $\dot{q}(t)$; hence the phase spaces are defined as $2f$-dimensional domains of p's and q's. To treat $2f$ variables as independent, $2f$ equations of motion must be furnished, twice as many as the Lagrange equations for f degrees of freedom $q_1, \ldots, q_f$. Moreover, each of the doubled set of differential equations must be first-order in time, in order that giving $q(0)$, $p(0)$ suffice for the unique determination of $q(t)$, $p(t)$. These will now be found to be characteristics of the hamiltonian formulation.

8.1 HAMILTON'S CANONICAL FORMULATION

To be noticed first is that the hamiltonian energy, $H = \sum_s p_s \dot{q}_s - L(q, \dot{q}, t)$ of (7.40), depends on the generalized velocities $\dot{q}_s$ at most through its dependence on the conjugate momenta, $p_s \equiv \partial L/\partial \dot{q}_s$. This is made evident when the derivative of H with respect to any $\dot{q}_\sigma$, holding not only all the q's, but also all the p's, constant, is formed:

$$\partial H/\partial \dot{q}_\sigma = p_\sigma - \partial L(q, \dot{q}, t)/\partial \dot{q}_\sigma = 0, \tag{8.1}$$

vanishing because of the definition of p_σ. Thus, despite the explicit appearance of the $\dot{q}$'s in the defining expression $H \equiv \sum_s p_s \dot{q}_s - L(q, \dot{q}, t)$, its variations with the $\dot{q}$'s can be ignored as nonexistent, except insofar as variations of the p's may occur. It must be possible to express the generalized energy as a function of q, p, t only, and after that is done the result $H(q, p, t)$ is called an "hamiltonian" function.

As an example, consider a system described by the lagrangian $L = \sum_i \frac{1}{2} m_i \dot{\mathbf{r}}_i^2 - V(\mathbf{r})$, which has momenta with components like $\partial L/\partial \dot{x}_i = m_i \dot{x}_i$. The hamiltonian function is formed from

$$\sum_i \mathbf{p}_i \cdot \dot{\mathbf{r}}_i - L = \boxed{\sum_i (\mathbf{p}_i^2/2m_i) + V(\mathbf{r}) \equiv H(\mathbf{r}, \mathbf{p})}\,, \tag{8.2}$$

through eliminating each $\dot{\mathbf{r}}_i$ in favor of $\mathbf{p}_i/m_i$.

For a particle described by polar coordinates, so that $L = \frac{1}{2} m \times (\dot{r}^2 + r^2 \dot{\varphi}^2) - V$ and $p_r = m\dot{r}$, $p_\varphi = mr^2\dot{\varphi}$, the hamiltonian is

$$H(r, \varphi, p_r, p_\varphi) = p_r^2/2m + p_\varphi^2/2mr^2 + V(r, \varphi). \tag{8.3}$$

The notation $p_\varphi \equiv L$ was used for the angular momentum earlier (4.15), and if this is reverted to here, the symbol L should not be confused with the lagrangian. In spherical coordinates r, ϑ, φ it is customary to write

$$\boxed{H = p_r^2/2m + \mathbf{L}^2/2mr^2 + V(r, \vartheta, \varphi),} \tag{8.4a}$$

where

$$\boxed{\mathbf{L}^2 \equiv p_\vartheta^2 + p_\varphi^2/\sin^2 \vartheta,} \tag{8.4b}$$

now with $p_\varphi = m(r \sin \vartheta)^2 \dot{\varphi}$, and $p_\vartheta = mr^2 \dot{\vartheta}$ as seen in (7.38). The quantity $\mathbf{L}^2/2mr^2$ is just the rotational kinetic energy, (4.18), transformed to spherical coordinates:

$$\tfrac{1}{2}m(v_\vartheta^2 + v_\varphi^2) = p_\vartheta^2/2mr^2 + p_\varphi^2/2mr^2 \sin^2 \vartheta,$$

since $p_\vartheta = mv_\vartheta r$ and $p_\varphi = mv_\varphi(r \sin \vartheta)$.

The Canonical Equations of Motion

The equations of motion in the variables q, p may be found by considering the variations of $H(q, p, t)$ with those variables. The connection $H = \sum_s p_s \dot{q}_s - L(q, \dot{q}, t)$ to the lagrangian formulation is used, and this gives first

$$\partial H(q, p, t)/\partial q_\sigma = -\partial L/\partial q_\sigma = -\dot{p}_\sigma,$$

according to the Lagrange equations of motion (7.33). Thus furnished are only f equations, for the time-evolutions of the variables $p_\sigma(t)$. Other f equations follow from

$$\partial H(q, p, t)/\partial p_\sigma = \dot{q}_\sigma .$$

The latter equations seem trivial in practice: in the example of the hamiltonian (8.2), they merely reaffirm relations like $\mathbf{p}_i/m_i = \dot{\mathbf{r}}_i$, and for (8.3) the relations $p_r/m = \dot{r}$ and $p_\varphi/mr^2 = \dot{\varphi}$. However, they perform the very important function of seeing to it that in a treatment of q *and* p as independent variables, the correct relationship of $\dot{q}$ to p is maintained during the motion. This will be demonstrated in more secure detail, by calling on the Variational Principle in the next subsection.

The outcome is that motion is treated as a time-evolution, $q(0)$, $p(0) \to q(t)$, $p(t)$, of $2f$ variables specifying its successive phases, being generated by some hamiltonian $H(q, p, t)$ according to the $2f$ equations of motion:

$$\boxed{\dot{q}_s = \partial H/\partial p_s, \qquad \dot{p}_s = -\partial H/\partial q_s \qquad (s = 1, \dots f).} \tag{8.5}$$

These are called Hamilton's "canonical equations" and the variable sets q and p are said to be each other's "canonical conjugates." The hamiltonian has the role here which is played by force functions in the Newtonian and d'Alembertian formulations, and by lagrangians in others.

It is simple to check that the canonical equations reduce to ones familiar in the Newtonian formulation in each of the examples of the preceding subsection. While the first of the equations (8.5) usually furnishes the connection between conjugate momentum and velocity, the second is some generalization of $\dot{p}_x = -\partial V/\partial x = F_x$.

The Variational Principle in Phase Spaces

The justification for regarding the q's and p's as equally independent variables, despite the derivability of $\dot{q}$ from $q(t)$, is more fully clarified by starting from the Variational Principle of (7.48). Now the motions between any chosen initial and final times, t_1 and t_2, will be represented by paths in the $2f$-dimensional phase space. Variations of the paths in such a space can be constructed by making $2f$ arbitrary choices of $\delta q(t) \equiv \delta q_1, \ldots, \delta q_f$ *and* of $\delta p(t) = \delta p_1, \ldots, \delta p_f$. They should be restricted by $\delta q(t_1) = 0$ and $\delta q(t_2) = 0$ in applying the Variational Principle.

The change to the phase space is completed when the "expenditure" integral is transformed to

$$I = \int_{t_1}^{t_2} L(q, \dot{q}, t)\, dt = \int_1^2 \left[\sum_s p_s\, dq_s - H(q, p, t)\, dt\right], \tag{8.6}$$

after the definition (7.40) of the hamiltonian is used. Upon the variations δq, δp in the phase space,

$$\delta I = \int_{t_1}^{t_2} dt \sum_s [\delta p_s \dot{q}_s + p_s\, \delta\dot{q}_s - (\partial H/\partial p_s)\, \delta p_s - (\partial H/\partial q_s)\, \delta q_s],$$

where

$$\int_{t_1}^{t_2} dt\, p_s\, \delta\dot{q}_s = \int_{t_1}^{t_2} dt[(d/dt)(p_s\, \delta q_s) - \dot{p}_s\, \delta q_s],$$

$$= [p_s\, \delta q_s]_{t_1}^{t_2} - \int_{t_1}^{t_2} dt \dot{p}_s\, \delta q_s .$$

The bracket in the last line vanishes for paths with $\delta q(t_1) = 0$ and $\delta q(t_2) = 0$. Consequently, the Variational Principle, $\delta I = 0$ in such path variations, yields

$$\int_{t_1}^{t_2} dt \sum_s [\delta p_s(\dot{q}_s - \partial H/\partial p_s) - \delta q_s(\dot{p}_s + \partial H/\partial q_s)] = 0.$$

This shows that variations δp_s as well as δq_s may be chosen arbitrarily if only the canonical equations are obeyed. This result was obtained even though the varied paths were restricted as to end points only by $\delta q(t_1) = \delta q(t_2) = 0$. That implies that the expenditure on the correct path is an extremum even as compared to paths with end points differing by $\delta p(t_1) \neq 0$ and $\delta p(t_2) \neq 0$. Plainly, the first-order canonical equations develop a unique path once only $q(t_1)$ and $p(t_1)$ are given; the Variational Principle uses instead a specification of $q(t_2)$ to replace that of $p(t_1)$.

Shown here has been the permissibility of treating both the q's and p's as independent variables if only the $2f$ canonical equations are used to determine their joint time-evolution during motion.

Considering variations of the form (8.6) calls attention to the possibility that variations of the generalized energy $H(q, p)$ need not be allowed for when dealing with a sufficiently complete description, one in which energy conservation can be taken for granted. Then

$$\delta I \to \delta \int_{q(t_1)}^{q(t_2)} \sum_s p_s \, dq_s = 0 \tag{8.7}$$

in path variations restricted to those in which all the alternative paths are required to have the same value for the conserved energy $H(q, p) = E$ at every phase point. The integral

$$A = \sum_s \int_{q(t_1)}^{q(t_2)} p_s \, dq_s \tag{8.8}$$

is called the "action" expended in the motion between the configurations $q(t_1)$ and $q(t_2)$, and (8.7) is referred to as a "Principle of Least Action." Explicit time-dependence can be excluded in energy conserving situations and then the simple expression $T = \frac{1}{2} \sum_s p_s \dot{q}_s$ (7.16) can be used for the kinetic energy; with this the action may be written

$$A = \int_{t_1}^{t_2} 2T \, dt, \tag{8.9}$$

another expression which is frequently to be seen in treatments of these matters (see also Exercises 8.7 and 8.8).

Hamiltonian Descriptions of Conservation Laws

A condition for the conservation of the generalized energy, H, follows directly from the canonical equations (8.5):

$$\frac{dH(q, p, t)}{dt} = \sum_s \left(\frac{\partial H}{\partial q_s} \dot{q}_s + \frac{\partial H}{\partial p_s} \dot{p}_s \right) + \frac{\partial H}{\partial t} = \frac{\partial H}{\partial t}. \tag{8.10}$$

Thus, as usual, explicit time-dependence, with its implications of disturbing external influences, must be excluded from the hamiltonian if there is to be energy conservation, $dH/dt = \partial H/\partial t = 0$.

All further conservation laws, arising from symmetries other than invariance to time-translation, can be built up from the conservation of any generalized momentum, p_i, which is conjugate to some ignorable coordinate, q_i. According to the canonical equations,

$$\dot{p}_i = -\partial H/\partial q_i = 0 \tag{8.11}$$

if H is invariant to changes of q_i. Generalized momentum conservation follows from ignorability of coordinates in the hamiltonian, just as it did from such ignorability in the lagrangian.

Since the canonical equations reduce to simple conservation conditions (8.11) for each ignorable coordinate, there is obvious advantage in using such coordinates as do become ignorable. They correspond to symmetries of the system. The introduction of such variables may require substitutions for whatever ones may be chosen initially, and systematic procedures for finding the appropriate transformations are developed next.

8.2 CANONICAL TRANSFORMATIONS

The transformations of lagrangian descriptions discussed in the last subsection of §7.4 have their counterpart in the hamiltonian formulation. It was seen in (7.63) that whenever a transformation is initiated by choosing some generator $F(q, Q, t)$, then a new hamiltonian, which may now be written

$$\mathscr{H}(Q, P, t) = H(q, p, t) + \partial F/\partial t, \tag{8.12}$$

arises to describe the same circumstances of motion as are described by $H(q, p, t)$. Here, $P_s = -\partial F/\partial Q_s$ of (7.62) is a new generalized momentum, conjugate to the new variable Q_s. The canonical equations for the new phase variables will be

$$\dot{Q}_s = \partial\mathscr{H}/\partial P_s, \qquad \dot{P}_s = -\partial\mathscr{H}/\partial Q_s, \tag{8.13}$$

as is evident from the connection (7.63) between hamiltonian and lagrangian, and the Lagrange equations (7.55). They are also to be expected from the invariance of the Variational Principle to the transformations, and the direct derivability of canonical equations from the principle as seen in the preceding section. Because the same canonical form (8.5) persists after the transformation to (8.13), the procedure is said to test "canonical invariance" and is called a "canonical transformation."

It was seen in (7.61) that if a given, successful description by an $L(q, \dot{q}, t)$ or $H(q, p, t)$ is to be transformed, then the generator $F(q, Q, t)$ introduces such Q's that

$$p_s = \partial F(q, Q, t)/\partial q_s. \tag{8.14a}$$

These, together with the definitions (7.62) of the new conjugate momenta

$$P_s = -\partial F(q, Q, t)/\partial Q_s, \tag{8.14b}$$

plainly have implicit in them relations

$$\begin{aligned} q_s &= q_s(Q_1, \ldots, Q_f; P_1, \ldots, P_f; t), \\ p_s &= p_s(Q_1, \ldots, Q_f; P_1, \ldots, P_f; t). \end{aligned} \tag{8.15}$$

A canonical transformation* of the *phase* space is thus generated, from one spanned by q's and p's to one having Q's and P's as the variables.

The possible phase space transformations (8.15) treat the coordinates and momenta on an equal footing and may include interchanges among them. Thus the special choice of generator $F = \sum_\sigma q_\sigma Q_\sigma$ leads to $p_s = Q_s$ and $P_s = -q_s$, by (8.14), and there is merely a renaming of q_s, p_s as $-P_s$, Q_s, respectively. This emphasizes that the important distinctions among phase-space variables are the conjugacies of pairs, rather than which are called "generalized coordinates" and which "generalized momenta." It is generally conceded, however, that the name "coordinate" is more appropriate for quantities that measure motion by actually changing in the course of it, and so their "conservation" would merely correspond to a lack of motion in the degree of freedom (and then the "variable" need not have been introduced in the first place!) There is more point in considering "conserved coordinates" when their conjugates are also conserved, as will be seen in §8.5.

A standard example of a phase space transformation is one that applies to a linear oscillator, having $H(x, p) = (p^2/2m) + \frac{1}{2}m\omega_0^2 x^2$, and is generated by† $F(x, Q) = \frac{1}{2}m\omega_0 x^2 \cot Q$. Then (8.14*a*, *b*) give

$$p = \partial F/\partial x = m\omega_0 x \cot Q, \qquad P = m\omega_0 x^2/2 \sin^2 Q.$$

The transformation relations (8.15) become

$$x = (2P/m\omega_0)^{1/2} \sin Q, \qquad p = (2m\omega_0 P)^{1/2} \cos Q$$

* Sometimes called a "contact" transformation because of certain geometrical properties of no great interest here.

† The development of (8.104) will show how this choice has been suggested.

in this case, and the new hamiltonian is

$$\mathscr{H} = H = \omega_0 P \cos^2 Q + \omega_0 P \sin^2 Q = \omega_0 P.$$

This is an ordinary constant energy, $H = E$, because t is ignorable (8.10), and $P = E/\omega_0$ is also expected to be conserved because Q does not occur in $\mathscr{H}$. The canonical equations (8.13) reduce to $\dot{Q} = \partial \mathscr{H}/\partial P = \omega_0$ and are trivial to solve for $Q = \omega_0 t + \varphi_0$. The spatial solution

$$x = \left(\frac{2E}{m\omega_0^2}\right)^{1/2} \sin(\omega_0 t + \varphi_0)$$

of (3.10) and (3.12) is an immediate consequence. This example illustrates the fact that a canonical transformation may be used to reduce the solution of equations of motion to triviality. On the other hand, the problem of finding the requisite transformation-generator F may be the more difficult! The possibilities will be explored in §8.5, as part of the Hamilton-Jacobi formulation.

The "Phase Function" as Generator

Since the transformations initiated by choosing generators of the form $F(q, Q, t)$ result in relations like (8.15), connecting all the variables q, p, Q, P, it should be possible to modify the procedure to the use of generators expressed as functions of others of those variables. The most widely used generators are of the type $S(q, P, t)$, sometimes called "Hamilton's Principal Functions." It will also turn out appropriate to call them "phase functions," instead (§8.6).

The elimination of Q's in favor of P's in the transformation-generating procedure may be carried out by what is called a "Legendre transformation" (of the canonical transformation!), much used in thermodynamics. An example of a Legendre transformation was provided when $\dot{q}$'s were eliminated in favor of p's in going from the Lagrange to the Hamilton description; the sum $\sum_s p_s \dot{q}_s$ was subtracted from $L(q, \dot{q}, t)$, with the result a function $-H(q, p, t)$, as demonstrated at the beginning of §8.1. The result depends on the fact that the new variables, p_s, are related to the $\dot{q}_s$ in the form $p_s = \partial L/\partial \dot{q}_s$. For the problem set now, the proposed new variables, P_s, are related to the Q_s in the form $P_s = -\partial F/\partial Q_s$ (8.14), and so

$$F(q, Q, t) + \sum_s P_s Q_s = S(q, P, t) \tag{8.16}$$

is used to define the new generator in terms of the old one, F. That $S(q, P, t)$

is indeed independent of the Q's for given values of q, P is readily checked by noting that the partial derivatives of (8.16) with respect to the Q's vanish.

In the resulting canonical transformation procedure, such phase variable correspondences as (8.15) are implicit in

$$\partial S/\partial q_s = p_s \qquad \text{and} \qquad \partial S/\partial P_s = Q_s, \tag{8.17}$$

in place of (8.14). The new hamiltonian (8.12) is now related to the old by

$$\mathscr{H}(Q, P, t) = H(q, p, t) + \partial S/\partial t, \tag{8.18}$$

in the transformations initiated by choices of $S(q, P, t)$.

The procedure may be illustrated for the familiar cartesian-to-polar coordinate transformation, $q_{1,2} \equiv x, y \to Q_{1,2} \equiv r, \varphi$, in which

$$r = \partial S/\partial p_r = (x^2 + y^2)^{1/2}, \qquad \varphi = \partial S/\partial p_\varphi = \tan^{-1}(y/x)$$

are supposed to hold. A generating function with which they do is

$$S(x, y, p_r, p_\varphi) = p_r(x^2 + y^2)^{1/2} + p_\varphi \tan^{-1}(y/x),$$

and then the remaining relations (8.17) yield

$$p_x = \partial S/\partial x = (xp_r/r) - (yp_\varphi)/r^2, \qquad p_y = \partial S/\partial y = (yp_r/r) + (xp_\varphi)/r^2.$$

These give correct expressions for the new generalized momenta:

$$P_1 \equiv p_r = (xp_x + yp_y)/r, \qquad P_2 \equiv p_\varphi = xp_y - yp_x.$$

Meanwhile, $\partial S/\partial t = 0$ in (8.18), and therefore the new hamiltonian is to be formed merely by substituting the polar variables for the cartesian ones, as in (8.3). The example may be regarded as a check that the transformation to the polar coordinates is indeed a canonical one.

It will be important for the next considerations to notice that the special choice

$$S_1 = \sum_s q_s P_s \tag{8.19}$$

generates the "identity transformation"—i.e. the relations (8.17) become $P_s = p_s$ and $q_s = Q_s$—and there is really no transformation at all.

Infinitesimal Transformations

Much of what can be learned from making variable transformations can already be found by making the changes "infinitesimal." Just what is meant by the latter is more precisely defined by the procedure to be adopted here. The essentials of such procedures have already been demonstrated in the treatment of the "variations," starting from the d'Alembert formulation in §7.1 and continuing in the Variational formulations. Those applications incidentally showed that valuable insights can be gained by trying infinitesimal changes.

Canonical transformations differing infinitesimally from the identity transformation by S_1 may be implemented by making choices for a new generator $G(q, p, t)$ such that

$$S(q, P, t) = S_1(q, P) + \varepsilon G(q, P, t). \tag{8.20}$$

Here, ε is an arbitrary parameter to be chosen small enough to make the ensuing procedure valid even if this takes $\varepsilon \to 0$ in the end. It is not incorporated into the arbitrary function G, but is to be separately chosen, just so that finite choices of G may be freely made, yet the additions to S_1 remain small for $\varepsilon \to 0$. The first of the relations (8.17) shows that

$$p_s = \partial S/\partial q_s = P_s + \varepsilon\, \partial G/\partial q_s, \tag{8.21}$$

and hence

$$\begin{aligned} G(q, P, t) &= G(q, p, t) + \sum_s (P_s - p_s)(\partial G/\partial p_s)_{P=p} + \cdots \\ &= G(q, p, t) - \varepsilon \sum_s (\partial G/\partial q_s)(\partial G/\partial p_s) + \cdots. \end{aligned}$$

Treating ε as infinitesimal shall mean dropping terms proportional to ε^2, and retaining only ones proportional to $\varepsilon^0 = 1$ or to ε; hence $G(q, P, t) \to G(q, p, t)$ in (8.20) and

$$S = \sum_s q_s P_s + \varepsilon G(q, p, t), \tag{8.22}$$

for which only expressions G in q, p, t are still to be constructed.

Besides (8.21), the procedure (8.17) yields

$$Q_s = \partial S/\partial P_s = q_s + \varepsilon\, \partial G/\partial p_s, \tag{8.23}$$

with the last derivative predicated on the fact that p_s and P_s need not be distinguished in terms already small $\sim\varepsilon$; that can be established more carefully in the same way as was (8.22). The consequent small changes,

$$\delta q_s \equiv Q_s - q_s \quad \text{and} \quad \delta p_s \equiv P_s - p_s, \tag{8.24a}$$

of the variables are given by

$$\delta q_s = \varepsilon\, \partial G/\partial p_s \quad \text{and} \quad \delta p_s = -\varepsilon\, \partial G/\partial q_s \tag{8.24b}$$

for any choice of $G(q, p, t)$. The corresponding change of the hamiltonian is

$$\delta H = \mathscr{H} - H = \varepsilon\, \partial G/\partial t. \tag{8.25}$$

The relations (8.24) and (8.25) together constitute the procedure for generating infinitesimal canonical transformations.

The Generators of Space-Time Shifts

Most immediately evident is the similarity in form of the infinitesimal transformation equations (8.24*b*) to the canonical equations of motion (8.5). Indeed, if the generator is chosen to be just $G \equiv H(q, p, t)$ and ε is chosen to be some time interval, $\varepsilon \equiv \delta t$, then the change of variables is by $\delta q_s = \dot{q}_s \delta t$ and $\delta p_s = \dot{p}_s\, \delta t$, just equal to the actual displacements in the motion during δt. Thus, motion itself may be regarded as a series of canonical transformations generated by the hamiltonian. There is a succession of transitions described as infinitesimal variable transformations,

$$q(t), p(t) \to Q \equiv q(t) + \dot{q}\, \delta t \approx q(t + \delta t), P \equiv p(t + \delta t), \tag{8.26a}$$

recalling the picture first pointed out in (2.9). At each step, the hamiltonian changes according to (8.25),

$$H(q, p, t) \to \mathscr{H} = H + \delta t(\partial H/\partial t) \approx H(q, p, t + \delta t), \tag{8.26b}$$

in readiness for generating the next infinitesimal transition in the motion.

The same transformation (8.26a), generated by $\varepsilon G \equiv \delta t H(q, p, t)$, can obviously be reinterpreted as shifting the times at which the entire motion takes place, back to times earlier by δt (when this is positive). Clearly $Q(t) \equiv q(t + \delta t)$ and $P(t) \equiv p(t + \delta t)$ can represent the same phases of motion as do $q(t)$ and $p(t)$, but passed through at earlier times, as indicated in Figure 8.1.

The generators of displacements in space are equally easy to construct. Suppose the system in motion is bodily translated in ordinary space, and every position $\mathbf{r}_i$ in it by $\Delta\mathbf{r}$. This can be construed as a transformation of the phase variables,

$$\mathbf{r}_i(t) \to \mathbf{R}_i(t) \equiv \mathbf{r}_i(t) + \Delta\mathbf{r}, \tag{8.27a}$$

$$\mathbf{p}_i(t) \to \mathbf{P}_i(t) \equiv \mathbf{p}_i(t),$$

as indicated in Figure 8.2 for one cartesian coordinate. In these terms, the generating equations (8.24) take the forms

$$\delta x_i = \varepsilon \frac{\partial G}{\partial p_{ix}} = \Delta x, \qquad \delta y_i = \varepsilon \frac{\partial G}{\partial p_{iy}} = \Delta y, \qquad \delta z_i = \varepsilon \frac{\partial G}{\partial p_{iz}} = \Delta z, \tag{8.28}$$

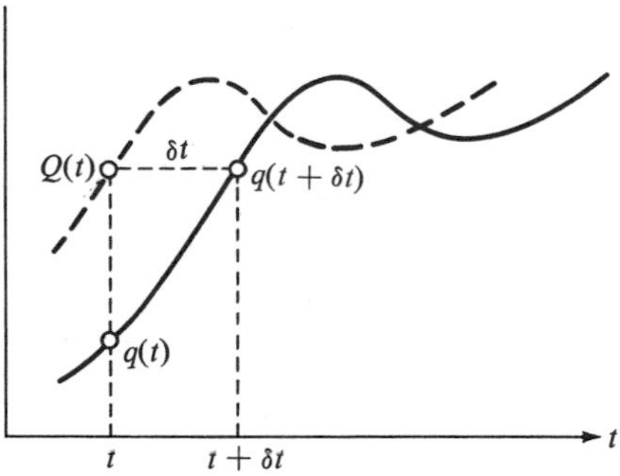

Figure 8.1

and $\boldsymbol{\delta}\mathbf{p}_i = -\varepsilon\nabla_i G(\mathbf{r}, \mathbf{p}) = 0$. Then, if the generator is given the form $\varepsilon G \equiv \Delta\mathbf{r}\cdot\mathbf{G}$, the vector $\mathbf{G}$ must be

$$\mathbf{G} = \sum_i \mathbf{p}_i, \tag{8.27b}$$

equal to the total linear momentum. Thus the total momentum is a generator of an over-all infinitesimal translation in space, just as the hamiltonian generates translation in time.

The displacement may instead be taken to be an over-all reorientation of the system by an angle $\delta\varphi$ around an arbitrarily chosen axis, $\hat{\boldsymbol{\omega}}$. Then the vectors giving the phases of the motion suffer the infinitesimal displacements

$$\boldsymbol{\delta}\mathbf{r}_i = \boldsymbol{\delta\omega}\times\mathbf{r}_i, \qquad \boldsymbol{\delta}\mathbf{p}_i = \boldsymbol{\delta\omega}\times\mathbf{p}_i, \tag{8.29a}$$

where $\boldsymbol{\delta\omega} \equiv \hat{\boldsymbol{\omega}}\delta\varphi$, if the origin of the vectors is put on the rotation axis. The generator needed in forms like (8.28) must now have a term $(\boldsymbol{\delta\omega}\times\mathbf{r}_i)_x p_{ix}$ to give δx_i, and a total $\varepsilon G = \sum_i (\boldsymbol{\delta\omega}\times\mathbf{r}_i)\cdot\mathbf{p}_i$ to yield all the $\boldsymbol{\delta}\mathbf{r}_i$. This total also suffices to give the $\boldsymbol{\delta}p_i = -\varepsilon\nabla_i G$, as is particularly clear when εG is rewritten:

$$\varepsilon G = \sum_i \boldsymbol{\delta\omega}\cdot(\mathbf{r}_i\times\mathbf{p}_i) = -\sum_i (\boldsymbol{\delta\omega}\times\mathbf{p}_i)\cdot\mathbf{r}_i.$$

A vector $\mathbf{G}$ may be defined so that $\varepsilon G \equiv \boldsymbol{\delta\omega}\cdot\mathbf{G}$, and then

$$\mathbf{G} = \sum_i \mathbf{r}_i\times\mathbf{p}_i = \mathbf{L}, \tag{8.29b}$$

the total angular momentum about the chosen momental center. The angular momentum is thus a generator of infinitesimal rotations.

Perhaps the simplest generator is one that displaces some one degree of freedom, q_s by δq_s. Then, if $\varepsilon = \delta q_s$ is chosen,

$$G = p_s \tag{8.30}$$

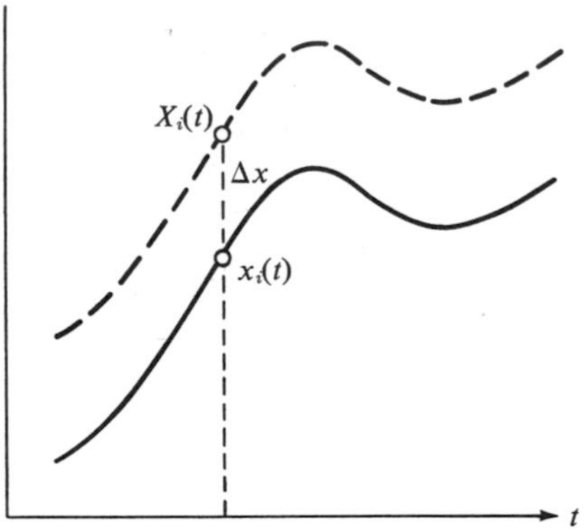

Figure 8.2

gives $\delta q_s = \varepsilon\, \partial G/\partial p_s$ correctly. This throws additional light on the property of canonical "conjugacy." The conjugate of q_s generates infinitesimal displacements of q_s. This has proved a valuable means means of *identifying* the variables it is important to introduce into theories under development.

Invariance Tests by Infinitesimal Transformation

The canonical transformations were originally introduced as merely formal redescriptions of a given motion, by $Q(t)$, $P(t)$, $\mathscr{H}(Q, P, t)$ in place of $q(t)$, $p(t)$, $H(q, p, t)$. They are then mere changes of viewpoint, akin to shifts of reference frame in ordinary space-time, without physical effect on the circumstances being viewed. However, the variable shifts may be *set* equal to actual physical operations,* as suggested by examples in the preceding subsection—for example, the translation of the system in motion to a new location in space. Such treatments offer means of testing the effects of the physical changes, as did the result (7.67) for the change of expenditure due to altering the path of motion. In particular, the conditions for the *invariance* of motion to the different physical circumstances may thus be found. The approach provided by the infinitesimal transformation procedure will not lead to new physical conclusions not obtainable by the former method, yet it is worth reviewing. It will implement an expectation that physical changes to which a motion is invariant ought to be equivalent to mere changes of viewpoint. Such an expectation is a generalization from regarding a bodily shift in space of an entire, isolated system as equivalent to an opposite shift of reference frame.

---

* This is sometimes described as a change from a "passive" to an "active" transformation.

To see how the invariance tests may be couched in the language of the infinitesimal transformations, consider first the space translations discussed in connection with Figure 8.2. Invariance to the shift implies that the displaced path is a physically possible one and that $\mathbf{R}_i(t) = \mathbf{r}_i(t) + \mathbf{\Delta r}$, $\mathbf{P}_i(t) = \mathbf{p}_i(t)$ is another solution of the same equations of motion, with the same hamiltonian, as are satisfied by $\mathbf{r}_i(t)$, $\mathbf{p}_i(t)$. The values $H(\mathbf{r}(t), \mathbf{p}(t), t) = H(\mathbf{R}(t), \mathbf{P}(t), t)$ must match at every moment t if the equations are to be the same; hence the invariance requirement is

$$\Delta H \equiv H(\mathbf{r} + \mathbf{\Delta r}, \mathbf{p}, t) - H(\mathbf{r}, \mathbf{p}, t) = 0. \tag{8.31a}$$

This is already sufficient for seeing how the momentum conservation requirement (7.70) arises, since $\Delta H = \sum_i \nabla_i H \cdot \mathbf{\Delta r}$ and the canonical equations give $\nabla_i H = -\dot{\mathbf{p}}_i$; then $\Delta H = -\mathbf{\Delta r} \cdot \sum_i \dot{\mathbf{p}}_i = 0$ as anticipated. However, it is desired that this be put into the language of the canonical transformations, when $\mathbf{R}_i \equiv \mathbf{r}_i + \mathbf{\Delta r}_i$, $\mathbf{P}_i \equiv \mathbf{p}_i$ are treated as new canonical variables and $H(\mathbf{R}, \mathbf{P}, t)$ is identified with $\mathscr{H}(\mathbf{R}, \mathbf{P}, t)$. The particular transformation here is an infinitesimal one, with $\boldsymbol{\delta}\mathbf{r}_i = \mathbf{\Delta r}$ and $\boldsymbol{\delta}\mathbf{p}_i = 0$ for every i. In this language, the invariance requirement (8.31a) is rewritten

$$\delta H \equiv H(\mathbf{r} + \boldsymbol{\delta}\mathbf{r}, \mathbf{p}, t) - H(\mathbf{r}, \mathbf{p}, t) = 0 \tag{8.31b}$$

(that is, as an invariance of the hamiltonian to an infinitesimal canonical transformation).

Invariance to any set of variable shifts $\delta q_1, \ldots, \delta q_f$ and $\delta p_i, \ldots, \delta p_f$ is reducible to requiring an invariance of the hamiltonian:

$$\delta H \equiv H(q + \delta q, p + \delta p, t) - H(q, p, t) = 0. \tag{8.32}$$

In general,

$$\begin{aligned} \delta H &= \sum_s [(\partial H/\partial q_s)\, \delta q_s + (\partial H/\partial p_s)\, \delta p_s] \\ &= \sum_s [-\dot{p}_s\, \delta q_s + \dot{q}_s\, \delta p_s] \end{aligned} \tag{8.33}$$

follows from the canonical equations. Then the conclusions drawn from $\delta H = 0$ depend on the particular nature of the shifts δq, δp to which there is supposed to be such invariance.

The transformation $q, p \to Q \equiv q + \delta q$, $P \equiv p + \delta p$ is a canonical one, since Q, P are to obey canonical equations of motion, with

$$\mathscr{H}(Q, P, t) \equiv H(Q, P, t) = H(q + \delta q, p + \delta p, t). \tag{8.34a}$$

It must therefore be possible to generate it by the procedure (8.24) (8.25) with some choice of $G(q, p)$. This generator cannot contain the time explicitly for transformations to which there is invariance, since for these transformations

$$\delta H \equiv \mathscr{H}(Q, P, t) - H(q, p, t) = \varepsilon\, \partial G/\partial t = 0 \tag{8.34b}$$

is to be required. With δq, δp being generated by $G(q, p)$, the expression (8.33) becomes

$$\delta H = -\varepsilon \sum_s [(\partial G/\partial p_s)\dot{p}_s + (\partial G/\partial q_s)\dot{q}_s] = -\varepsilon dG/dt. \tag{8.35}$$

Thus the invariance $\delta H = 0$ leads to $\dot{G} = 0$, and it is learned that transformations to which the hamiltonian is invariant are always generated by quantities $G(q, p)$, which are *conserved* during the motion. Conversely, any combination of canonical variables which forms a constant of the motion generates a transformation to which the motion is invariant.

Examples of infinitesimal generators which are constants of the motion are provided by invariances to the space-time displacements discussed in the preceding subsection. The total momentum (8.27*b*) generates spatial translations and is conserved in motions invariant to the translations. The angular momentum (8.29*b*) generates infinitesimal rotations and is conserved in motions invariant to the reorientations. Likewise, the hamiltonian energy is conserved in motions invariant to the displacements in time which it generates.

The requirement for invariance to the shift of times indicated in Figure 8.1 might be initiated in a way which gives it a form different from the purportedly general one (8.32). For the same phases of motion $Q(t) = q(t + \delta t)$, $P(t) = p(t + \delta t)$ to be attained at the two different times, the canonical equations of motion must have

$$H(Q(t), P(t), t) = H(q(t + \delta t), p(t + \delta t), t + \delta t) \equiv H(Q(t), P(t). t + \delta t).$$

Thus required is $\delta H \equiv \delta t(\partial H/\partial t) = 0$—that is, an invariance to a shift in just the explicit time-dependence, already seen in (8.10) as the condition for energy conservation. On the other hand,

$$\delta q = q(t + \delta t) - q(t) = \dot{q}\delta t \qquad \text{and} \qquad \delta p = \dot{p}\delta t,$$

and hence the change of hamiltonian (8.33) thus generated is

$$\delta H = \sum_s [-\dot{p}_s \dot{q}_s \delta t + \dot{q}_s \dot{p}_s \delta t] = 0,$$

and so the condition in the form (8.32) is also satisfied. The reason this happens, of course, is that what is really demanded for invariance to time-displacements is a constancy of the hamiltonian itself. Not only must the hamiltonian values be the same at the two points marked $Q(t)$ and $q(t + \delta t)$ in Figure 8.1 (with $\delta H \equiv (\partial H/\partial t)\delta t = 0$), but also at the two points marked $Q(t)$ and $q(t)$; that is, δH due to $\delta q = Q(t) - q(t)$ and $\delta p = P(t) - p(t)$ must vanish, and indeed the hamiltonian energy must be the same on all points of both paths. These findings emphasize that the energy conservation is *not* essential for the invariances discussed above, for linear or angular momenta to be conserved.

8.3 POISSON BRACKETS

The questions of invariance in the preceding section turned on the requirement (8.32) on the change of hamiltonian $\delta H(q, p, t)$ generated by an infinitesimal change in the variables: δq, δp. There is interest in the change thus produced in any function $F(q, p, t)$ of the canonical variables,

$$\begin{aligned}\delta F &= \sum_s \left[\left(\frac{\partial F}{\partial q_s}\right)\delta q_s + \left(\frac{\partial F}{\partial p_s}\right)\delta p_s\right] \\ &= \varepsilon \sum_s \left[\left(\frac{\partial F}{\partial q_s}\right)\left(\frac{\partial G}{\partial p_s}\right) - \left(\frac{\partial F}{\partial p_s}\right)\left(\frac{\partial G}{\partial q_s}\right)\right] \equiv \varepsilon[F, G],\end{aligned} \tag{8.36}$$

when δq, δp are generated by $G(q, p, t)$ as in (8.24). The final bracket symbol $[F, G]$ is *defined* to stand for the indicated combination of derivatives. Such a combination is called a "Poisson bracket" of the pair of quantities F and G, each of which may be any function of the canonical variables. Poisson brackets of various pairs of canonical functions form significant elements in expressions of Hamiltonian Theory. They derive importance from their possession of "canonical invariance." This means that a Poisson bracket has the same value whether the derivatives are taken with respect to a set q, p or with respect to any other set Q, P which is a canonical transform of q, p. The basis of this assertion will be discussed in the next section; here, Poisson bracket expressions on a given basis q, p will be investigated.

Fundamental Properties of Poisson Brackets

It is immediately obvious that

$$[G, F] = -[F, G] \qquad \text{and} \qquad [F, F] = 0. \tag{8.37}$$

Any pair of functions for which $[F, G] = 0$ occurs will be said to "commute" with each other "in a Poisson bracket sense." Thus, any function "Poisson-commutes" with itself.

When either F of G is chosen to be one of the conjugate variables q_s or p_s, relations like

$$[q_s, G] = \partial G/\partial p_s \qquad \text{and} \qquad [p_s, G] = -\partial G/\partial q_s \tag{8.38}$$

develop. If εG is used to generate an infinitesimal transformation, then the brackets (8.38), multiplied by ε, given the infinitesimal variables changes δq_s and δp_s, respectively, as comparison to (8.24) shows.

The so-called "fundamental Poisson brackets" emerge when both members of each bracket are chosen from the basic phase-space variables:

$$\boxed{[q_s, q_\sigma] = 0, \qquad [p_s, p_\sigma] = 0, \qquad [q_s, p_\sigma] = \delta_{s\sigma}}\,, \tag{8.39}$$

The last, two-index, symbol here is called a "Kronecker delta" and is defined to be $\delta_{s\sigma} = 0$ for $\sigma \neq s$, while $\delta_{s\sigma} = 1$ for $\sigma = s$.

Poisson Bracketing with the Hamiltonian

The results (8.38) show how variations of any function with a given canonical variable may be represented by a Poisson bracket with the conjugate of that variable. Variations with time may be represented by a Poisson-bracketing with the variable "conjugate" to time, the hamiltonian itself:

$$[F, H] \equiv \sum_s \left(\frac{\partial F}{\partial q_s}\frac{\partial H}{\partial p_s} - \frac{\partial F}{\partial p_s}\frac{\partial H}{\partial q_s}\right) = \sum_s \left(\frac{\partial F}{\partial q_s}\dot{q}_s + \frac{\partial F}{\partial p_s}\dot{p}_s\right)$$

follows from the canonical equations. The last expression is just the time rate of change of F arising from its dependence on the motion $q(t)$, $p(t)$. It differs from the total time derivative only if $F(q, p, t)$ also depends on the time explicitly—that is,

$$\boxed{dF(q, p, t)/dt = [F, H] + \partial F/\partial t}\,. \tag{8.40}$$

Notice that this is consistent with $dH/dt = \partial H/\partial t$ (8.10), since $[H, H] \equiv 0$.

When F is chosen to be q_s or p_s in (8.40),

$$[q_s, H] = \dot{q}_s \qquad \text{and} \qquad [p_s, H] = \dot{p}_s. \tag{8.41a}$$

When $G \equiv H$ is chosen for the generator in (8.38),

$$[q_s, H] = \partial H/\partial p_s \qquad \text{and} \qquad [p_s, H] = -\partial H/\partial q_s. \tag{8.41b}$$

Thus the canonical equations of motion are implicit in the properties of Poisson brackets. Since the expressions (8.41*b*) are special cases of the infinitesimal transformation relations (8.38), they reiterate the earlier finding (8.26) that the hamiltonian generates the infinitesimal transitions in a motion.

Poisson-bracketing with the hamiltonian of the transformation generators $G(q, p, t)$ has interest because, according to (8.36), this gives the change δH in the hamiltonian caused by the substitution $q \to q + \delta q$, $p \to p + \delta p$:

$$\delta H = \varepsilon[H, G] = H(q + \delta q, p + \delta p, t) - H(q, p, t). \tag{8.42}$$

It is just the vanishing of this, $\delta H = 0$, which forms the requirement (8.32) for invariance to the variable shift. Now

$$[G(q, p), H(q, p, t)] = dG(q, p)/dt = 0 \tag{8.43}$$

in these circumstances, reiterating the finding (8.35) that it is constants of the motion which generate transformations to which the motion is invariant. The *new* conclusion here is that the constants of the motion "Poisson-commute" with the hamiltonian.

Indeed, it is immediately evident from (8.40) that any combination $F(q, p)$ of the canonical variables, not explicitly time-dependent, will be conserved during motion generated by an hamiltonian with which it Poisson-commutes:

$$dF(q, p)/dt = [F(q, p), H(q, p)] = 0. \tag{8.44}$$

Every conserved constant in the motion Poisson-commutes with the hamiltonian.

Poisson-Bracketing with Angular Momenta

The basic Poisson brackets (8.39) may be supplemented with ones of almost equal importance, formed with the angular momentum vectors $\mathbf{L}_i = \mathbf{r}_i \times \mathbf{p}_i$.

Relations like $[x_i, L_{ix}] = 0$, $[x_i, L_{iy}] = (\partial/\partial p_{ix})(zp_x - xp_z)_i = z_i$, and $[x_i, L_{iz}] = -y_i$ follow from (8.38). These three may be gathered into a single expression by representing an arbitrary component of $\mathbf{L}_i$ as its projection on an appropriately defined unit vector, $\hat{\boldsymbol{\omega}}$. Then

$$\hat{\boldsymbol{\omega}} \cdot \mathbf{L}_i = \hat{\omega}_x L_{ix} + \hat{\omega}_y L_{iy} + \hat{\omega}_z L_{iz}$$

and

$$[x_i, \hat{\boldsymbol{\omega}} \cdot \mathbf{L}_i] = \hat{\omega}_x \cdot 0 + \hat{\omega}_y z_i - \hat{\omega}_z y_i = (\hat{\boldsymbol{\omega}} \times \mathbf{r}_i)_x .$$

Moreover, $\mathbf{L}_i$ here may be replaced by the total angular momentum $\mathbf{L} = \sum_i \mathbf{L}_i$ without affecting the result, since only derivatives with respect to the variables of the ith "particle" are involved. Plainly, a general result is

$$[\mathbf{r}_i, \hat{\boldsymbol{\omega}} \cdot \mathbf{L}] = \hat{\boldsymbol{\omega}} \times \mathbf{r}_i \tag{8.45}$$

and this is already implicit in the fact that $\boldsymbol{\delta\omega} \cdot \mathbf{G} \equiv \delta\varphi\hat{\boldsymbol{\omega}} \cdot \mathbf{L}$ of (8.29) generates the rotation $\boldsymbol{\delta}\mathbf{r}_i = \delta\varphi\hat{\boldsymbol{\omega}} \times \mathbf{r}_i$. It is equally plain that

$$[\mathbf{p}_i, \hat{\boldsymbol{\omega}} \cdot \mathbf{L}] = \hat{\boldsymbol{\omega}} \times \mathbf{p}_i. \tag{8.46}$$

Thus, Poisson-bracketing of the position or momentum vectors $\mathbf{r}_i$, $\mathbf{p}_i$ with a component $\hat{\boldsymbol{\omega}} \cdot \mathbf{L}$ of the angular momentum along an axis $\hat{\boldsymbol{\omega}}$ produces what might be called "rotational derivatives," $\boldsymbol{\delta}\mathbf{r}_i/\delta\varphi$ or $\boldsymbol{\delta}\mathbf{p}_i/\delta\varphi$, about $\hat{\boldsymbol{\omega}}$.

The angular momentum vector $\mathbf{L}_i = \mathbf{r}_i \times \mathbf{p}_i$ may itself be rotated in a similar way. The component of $[\mathbf{L}, \hat{\boldsymbol{\omega}} \cdot \mathbf{L}]$ on an arbitrary direction $\mathbf{e}$ is

$$[\mathbf{e} \cdot \mathbf{L}, \hat{\boldsymbol{\omega}} \cdot \mathbf{L}] = \sum_i \left\{\frac{\partial}{\partial \mathbf{r}_i}(\mathbf{e} \cdot \mathbf{L}_i)\frac{\partial}{\partial \mathbf{p}_i}(\hat{\boldsymbol{\omega}} \cdot \mathbf{L}) - \frac{\partial}{\partial \mathbf{p}_i}(\mathbf{e} \cdot \mathbf{L}_i) \cdot \frac{\partial}{\partial \mathbf{r}_i}(\hat{\boldsymbol{\omega}} \cdot \mathbf{L})\right\},$$

where a notation like that used in (7.49*b*), for the gradient types of derivatives, has again been found most convenient. The derivatives here can themselves be equated to Poisson brackets via (8.38), and then

$$\begin{aligned}[\mathbf{e} \cdot \mathbf{L}, \hat{\boldsymbol{\omega}} \cdot \mathbf{L}] &= -\sum_i \{[\mathbf{p}_i, \mathbf{e} \cdot \mathbf{L}_i] \cdot [\mathbf{r}_i, \hat{\boldsymbol{\omega}} \cdot \mathbf{L}] - [\mathbf{r}_i, \mathbf{e} \cdot \mathbf{L}_i] \cdot [\mathbf{p}_i, \hat{\boldsymbol{\omega}} \cdot \mathbf{L}]\} \\ &= -\sum_i \{(\mathbf{e} \times \mathbf{p}_i) \cdot (\hat{\boldsymbol{\omega}} \times \mathbf{r}_i) - (\mathbf{e} \times \mathbf{r}_i) \cdot (\hat{\boldsymbol{\omega}} \times \mathbf{p}_i)\},\end{aligned}$$

with the last line following from (8.45) and (8.46). Manipulations of the multiple products give

$$\begin{aligned}[\mathbf{e} \cdot \mathbf{L}, \hat{\boldsymbol{\omega}} \cdot \mathbf{L}] &= -\mathbf{e} \cdot \sum_i \{\mathbf{p}_i \times (\hat{\boldsymbol{\omega}} \times \mathbf{r}_i) - \mathbf{r}_i \times (\hat{\boldsymbol{\omega}} \times \mathbf{p}_i)\} \\ &= +\mathbf{e} \cdot \sum_i \{\mathbf{r}_i(\hat{\boldsymbol{\omega}} \cdot \mathbf{p}_i) - \mathbf{p}_i(\hat{\boldsymbol{\omega}} \cdot \mathbf{r}_i)\} \\ &= \mathbf{e} \cdot \left\{\hat{\boldsymbol{\omega}} \times \left(\sum_i \mathbf{r}_i \times \mathbf{p}_i\right)\right\} \equiv \mathbf{e} \cdot (\hat{\boldsymbol{\omega}} \times \mathbf{L}).\end{aligned} \tag{8.47}$$

Since the direction $\mathbf{e}$ was chosen arbitrarily, the result

$$[\mathbf{L}, \hat{\boldsymbol{\omega}} \cdot \mathbf{L}] = \hat{\boldsymbol{\omega}} \times \mathbf{L} \tag{8.48}$$

is valid, and this is the expected rotation: $\boldsymbol{\delta}\mathbf{L}/\delta\varphi$.

The expression (8.47) vanishes for parallel components ($\mathbf{e} \cdot \hat{\boldsymbol{\omega}} = 1$) of the angular momentum factors, as should any Poisson bracket of a quantity with itself. When projections along the perpendicular x- and y-axis are considered ($\mathbf{e} \equiv \mathbf{i}$, $\hat{\boldsymbol{\omega}} \equiv \mathbf{j}$), the result is

$$[L_x, L_y] = (\mathbf{i} \times \mathbf{j}) \cdot \mathbf{L} = L_z,$$

and, more generally,

$$\boxed{[L_k, L_l] = L_m} \tag{8.49}$$

when the cartesian components are put in "cyclic" order ($k, l, m \equiv x, y, z$ or y, z, x or z, x, y). These are the basic Poisson brackets of angular momenta. They reveal the important fact that no more than one cartesian component of an angular momentum can serve as a conjugate variable in any canonical formulation. Any pair of canonical momenta must Poisson-commute according to the basic properties (8.39), and (8.49) shows that the angular momentum components do not Poisson-commute.

Each component of the angular momentum does Poisson-commute with the magnitude $L^2 = L_x^2 + L_y^2 + L_z^2 \equiv \sum_k L_k^2$:

$$\begin{aligned}
[L^2, \hat{\boldsymbol{\omega}} \cdot \mathbf{L}] &= \sum_i \left\{\left(\frac{\partial}{\partial \mathbf{r}_i} L_i^2\right) \cdot \left(\frac{\partial}{\partial \mathbf{p}_i} \hat{\boldsymbol{\omega}} \cdot \mathbf{L}\right) - \left(\frac{\partial}{\partial \mathbf{p}_i} L_i^2\right) \cdot \left(\frac{\partial}{\partial \mathbf{r}_i} \hat{\boldsymbol{\omega}} \cdot \mathbf{L}\right)\right\} \\
&= \sum_{i,k} 2L_{ik}\left\{\left(\frac{\partial}{\partial \mathbf{r}_i} L_{ik}\right) \cdot \left(\frac{\partial}{\partial \mathbf{p}_i} \hat{\boldsymbol{\omega}} \cdot \mathbf{L}\right) - \left(\frac{\partial}{\partial \mathbf{p}_i} L_{ik}\right) \cdot \left(\frac{\partial}{\partial \mathbf{r}_i} \hat{\boldsymbol{\omega}} \cdot \mathbf{L}\right)\right\} \\
&= 2\sum_i \mathbf{L}_i \cdot [\mathbf{L}_i, \hat{\boldsymbol{\omega}} \cdot \mathbf{L}] = 2\sum_i \mathbf{L}_i \cdot (\hat{\boldsymbol{\omega}} \times \mathbf{L}_i) = 0.
\end{aligned}$$

Thus

$$[L^2, \hat{\boldsymbol{\omega}} \cdot \mathbf{L}] = 0,$$

the commutation property in question. It may also be written as a vector equation

$$\boxed{[L^2, \mathbf{L}] = 0}, \tag{8.50}$$

since $\hat{\boldsymbol{\omega}}$ could be chosen arbitrarily.

8.4 CANONICAL INVARIANTS

Certain generalities about phase space descriptions of motion have interest which extends beyond the domain of Classical Mechanics "proper." The generalities in question are associated with the canonical invariance of various quantities and relations, forms that are equally valid regardless of which particular set of canonical variables is used in their expression. Specific examples will be reviewed in this section.

The fundamental canonical equations of motion themselves have canonical invariance, as exhibited by (8.5) and (8.13). They have it by definition, actually, since what constitutes a canonical transformation $q, p, H \to Q, P, \mathscr{H}$ has been defined as one that leads to the same forms for the canonical equations expressed in either the old or new variables.

The Invariance of Poisson Brackets

It has already been mentioned that Poisson bracket expressions are canonical invariants. This means in detail that

$$[F, G]_{q,p} \equiv \sum_s \left(\frac{\partial F}{\partial q_s}\frac{\partial G}{\partial p_s} - \frac{\partial F}{\partial p_s}\frac{\partial G}{\partial q_s} \right)$$

$$= \sum_\sigma \left(\frac{\partial F}{\partial Q_\sigma}\frac{\partial G}{\partial P_\sigma} - \frac{\partial F}{\partial P_\sigma}\frac{\partial G}{\partial Q_\sigma} \right) \equiv [F, G]_{Q,P} \qquad (8.51)$$

when the q, p and the Q, P are related by a canonical transformation like (8.14) or (8.17). The proof here will be restricted to the infinitesimal transformations (8.24), on the basis that the finite transformations may be approached by infinitesimal steps. More complete proofs have been constructed but are perhaps no more illuminating.

As a first step, consider the "fundamental" Poisson brackets (8.39). The similar brackets containing Q's and P's in place of the q's and p's are constructed with the help of the derivatives

$$\frac{\partial Q_i}{\partial q_s} = \delta_{is} + \varepsilon \frac{\partial^2 G}{\partial q_s \, \partial p_i}, \qquad \frac{\partial Q_i}{\partial p_s} = \varepsilon \frac{\partial^2 G}{\partial p_s \, \partial p_i},$$

$$\frac{\partial P_j}{\partial q_s} = -\varepsilon \frac{\partial^2 G}{\partial q_s \, \partial q_j}, \qquad \frac{\partial P_j}{\partial p_s} = \delta_{js} - \varepsilon \frac{\partial^2 G}{\partial p_s \, \partial q_j},$$

which follow from (8.24). The insertion of these into the defining expression (8.51) of a Poisson bracket yields, for example,

$$[Q_i, P_j]_{q,p} = \sum_s \left\{ \delta_{is}\,\delta_{js} + \varepsilon\,\delta_{js} \frac{\partial^2 G}{\partial q_s \, \partial p_i} - \varepsilon\,\delta_{is} \frac{\partial^2 G}{\partial p_s \, \partial q_j} \right\},$$

after dropping terms proportional to ε^2, as is required for a consistent application of the infinitesimal transformation procedure. In the sum of terms, only those with $s = i$ or $s = j$ survive, and therefore

$$[Q_i, P_j]_{q,p} = \delta_{ij} + \varepsilon \left(\frac{\partial^2 G}{\partial q_j \, \partial p_i} - \frac{\partial^2 G}{\partial p_i \, \partial q_j} \right).$$

The parenthesis here vanishes identically and the fundamental Poisson bracket result,

$$[Q_i, P_j] = \delta_{ij}, \tag{8.52a}$$

follows. The canonically invariant transforms of the others in (8.39),

$$[Q_i, Q_j] = 0 \qquad \text{and} \qquad [P_i, P_j] = 0, \tag{8.52b}$$

can be obtained in similar ways.

Next consider the general Poisson bracket (8.51). Its invariance can be proved from the fundamental brackets (8.52) without further resort to the transformation procedures. The bracket $[F, G]_{q,p}$ may be constructed with G treated as a function of the new variables Q, P; that is,

$$\frac{\partial G}{\partial q_s} = \sum_\sigma \left\{ \frac{\partial G}{\partial Q_\sigma} \frac{\partial Q_\sigma}{\partial q_s} + \frac{\partial G}{\partial P_\sigma} \frac{\partial P_\sigma}{\partial q_s} \right\},$$

$$\frac{\partial G}{\partial p_s} = \sum_\sigma \left\{ \frac{\partial G}{\partial Q_\sigma} \frac{\partial Q_\sigma}{\partial p_s} + \frac{\partial G}{\partial P_\sigma} \frac{\partial P_\sigma}{\partial p_s} \right\}.$$

Substitution into (8.51) is readily seen to yield, after some rearrangement of terms,

$$[F, G]_{q,p} = \sum_\sigma \left\{ \frac{\partial G}{\partial Q_\sigma} [F, Q_\sigma]_{q,p} + \frac{\partial G}{\partial P_\sigma} [F, P_\sigma]_{q,p} \right\}. \tag{8.53}$$

A new expression for $[F, Q_\sigma]_{q,p} = -[Q_\sigma, F]_{q,p}$ can be obtained by replacing F with Q_σ and G with F in (8.53):

$$[Q_\sigma, F]_{q,p} = \sum_\mu \left\{ \frac{\partial F}{\partial Q_\mu} [Q_\sigma, Q_\mu]_{q,p} + \frac{\partial F}{\partial P_\mu} [Q_\sigma, P_\mu]_{q,p} \right\},$$

which reduces to

$$[Q_\sigma, F] = \partial F/\partial P_\sigma \tag{8.54a}$$

because of (8.52). The last is just the canonical transform of the first type of equation (8.38), and the second can be obtained similarly,

$$[P_\sigma, F] = -\partial F/\partial Q_\sigma, \tag{8.54b}$$

thus exhibiting a canonical invariance of the forms (8.38). Putting these results into (8.53) yields

$$[F, G]_{q,p} = [F, G]_{Q,P}, \tag{8.55}$$

just the conclusion sought. Because of the invariance, specification of the variables as by the subscripts here is unnecessary.

Phase Volumes

Of particular interest to Statistical Mechanics is the invariance of the volume elements of phase space:

$$d\Omega \equiv dq_1\, dq_2 \cdots dq_f\, dp_1\, dp_2 \cdots dp_f = dQ_1 \cdots dQ_f\, dP_1 \cdots dP_f. \quad (8.56)$$

This derives importance from a frequently employed hypothesis: lacking any further information, the chance that a system has its variables within an interval $d\Omega$, in a "uniformly available" part of the phase space, is proportional to the size of $d\Omega$. Plainly, such a hypothesis is tenable only if it is invariant to the particular variables used in the description.

The proof of the invariance (8.56) will only be illustrated for the case of one degree of freedom ($f = 1$); the complete proof is not essentially different in essence, and is considerably lengthier.

Define two-dimensional vectors **dq** and **dp** along the appropriate orthogonal directions of the $f = 1$ phase "plane." Then $d\Omega \equiv dq\, dp = |\mathbf{dq} \times \mathbf{dp}|$. This is to be expressed in terms of the new variables Q, P:

$$\begin{aligned} \mathbf{dq}(Q, P) &= (\partial q/\partial Q)_P\, \mathbf{dQ} + (\partial q/\partial P)_Q\, \mathbf{dP}, \\ \mathbf{dp}(Q, P) &= (\partial p/\partial Q)_P\, \mathbf{dQ} + (\partial p/\partial P)_Q\, \mathbf{dP}, \end{aligned} \quad (8.57)$$

where $\mathbf{dQ} \cdot \mathbf{dP} = 0$ because of what is meant by partial derivatives. It follows that

$$d\Omega = |\mathbf{dq} \times \mathbf{dp}| = \left| \left(\frac{\partial q}{\partial Q}\right)_P \left(\frac{\partial p}{\partial P}\right)_Q - \left(\frac{\partial q}{\partial P}\right)_Q \left(\frac{\partial p}{\partial Q}\right)_P \right| |\mathbf{dQ} \times \mathbf{dP}|. \quad (8.58)$$

The combination of partial derivatives here is just the fundamental Poisson bracket $[q, p] = 1$, and hence

$$d\Omega \equiv dq\, dp = dQ\, dP, \quad (8.59)$$

the anticipated invariance. Such ratios of volume elements as the combination of derivatives in (8.58) are known as "Jacobian determinants," in this case denoted by

$$\frac{\partial(q, p)}{\partial(Q, P)} \equiv \begin{vmatrix} (\partial q/\partial Q)_P & (\partial q/\partial P)_Q \\ (\partial p/\partial Q)_P & (\partial p/\partial P)_Q \end{vmatrix}. \quad (8.60)$$

The value of this determinant (9.65) is just the Poisson bracket $[q, p]$.

The proof given here depends on the canonical invariance of the fundamental Poisson brackets, proved in the preceding subsection for infinitesimal transformations. In the $f = 1$ case, the proof for finite transformations is also quite brief. According to (8.17) the canonical transformation may be generated by some $S(q, P, t)$ with

$$p = (\partial S/\partial q)_P, \qquad Q = (\partial S/\partial P)_q.$$

This gives p as a function of q, P, and so

$$\left(\frac{\partial p}{\partial P}\right)_Q = \left(\frac{\partial^2 S}{\partial q^2}\right)\left(\frac{\partial q}{\partial P}\right)_Q + \frac{\partial^2 S}{\partial P\,\partial q}, \qquad \left(\frac{\partial p}{\partial Q}\right)_P = \frac{\partial^2 S}{\partial q^2}\left(\frac{\partial q}{\partial Q}\right)_P.$$

Substituting these expressions into the above Jacobian results in a cancellation of two terms, leaving

$$[q, p]_{Q,P} = \left(\frac{\partial q}{\partial Q}\right)_P \frac{\partial^2 S}{\partial P\,\partial q} = \left(\frac{\partial q}{\partial Q}\right)_P \left(\frac{\partial Q}{\partial q}\right)_P.$$

Now the expression (8.57) for the differential elements shows that

$$(\partial Q/\partial q)_P = 1/(\partial q/\partial Q)_P,$$

and hence $[q, p] = 1$, the anticipated result.

An important example of the phase-volume invariance,

$$dx\,dy\,dp_x\,dp_y = dr\,d\varphi\,dp_r\,dp_\varphi,$$

is easy to check directly. Reference to page 196 leads to $dx\,dy = dr \cdot r\,d\varphi$ and $dp_x\,dp_y = r^{-1}\,dp_r\,dp_\varphi$, showing that configuration space elements $dq_1 \cdots dq_f$ and momentum space elements $dp_1 \cdots dp_f$ are not in general individually invariant, but only their product is.

Liouiville's Theorem

An important generalization known as Liouiville's Theorem follows from the invariance of phase-volume elements. The contexts in which the theorem acquires meaning should be described first.

The instantaneous state of motion, or phase, of any system is known completely only when the values of all the variables $q_1(t), \ldots, q_f(t)$, $p_1(t), \ldots, p_f(t)$ are given; that is, a "system-point" is definitely located in the phase space. Once it is so located at some initial time, then its future locations are as definitely determinable, from the equations of motion. As time goes on, the system-point describes a well-defined path in the phase space.

In some problems the given information may not be complete enough to locate a unique phase-space point. For example, it may only be known that the position coordinates of a set of particles are within some vessel, and that the momenta correspond to some given total energy. Then it is known only that the system-point is somewhere within an "available" volume of the phase space. A variety of motions are judged equally likely, one for each point from which the motion might have started in consistency with the data. All these equally possible motions should be considered in coming to conclusions about the possible results. It is as if a whole "ensemble" of equivalent, independent systems were to be treated, all governed by the same equations of motion but differing as to details of the starts given to their motions.

It is in this context that the statistical hypothesis mentioned just below (8.56) is made. It is judged that the chance of finding a system-point within a given element of the available phase-volume is proportional to the size of the element—that is, that this size of continuous volume is proportional to the number of the equally likely (so-judged!) system-points it contains. To be tenable, such an assumption must be possible to make at any phase of the motions, and will be possible only if the system-points forming a given initial phase-volume continue to occupy a volume of the same size as the motions continue. That this is indeed so, under the regime of canonical equations of motion, is just the burden of the Liouiville theorem.

In proving the theorem, a continuum of dN system-points of the ensemble, forming a simply connected volume-element $d\Omega$ at some initial instant, is considered. As the motions of the systems proceed, the points continue to form a simply connected volume, since the path of each is governed by differential equations demanding continuity of solution. There can never be escape of a point through the continuous volume surface, since that would entail a crossing of paths, whereas the equations of motion determine a unique path from any phase-space point.

The displacements of the dN system-points in $d\Omega$, which will occur during the physical motions, bring them to new locations which are just canonical transformations of the initial points; that motion is equivalent to canonical transformations generated by the hamiltonian was found in leading up to (8.26). It follows that the new volumes occupied by the dN points are just canonical transforms of $d\Omega$, and this was shown to be invariant in the preceding subsection. Thus the dN points will continue to form a volume of the size of $d\Omega$, even though it may move in the space and may suffer changes of shape.

The Liouiville theorem may be expressed formally in terms of a *density* distribution of the ensemble of system-points over the phase space:

$$D(q, p, t) \equiv dN/d\Omega. \tag{8.61}$$

Some choice is to be made for the initial ensemble distribution, $D(q, p, 0)$, depending on the data at hand (for example, $D(q, p, 0) = 0$ at "unavailable" points q, p when "availability" is understood as indicated above). The theorem has it that the $d\Omega$ occupied by the given dN is independent of time (motion), and hence that $dD/dt = 0$. Expressed as in (8.40),

$$dD/dt = [D, H] + \partial D/\partial t = 0. \tag{8.62}$$

This must hold for any distribution D which may be adopted as representative of whatever information has been given.

An important class of cases is expected to have $\partial D/\partial t = 0$—that is, those cases for which the chance of finding the system in any given phase q, p of motion is expected to be independent of time. Such systems are said to be in a steady state of "equilibrium." The distribution $D(q, p)$ of phase-points at which such a system might be found, if trial were possible and made, must "Poisson-commute" with the hamiltonian,

$$[D(q, p), H] = 0, \tag{8.63}$$

and D is a constant of the motion (8.44). Expressions containing q, p in combinations forming constants of the motion are therefore used to represent the ensemble distributions, $D(q, p)$, of systems in equilibrium.

An Example of a "Phase-Space Spectrum"

The abstractions of this section will now be leavened by a specific example of how statistical considerations of the type alluded to may be used.

The nucleus of a beryllium atom, whenever it is formed with a certain excess of "excitation" energy, explodes into three fragments: one is a "neutron" of mass m, and each of the other two is an "alpha-particle" of mass $4m$. The question is how the energy released, W, is shared among the fragments when the process is free to occur in any way consistent with energy and momentum conservation.

If $\mathbf{p}$ is the momentum acquired by the neutron and $\mathbf{P}_1$, $\mathbf{P}_2$ are the momenta of the alpha-particles (in a given instance), then

$$\begin{aligned} W &= p^2/2m + (P_1^2 + P_2^2)/8m. \\ 0 &= \mathbf{p} + \mathbf{P}_1 + \mathbf{P}_2 . \end{aligned} \tag{8.64}$$

It is convenient to work with the total (CM) momentum of the two alpha-particles $\mathbf{P}_1 + \mathbf{P}_2 = -\mathbf{p}$, and with their "internal" (5.13) momentum, $\mathbf{P} = \frac{1}{2}(\mathbf{P}_1 - \mathbf{P}_2)$. The momentum variables $\mathbf{p}$ and $\mathbf{P}$ are then subject only to the energy conservation condition,

$$W = 9p^2/16m + P^2/4m. \tag{8.65}$$

The neutron momentum may according to this have any magnitude between $p = 0$ and $p = p_0 \equiv \frac{4}{3}(mW)^{1/2}$.

Since **p** and **P** are unrestricted by the momentum conservation condition, and their directions are totally unrestricted (it is rather $\mathbf{P}_{1,2} = -\frac{1}{2}\mathbf{p} \pm \mathbf{P}$ which are restricted, for a given **p** and **P**), the process can presumably take place with these variables falling into any part of a momentum-space volume* proportional to

$$\Phi = \int 4\pi p^2 \, dp \int 4\pi P^2 \, dP. \tag{8.66a}$$

A finite range of integration $\Delta P = \int dP$, *for a given p-value*, would be permissible only if there were a range of possibilities $\Delta W = P\,\Delta P/2m$ for the energy release, because of the energy conservation condition (8.65). It is results in the limit $\Delta W \to 0$ which will be of interest.

For vanishingly small ΔW, the "available phase volume" is proportional to

$$\Phi = 32\pi^2 m \,\Delta W \int_0^{p_0} P p^2 \, dp = 3\pi^3 m p_0^4 \,\Delta W, \tag{8.66b}$$

since $P = \frac{3}{2}(p_0^2 - p^2)^{1/2}$ is required for the energy conservation. What has more interest is the *fraction df* of the possible processes which can be expected to occur with, say, the neutron momentum magnitude p in a range dp:

$$df = 48\pi^2 m \,\Delta W \, p^2 (p_0^2 - p^2)^{1/2} \, dp/\Phi. \tag{8.67}$$

The factor ΔW cancels out when the above result for Φ is used; hence a conclusion

$$df/dp = (16/\pi p_0^4) p^2 (p_0^2 - p^2)^{1/2} \tag{8.68}$$

is reached. Such a result is known as a "statistical" or "phase-space" spectrum. It presents the expectations for the fractions of a very large number of events which will emit neutrons having the various possible momenta, $0 < p < p_0$. It can be seen that most of the neutrons are expected to acquire the momentum $(\frac{2}{3})^{1/2} p_0$, and vanishingly small numbers to have $p \approx 0$ or $p \approx p_0$. Of course, such expectations are relative to what phases have been initially taken as equally likely.

---

* It is thus being assumed that the momentum-space factors of the phase-volume elements are invariant throughout the motions after the explosion, *separately* from the ordinary-space volume elements. This corresponds to treating the particles as having some well-defined individual linear momenta as soon as they are formed at the point of explosion, and as having those momenta uninfluenced by the relative particle-positions thereafter (there is no "final-state interaction").

The *energy* spectrum corresponding to (8.68) is obtained by substituting $E = p^2/2m$ and $dE = p\,dp/m$:

$$df/dE = (81/8\pi W^2)[E(8W/9 - E)]^{1/2}. \tag{8.69}$$

Plotted against E, this has a semielliptical shape ranging between zeroes at $E = 0$ and $E = E_0 = 8W/9$; its maximum is in the middle of the range, at $E = \frac{1}{2}E_0 = 4W/9$. When observations conform to such "statistical" expectations, it is concluded that there is no appreciable interaction, among the dispersing fragments, to disturb the randomness of emissions restricted only by energy and momentum conservation.

8.5 THE HAMILTON–JACOBI FORMULATION

It is in principle desirable to choose such a set of canonical variables that as many as possible become ignorable. For each ignorable coordinate q_i, a canonical equation $\dot{p}_i = -\partial H/\partial q_i = 0$ becomes trivial to solve, and a conservation relation, $p_i =$ constant in the motion, is formed. The value of conservation relations for obtaining descriptions of motion has been demonstrated repeatedly; for example, the motion in a central force field was derived in §4.2 from just the conservation of energy E and of angular momentum $\mathbf{L}$.

Variables which become ignorable may be sought with the help of the canonical transformation-generating procedure of (8.17) and (8.18). This will be found powerful enough to enable introducing new canonical variables Q, P, *all* of which become ignorable. That is, the new hamiltonian $\mathscr{H}$ will contain none of them, so that the canonical equations of motion reduce to triviality,

$$\dot{Q}_s = \partial\mathscr{H}/\partial P_s = 0 \qquad \text{and} \qquad \dot{P}_s = -\partial\mathscr{H}/\partial Q_s = 0, \tag{8.70}$$

and each of the $2f$ "variables" $Q_1, \ldots, Q_f, P_1, \ldots, P_f$ is a constant of the motion. Indeed, the transformation $H(q, p, t) \to \mathscr{H}$ may be made such that any initial explicit time-dependence is also eliminated and a gauge in which $\mathscr{H} \equiv 0$ may be chosen. This reduction of the equations of motion to merely $\dot{Q}_s = 0$, $\dot{P}_s = 0$, changes the problem to one of finding a transformation-generator $S(q, P, t)$ which achieves that result. This constitutes the "Hamilton–Jacobi" approach to the solution of problems in Mechanics.

The General Hamilton–Jacobi Equation

The transformation-generating procedure of (8.17) and (8.18) requires adopting a generator $S(q, P, t)$ that then relates any given set of conjugate variables q, p to a new set Q, P via $p_s = \partial S/\partial q_s$ and $Q_s = \partial S/\partial P_s$. Meanwhile, a new hamiltonian $\mathscr{H}(Q, P, t) = H(q, p, t) + \partial S/\partial t$ is formed. It becomes obvious that a generator that is to produce $\mathscr{H} \equiv 0$ will be subject to the differential equation

$$H(q, \partial S/\partial q, t) + \partial S/\partial t = 0, \tag{8.71}$$

formed by substituting a differential $\partial S/\partial q_s$ for each p_s occurring in the initial hamiltonian. This is called the general, or "time-dependent," Hamilton–Jacobi differential equation.

The special generating functions $S(q, P, t)$ that produce $\mathscr{H} \equiv 0$, and thereby recast the description into terms of constants-in-the-motion Q_s, P_s, acquire a significance that is more nearly physical than their role as mathematical devices for variable substitution. Consider the change undergone by $S(q, P, t)$ during a time-interval, dt, of actual motion:

$$dS(q, P, t) = \sum_s (\partial S/\partial q_s)\, dq_s + (\partial S/\partial t)\, dt,$$

since $dP_s = \dot{P}_s\, dt = 0$ during the motion (8.70). Because $\partial S/\partial q_s = p_s$ and $\partial S/\partial t = -H$ when $\mathscr{H} \equiv 0$,

$$dS = \sum_s p_s\, dq_s - H\, dt \equiv L\, dt = dI \tag{8.72}$$

is an element of what was called the "expenditure" during motion, in the form already seen in (8.6).

A first point to notice about the Hamilton–Jacobi equation (8.71) is that, although a function $S(q, P, t)$ of not only the q's and t, but also of the P's, is to be found, the equation puts restrictions only on the way S must depend on the q's and the time. Thus, arbitrariness is still left in the way that $P_1, \ldots, P_f$ will enter into $S(q, P, t)$ despite the highly restricted objective of the transformation, namely an $\mathscr{H} \equiv 0$. This may not be entirely unexpected. First, the P's are to be constants in the motion and so may be equated to the various arbitrary integration constants that will characterize a general solution of such a differential equation as (8.71). Second, any combination of constants of the motion is another constant of the motion; hence a remaining wide variety of specific choices for the $P_1, \ldots, P_f$ is only to be expected.

How the Hamilton–Jacobi equation helps determine a generator $S(q, P, t)$ for any specific case is best demonstrated in connection with a definite example. This will avoid the excessive abstraction involved in referring to contingencies which may arise in various cases without a specific one in mind, yet will not prevent the drawing of sufficiently general conclusions. The example to be treated is the ($f = 2$)-dimensional configuration problem of a particle in a central-force field (§4.2). It will have the advantage of showing how the use of the conservation of E and L in that problem fits into the present scheme, with its objective of making maximum use of just such conservation relations.

The initial hamiltonian for the central-force problem is just the one (8.3), with a potential $V(r)$. When $p_r = \partial S/\partial r$ and $p_\varphi = \partial S/\partial \varphi$ are substituted for the conjugate momenta, the Hamilton–Jacobi equation

$$\frac{1}{2m}\left(\frac{\partial S}{\partial r}\right)^2 + \frac{1}{2mr^2}\left(\frac{\partial S}{\partial \varphi}\right)^2 + V(r) + \frac{\partial S}{\partial t} = 0 \tag{8.73}$$

is formed. It is sometimes feasible, for such a differential equation, to find a solution which has a "separable" form as regards the variables r, φ, t:

$$S = S_r(r, P_1, P_2) + S_\varphi(\varphi, P_1, P_2) + S_t(t, P_1, P_2). \tag{8.74}$$

Although it may seem that only special solutions could have such separability, the final result will here turn out to be recognizable as a general solution. Care has been taken to indicate in (8.74) that new "conjugate momenta" are in the end to be identified in $S(r, \varphi, P_1, P_2, t)$, although they do not appear explicitly in the Hamilton–Jacobi equation. The P's have the role of integration constants as regards the differential equation.

For solutions of the form (8.74), the differential equation first yields

$$\frac{\partial}{\partial t} S_t(t, P_1, P_2) = -\frac{1}{2m}\left[\left(\frac{\partial S_r}{\partial r}\right)^2 + \frac{1}{r^2}\left(\frac{\partial S_\varphi}{\partial \varphi}\right)^2\right] - V(r), \tag{8.75}$$

with the time variable excluded from the right side. On the other hand, the variables r, φ are excluded from the left side, and therefore the two sides can remain equal for all r, φ, t only if each side has no resultant dependence on these variables, but equals some "integration constant" $\alpha_1(P_1, P_2)$. Then $\partial S_t/\partial t = \alpha_1$ yields

$$S_t(t, P_1, P_2) = \alpha_1(P_1, P_2)t + \alpha_0(P_1, P_2), \tag{8.76}$$

where α_0 is a second "integration constant" as regards dependence on r, φ, t.

The equation of (8.75) to $\alpha_1(P_1, P_2)$ also yields

$$p_\varphi = \partial S_\varphi/\partial \varphi = \pm r[-2m(\alpha_1 + V) - (\partial S_r/\partial r)^2]^{1/2}, \tag{8.77}$$

and, again, both sides of this can have no resultant dependence on r, φ, t, and so must be equal to still another "integration constant," $\alpha_2(P_1, P_2)$. As a result,

$$S_\varphi(\varphi, P_1, P_2) = \alpha_2(P_1, P_2)\varphi. \tag{8.78}$$

Another "integration constant" could be added to this, as α_0 was to S_t of (8.76); however, such an additive constant can be amalgamated into $\alpha_0(P_1, P_2)$ in the final expression for $S = S_r + S_\varphi + S_t$.

Finally, S_r is to be obtained from the equation of (8.77) to $\alpha_2(P_1, P_2)$:

$$p_r = \partial S_r/\partial r = \pm[-2m(\alpha_1 + V) - \alpha_2^2/r^2]^{1/2}. \tag{8.79}$$

Thus found is a solution of the Hamilton–Jacobi equation:

$$S(r, \varphi, t, \alpha_0, \alpha_1, \alpha_2) = \alpha_0 + \alpha_1 t + \alpha_2 \varphi + \int^r p_r(r, \alpha_1, \alpha_2)\, dr. \tag{8.80}$$

This may be recognized as having the requisite number of arbitrary constants $\alpha_0, \alpha_1, \alpha_2$ to be a general solution of a differential equation which is of *first* order in differentials with respect to *three* variables r, φ, t.

A conclusion that can be drawn at this point is that a general solution of the Hamilton–Jacobi equation will contain f arbitrary integration constants if it pertains to a problem with f degrees of freedom $q_1, \ldots, q_f$—that is, $S(q_1, \ldots, q_f; \alpha_1, \ldots, \alpha_f; t)$—since the equation will contain f first-order differentials, $\partial S/\partial q_s$. Moreover, an $(f+1)$st constant, α_0, may always be added to S as should have been obvious from the beginning, since the procedure involves only derivatives of S. The α's may be arbitrary combinations of the new "conjugate momenta," $P_1, \ldots, P_f$, since these are to be constants in the motion. Application of the procedure has left a freedom of choice as to what the f relationships $\alpha_s(P_1, \ldots, P_f)$ may be. A specific choice for the set of relations will define the new "variables" $P_1, \ldots, P_f$ and also their conjugates, through $Q_s = (\partial/\partial P_s)S(q, \alpha(P), t)$. Of course, it is desirable to make the choices in such a way that the quantities Q, P have as immediate a physical significance as possible.

The last step may be illustrated by completing the above example. Since $H(r, \varphi, p_r, p_\varphi)$ is not explicitly time-dependent, it is a conserved energy $H = E$ (see (8.10) and (7.41)). Moreover, $H = -\partial S/\partial t = -\alpha_1$, and hence the constant of motion α_1 can be identified with the negative of the conserved energy:

$$\alpha_1(P_1, P_2) = -E. \tag{8.81}$$

From $p_\varphi = \partial S_\varphi/\partial\varphi = \partial S/\partial\varphi$,

$$\alpha_2(P_1, P_2) = p_\varphi \equiv L \tag{8.82}$$

is identified with the conserved angular momentum. Thus, depending on the way the relationships of P_1, P_2 to α_1, α_2 are adopted, they may be any combinations of E and L: $P_1(E, L)$ and $P_2(E, L)$, conserved because E and L are conserved. Physically meaningful choices are desirable, and letting

$$P_1 \equiv E \qquad \text{and} \qquad P_2 \equiv L, \tag{8.83}$$

as would follow from adopting the relationships $\alpha_1(P_1, P_2) \equiv -P_1$ and $\alpha_2(P_1, P_2) \equiv P_2$, seems as good a set of choices as any. The generator (8.80) can be written

$$S = -Et + L\varphi \pm \int^r dr[2m(E - V) - L^2/r^2]^{1/2}, \tag{8.84}$$

with an arbitrary additive constant (α_0) implicit in the indefinite integration that remains.

The new "variables" Q_1, Q_2 conjugate to $P_1 \equiv E$ and $P_2 \equiv L$ are still to be found, from relations $Q_s = \partial S/\partial P_s$. Conjugate to $P_1 \equiv E$ is

$$Q_1 = \partial S/\partial E = -t + \int^r dr/\dot{r}, \tag{8.85}$$

with $\dot{r} = \pm[2(E - V - L^2/2mr^2)/m]^{1/2}$, being identifiable as the radial velocity (see (4.19) and (4.21)). Conjugate to $P_2 \equiv L$ is

$$Q_2 = \partial S/\partial L = \varphi - (L/m) \int^r dr/r^2\dot{r}. \tag{8.86}$$

Both Q_1 and Q_2 are also supposed to be constants in the motion and that they indeed are is readily checked. Since $dr/\dot{r} = dt$, the expression (8.85) is merely the identity: $\int dt = t +$ arbitrary constant (Q_1). Since $L/mr^2 = \dot{\varphi}$, the expression (8.86) is equivalent to $\int d\varphi = \varphi +$ arbitrary constant ($-Q_2$). Thus Q_1, Q_2 turn out to be the remaining two arbitrary constants (besides E, L) to be expected in the general solution of the two-dimensional problem. They help in fitting the results to whatever initial conditions are given the motion and are therefore constants characteristic of a specific start just as are E and L. Their possible role will be explored further below.

Hamilton–Jacobi Mechanics

The discussion of the Hamilton–Jacobi equation in the preceding subsection has made it clear that it offers as complete a method for arriving at descriptions of motion as do any equations of motion. The final results (8.85) and (8.86), of the specific problem treated, have implicit in them the radius as function of time, $r(t)$, and the orbit $r(\varphi)$. The form in which (8.86) contains the orbit

$$\varphi(r) = (L/m)\int^r dr/r^2\dot{r} + \text{arbitrary constant } (Q_2) \tag{8.87}$$

is practically identical with the form (4.21) obtained in the Newtonian formulation. The time-dependence implicit in (8.85),

$$t = \int^{r} dr/\dot{r} + \text{arbitrary constant } (-Q_1), \tag{8.88}$$

is an equally acceptable alternative to the form (1.62),

$$t = \int^{\varphi} d\varphi/\dot{\varphi} + \text{arbitrary constant},$$

used in the Newtonian formulation. The method has in this case amounted to using the approach of §4.2 (that is, starting with "first integrals," for E and L, of the Newtonian equations).

As a basic formulation of Mechanics, the Hamilton–Jacobi theory requires the adoption of a set of coordinates $q \equiv q_1, \ldots, q_f$, as does the Lagrange formulation, together with an hamiltonian functional form, as does the canonical formulation. The single Hamilton–Jacobi equation then plays the role of all the equations of motion.

The Hamilton–Jacobi function $S(q, P, t)$ may be regarded as the generator of a transformation $q(0), p(0) \leftrightarrow q(t), p(t)$, relating the initial phase of the motion to one obtaining a finite time t later, just as the generator (8.22) with $G = H$,

$$S'(q, P', t) = \sum_s q_s P'_s + H(q, p, t)\, \delta t, \tag{8.89}$$

was found to generate infinitesimal steps (8.26) of the motion, with $q(0)$, $p(0) \rightarrow Q'(t) = q(\delta t)$, $P'(t) = p(\delta t)$. However, the Hamilton–Jacobi transformation treats $q(t)$, $p(t)$ as the "old" variables, and is more directly adaptable to a definition $Q \equiv q(0)$, $P \equiv p(0)$ of the "new" ones. Thus, $Q_2 = \varphi(0)$ in (8.87) if the lower limit of the integration is taken to be $r = r(0)$. Plainly, $q(0)$ and $p(0)$ are constants for a specific motion, such as the Hamilton–Jacobi "variables" Q, P must be, and may be treated as "conserved" quantities in place of, say, $E = H(q(0), p(0))$ and $L(q(0), p(0))$. The transformation relations with an Hamilton–Jacobi generator $S(q, P, t) = S(q(t), p(0), t)$ take forms

$$p_s(t) = \partial S(q, p(0), t)/\partial q_s, \qquad Q_s = q_s(0) = \partial S(q, P, t)/\partial P, \tag{8.90}$$

which have implicit in them the solutions

$$q_s(t, q(0), p(0)), \qquad p_s(t, q(0), p(0)), \tag{8.91}$$

of what was called the "prototype" problem of Mechanics (page 37).

The Time-Independent Hamilton–Jacobi Equation

The general Hamilton–Jacobi equation (8.71) can be partially integrated immediately for the important cases in which the hamiltonian $H(q, p)$ is not explicitly time-dependent. This corresponds to the cases of (generalized) energy conservation, as follows from (8.10), and it is appropriate to set $H(q, p) = E$. Now

$$\frac{\partial}{\partial t} S(q, P, t) = -E \rightarrow S(q, P, t) = -Et + W(q, P), \tag{8.92}$$

with a new function $W(q, P)$ taking over as transformation-generator in the sense that (8.17) becomes

$$p_s = \partial W/\partial q_s \qquad \text{and} \qquad Q_s = \partial W/\partial P_s. \tag{8.93a}$$

The general Hamilton–Jacobi equation is now replaced by a time-independent one,

$$H(q, \partial W/\partial q) = E, \tag{8.93b}$$

constituting the Hamilton–Jacobi formulation for energy-conserving systems.

The same transformation procedure (8.93) would have been the result if the original objective had been set at making only the new "coordinates" Q ignorable, and not the new momenta, P, in a time-independent transformation generated by a function called $W(q, P)$, of an energy-conserving system with $H(q, p) = E$. Then a new hamiltonian $\mathscr{H}(P) = H + \partial W/\partial t = H = E(P)$ is expected and canonical equations

$$\dot{P}_s = -\partial E(P)/\partial Q_s = 0, \qquad \dot{Q}_s = \partial E(P)/\partial P_s \equiv v_s(P), \tag{8.94}$$

with $v_s(P)$ a constant in the motion because the P's are. It is now plain that the time-independent transformation procedure has (8.93) as the appropriate version of (8.17) and (8.18).

Since $dW/dt = \sum_s(\partial W/\partial q_s)\dot{q}_s$ during the motion with each P_s constant, (8.72) is here replaced by

$$dW = \sum_s p_s\, dq_s, \tag{8.95}$$

which is just an element of the "action" (8.8) being expended during the motion. The same result follows from (8.72) with $S = W - Et = W - Ht$.

A solution of the time-independent Hamilton–Jacobi equation will lead to a description of motion through

$$Q_s(t) = \partial W(q, P)/\partial P_s = v_s(P)t + c_s(P), \tag{8.96}$$

as follows from (8.94). Implicit here are $q_s(t, P, c)$ with the P's and c's serving as the $2f$ arbitrary integration constants that will enable fitting the solution to any given initial conditions.

One of the simplest illustrations of the role played by the variables $Q = \nu t + c$ is provided by the linear harmonic oscillator.* The Hamilton–Jacobi equation for this case is

$$H = \frac{1}{2m}\left(\frac{\partial W}{\partial x}\right)^2 + \tfrac{1}{2}m\omega_0^2 x^2 = E(P), \tag{8.97}$$

which yields

$$W(x, P) = \int p\,dx = \pm \int dx[2mE(P) - (m\omega_0 x)^2]^{1/2}. \tag{8.98}$$

A definition for the constant of motion P is still to be chosen A customary choice is to make it the constant action (8.8), which is expended by the one degree of freedom in one full period of the motion, meaning by this the cycle during which x varies from one limit $x = -(2E/m\omega_0^2)^{1/2} \equiv -a$ at which $p = 0$, to the other at $x = +a$, and back again to $x = -a$. Thus

$$P \equiv \oint p\,dx = 2\int_{-a}^{a} dx[(m\omega_0)^2(a^2 - x^2)]^{1/2} = 2\pi E(P)/\omega_0. \tag{8.99}$$

The new hamiltonian (8.97) is thus

$$E(P) = \omega_0 P/2\pi, \tag{8.100}$$

and the canonical equations (8.94) are

$$\dot{P} = -\partial E/\partial Q = 0, \qquad \dot{Q} = \partial E/\partial P \equiv \nu(P) = \omega_0/2\pi. \tag{8.101}$$

Now the P-dependence of the generator (8.98) can be made explicit in

$$W(x, P) = \int dx[m\omega_0(P/\pi - m\omega_0 x^2)]^{1/2}, \tag{8.102a}$$

and hence

$$\begin{aligned} Q = \partial W/\partial P &= (m\omega_0/2\pi)\int dx/[(m\omega_0 P/\pi)(1 - \pi m\omega_0 x^2/P)]^{1/2} \\ &= (2\pi)^{-1}\sin^{-1}(\pi m\omega_0/P)^{1/2}x. \end{aligned} \tag{8.102b}$$

Thus

$$x = (P/\pi m\omega_0)^{1/2}\sin 2\pi Q = a\sin 2\pi(\nu t + c), \tag{8.103}$$

---

* This has already been reviewed, from a different starting point, on pages 194–195.

and $Q = \nu t + c$ plays the role of a "phase-angle" of oscillation when it is conjugate to the action expended in one oscillation.

The example here reveals how the transformation-generator $F(x, Q)$ used for illustration on page 194 was suggested. The connection between generators of the type $F(x, Q)$ and the type $S(x, P) = W(x, P)$ is, according to (8.16),

$$F(x, Q) = W(x, P) - PQ.$$

With $W(x, P)$ of (8.102a) integrated out,

$$F(x, Q) = \tfrac{1}{2} m\omega_0 x \left[\frac{P}{\pi m\omega_0} - x^2\right]^{1/2} + \frac{P}{2\pi} \sin^{-1}\left(\frac{\pi m\omega_0}{P}\right)^{1/2} x - PQ,$$

in which the last two terms cancel according to $Q(x, P)$ of (8.102b). When the same relation, in the form

$$P = \pi m\omega_0 (x/\sin 2\pi Q)^2,$$

is used to eliminate P,

$$F(x, Q) = \tfrac{1}{2} m\omega_0 x^2 [\sin^{-2} 2\pi Q - 1]^{1/2} = \tfrac{1}{2} m\omega_0 x^2 \cot 2\pi Q. \quad (8.104)$$

This is the same as the generator used on page 194 except for the rescalings $2\pi Q \to Q$ and $P/2\pi \to P$.

Action and Angle Variables

Canonical transformation to variables Q, P such that P becomes an action expended by some one degree of freedom during a cycle of its motion, as in (8.99), is customary whenever such a quantity is definable (that is, whenever the motion *has* periodic cycles). The coordinate Q conjugate to such an "action variable" is called an "angle variable," since a common case of cyclic motion is one in which some angle completes an orbit, 2π, in each cycle. The types of motion in which the action and angle variables prove most useful are ones called "multiply periodic"; in these, each cyclic degree of freedom, q_s, repeats its cycle of values in some period, T_s, which may be different for different degrees of freedom. The solar system, with each of its planets having a different "year," provides a case in point. Another example is provided by the anisotropic oscillator (page 85), having some "Lissajou figure" as orbit. When two of the periods are equal, as for the x and y degrees of freedom of an *isotropic* oscillator, they are said to have "degenerated" to one value.

There is a standard notation for action and angle variables: J_s instead of P_s and w_s instead of Q_s. By definition,

$$J_s = \oint p_s \, dq_s, \tag{8.105}$$

where q_s is a degree of freedom that goes through a cycle of values defining the range of the integration. The integration produces a constant J_s, which is taken to be an end result of an Hamilton–Jacobi transformation by a generator $W(q, J)$, a solution of the differential equation (8.93*b*). Then the angle variable conjugate to J_s must be

$$w_s = \partial W / \partial J_s, \tag{8.106}$$

according to (8.93*a*). As in (8.94), the canonical equations of motion give

$$\dot{w}_s = \partial H(J)/\partial J_s = \nu_s(J), \qquad \text{constant,} \tag{8.107}$$

where $H(J) = E$ is the transformed hamiltonian.

The constant ν_s is just the frequency, $= T_s^{-1}$, of the degree of freedom q_s. That can be seen by examining the change, Δw_s, which occurs in $w_s = \nu_s t$ + constant, during one cycle of q_s:

$$\Delta w_s \equiv \oint (\partial w_s/\partial q_s) \, dq_s = \oint (\partial^2 W/\partial q_s \, \partial J_s) \, dq_s .$$

Since the J's are constants during the motion $q_s(t)$, the last integral can be evaluated as

$$\Delta w_s = \frac{\partial}{\partial J_s} \oint \frac{\partial W}{\partial q_s} \, dq_s = \frac{\partial}{\partial J_s} \oint p_s \, dq_s = 1.$$

Thus $\Delta w_s = \nu_s \, \Delta t$ is just one unit for a period $\Delta t = T_s$ of q_s, making $\nu_s T_s = 1$ and

$$\nu_s = 1/T_s,$$

as asserted. The formula (8.107),

$$\nu_s = \partial H(J)/\partial J_s, \tag{8.108}$$

is an expression for the frequency which is of some historical significance in the development of the quantum mechanics of orbital motions.

The Kepler Problem

Putting the problem of elliptical orbits in an "inverse square" force field into terms of action and angle variables has helped in the understanding of the connections between classical and quantum-mechanical descriptions (the Correspondence Principle).

Written in a way appropriate to the motion of a point charge $-e$ in the field of a fixed point charge $+Ze$, the hamiltonian (8.4) becomes

$$H = \frac{p_r^2}{2m} + \frac{p_\vartheta^2}{2mr^2} + \frac{p_\varphi^2}{2mr^2 \sin^2 \vartheta} - \frac{Ze^2}{r} = E. \tag{8.109a}$$

This is to be converted into a Hamilton–Jacobi equation for a generator $W(r, \vartheta, \varphi, J_r, J_\vartheta, J_\varphi)$, which will be sought in the separable form

$$W = W_r(r) + W_\vartheta(\vartheta) + W_\varphi(\varphi), \tag{8.109b}$$

so that it is substitutions of

$$p_r = \partial W_r/\partial r, \qquad p_\vartheta = \partial W_\vartheta/\partial \vartheta, \qquad p_\varphi = \partial W_\varphi/\partial \varphi$$

into $H(r, \vartheta, p_r, p_\vartheta, p_\varphi)$, which yield the appropriate differential equation. Then it follows that

$$p_\varphi^2 = \left(\frac{\partial W_\varphi}{\partial \varphi}\right)^2 = r^2 \sin^2 \vartheta \left[2m\left(E + \frac{Ze^2}{r}\right) - \left(\frac{\partial W_r}{\partial r}\right)^2 - \frac{1}{r^2}\left(\frac{\partial W_\vartheta}{\partial \vartheta}\right)^2\right]$$

must remain constant during the motions of r, ϑ, φ; the constant p_φ is just the z-component, L_z, of the conserved angular momentum, and the corresponding action variable is

$$J_\varphi \equiv \oint p_\varphi \, d\varphi = 2\pi L_z. \tag{8.110}$$

Next, it becomes clear that

$$\left(\frac{\partial W_\vartheta}{\partial \vartheta}\right)^2 + \frac{L_z^2}{\sin^2 \vartheta} = r^2\left[2m\left(E + \frac{Ze^2}{r}\right) - \left(\frac{\partial W_r}{\partial r}\right)^2\right]$$

must be constant expressions, to be identified with the square, L^2, of the conserved angular momentum. Thus the second action variable,

$$J_\vartheta = \oint p_\vartheta \, d\vartheta = \oint [L^2 - L_z^2/\sin^2 \vartheta]^{1/2} \, d\vartheta,$$

is to be found from an integration in which L^2, L_z are constants. The integration is most easily carried out over an angle, ψ, in the plane of the orbit, so defined that $r\dot{\psi}$ is the velocity component normal to **r**:

$$\begin{aligned} r^2\dot{\psi}^2 &= v^2 - \dot{r}^2 \\ &= r^2\dot{\vartheta}^2 + r^2\dot{\varphi}^2 \sin^2 \vartheta \\ &= (p_\vartheta \dot{\vartheta} + p_\varphi \dot{\varphi})/m. \end{aligned}$$

Clearly,

$$L^2 = m^2r^4(\dot{\vartheta}^2 + \sin^2 \vartheta\dot{\varphi}^2) = m^2r^4\dot{\psi}^2$$

and

$$L\dot{\psi} = p_\vartheta\dot{\vartheta} + p_\varphi \dot{\varphi},$$

a form to be expected from (7.16). Thus

$$J_\vartheta \equiv \oint p_\vartheta \dot{\vartheta}\, dt = \oint (L d\psi - L_z\, d\varphi) = 2\pi(L - L_z). \tag{8.111}$$

Finally,

$$J_r \equiv \oint p_r\, dr$$

must be found from an integration of

$$p_r = \frac{\partial W_r}{\partial r} = \pm\left[2m\left(E + \frac{Ze^2}{r} - \frac{L^2}{2mr^2}\right)\right]^{1/2} \tag{8.112}$$

The range of the integration is from perihelion to aphelion and back; these extreme radii are just roots of the last expression, since the radial momentum vanishes at those "turning points." Sommerfeld invented elegant methods for carrying out such definite integrations, but reference to standard tables of integrals suffices, and the result

$$J_r = -2\pi L + \pi Ze^2[2m/(-E)]^{1/2}$$

is obtained. It has already been found that $2\pi L = J_\vartheta + J_\varphi$, and so the solution of the last expression for the energy yields

$$E = -\frac{2\pi^2 mZ^2e^4}{(J_r + J_\vartheta + J_\varphi)^2}. \tag{8.113}$$

The importance of this way of expressing the energy stems from the fact that it is only necessary to assume that the action expended per cycle by each degree of freedom is an integer multiple of "Planck's constant," h, so that each $J_s = n_s h$, in order to obtain the empirical "Balmer formula" for the energies found for hydrogen-like atoms:

$$E_n = -\frac{2\pi^2 mZ^2e^4}{n^2h^2}, \tag{8.114}$$

when $n = n_r + n_\vartheta + n_\varphi = 1, 2, 3, \ldots$. The hypothesis $\oint p_s\, dq_s = n_s h$ characterizes the "old quantum theory" but is only a "semiclassical approximation" to modern quantum mechanics.

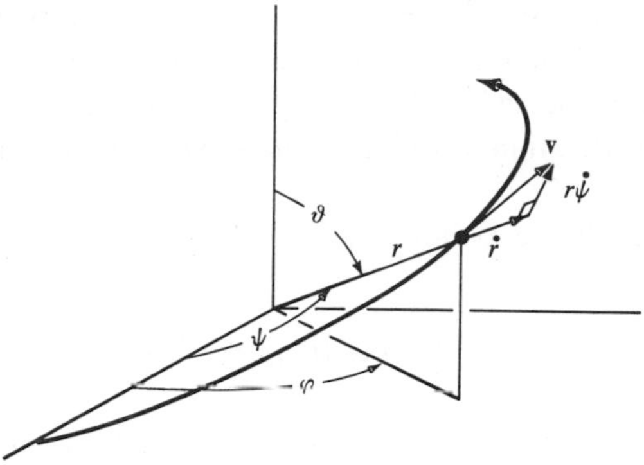

Figure 8.3

As indicated in (8.108), the frequencies with which the degrees of freedom go through their cycles are to be obtained from derivatives of the energy with respect to the corresponding action "variables." Here the three frequencies "degenerate" into the single one

$$\nu = \frac{4\pi^2 m Z^2 e^4}{(J_r + J_\vartheta + J_\varphi)^3} = \frac{(-2E)^{3/2}}{2\pi Z e^2 m^{1/2}}.$$

When it is recalled that the semimajor radius a of the elliptical orbit is related to the energy by $E = -Ze^2/2a$ (4.27), the expression for the period $T = 1/\nu$ may be written:

$$T = 2\pi(m/Ze^2)^{1/2} a^{3/2}.$$

This is just Kepler's Third Law (4.28) again.

8.6 THE "WAVE OPTICS" OF CLASSICAL TRAJECTORIES

The name "phase function" was suggested for $S(q, P, t)$ on page 195 because assigning a specific value to it may be interpreted as focusing attention on a specific "phase" in a "configuration wave" which may be associated with possible motions of an ensemble of identical systems. The picture here being alluded to has played no essential role in Classical Mechanics "proper," but has helped clarify relationships among rather disparate theories. One example is provided by the relation between rays of light (regarded as trajectories of light "corpuscles") and "wave optics." Another is the connection between classical particle mechanics and the Wave Mechanics of particles. Neither example can be explored in depth at this stage (see §15.3), but some acquaintance with the configuration-wave picture can now be acquired. Its consideration will be restricted to the case of energy-conserving systems.

Consider all the points q $(\equiv q_1, \ldots, q_f)$ of a configuration space at which

$$S(q, P, t) = W(q, P) - Et, \tag{8.115}$$

of (8.92), has a constant value at a given moment, and with given values for all the conserved quantities $E, P(\equiv P_1, \ldots, P_f)$. The set of points q making $W(q, P) = S_1 + Et_1$ (that is, with specific values $S = S_1$ and $t = t_1$) form a (hyper)surface in the configuration space. Such surfaces are "propagated" in the space in the sense that $S = S_1$ may also be assigned for $t = t_1 + \Delta t$, and the new surface thus defined may be regarded as a new "location" for the initial surface, acquired during the time interval Δt. A schematic diagram of such a moving surface is shown in Figure 8.4. Each surface may be regarded as defining a "wave front," a name which is given to surfaces of constant "phase" in the terminology of wave propagation (Chapter 14). A more explicit example is provided by the central force problem of (8.84), which yields

$$W(r, \varphi, E, L) = \int^r p_r\, dr + L\varphi = S + Et.$$

For any given value $W_1 = S_1 + Et_1$, this defines a curve $r(\varphi)$ in the configuration plane of r, φ, which is one of the "wave" fronts in question.

There remain to be found the connections between the configuration waves and the trajectories actually followed by the members of the ensemble, in the same configuration space. They are most transparent for the case in which the system considered, and so each member of the ensemble, is a single particle moving in a configuration space of its cartesian coordinates x, y, z, under the influence of a potential $V(x, y, z)$. The direction of the particle's trajectory at any point is given by the momentum vector $\mathbf{p}$, having the conjugates of x, y, z as its cartesian components. The phases $W(x, y, z) = S + Et$ of the associated configuration wave fronts are to be found as solutions of the Hamilton–Jacobi equations (8.93),

$$\begin{aligned} E &= \frac{1}{2m}\left[\left(\frac{\partial W}{\partial x}\right)^2 + \left(\frac{\partial W}{\partial y}\right)^2 + \left(\frac{\partial W}{\partial z}\right)^2\right] + V(x, y, z) \\ &\equiv |\nabla W|^2/2m + V, \end{aligned} \tag{8.116}$$

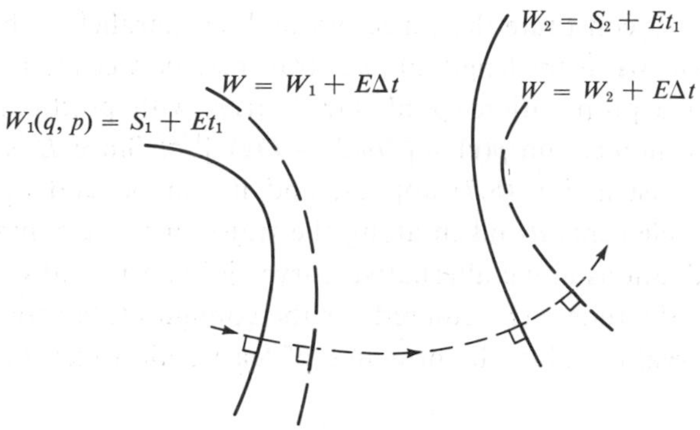

Figure 8.4

with $\nabla W = \mathbf{p}$. Now it is known from the consideration of the "equipotentials" in (2.36) that the surfaces of the constant $W(x, y, z)$ will be normal to the gradients of W. Thus a first relation between the configuration wave fronts and the trajectories is that the latter pass through the wave fronts perpendicularly, $\mathbf{p} = \nabla W$ being normal to these. This finding also holds in general: the configuration-wave fronts are propagated in such a way that they stay normal to the trajectories of the system-points in the configuration space. A sample system-trajectory is indicated in Figure 8.4, by the curve with arrows attached.

The next question is concerned with the relations between the propagation velocities of the wave fronts (these may differ at different points of a front) and the speeds of the system-points along their trajectories. These are again simple in the case of the single-particle systems, where the speeds of the particles are $v = [2(E - V)/m]^{1/2}$. The configuration-wave propagation speed u along some one trajectory can be found as $u = ds/dt$, where ds is a normal distance between two wave fronts given by $W_1 = S_1 + Et_1$ and $W = W_1 + dW = S_1 + E(t_1 + dt)$. Then $dW = Edt = |\nabla W|ds$ and

$$u = E/|\nabla W| = E/p = E/mv. \tag{8.117}$$

It turns out, since the total energy $E = \frac{1}{2}mv^2 + V$ is constant, that the wave-propagation velocity u is *inversely* proportional to the particle's speed for a given energy. The wave front and the particle are *not* propagated together. For a free particle, which has the *constant* velocity $v = (2E/m)^{1/2}$, the wave-propagation speed is $u = \frac{1}{2}v$.

Consider now the action (8.95) which the particle has expended between positions $\mathbf{r}_0$ and $\mathbf{r}$ on its path. This may be evaluated as

$$W(\mathbf{r}, \mathbf{p}_0) = \int_{\mathbf{r}_0}^{\mathbf{r}} \mathbf{p} \cdot d\mathbf{r} = \int_0^{s(\mathbf{r})} p\, ds = E \int_0^{s(\mathbf{r})} ds/u(\mathbf{r}). \tag{8.118}$$

The elements ds constitute the trajectory and are parallel to the momentum at each instant; $s(\mathbf{r})$ is the length of path traversed between $\mathbf{r}_0$ and $\mathbf{r}$. The configuration-wave propagation speed, $u(\mathbf{r})$, varies with position, inversely to the particle's momentum $p(\mathbf{r}) = [2m(E - V(\mathbf{r}))]^{1/2}$. Since E is constant, the principle of least action (8.7) applies, and it can be said that the action evaluated for elements ds taken along the trajectory is a minimum as compared to evaluations along alternative curves joining $\mathbf{r}_0$ and $\mathbf{r}$. The quantity $ds/u(\mathbf{r})$ is just the time $dt(\mathbf{r})$ required for the configuration wave to pass over the path interval ds. Thus the principle of least action may be expressed as

$$\delta \int \frac{ds}{u(\mathbf{r})} = \delta \int dt(\mathbf{r}) = 0. \tag{8.119}$$

In the theory of geometrical optics, this is known as "Fermat's Principle of Least Time" (§15.3). For particles, it is *not* the time required for the particle to traverse the trajectory that is thus minimized, but rather the time required for the configuration wave to do so. (However, in the special case of the massless "corpuscles of light," the speeds of the free corpuscles and of their configuration wave coincide!)

EXERCISES

8.1. Consider the canonical equations generated by the hamiltonian (8.4).

(a) Show how they lead to the expressions (1.39) for the accelerations of the spherical coordinates.

(b) Show how the conservation of the *total magnitude* of angular momentum follows from them when the field is a central one.

8.2. To see an example of an explicitly time-dependent hamiltonian function, $H(\varphi, p, t)$, construct one for the system of Figure 7.3 or Figure 7.4.

8.3. For a free particle, the hamiltonian (7.41) is simply $H = T$, and the canonical equations give $\dot{p}_s = -\partial T/\partial q_s$. In the Lagrangian formulation, $L = T$ in (7.32) and the equations of motion (7.33) give

$$(d/dt)(\partial L/\partial \dot{q}_s) \equiv \dot{p}_s = +\partial T/\partial q_s.$$

How are these two formulas for $\dot{p}_s$ to be reconciled?

8.4. The motion of a mass-particle m in a central field $V(r)$ is to be described in a frame which is rotating about an axis through the force-center, $r = 0$, with a constant angular velocity $\boldsymbol{\omega}$, relative to an inertial frame.

(a) Show that the generalized momentum **P**, conjugate to the position vector $\mathbf{R}(t)$ ($=\mathbf{r}$, but to be resolved on the rotating frame) is

$$\mathbf{P} = m[\mathbf{V} + \boldsymbol{\omega} \times \mathbf{R}],$$

where **V** is the velocity relative to the rotating frame. How do its values compare to those of the "kinetic" momentum, $\mathbf{p} = m\mathbf{v}$, in the inertial frame?

(b) Construct the hamiltonian $\mathscr{H}(\mathbf{R}, \mathbf{P}) = \mathbf{P} \cdot \mathbf{V} - L$, where $L = \frac{1}{2}mv^2 - V(r)$. Show that $\mathscr{H}$ has the conserved value $E - \boldsymbol{\omega} \cdot \mathbf{L}$, where E is the energy, and **L** the conserved angular momentum, in the inertial frame (but note that $E - \boldsymbol{\omega} \cdot \mathbf{L} \neq \frac{1}{2}mV^2 + V(R)$!)

(c) Show that the canonical equations generated by $\mathscr{H}(\mathbf{R}, \mathbf{P})$ check with the expectations (2.37).

8.5. Suppose that a uniform magnetic field **B** (see Exercise 7.12) is imposed on a point-charge moving in a central potential field $V(r)$.

(a) Find the generalized momentum, **p**, which is conjugate to **r**, and construct the hamiltonian $H(\mathbf{r}, \mathbf{p})$.

(b) Compare the canonical equations generated by $H(\mathbf{r}, \mathbf{p})$ to the findings of Exercise 7.12(a).

(c) Show that in a weak magnetic field, such that effects proportional to B^2 may be neglected,

$$H(\mathbf{r}, \mathbf{p}) = p^2/2m + V(r) - \boldsymbol{\mu} \cdot \mathbf{B},$$

where $\boldsymbol{\mu} \equiv (q/2mc)(\mathbf{r} \times \mathbf{p})$. (See Exercise 7.12. Here, $-\boldsymbol{\mu} \cdot \mathbf{B}$ is frequently treated as a "perturbation," of motions to be regarded as "internal" to the system. See Exercises 8.6 and 8.14.)

8.6. Draw on results from the two preceding exercises for the following conclusions.

(a) The conjugate momentum of a particle in a uniform magnetic field simulates that for a motion relative to a frame rotating with the so-called "Larmor precession frequency," $\omega_L \equiv qB/2mc$.

(b) However, the *hamiltonian* found in Exercise 8.5 simulates one in the rotating frame only for weak fields, when effects proportional to ω_L^2 may be neglected.

(c) The angular momentum vector, conserved when the field is absent, "precesses" with the Larmor frequency in the presence of the weak field.

(d) The motion in a strong field differs from motion relative to the rotating frame because of the absence of "centrifugal effects"; the magnetic force simulates only the "Coriolis effects," as should have already been evident from the equations of motion found in Exercise 7.12(a).

8.7. The Principle of Least Action, discussed in connection with (8.7), actually has a more general provenance than there indicated; it is also satisfied if the varied paths are allowed to have varied end times (that is, $\Delta t_{1,2} \neq 0$ as in Figure 7.6, *but* $\Delta q(t_{1,2}) = 0$ as in Figure 7.5). Show that this is so because the energy $H = E$ is to be kept at a steady, constant value. (Show that the action integral can be written

$$A = \int_{t_1}^{t_2} dtL + H(t_2 - t_1),$$

and use the result (7.67), properly modified.)

8.8. The Principle of Least Action (8.7) has the "Jacobi form" when reference to time is eliminated, so that it can "in principle" yield *orbits* directly. That is done by substituting $T = \frac{1}{2}(ds/dt)^2$ into (8.9), after defining $ds^2 \equiv \sum \mu_{s\sigma}\, dq_s\, dq_\sigma$. (See (7.13). Then ds is called a "line element" in an f-dimensional "curved space," having $\mu_{s\sigma}(q)$ as "metric.") Show that the principle may be written

$$\delta \int \sqrt{T}\, ds = 0.$$

What is the explicit expression for ds^2 in the case of a single particle described by polar coordinates?

8.9. Check that the transformation to the relative and centroidal coordinates of two mass-particles, introduced in §5.1, is a canonical one by constructing an appropriate generator.

8.10. Find the suitable interpretation for the transformation generated by

$$S = (x \cos \omega t + y \sin \omega t)P_X + (-x \sin \omega t + y \cos \omega t)P_Y + zP_Z.$$

(Compare Exercise 8.4. Note that $S = \mathbf{P} \cdot \mathbf{R} = \mathbf{p} \cdot \mathbf{r}$ in value.)

8.11. Show that for $t \to \delta t \to 0$, the generator of the preceding exercise differs from one yielding an identity transformation by

$$\delta t G = -\boldsymbol{\omega} \delta t \cdot (\mathbf{r} \times \mathbf{p}),$$

and investigate the results of using this G in (8.24).

8.12. Prove the following Poisson bracket relations among the functions $A(q, p, t)$, $B(q, p, t)$, and $C(q, p, t)$:

$$[A, B + C] = [A, B] + [A, C],$$
$$[A, BC] = [A, B]C + B[A, C],$$
$$[A, [B, C]] + [B, [C, A]] + [C, [A, B]] = 0.$$

The last relation is known as a "Jacobi identity" (compare Exercise 1.7).

8.13. The results of the preceding exercise prove useful in such investigations as the following, for the system of Exercise 8.5.

(a) Find the connection between the Poisson bracket $[v_x, v_y]$ and the magnetic field **B**.

(b) Check that it is Poisson-bracketing with the moment, $\mathbf{r} \times \mathbf{p}$, of the *conjugate* momentum, rather than with the "kinetic" angular momentum, $\mathbf{r} \times m\mathbf{v}$, that rotates $\mathbf{p}$ as in (8.46).

8.14. Construct the Hamilton–Jacobi equation for the system of Exercise 8.5 with $q = -e$, in spherical coordinates.

(a) Show that the equation does *not* have separable solutions, $W = W_r + W_\vartheta + W_\varphi$, unless effects proportional to B^2 are neglected and the approximation of Exercise 8.5(c) is used.

(b) Show that the approximate ("weak-field") solution can be carried through exactly as for the "Kepler problem" of §8.5, and that the effect of the magnetic field is merely to replace E with $E - \omega_L L_z$, where $\omega_L = |eB/2mc|$. (Compare this with the effect of the rotation considered in Exercise 8.4(b).)

(c) Show that the effect of the small magnetic field (called the "weak-field Zeeman effect") on the "hydrogen-like atom" considered in the above "Kepler problem" is to remove one degree of degeneracy in the frequencies. (Whereas $\nu_r = \nu_\vartheta$ are unchanged, ν_φ is changed by $\Delta\nu_\varphi = \omega_L/2\pi$.)

8.15. Consider the effect of a uniform electric field $\mathscr{E}$ on the "hydrogen-like" atomic system of (8.109*a*) (the so-called "Stark effect"). It may be represented by the augmented potential energy

$$V(r) = \frac{-Ze^2}{r} - e\mathscr{E}r\cos\vartheta\,,$$

if the polar axis is put parallel to $\mathscr{E}$. It will then be found the Hamilton–Jacobi equation is not separable in the spherical coordinates, but does become separable in the "confocal paraboloidal" coordinates defined by

$$q_1 = r(1 - \cos\vartheta), \qquad q_2 = r(1 + \cos\vartheta), \qquad q_3 = \varphi,$$

where $0 \leq q_{1,2} < \infty$ and $0 \leq q_3 < 2\pi$.

(a) By establishing connections to any definition of a parabola with which you are familiar, show that the surfaces of revolution defined by $q_1 =$ constant have intersections with planes through the polar axis ($q_3 =$ constant) which are parabolas with "focus" at the origin and opening out to $z = r\cos\theta \to +\infty$. (The surfaces $q_2 =$ constant are similar paraboloids opening out to $z = -\infty$.)

(b) Show that the hamiltonian may be written as

$$H = \frac{2(q_1 p_1^2 + q_2 p_2^2)}{m(q_1 + q_2)} + \frac{p_3^2}{2mq_1q_2} - \frac{2Ze^2}{q_1 + q_2} + \tfrac{1}{2}e\mathscr{E}(q_1 - q_2).$$

(c) Show that the action $J_3 = 2\pi p_3 = 2\pi L_z$ is a constant of the motion, and that $J_{1,2}$ are calculable from the integrals

$$J_{1,2} = \sqrt{\frac{m}{2}} \oint dq_{1,2} \left[E + \frac{Ze^2 \pm \lambda}{q_{1,2}} - \frac{L_z^2}{2mq_{1,2}^2} \mp \tfrac{1}{2}e\mathscr{E}q_{1,2}\right]^{1/2},$$

respectively. Here λ is an arbitrary constant, eventually determined by the starting conditions, which has replaced the total angular momentum L of the field-free case; its value for a given L_z helps determine the orientations of the motions relative to the field. (The reader may want to make a deeper study of these matters by comparing the results here with his knowledge about the $\mathscr{E} = 0$ case, reformulated in terms of the paraboloidal coordinates.)

CHAPTER 9

ROTATIONS AND FRAME TRANSFORMATIONS

The description of rotation will now be given more attention, in preparation for treating rigid body motions in the next chapter. A body's orientational configuration may quite obviously be described by specifying it for an orthogonal reference frame fixed in or to the body, a so-called "body frame." Rotations of such frames have already been considered in §1.4, where instantaneous orientations were specified by orthogonal unit vectors **I**, **J**, **K**(t), and their rates of rotation by an instantaneous angular velocity vector $\mathbf{\Omega}(t)$. Here, various amplifications of, and alternatives to, that description will be considered.

9.1 FINITE ROTATIONS

In the treatment of §1.4, the angular velocity $\mathbf{\Omega}$ determined the rate of change of each axis vector **I**, **J**, **K** in the expressions (1.41), and each of these is a special example of the velocity (1.42),

$$\dot{\mathbf{R}} = \mathbf{\Omega} \times \mathbf{R}, \tag{9.1}$$

imparted to an arbitrary point at **R**, fixed to the rotating frame. Thus, an "infinitesimal" displacement, $\mathbf{dR} = \mathbf{\Omega}\, dt \times \mathbf{R}$, is described, but not an integrated finite one, $\mathbf{R}' - \mathbf{R}$.

Finite Rotational Displacements

A description of a finite rotational displacement, about a given axis, may be attempted through a generalization of the description of the infinitesimal one in Figure 1.7. In Figure 9.1, a point at **R** is rigidly connected

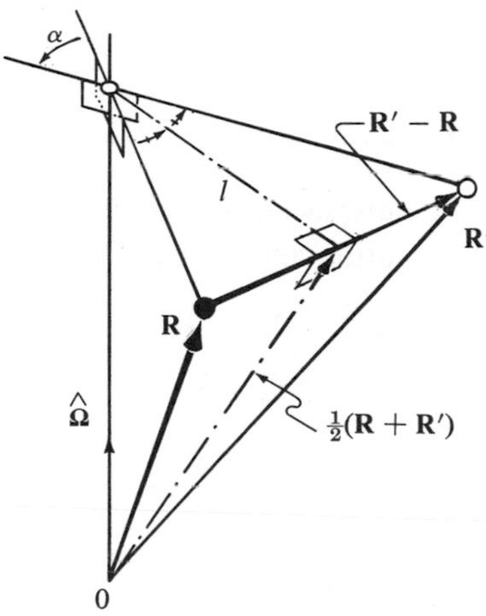

Figure 9.1

to a rotation center at $\mathbf{R} = 0$. A rotation through the angle α, about an axis steadily directed along $\hat{\mathbf{\Omega}}$, moves the point to $\mathbf{R}'$. The point's distance from every point of the rotation axis remains constant, and so the displacement vector $\mathbf{R}' - \mathbf{R}$ is in a plane normal to $\hat{\mathbf{\Omega}}$. With the magnitudes of **R** and $\mathbf{R}'$ equal, $\mathbf{R}' - \mathbf{R}$ is also perpendicular to the mean vector $\frac{1}{2}(\mathbf{R} + \mathbf{R}')$. The displacement is then normal to the plane containing the axis and the mean-vector and so $\mathbf{R}' - \mathbf{R}$ has the direction of $\hat{\mathbf{\Omega}} \times \frac{1}{2}(\mathbf{R} + \mathbf{R}')$. The magnitude $|\hat{\mathbf{\Omega}} \times \frac{1}{2}(\mathbf{R} + \mathbf{R}')| = l$ is just the length of the perpendicular dropped from the axis to $\mathbf{R}' - \mathbf{R}$, in the plane normal to the axis, and hence $\frac{1}{2}|\mathbf{R}' - \mathbf{R}| = l \tan \frac{1}{2}\alpha$. It follows that

$$\mathbf{R}' - \mathbf{R} = 2[\hat{\mathbf{\Omega}} \times \tfrac{1}{2}(\mathbf{R} + \mathbf{R}')] \tan \tfrac{1}{2}\alpha$$

gives both the direction and magnitude of the displacement vector correctly. There is involved in this expression a vector independent of the point being considered:

$$\mathbf{w} \equiv 2\hat{\mathbf{\Omega}} \tan \tfrac{1}{2}\alpha, \tag{9.2}$$

being purely a representation of the characteristics of the rotation (the axis $\hat{\mathbf{\Omega}}$ and the angle α).

The finite-rotation vector $\mathbf{w}$ allows expressing the rotational displacement by the so-called "Rodrigues formula,"

$$\mathbf{R}' - \mathbf{R} = \mathbf{w} \times \tfrac{1}{2}(\mathbf{R}' + \mathbf{R}). \tag{9.3}$$

This may be solved for $\mathbf{R}'$ explicitly by using expressions obtained from scalar and vector multiplications of $\mathbf{w}$ into (9.3) (compare Exercise 1.2):

$$\mathbf{R}' = \mathbf{R} + (1 + \tfrac{1}{4}w^2)^{-1}[\mathbf{w} \times (\mathbf{R} + \tfrac{1}{2}\mathbf{w} \times \mathbf{R})]. \tag{9.4}$$

It is easy to find that $1 + \frac{1}{4}w^2 = \cos^{-2} \frac{1}{2}\alpha$ and that

$$\mathbf{R}' - \mathbf{R} = (\hat{\mathbf{\Omega}} \times \mathbf{R}) \sin \alpha + [\hat{\mathbf{\Omega}} \times (\hat{\mathbf{\Omega}} \times \mathbf{R})](1 - \cos \alpha), \tag{9.5}$$

which is a resolution of the displacement vector on two orthogonal directions in a plane normal to the rotation axis. Since $|\hat{\mathbf{\Omega}} \times (\hat{\mathbf{\Omega}} \times \mathbf{R})| = |\hat{\mathbf{\Omega}} \times \mathbf{R}|$,

$$\begin{aligned} |\mathbf{R}' - \mathbf{R}|^2 &= 2|\hat{\mathbf{\Omega}} \times \mathbf{R}|^2(1 - \cos \alpha), \\ |\mathbf{R}' - \mathbf{R}| &= 2|\hat{\mathbf{\Omega}} \times \mathbf{R}| \sin \tfrac{1}{2}\alpha. \end{aligned} \tag{9.6}$$

The results (9.5) and (9.6) may be verified by the construction in the normal plane shown in Figure 9.2.

Compounding Finite Rotations

The difficulties behind the representation of the rotational displacement when the axis itself rotates will be better appreciated after successive rotations about different *fixed* axes are considered.

Let $\mathbf{R} \to \mathbf{R}' \to \mathbf{R}''$ with

$$\mathbf{R}' - \mathbf{R} = \mathbf{w}_1 \times \tfrac{1}{2}(\mathbf{R}' + \mathbf{R}) \qquad \text{and} \qquad \mathbf{R}'' - \mathbf{R}' = \mathbf{w}_2 \times \tfrac{1}{2}(\mathbf{R}'' + \mathbf{R}').$$

The vector sum of these gives

$$\mathbf{R}'' - \mathbf{R} = (\mathbf{w}_1 + \mathbf{w}_2) \times \tfrac{1}{2}\mathbf{R}' + \mathbf{w}_1 \times \tfrac{1}{2}\mathbf{R} + \mathbf{w}_2 \times \tfrac{1}{2}\mathbf{R}''.$$

The reference to the intermediate position $\mathbf{R}'$ can be eliminated from this expression with the help of the preceding ones, and then

$$\mathbf{R}'' - \mathbf{R} = \mathbf{w}_3 \times \tfrac{1}{2}(\mathbf{R}'' + \mathbf{R}),$$

with

$$\mathbf{w}_3 = \frac{\mathbf{w}_1 + \mathbf{w}_2 - \frac{1}{2}\mathbf{w}_1 \times \mathbf{w}_2}{1 - \frac{1}{4}\mathbf{w}_1 \cdot \mathbf{w}_2}. \tag{9.7}$$

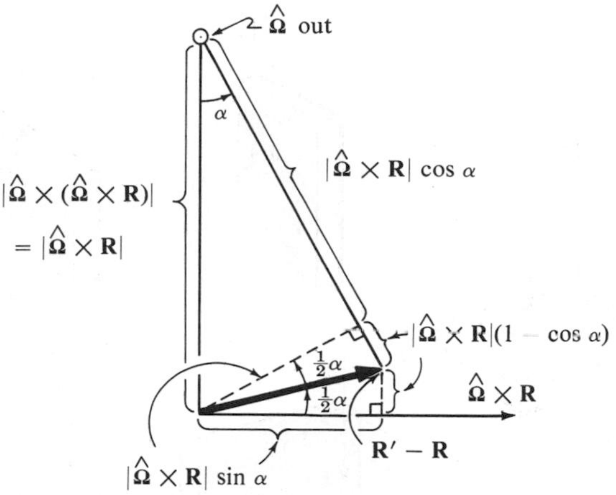

Figure 9.2

Thus the succession of two rotations, $\mathbf{w}_1$ followed by $\mathbf{w}_2$, is equivalent to a single rotation $\mathbf{w}_3$. It is easy to check that if the two rotations are about the same axis $\hat{\boldsymbol{\Omega}}_1 = \hat{\boldsymbol{\Omega}}_2 = \hat{\boldsymbol{\Omega}}$, then $\mathbf{w}_3 = 2\hat{\boldsymbol{\Omega}} \tan \frac{1}{2}(\alpha_1 + \alpha_2)$, as to be expected. Even then, $\mathbf{w}_3 \neq \mathbf{w}_1 + \mathbf{w}_2$, because of the "renormalization" by the denominator of (9.7).

The most important result made explicit by the representations here is that finite rotations about different axes do not "commute." The result depends on the order in which the individual rotations are carried out. If $\mathbf{w}_2$ were performed first, instead, and this were followed by $\mathbf{w}_1$, then the resultant

$$\mathbf{w}_3' = \frac{\mathbf{w}_2 + \mathbf{w}_1 - \frac{1}{2}\mathbf{w}_2 \times \mathbf{w}_1}{1 - \frac{1}{4}\mathbf{w}_2 \cdot \mathbf{w}_1} = \frac{\mathbf{w}_1 + \mathbf{w}_2 + \frac{1}{2}\mathbf{w}_1 \times \mathbf{w}_2}{1 - \frac{1}{4}\mathbf{w}_1 \cdot \mathbf{w}_2} \tag{9.8}$$

would not be the same as $\mathbf{w}_3$, except when the rotation axes are the same and $\mathbf{w}_1 \times \mathbf{w}_2 = 0$.

Figure 9.3 illustrates successive rotations of an object (a book), by 90° each, about the x and y axes, respectively ($\mathbf{w}_1 = 2\mathbf{i}$, $\mathbf{w}_2 = 2\mathbf{j}$). Starting from the orientation in (a), the orientation (c) is the result of a rotation $\mathbf{w}_1$ followed by $\mathbf{w}_2$; orientation (c′) is a result of a rotation $\mathbf{w}_2$ followed by $\mathbf{w}_1$. The orientation of the object at any phase may be specified by giving orthogonal unit vectors $\mathbf{I}$ and $\mathbf{J}$ of body axes fixed in it. Then $\mathbf{I}(a) = \mathbf{i}$ and $\mathbf{J}(a) = \mathbf{j}$, initially, and (9.4) gives the intermediate orientations (b) and (b') when $\mathbf{w}_1$ and $\mathbf{w}_2$ are respectively used:

$$\mathbf{I}(b) = \mathbf{i}, \qquad \mathbf{J}(b) = \mathbf{j} + \tfrac{1}{2}[2\mathbf{i} \times (\mathbf{j} + \mathbf{i} \times \mathbf{j})] = \mathbf{j} + \mathbf{k} - \mathbf{j} = \mathbf{k},$$

$$\mathbf{I}(b') = \mathbf{i} + \tfrac{1}{2}[2\mathbf{j} \times (\mathbf{i} + \mathbf{j} \times \mathbf{i})] = \mathbf{i} - \mathbf{k} - \mathbf{i} = -\mathbf{k}, \qquad \mathbf{J}(b') = \mathbf{j}.$$

Next, $\mathbf{w}_2$ may be applied to configuration (b) to give $\mathbf{I}(c) = -\mathbf{k}$, $\mathbf{J}(c) = \mathbf{i}$, and

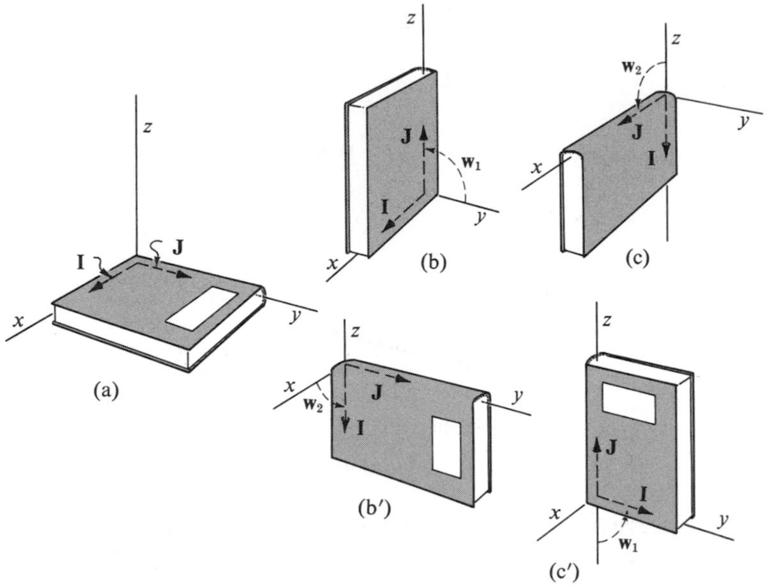

Figure 9.3

$\mathbf{w}_1$ applied to (b') gives $\mathbf{I}(c') = \mathbf{j}$, $\mathbf{J}(c') = \mathbf{k}$. These results for (c) and (c') also follow directly when the single rotations $\mathbf{w}_3 = 2(\mathbf{i} + \mathbf{j} - \mathbf{k})$ and $\mathbf{w}'_3 = 2(\mathbf{i} + \mathbf{j} + \mathbf{k})$, calculated from (9.7) and (9.8), are respectively applied to (a). Each of the rotations $\mathbf{w}_3$ and $\mathbf{w}'_3$ are by 120°.

The Infinitesimal Limit

The definition of the finite-rotation vector $\mathbf{w}$ of (9.2) was so normalized that for $\alpha \to d\alpha = \dot{\alpha}\, dt \to 0$, $\mathbf{w} \to 2\hat{\mathbf{\Omega}}(\frac{1}{2}\dot{\alpha}\, dt) = \mathbf{\Omega}\, dt$, where $\mathbf{\Omega} = \hat{\mathbf{\Omega}}\dot{\alpha}$ is the angular velocity of the infinitesimal rotation. The corresponding infinitesimal displacement becomes, according to (9.3),

$$\mathbf{R}' - \mathbf{R} = \mathbf{dR} = \mathbf{\Omega}\, dt \times \tfrac{1}{2}(2\mathbf{R} + \mathbf{dR}) \to \mathbf{\Omega}\, dt \times \mathbf{R},$$

just as expected from $\dot{\mathbf{R}} = \mathbf{\Omega} \times \mathbf{R}$ (9.1).

The important result which can now be obtained is that—despite the nonlinearity in the addition of, and the noncommutativity in the compounding of, the finite rotations—angular velocities can be added vectorially, and do commute. Thus, when $\mathbf{w}_1 \to \mathbf{\Omega}_1\, \Delta t$ and $\mathbf{w}_2 \to \mathbf{\Omega}_2\, \Delta t$, the expression (9.7) for the resultant yields

$$\mathbf{\Omega}_3 \equiv \lim_{\Delta t \to 0} \frac{\mathbf{w}_3}{\Delta t} = \lim \frac{\mathbf{\Omega}_1 + \mathbf{\Omega}_2 - \frac{1}{2}\mathbf{\Omega}_1 \times \mathbf{\Omega}_2\, \Delta t}{1 - \frac{1}{4}\mathbf{\Omega}_1 \cdot \mathbf{\Omega}_2\, \Delta t} = \mathbf{\Omega}_1 + \mathbf{\Omega}_2, \qquad (9.9)$$

and $\mathbf{w}'_3 \to \mathbf{\Omega}_2 + \mathbf{\Omega}_1 = \mathbf{\Omega}_1 + \mathbf{\Omega}_2$.

The vectors **w** help represent integrated effects of $\mathbf{\Omega}(t)$. They could be used to specify any instantaneous orientation of a body in that they can give the axis and angle of a single finite rotation by which the orientation could be reached. However, its awkward algebraic properties make such a line of development unnecessarily complex. Representations which are easier to develop will be introduced in the following sections. The representations by **w** were worth attention because they are closest to the representation of rotation *rates* by angular velocities, $\mathbf{\Omega}$. They show most clearly how commuting angular velocities can be consistent with noncommuting finite rotations. Meanwhile, they afford a general insight into the properties of finite rotations: that successive ones about different axes can be replaced by one rotation about a single axis, that they are not *vectorially* additive, and that they do not commute.

9.2 EULER ANGLES

The components of the angular velocity $\mathbf{\Omega}(t)$ may be regarded as "generalized velocities" (§7.2), corresponding to the three rotational degrees of freedom. However, they cannot easily be treated as just time-derivatives of any simply definable "generalized coordinates," except for rotation about a fixed axis, $\hat{\mathbf{\Omega}}$. In the latter case, an instantaneous orientation may be described by a single azimuthal angle $\varphi(t)$ about the axis, and $\Omega = \dot{\varphi}$ simply. When the rotation axis itself rotates, then the connection of $\mathbf{\Omega}(t)$ to any convenient orientation coordinates is more complicated, as will be seen.

Euler Coordinates

The most widely used coordinates for specifying instantaneous orientations of rotating axes are the Euler angles φ, ϑ, ψ indicated in Figure 9.4. This shows an arbitrary instantaneous orientation of "body axes" X, Y, Z, relative to a nonrotating frame x, y, z. The orientation plainly is arbitrary if, first, the Z-axis has been rotated from coincidence with z through an arbitrary angle ϑ. Any such rotation is equivalent to one about an axis which lies in both the x, y and X, Y planes, and hence forms their intersection (sometimes called the "nodal line"). The orientation of this line could also have been chosen arbitrarily, and may be specified by the angle φ it makes with the x, z plane. The X, Y plane has now been arbitrarily tilted up from the x, y plane; an arbitrary choice still exists for the orientation of the X, Y axes in the tilted plane. The choice will have been made if a third angle, ψ, between X-axis and nodal line, has been adopted.

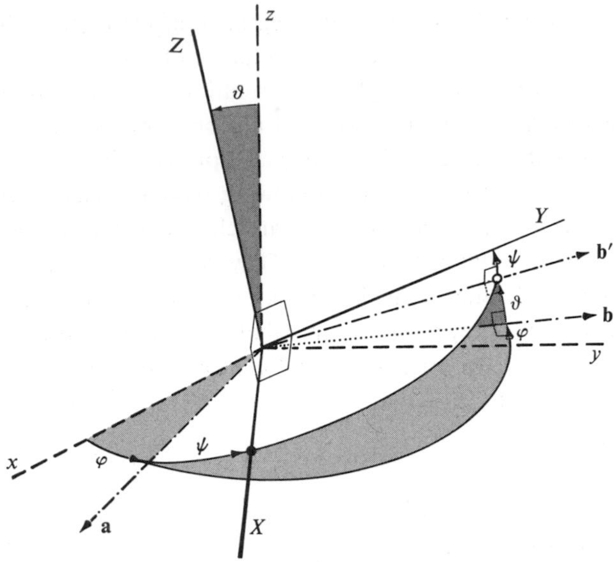

Figure 9.4

The result that an arbitrary orientation of a body can be specified by just three angles corresponds to a fact which may have been plain from the beginning: a rigid body has just three rotational degrees of freedom. That is why the rotation rate was fully specifiable by the three components of the angular velocity vector $\mathbf{\Omega}$, and any finite rotation by the three-component vector $\mathbf{w}$ of (9.2).

When considering resolutions of vectors on the two frames of Figure 9.4, it helps to make provisional use of the nodal line as a subsidiary axis, denoted by a unit vector $\mathbf{a}$. The axis normal to it in the x, y plane will be denoted $\mathbf{b}$, while the one in the body X, Y plane is denoted $\mathbf{b}'$.

The connection of the Euler specification to that by the body axis vectors $\mathbf{I}$, $\mathbf{J}$, $\mathbf{K}$, or to that in terms of an arbitrary body point $\mathbf{R} = \mathbf{I}X + \mathbf{J}Y + \mathbf{K}Z$, may be obtained by considering the components of these vectors on the nonrotating "space" axes, $\mathbf{i}$, $\mathbf{j}$, $\mathbf{k}$. Thus, reference to Figure 9.4 easily shows that

$$\mathbf{I} = \mathbf{a} \cos \psi + \mathbf{b}' \sin \psi \tag{9.10a}$$

and

$$\left.\begin{aligned} \mathbf{a} &= \mathbf{i} \cos \varphi + \mathbf{j} \sin \varphi, \\ \mathbf{b}' &= \mathbf{b} \cos \vartheta + \mathbf{k} \sin \vartheta \\ &= (\mathbf{j} \cos \varphi - \mathbf{i} \sin \varphi) \cos \vartheta + \mathbf{k} \sin \vartheta, \end{aligned}\right\} \tag{9.11}$$

and hence

$$\mathbf{I} = \mathbf{i}(\cos\varphi\cos\psi - \sin\varphi\cos\vartheta\sin\psi)$$
$$+\mathbf{j}(\sin\varphi\cos\psi + \cos\varphi\cos\vartheta\sin\psi) + \mathbf{k}\sin\vartheta\sin\psi. \qquad (9.10b)$$

The use of the subsidiary axes has obviated the necessity of using spherical trigonometric formulas, as in direct evaluations of direction cosines like $\mathbf{I}\cdot\mathbf{i}$. However, much more elegant methods of obtaining such resolutions as (9.10*b*) will be found in the next section. Those methods will also afford the best way to discern the connection between the Euler description and that by the finite-rotation vector **w** of (9.2).

Connection of Ω to the Euler Angular Velocities

Rates of change of the Euler angles correspond to a resultant angular velocity that is most immediately representable by

$$\mathbf{\Omega} = \mathbf{k}\dot{\varphi} + \mathbf{a}\dot{\vartheta} + \mathbf{K}\dot{\psi}. \qquad (9.12)$$

The vector additivity used here was justified in (9.9). The component terms of (9.12) are not all orthogonal, since, although $\mathbf{k}\cdot\mathbf{a} = \mathbf{K}\cdot\mathbf{a} = 0$, $\mathbf{k}\cdot\mathbf{K} = \cos\vartheta \neq 0$ in general. A resolution on the orthogonal body axes,

$$\mathbf{\Omega} = \mathbf{I}\Omega_X + \mathbf{J}\Omega_Y + \mathbf{K}\Omega_Z, \qquad (9.13)$$

will prove more useful. Reference to Figure 9.4 shows that

$$\left.\begin{aligned} \mathbf{k}\,\dot{\varphi} &= \dot{\varphi}(\mathbf{K}\cos\vartheta + \mathbf{b}'\sin\vartheta) \\ &= \mathbf{K}\dot{\varphi}\cos\vartheta + \dot{\varphi}\sin\vartheta(\mathbf{J}\cos\psi + \mathbf{I}\sin\psi), \\ \mathbf{a}\dot{\vartheta} &= \dot{\vartheta}\,(\mathbf{I}\cos\psi - \mathbf{J}\sin\psi), \end{aligned}\right\} \qquad (9.14)$$

and hence that

$$\left.\begin{aligned} \Omega_X &= \dot{\varphi}\sin\vartheta\sin\psi + \dot{\vartheta}\cos\psi, \\ \Omega_Y &= \dot{\varphi}\sin\vartheta\cos\psi - \dot{\vartheta}\sin\psi, \\ \Omega_Z &= \dot{\varphi}\cos\vartheta + \dot{\psi}. \end{aligned}\right\} \qquad (9.15)$$

It is equally easy, with the help of (9.11), to find the resolution on the non-rotating axes,

$$\left.\begin{aligned} \Omega_x &= \dot{\vartheta}\cos\varphi + \dot{\psi}\sin\vartheta\sin\varphi, \\ \Omega_y &= \dot{\vartheta}\sin\varphi - \dot{\psi}\sin\vartheta\cos\varphi, \\ \Omega_z &= \dot{\varphi} + \dot{\psi}\cos\vartheta. \end{aligned}\right\} \qquad (9.16)$$

A disadvantage of the Euler description examined here is that the angles φ, ϑ, ψ do not have as easily defined a "covariance" with frame changes as do components of a vector for example. That will be remedied in the next section, with a description of an orientational configuration by a single "entity" (to be given in (9.49)) having a well-defined covariance (like that shown in (9.115)), and in such respects analogous to the representation of the three translational degrees of freedom of a point by a single vector "entity" like **r**. Moreover, the meaning of "covariance" will be clarified.

9.3 ROTATIONAL TRANSFORMATION MATRICES

Another way of representing frame orientations will be given extended attention because it serves to introduce an important mode of mathematical expression, the matrix algebra. The representation may be developed by considering the immediate problem to be one of specifying the orthogonal unit vectors **I**, **J**, **K**, which point out the body axes, relative to a fixed space-frame **i**, **j**, **k**. This may be done by simply giving the three cartesian components of each of the three unit vectors, in projection forms like those of (1.16):

$$\left.\begin{matrix} \mathbf{I}\cdot\mathbf{i}, & \mathbf{I}\cdot\mathbf{j}, & \mathbf{I}\cdot\mathbf{k}, \\ \mathbf{J}\cdot\mathbf{i}, & \mathbf{J}\cdot\mathbf{j}, & \mathbf{J}\cdot\mathbf{k}, \\ \mathbf{K}\cdot\mathbf{i}, & \mathbf{K}\cdot\mathbf{j}, & \mathbf{K}\cdot\mathbf{k}. \end{matrix}\right\} \tag{9.17}$$

Of these nine cosines, only three are independent because of the orthogonality of each set of unit vectors, as will be seen, and that again corresponds to the existence of just three orientational degrees of freedom. Although six are redundant, it yet proves economical to work with the whole array of nine quantities because this can be treated as a single algebraic entity, a "matrix."

Linear Vector Transformations

Whereas the representations discussed in the preceding sections are specifically designed for three dimensions, the matrix representations are readily adapted to any number, and may as well be discussed here for an arbitrary number of dimensions, n. Of course, the ordinary three-dimensional space has the greatest importance, but mappings on various types of multidimensional vector spaces have also proved useful for physical descriptions.

An n-dimensional space will here be taken to be one in which as many as n mutually orthogonal directions may be defined. A set of them may be pointed out by unit vectors $\mathbf{e}_1, \mathbf{e}_2, \ldots, \mathbf{e}_n$ having "orthonormality" properties like those in (1.12):

$$\mathbf{e}_k \cdot \mathbf{e}_l = \delta_{kl} \qquad (\equiv 1 \text{ for } k = l; \equiv 0 \text{ otherwise}). \tag{9.18}$$

The set is complete enough to define all the vectors

$$\mathbf{r} = \mathbf{e}_1 x_1 + \mathbf{e}_2 x_2 + \cdots + \mathbf{e}_n x_n, \tag{9.19a}$$

formed with any choices for the components $x_1, x_2, \ldots, x_n$. Resolutions into components like this may be written more economically as

$$\mathbf{r} = \sum_{k=1}^{n} \mathbf{e}_k x_k \equiv \mathbf{e}_k x_k, \tag{9.19b}$$

the summation sign being left off in the final expression in agreement with a "summation convention," by which the mere repetition of an index (in a product) implies that the index is to be summed on.

It is not any particular set of components $x_k = \mathbf{e}_k \cdot \mathbf{r}$ which is held to characterize a given vector $\mathbf{r}$, since any alternative set of n mutually orthogonal directions, $\{\mathbf{e}_l'\} \equiv \mathbf{e}_1', \mathbf{e}_2', \ldots, \mathbf{e}_n'$, arbitrarily rotated with respect to $\{\mathbf{e}_k\}$ $\equiv \mathbf{e}_1, \mathbf{e}_2, \ldots, \mathbf{e}_n$, may be used for the resolution of the vector:

$$\mathbf{r} = \mathbf{e}_k x_k = \mathbf{e}_l' x_l'. \tag{9.20}$$

It is rather the relations,

$$x_j' = \mathbf{e}_j' \cdot \mathbf{r} = (\mathbf{e}_j' \cdot \mathbf{e}_k) x_k, \tag{9.21}$$

which develop among components on relatively rotated bases that characterize vector behavior (its "covariance"). Thus a vector $\mathbf{r}$ is an entity that may be assigned alternative sets of components, $\{x_k\}$ or $\{x_l'\}$ or $\{x_m''\}$, and so on, on alternative bases, $\{\mathbf{e}_k\}$ or $\{\mathbf{e}_l'\}$ or $\{\mathbf{e}_m''\}$, and so on, with the sets of components having linear transformation interrelationships of the type (9.21). The array of n^2 transformation coefficients (each a cosine),

$$\mathbf{e}_j' \cdot \mathbf{e}_k \equiv a_{jk}, \tag{9.22}$$

is a generalization of the array (9.17), which was to be investigated. Besides its employment for specifying the orientation of one frame relative to another, the array has now been found useful for defining the covariance of (vector) components in the frame change.

Square Matrices

Each of the elements a_{jk} of the array, being a scalar product of unit vectors, is just a direction cosine that helps measure the size of a rotation. It is the cosine magnitudes that characterize a rotation, rather than the particular basis vectors $\{\mathbf{e}_k\}$ and $\{\mathbf{e}'_j\}$ that give possible starting and ending frame orientations. The same array of numbers a_{jk} might equally well be equated to cosines $\mathbf{e}''_j \cdot \mathbf{e}'''_k$, measuring the rotation of a *third* basis, $\{\mathbf{e}''_j\}$, to a fourth one, $\{\mathbf{e}'''_k\}$, making the same angles with $\{\mathbf{e}''_j\}$ as do the $\{\mathbf{e}_k\}$ with $\{\mathbf{e}'_j\}$. The arrays of cosines like a_{jk} may be treated as n^2-component entities that have one-to-one correspondence with the possible rotations. Each such entity is called a rotational transformation *matrix*, and is said to possess n^2 *matrix elements*, a_{jk}. Matrices will here be symbolized by sans-serif letters like a, and whenever a detailed specification of the elements of one is desired, they are arranged into an "$n \times n$" square array:

$$\mathsf{a} = \begin{pmatrix} a_{11} & a_{12} & \cdots & a_{1n} \\ a_{21} & a_{22} & \cdots & a_{2n} \\ \vdots & \vdots & \vdots & \vdots \\ a_{n1} & a_{n2} & \cdots & a_{nn} \end{pmatrix}. \tag{9.23}$$

The ordering conventions are such that $a_{j1}, a_{j2}, \ldots, a_{jn}$ are the elements of the jth *row*, and $a_{1k}, a_{2k}, \ldots, a_{kn}$ form the kth *column*. Sometimes the numerical indices like j or $k = 1, 2 \ldots, n$ are replaced by such as j or $k = x, y, z$ or X, Y, Z, when these become appropriate.

The simplest nontrivial examples of rotational transformation matrices are provided by two-dimensional frame rotations like the one indicated in Figure 9.5. Such linear transformation relations as

$$\begin{aligned} x'_1 &= x_1 \cos\varphi + x_2 \sin\varphi, \\ x'_2 &= -x_1 \sin\varphi + x_2 \cos\varphi, \end{aligned} \tag{9.24}$$

are generated by rotation through an angle φ, and the corresponding matrix of the coefficients is

$$\mathsf{a}(\varphi) = \begin{pmatrix} \cos\varphi & \sin\varphi \\ -\sin\varphi & \cos\varphi \end{pmatrix}, \tag{9.25}$$

there being one such matrix for each possible value $0 \leq \varphi < 2\pi$. the single parameter φ of the two-dimensional rotations is replaced by three parameters in three-dimensional rotations. Thus, the Euler angles φ, ϑ, ψ can be used to express the rotational transformation matrix relating the body and space frames of Figure 9.4; one will be presented in (9.49), after an elegant way of obtaining it is developed.

Orthonormal Matrices

The rotational transformation matrices are of a special type, having certain interrelationships among their elements. That is particularly clear

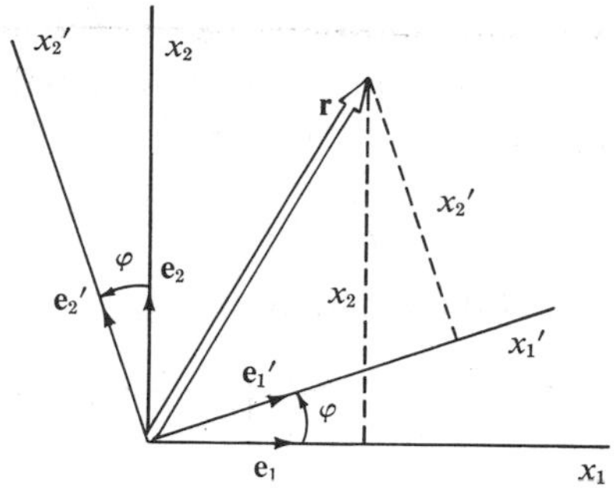

Figure 9.5

in the case of the array (9.17) when it is displayed in the equivalent alternative ways:

$$\begin{pmatrix} \mathbf{I}\cdot\mathbf{i} & \mathbf{I}\cdot\mathbf{j} & \mathbf{I}\cdot\mathbf{k} \\ \mathbf{J}\cdot\mathbf{i} & \mathbf{J}\cdot\mathbf{j} & \mathbf{J}\cdot\mathbf{k} \\ \mathbf{K}\cdot\mathbf{i} & \mathbf{K}\cdot\mathbf{j} & \mathbf{K}\cdot\mathbf{k} \end{pmatrix} = \begin{pmatrix} I_x & I_y & I_z \\ J_x & J_y & J_z \\ K_x & K_y & K_z \end{pmatrix} = \begin{pmatrix} i_X & j_X & k_X \\ i_Y & j_Y & k_Y \\ i_Z & j_Z & k_Z \end{pmatrix}. \tag{9.26}$$

The squares of the elements in any one row or in any one column, add up to unity because each vector **i**, **j**, **k**, **I**, **J**, **K** has unit magnitude. Moreover, the sum of the products of corresponding elements from any pair of rows, or from any pair of columns, vanishes because of the orthogonalities $\mathbf{i}\cdot\mathbf{j}=\mathbf{j}\cdot\mathbf{k}=\mathbf{k}\cdot\mathbf{i}=0$ and $\mathbf{I}\cdot\mathbf{J}=\mathbf{J}\cdot\mathbf{K}=\mathbf{K}\cdot\mathbf{I}=0$. In the more general rotational transformation matrix formed of the elements $a_{jk}=\mathbf{e}_j'\cdot\mathbf{e}_k$ (9.22), the kth column consists of the components of the unit vector $\mathbf{e}_k$ on the basis $\{\mathbf{e}_l'\}$:

$$\mathbf{e}_k = \mathbf{e}_j'(\mathbf{e}_j'\cdot\mathbf{e}_k) \equiv \mathbf{e}_j' a_{jk}, \tag{9.27}$$

when resolved like **r** of (9.20) and (9.21). Consequently

$$a_{1k}^2 + a_{2k}^2 + \cdots + a_{nk}^2 = 1 \tag{9.28a}$$

because $\mathbf{e}_k\cdot\mathbf{e}_k=1$, and

$$a_{1k}a_{1l} + a_{2k}a_{2l} + \cdots + a_{nk}a_{nl} = 0 \tag{9.28b}$$

because $\mathbf{e}_k\cdot\mathbf{e}_l=0$. All the relations of this kind may be gathered into the expression

$$\Big(\sum_j\Big) a_{jk}a_{jl} = \delta_{kl} \qquad (k, l = 1, 2, \ldots, n), \tag{9.29}$$

representing a property characteristic of any matrix which transforms one orthonormal basis into another. Any matrix obeying such conditions is called an *orthonormal* matrix.

Basically, the orthonormality property of rotational transformation matrices follows from the fact that a rotation, by definition, leaves the length of every vector unchanged. Consider the linear transformation

$$x'_j = a_{jk} x_k . \tag{9.30}$$

If it is to leave the length of the vector invariant,

$$r^2 \equiv \mathbf{r} \cdot \mathbf{r} = \sum_j (x'_j)^2 = \sum_j \Big(\sum_k a_{jk} x_k\Big)\Big(\sum_l a_{jl} x_l\Big)$$
$$= \sum_{k,l} x_k x_l \Big(\sum_j a_{jk} a_{jl}\Big)$$

must equal $\sum_k x_k^2$. This will be so, regardless of the vector's components, only if (9.29) holds; it is necessary to make the terms with $k \neq l$ vanish and to provide unit coefficients for the squared terms.

To have the orthonormality property, the matrix elements must obey n conditions of the type (9.28*a*), and also $n(n-1)/2$ conditions like (9.28*b*), since this is the number of independent pairs that can be formed of n columns. Thus the n^2 elements must obey $n + \frac{1}{2}n(n-1) = \frac{1}{2}n(n+1)$ conditions and $n^2 - \frac{1}{2}n(n+1) = \frac{1}{2}n(n-1)$ choices for elements are left open. That accounts for the fact that the $n = 2$ representation, $\mathbf{a}(\varphi)$ of (9.25), contained just one free parameter φ, and that just $\frac{1}{2} \cdot 3(3-1) = 3$ Euler angle degrees of freedom exist in the three dimensional rotations.

It should also be pointed out that not only each column, but also each *row* of the matrix elements $a_{jk} = \mathbf{e}'_j \cdot \mathbf{e}_k$, considered above, form the components of a unit vector. Thus $a_{j1}, a_{j2}, \ldots, a_{jn}$ are the components of $\mathbf{e}'_j$ on the basis $\{\mathbf{e}_k\} \equiv \mathbf{e}_1, \mathbf{e}_2, \ldots, \mathbf{e}_n$. This means that the orthonormality conditions may also be written as

$$\Big(\sum_l\Big) a_{jl} a_{kl} = \delta_{jk} . \tag{9.31}$$

These are not new conditions, additional to (9.29), for they can be shown to follow from (9.29), and vice versa, as will be seen most easily after (9.59).

Matrix Products

Suppose that a frame is subjected to two successive rotations, from an orientation $\{\mathbf{e}_l\}$ to a second, $\{\mathbf{e}'_k\}$, and then to a third, $\{\mathbf{e}''_j\}$, so that the components of a given vector $\mathbf{r}$ are transformed according to

$$x'_k = a_{kl} x_l \qquad \text{and} \qquad x''_j = b_{jk} x'_k . \tag{9.32}$$

Clearly

$$x''_j = b_{jk} a_{kl} x_l = c_{jl} x_l \tag{9.33}$$

if

$$c_{jl} = b_{jk}\, a_{kl} \qquad \Big(\equiv \sum_k b_{jk}\, a_{kl}\Big). \tag{9.34}$$

Thus two successive rotations, $\{\mathbf{e}_l\} \to \{\mathbf{e}'_k\}$ and $\{\mathbf{e}'_k\} \to \{\mathbf{e}''_j\}$, are replaced by a single, resultant rotation, $\{\mathbf{e}_l\} \to \{\mathbf{e}''_j\}$. The individual rotations are represented by matrices **a** and **b** and the resultant transformation matrix may be treated as the product of these matrices,

$$\mathbf{c} = \mathbf{ba}, \tag{9.35}$$

if the elements c_{jl} of such a product matrix are defined as the combinations (9.34) of elements of the factor matrices. The expression (9.35) is a matrix equation, having the n^2 components (9.34).

When each matrix is written out as an array of the type (9.23), then the product equation (9.35) may be displayed as

$$\begin{pmatrix} b_{11} & b_{12} & \cdots & b_{1n} \\ b_{21} & b_{22} & \cdots & b_{2n} \\ \vdots & \vdots & \vdots & \vdots \\ b_{n1} & b_{n2} & \cdots & b_{nn} \end{pmatrix} \begin{pmatrix} a_{11} & a_{12} & \cdots & a_{1n} \\ a_{21} & a_{22} & \cdots & a_{2n} \\ \vdots & \vdots & \vdots & \vdots \\ a_{n1} & a_{n2} & \cdots & a_{nn} \end{pmatrix} = \begin{pmatrix} c_{11} & c_{12} & \cdots & c_{1n} \\ c_{21} & c_{22} & \cdots & c_{2n} \\ \vdots & \vdots & \vdots & \vdots \\ c_{n1} & c_{n2} & \cdots & c_{nn} \end{pmatrix}. \tag{9.36}$$

The successive elements $c_{j1}, \ldots, c_{jn}$ in the jth row of the product matrix are successively obtained by multiplying the jth row of **b** into the successive columns of **a**, as dictated by the definition (9.34). A simple example of a matrix product is provided by two successive rotations of the type indicated in Figure 9.5, through angles φ_1 and φ_2 respectively:

$$\begin{pmatrix} \cos\varphi_2 & \sin\varphi_2 \\ -\sin\varphi_2 & \cos\varphi_2 \end{pmatrix} \begin{pmatrix} \cos\varphi_1 & \sin\varphi_1 \\ -\sin\varphi_1 & \cos\varphi_1 \end{pmatrix} = \begin{pmatrix} \cos(\varphi_1+\varphi_2) & \sin(\varphi_1+\varphi_2) \\ -\sin(\varphi_1+\varphi_2) & \cos(\varphi_1+\varphi_2) \end{pmatrix}, \tag{9.37}$$

since $\cos\varphi_2 \cos\varphi_1 + \sin\varphi_2(-\sin\varphi_1) = \cos(\varphi_2+\varphi_1)$, and so on. The order of the factor matrices does not matter in this simple example because the two rotations are about the same axis. In general, however,

$$(\mathbf{ab})_{jl} = a_{jk}\, b_{kl} \neq (\mathbf{ba})_{jl} = b_{jk}\, a_{kl},$$

or

$$\mathbf{ab} \neq \mathbf{ba}. \tag{9.38}$$

An example will be shown in (9.50). Algebra with such a property is called "noncommutative," and the factor matrices **a** and **b** are said not to "commute," in general.

Transposed Matrices

It was a bit cavalier to assume above that the product **c** of two rotation matrices **a** and **b** will again be a rotation matrix. Such a matrix must be of a restricted type, having an orthonormality property like (9.29). That the product of two matrices obeying orthonormality conditions also obeys an orthonormality condition can indeed be proved in terms of statements like (9.29) of such conditions. However, recasting the latter into matrix equations makes the expressions briefer. The conditions may be represented by a matrix equation if the "transpose" $\tilde{\mathbf{a}}$ of any matrix **a** is defined. By the transpose will be meant a matrix formed from **a** by interchanging its rows and columns, so that

$$(\tilde{\mathbf{a}})_{kl} = a_{lk}. \tag{9.39}$$

Then, since $\sum_j a_{jk} a_{jl} = \sum_j (\tilde{\mathbf{a}})_{kj} a_{jl}$, this qualifies as an element of the product $(\tilde{\mathbf{a}}\mathbf{a})_{kl}$, according to the multiplication rule (9.34). Now the orthonormality condition on **a** may be written

$$\tilde{\mathbf{a}}\mathbf{a} = \mathbf{1}, \tag{9.40}$$

if the right side is the unit matrix

$$\mathbf{1} \equiv \begin{pmatrix} 1 & 0 & \cdots & 0 \\ 0 & 1 & \cdots & 0 \\ \vdots & \vdots & & \vdots \\ 0 & 0 & \cdots & 1 \end{pmatrix}, \tag{9.41}$$

with elements $(\mathbf{1})_{kl} = \delta_{kl}$.

It is easy to see that for any product of matrices, $\mathbf{f} = \mathbf{abc}\cdots$, the transpose of the resultant is

$$\tilde{\mathbf{f}} = \widetilde{(\mathbf{abc}\cdots)} = \cdots \tilde{\mathbf{c}}\tilde{\mathbf{b}}\tilde{\mathbf{a}}, \tag{9.42}$$

a product of the individual transposes taken in reverse order. The proof may be illustrated for the case of three factors:

$$\begin{aligned}(\tilde{\mathbf{f}})_{kl} &= (\mathbf{abc})_{lk} = a_{lm} b_{mn} c_{nk} = \tilde{a}_{ml} \tilde{b}_{nm} \tilde{c}_{kn} \\ &= \tilde{c}_{kn} \tilde{b}_{nm} \tilde{a}_{ml} = (\tilde{\mathbf{c}}\tilde{\mathbf{b}}\tilde{\mathbf{a}})_{kl}.\end{aligned}$$

This development stresses the fact that conforming to the rules of matrix multiplication (9.34) always requires summation over repeated *contiguous*, inner indices when the factors are written in the proper order. The transpose of the orthonormality condition (9.40) yields

$$\tilde{\mathbf{a}}\mathbf{a} = \tilde{\mathbf{1}} \equiv \mathbf{1}, \tag{9.43}$$

making no change because the unit matrix is symmetrical, and so $\tilde{\mathbf{1}} = \mathbf{1}$.

It can now be easily checked that the product of two orthonormal matrices is indeed an orthonormal matrix, interpretable as providing a resultant rotational transformation. For, if $\tilde{\mathbf{a}}\mathbf{a} = \mathbf{1}$ and $\tilde{\mathbf{b}}\mathbf{b} = \mathbf{1}$, then

$$\tilde{\mathbf{c}}\mathbf{c} = (\widetilde{\mathbf{ba}})\mathbf{ba} = \tilde{\mathbf{a}}\tilde{\mathbf{b}}\mathbf{ba} = \tilde{\mathbf{a}}\mathbf{a} = \mathbf{1}. \tag{9.44}$$

Use has been made here of the obvious fact that the product of the unit matrix with any matrix is again the matrix, for $(\mathbf{1}.\mathbf{a})_{kl} = \delta_{km} a_{ml} = a_{kl}$ and $(\mathbf{a}.\mathbf{1})_{kl} = a_{km}\delta_{ml} = a_{kl}$.

Matrix Addition

To complete an algebra of matrices, a sum, $\mathbf{s} = \mathbf{a} + \mathbf{b}$, of two matrices may be defined. The addition and subtraction of matrices have minor roles as compared to matrix multiplication, but occasions arise in which they may be put to use. It is found that a definition in which the sum matrix has elements, $s_{kl} = a_{kl} + b_{kl}$, which are just sums of the corresponding elements of the matrices being added, is useful. A simple illustration is provided when a matrix expressing the difference from identity of an infinitesimal rotation is separated out. Thus, when $\varphi \to \delta\varphi \to 0$ in the two-dimensional matrix (9.25), $\cos\delta\varphi = 1$ and $\sin\delta\varphi \approx \delta\varphi$; hence

$$\begin{aligned}\mathbf{a}(\delta\varphi) &= \begin{pmatrix} 1 & \delta\varphi \\ -\delta\varphi & 1 \end{pmatrix} = \begin{pmatrix} 1 & 0 \\ 0 & 1 \end{pmatrix} + \begin{pmatrix} 0 & \delta\varphi \\ -\delta\varphi & 0 \end{pmatrix} \\ &= \mathbf{1} + i\,\delta\varphi\sigma_2,\end{aligned} \tag{9.45a}$$

where σ_2 is a widely used notation for the matrix

$$\sigma_2 = \begin{pmatrix} 0 & -i \\ i & 0 \end{pmatrix}. \tag{9.45b}$$

Implicit in these considerations is the conclusion that the product $\zeta\mathbf{a}$ of a matrix with any number ζ, real or complex, is a matrix with elements $(\zeta\mathbf{a})_{kl} = \zeta a_{kl}$; that is, every element of $\mathbf{a}$ is multiplied by the number.

The Three-Dimensional Representation

The matrix $\mathbf{a}(\varphi, \vartheta, \psi)$, which corresponds to the rotation of the body axes in Figure 9.4, may be constructed as a product of three rotations,

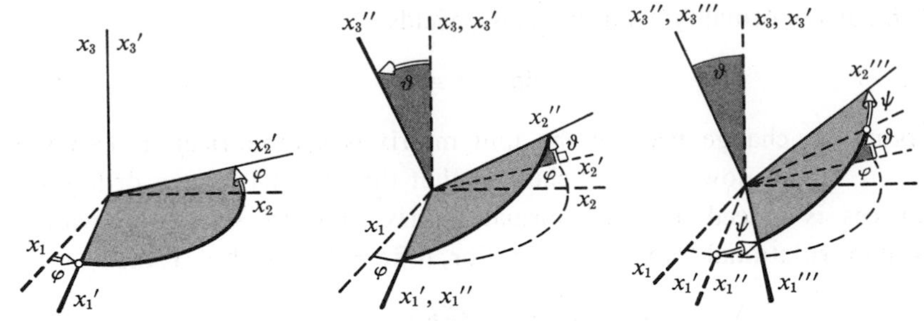

Figure 9.6

since the final orientation may be reached in the three steps shown in Figure 9.6. In the first step, the body axes are rotated about the space $z \equiv x_3$ axis through an angle φ, to axes labeled $x_1', x_2', x_3' \equiv x_3$. A given vector will have components on the two sets of axes having relations $x_\kappa' = a_{kl}(\varphi)x_l$, which are the same as the two-dimensional relations (9.24) except that they are supplemented by $x_3' = x_3$; hence the matrix of transformation coefficients is

$$\mathbf{a}(\varphi) = \begin{pmatrix} \cos\varphi & \sin\varphi & 0 \\ -\sin\varphi & \cos\varphi & 0 \\ 0 & 0 & 1 \end{pmatrix}. \tag{9.46}$$

The second step is a rotation through ϑ about the nodal line, forming the x_1' axis, to an orientation labeled as the $x_1'' \equiv x_1'$, x_2'', x_3'' axes. The relations $x_j'' = b_{jk}(\vartheta)x_k'$ are

$$x_1'' = x_1', \qquad x_2'' = x_2' \cos\vartheta + x_3' \sin\vartheta,$$
$$x_3'' = -x_2' \sin\vartheta + x_3' \cos\vartheta,$$

and hence

$$\mathbf{b}(\vartheta) = \begin{pmatrix} 1 & 0 & 0 \\ 0 & \cos\vartheta & \sin\vartheta \\ 0 & -\sin\vartheta & \cos\vartheta \end{pmatrix}. \tag{9.47}$$

Finally, there is a rotation $x_1'', x_2'', x_3'' \rightarrow x_1''', x_2''', x_3''' \equiv x_3''$, which is a rotation about a "z-axis" again, and so $x_i''' = a_{ij}(\psi)x_j''$ with a matrix $\mathbf{a}(\psi)$, which is the same as (9.46) except for the replacement of the angle φ by ψ. Thus

$$x_i''' = a_{ij}(\psi)b_{jk}(\vartheta)a_{kl}(\varphi)x_l \equiv a_{il}(\varphi, \vartheta, \psi)x_l,$$

in which

$$\mathbf{a}(\varphi, \vartheta, \psi) = \mathbf{a}(\psi)\mathbf{b}(\vartheta)\mathbf{a}(\varphi). \tag{9.48}$$

The factor arrays can be multiplied out to yield

$$\mathbf{a}(\varphi, \vartheta, \psi) = \begin{pmatrix} \cos\varphi\cos\psi - \cos\vartheta\sin\varphi\sin\psi & \sin\varphi\cos\psi + \cos\vartheta\cos\varphi\sin\psi & \sin\vartheta\sin\psi \\ -\cos\varphi\sin\psi - \cos\vartheta\sin\varphi\cos\psi & -\sin\varphi\sin\psi + \cos\vartheta\cos\varphi\cos\psi & \sin\vartheta\cos\psi \\ \sin\vartheta\sin\varphi & -\sin\vartheta\cos\varphi & \cos\vartheta \end{pmatrix}, \quad (9.49)$$

exhibiting the connection between description by the array (9.17) and the Euler angles. The first row has already been obtained in (9.10). The orthonormality of each of the factor matrices is evident and can also be checked for their product.

The order of the factors in (9.48) must be maintained with care, since, for example,

$$\mathbf{a}(\varphi)\mathbf{b}(\vartheta) \neq \mathbf{b}(\vartheta)\mathbf{a}(\varphi). \quad (9.50)$$

The lack of commutation stems from the fact that the two matrices $\mathbf{a}(\varphi)$ and $\mathbf{b}(\vartheta)$ refer to different rotation axes, x_3 and x_1', respectively. Notice that for $\vartheta = 0$, $\mathbf{a}(\varphi, 0, \psi) = \mathbf{a}(\varphi + \psi)$ is symmetrical in φ and ψ because $\mathbf{a}(\varphi)$ and $\mathbf{a}(\psi)$ then commute ($x_3'' \equiv x_3$).

Column and Row Vectors

The formalism may be expanded in such a way that all the relations involved can be written as matrix equations. For the relations $x_k' = a_{kl}x_l$, this is done by defining "column matrices"—arrays in terms of which the totality of relations may be written as

$$\begin{pmatrix} x_1' \\ x_2' \\ \vdots \\ x_n' \end{pmatrix} = \begin{pmatrix} a_{11}x_1 + a_{12}x_2 + \cdots + a_{1n}x_n \\ a_{21}x_1 + a_{22}x_2 + \cdots + a_{2n}x_n \\ \vdots \\ a_{n1}x_1 + a_{n2}x_2 + \cdots + a_{nn}x_n \end{pmatrix} = \begin{pmatrix} a_{11} & a_{12} & \cdots & a_{1n} \\ a_{21} & a_{22} & \cdots & a_{2n} \\ \vdots & \vdots & & \vdots \\ a_{n1} & a_{n2} & \cdots & a_{nn} \end{pmatrix} \begin{pmatrix} x_1 \\ x_2 \\ \vdots \\ x_n \end{pmatrix} \quad (9.51)$$

The multiplication of a square matrix into a column as in the last line is defined in such a way that it yields the column of sums in the first line. This plainly makes such products equivalent to products of square matrices, if a column is regarded as a square matrix having only zero elements in all but one column.

A column amounts to a display of the components of a vector on some basis. Both the vectors involved in (9.51) are the same one, $\mathbf{r}$, but, on the left, the column displays this vector on the basis $\mathbf{e}_1' \cdots \mathbf{e}_n'$, rather than its components on $\mathbf{e}_1, \ldots, \mathbf{e}_n$; hence it will there be denoted $(\mathbf{r})'$ instead of $\mathbf{r}$. In these notations, the relation may be written as the n-component equation of column vectors,

$$(\mathbf{r})' = \mathbf{ar}, \tag{9.52}$$

with the understanding that the "operation" of a square matrix on a column produces a column.

Row matrix displays of vector components also prove useful if they are regarded as transposes of columns thus:

$$\tilde{\mathbf{r}} \equiv (x_1, x_2, \ldots, x_n). \tag{9.53}$$

The multiplication of this into a column vector $\mathbf{R}$ produces a single number

$$\tilde{\mathbf{r}}\mathbf{R} = (x_1, x_2, \ldots, x_n)\begin{pmatrix} X_1 \\ X_2 \\ \vdots \\ X_n \end{pmatrix} = x_1 X_1 + \cdots + x_n X_n \equiv (\mathbf{r} \cdot \mathbf{R}), \tag{9.54}$$

according to the rules for matrix products, that is, if a row is treated as a square matrix having only zero elements in all rows except one and a single number like scalar $(\mathbf{r} \cdot \mathbf{R})$ is treated as a square matrix having only one nonvanishing element. Notice that the product of a row $\tilde{\mathbf{r}}$ into a column $\mathbf{R}$ is a simple scalar product only if the components displayed in both are projections of $\mathbf{r}$ and $\mathbf{R}$ on the same set of axes. There will be little use for products like $\tilde{\mathbf{r}}(\mathbf{r})'$ of vectors on different bases. Because $\mathbf{R} \cdot \mathbf{r} = \mathbf{r} \cdot \mathbf{R}$,

$$\tilde{\mathbf{R}}\mathbf{r} = \tilde{\mathbf{r}}\mathbf{R} = (\widetilde{\tilde{\mathbf{r}}\mathbf{R}}). \tag{9.55}$$

That is, the scalar product is symmetric.

The representability of a scalar product by the multiplication of a row vector into a column offers means for a concise derivation of the important orthonormality relations (9.40). As before, the derivation stems from the length invariance of a vector in rotations, or $(\mathbf{r})' \cdot (\mathbf{r})' = \mathbf{r} \cdot \mathbf{r} \equiv r^2$. In matrix notation, this equality is written $(\tilde{\mathbf{r}})'(\mathbf{r})' = \tilde{\mathbf{r}}\mathbf{r}$, and hence from (9.52)

$$(\tilde{\mathbf{r}})'(\mathbf{r})' = (\widetilde{\mathbf{ar}})\mathbf{ar} = \tilde{\mathbf{r}}\,\tilde{\mathbf{a}}\mathbf{ar} \rightarrow \tilde{\mathbf{r}}\mathbf{r}$$

if $\tilde{\mathbf{a}}\mathbf{a} = \mathbf{1}$. It is being taken for granted here that the rule of reversing the order of the factors when performing transpositions extends to $(\widetilde{\mathbf{ar}}) = \tilde{\mathbf{r}}\,\tilde{\mathbf{a}}$; that is readily verified by writing out the column $\mathbf{ar}$ and transposing it into a row.

The converse multiplication of columns into rows yields square matrices as in

$$\mathbf{R}\tilde{\mathbf{r}} = \begin{pmatrix} X_1 \\ X_2 \\ \vdots \\ X_n \end{pmatrix} (x_1 x_2 \cdots x_n) = \begin{pmatrix} (X_1 x_1) & (X_1 x_2) & \cdots & (X_1 x_n) \\ (X_2 x_1) & (X_2 x_2) & \cdots & (X_2 x_n) \\ \vdots & \vdots & & \vdots \\ (X_n x_1) & (X_n x_2) & \cdots & (X_n x_n) \end{pmatrix}, \quad (9.56)$$

according to the rules for products developed here. The elements of the product matrix here form the n^2 components of what is known as an "n-dimensional tensor of the second rank," to be discussed further below. Notice that $(\widetilde{\mathbf{R}\tilde{\mathbf{r}}}) = \mathbf{r}\tilde{\mathbf{R}} \neq \tilde{\mathbf{r}}\mathbf{R} = (\mathbf{r} \cdot \mathbf{R})$.

Inverse Matrices

Such linear relationships as $x_k' = a_{kl} x_l$ may usually be inverted (solved for the x_l as linear combinations of the x_k'). The coefficients in the result form the *inverse* $\mathbf{a}^{-1}$ of the matrix $\mathbf{a}$ thus:

$$x_l = a_{lk}^{-1} x_k' \rightarrow \mathbf{r} = \mathbf{a}^{-1}(\mathbf{r})'. \quad (9.57)$$

The matrix equation here appears as a consequence of

$$\mathbf{a}^{-1}\mathbf{a} = \mathbf{1}, \quad (9.58)$$

when the matrix $\mathbf{a}^{-1}$ is multiplied into $(\mathbf{r})' = \mathbf{ar}$ from the left. Multiplying (9.58) by $\mathbf{a}$ yields $\mathbf{aa}^{-1}\mathbf{a} = \mathbf{a}$, hence $\mathbf{aa}^{-1} = \mathbf{1}$ is also valid.

The elements of the inverse matrix are simple to find when it is an orthonormal matrix that is to be inverted, for the orthonormality property $\tilde{\mathbf{a}}\mathbf{a} = \mathbf{1}$ demonstrates that

$$\mathbf{a}^{-1} = \tilde{\mathbf{a}} \quad (9.59)$$

and each $a_{lk}^{-1} = a_{kl}$ simply. This conjoined with (9.58) shows that $\mathbf{a}\tilde{\mathbf{a}} = \mathbf{1}$ is also valid for orthonormal matrices, and this corresponds to deriving the relations (9.31) from (9.29), as mentioned in connection with (9.31). The equivalence between inverse and transpose for orthonormal matrices is particularly transparent in such forms as $a_{jk} = \mathbf{e}'_j \cdot \mathbf{e}_k$ (9.22). From the vector resolutions (9.20), the expressions

$$\begin{aligned} x'_j &= (\mathbf{e}'_j \cdot \mathbf{e}_k)x_k \leftrightarrow x_k = (\mathbf{e}_k \cdot \mathbf{e}'_j)x'_j, \\ (\mathbf{a}^{-1})_{kj} &= \mathbf{e}_k \cdot \mathbf{e}'_j = a_{jk} = (\tilde{\mathbf{a}})_{kj} \end{aligned} \tag{9.60}$$

are obvious.

The process of inverting the product $\mathbf{f} = \mathbf{abc}\cdots$ of several matrices becomes plain from the fact that

$$\cdots \mathbf{c}^{-1}\mathbf{b}^{-1}\mathbf{a}^{-1}\mathbf{abc}\cdots = \cdots \mathbf{c}^{-1}\mathbf{b}^{-1}\mathbf{bc}\cdots = \cdots \mathbf{c}^{-1}\mathbf{c}\cdots = \mathbf{1}.$$

Thus

$$\mathbf{f}^{-1} = (\mathbf{abc}\cdots)^{-1} = \cdots \mathbf{c}^{-1}\mathbf{b}^{-1}\mathbf{a}^{-1}, \tag{9.61}$$

with a reversal in the order of the factor matrices just as in the transposition process (9.42).

Finding the inverse of a matrix $\mathbf{a}$ which is *not* orthonormal, so that the relation (9.59) is not available, can become quite laborious. The most direct approach is to solve the simultaneous linear equations implicit in $\mathbf{aa}^{-1} = \mathbf{1}$ for each of the unknown elements a_{kl}^{-1} in terms of the given elements a_{jk}. The n equations for $a_{1l}^{-1}, a_{2l}^{-1}, \ldots, a_{nl}^{-1}$ are

$$\begin{matrix} a_{11}a_{1l}^{-1} + a_{12}a_{2l}^{-1} + \cdots + a_{1n}a_{nl}^{-1} = \delta_{1l} = 0, \\ a_{21}a_{1l}^{-1} + a_{22}a_{2l}^{-1} + \cdots + a_{2n}a_{nl}^{-1} = \delta_{2l} = 0, \\ \vdots \\ a_{l1}a_{1l}^{-1} + a_{l2}a_{2l}^{-1} + \cdots + a_{ln}a_{nl}^{-1} = \delta_{ll} = 1, \\ \vdots \\ a_{n1}a_{1l}^{-1} + a_{n2}a_{2l}^{-1} + \cdots + a_{nn}a_{nl}^{-1} = \delta_{nl} = 0. \end{matrix} \tag{9.62}$$

It is well known that the solutions of such simultaneous linear equations, when they exist, can be written as ratios of "determinants," which will be discussed next.

The Determinants of Matrices

A determinant like

$$|\mathbf{a}| \equiv \begin{vmatrix} a_{11} & a_{12} & \cdots & a_{1n} \\ a_{21} & a_{22} & \cdots & a_{2n} \\ \vdots & \vdots & & \vdots \\ a_{n1} & a_{n2} & \cdots & a_{nn} \end{vmatrix} \tag{9.63}$$

stands for a single number, rather than for a collection of n^2 component numbers as does the matrix **a**. The value of the determinant is by definition to be obtained as the sum of products

$$|\mathbf{a}| = \sum_P (-)^P a_{1,P1} a_{2,P2} \cdots a_{n,Pn}. \tag{9.64}$$

Here, P symbolizes any one of the $n! = n(n-1) \cdots 2 \cdot 1$ different possible permutations of the n integers $1, 2, \ldots, n$. With Pk standing for the integer which replaces k of $1, 2, \ldots, n$ after a given permutation of these, each term of (9.64) patently contains as a factor one element from every row (or every column) of the array. The symbol $(-)^P$ stands for $+1$ when P is an *even* permutation—that is, one obtained by making an even number of pair interchanges ("transpositions") among $1, 2, \ldots, n$; $(-)^P = -1$ for the odd permutations. The sum is to contain a term for each one of the possible permutations and so has $n!$ terms. In the 2×2 case,

$$\begin{vmatrix} a_{11} & a_{12} \\ a_{21} & a_{22} \end{vmatrix} = a_{11}a_{22} - a_{12}a_{21}, \tag{9.65}$$

and the 3×3 determinant has the value

$$\begin{vmatrix} a_{11} & a_{12} & a_{13} \\ a_{21} & a_{22} & a_{23} \\ a_{31} & a_{32} & a_{33} \end{vmatrix} = \begin{matrix} (a_{11}a_{22}a_{33} + a_{12}a_{23}a_{31} + a_{13}a_{21}a_{32}) \\ -(a_{11}a_{23}a_{32} + a_{13}a_{22}a_{31} + a_{12}a_{21}a_{33}). \end{matrix} \tag{9.66}$$

For larger n, a less direct approach to the evaluation becomes more efficient: an "expansion into minors."

By a *minor* determinant of an element a_{kl} is meant an $(n-1) \times (n-1)$ determinant formed from the $n \times n$ determinant $|\mathbf{a}|$ by striking out the row and the column that contain a_{kl} (the kth row and lth column); it will here be denoted $|\mathbf{a}|^{kl}$. Laplace found that the relations

$$\sum_i (-)^{i+l} |\mathbf{a}|^{il} a_{im} = |\mathbf{a}| \, \delta_{lm} = \sum_i (-)^{i+l} |\mathbf{a}|^{li} a_{mi} \tag{9.67}$$

hold for every choice of the pair of integers l, m. Thus the choice $l = 1$, $m = 2$ in the 2×2 case leads to

$$\sum_{i=1,2} (-)^{i+1} |\mathbf{a}|^{i1} a_{i2} = +a_{22}a_{12} - a_{12}a_{22} \equiv 0,$$

since $|\mathbf{a}|^{11} = a_{22}$ and $|\mathbf{a}|^{21} = a_{12}$. Choosing $l = m$ gives an expression for $|\mathbf{a}|$ which constitutes an expansion into minors; thus $l = m = 1$ in the 3×3 case gives

$$\begin{aligned} |\mathbf{a}| &= |\mathbf{a}|^{11} a_{11} - |\mathbf{a}|^{21} a_{21} + |\mathbf{a}|^{31} a_{31} \\ &= a_{11} \begin{vmatrix} a_{22} & a_{23} \\ a_{32} & a_{33} \end{vmatrix} - a_{21} \begin{vmatrix} a_{12} & a_{13} \\ a_{32} & a_{33} \end{vmatrix} + a_{31} \begin{vmatrix} a_{12} & a_{13} \\ a_{22} & a_{23} \end{vmatrix}, \end{aligned}$$

which can readily be seen to agree with (9.66). The fewest minors need actual evaluation when the integer $l = m$ is so chosen that the column with a_{im} (or the row with a_{mi}) has as many zero elements as possible.

The handling of determinants can be eased by taking advantage of various other generalities which have been shown to follow from the definition (9.64). Thus the value of a determinant is patently unchanged when its rows and columns are interchanged:

$$|\tilde{\mathbf{a}}| = |\mathbf{a}|. \tag{9.68}$$

The sign of the value is reversed when any one pair of rows (or of columns) is interchanged; this has the immediate consequence that $|\mathbf{a}| = 0$ if any two rows (or two columns) of $|\mathbf{a}|$ are identical within a common factor. When any *one* row (or column) is multiplied by a constant, the whole value of the determinant is multiplied by the same constant. This contrasts with the definition of the constant multiple of a *matrix*, by which $(\zeta\mathbf{a})_{kl} = \zeta a_{kl}$; that is, every element of the matrix is multiplied by the constant ζ, as if this itself were a diagonal matrix with elements $\zeta_{kl} = \zeta\delta_{kl}$. Accordingly,

$$|\zeta\mathbf{a}| = \zeta^n\,|\mathbf{a}|. \tag{9.69}$$

The value of a determinant is unchanged if to any row of it, say $\{a_{il}\} \equiv a_{i1}, a_{i2}, \ldots, a_{in}$, is added any linear combination of the other rows—that is, when to a_{il} is added some $\sum_{j\neq i} \zeta_j a_{jl}$, with the set $\{\zeta_j\}$ the same for every l. This property is frequently used to reduce to zero as many elements of some one row as possible, in preparation for as brief an expansion into minors as possible.

If $\mathbf{c} = \mathbf{ba}$ is a product of two matrices, then its determinant is the product of the determinants of the factor matrices, thus:

$$|\mathbf{c}| = |\mathbf{b}|\,.\,|\mathbf{a}|. \tag{9.70}$$

The order of the determinantal factors is immaterial, so $|\mathbf{c}'| = |\mathbf{c}|$ even when $\mathbf{c}' = \mathbf{ab} \neq \mathbf{ba} = \mathbf{c}$. It then follows from $|\mathbf{a}^{-1}\mathbf{a}| = |\mathbf{1}| = 1$ that

$$|\mathbf{a}^{-1}| = \frac{1}{|\mathbf{a}|} = |\mathbf{a}|^{-1}. \tag{9.71}$$

Since $\mathbf{a}^{-1} = \tilde{\mathbf{a}}$ for an *orthonormal* matrix,

$$|\mathbf{a}|^{-1} = |\tilde{\mathbf{a}}| = |\mathbf{a}| \qquad \text{and} \qquad |\mathbf{a}|^2 = 1. \tag{9.72}$$

Thus the orthonormal transformation matrices have determinants $|\mathbf{a}| = +1$ or -1. Any matrix with $|\mathbf{a}|^2 = 1$ is said to be "unimodular."

Determinants were originally defined because of their occurrence in the expression of solutions of simultaneous linear equations. For example, the equations

$$\begin{aligned} a_{11}x_1 + a_{12}x_2 + \cdots + a_{1n}x_n &= h_1, \\ a_{21}x_1 + a_{22}x_2 + \cdots + a_{2n}x_n &= h_2, \\ \vdots \qquad \vdots \qquad \quad \vdots \quad &\quad \vdots \\ a_{n1}x_1 + a_{n2}x_2 + \cdots + a_{nn}x_n &= h_n, \end{aligned} \tag{9.73}$$

have the solution

$$x_k = |\mathbf{h}(k)|/|\mathbf{a}|, \qquad k = 1, 2, \ldots, n, \tag{9.74}$$

where $\mathbf{h}(k)$ is a matrix formed from the matrix $\mathbf{a}$ by replacing its kth column by a column consisting of the "inhomogeneities," $h_1, h_2, \ldots, h_n$, in the equations. The determinant $|\mathbf{h}(k)|$ may be expanded in terms of the minors of the column of h's, and these minors are obviously the same as the minors $|\mathbf{a}|^{1k}$, $|\mathbf{a}|^{2k}, \ldots, |\mathbf{a}|^{nk}$ of the displaced column of elements $a_{1k}, a_{2k}, \ldots, a_{nk}$ of $|\mathbf{a}|$. Then

$$x_k = |\mathbf{a}|^{-1} \sum_i (-)^{i+k} |\mathbf{a}|^{ik} h_i, \tag{9.75}$$

a form known as "Cramer's rule."

In the special case of the linear equations (9.62), which are to be solved for the elements a_{lk}^{-1}, the unknowns are $x_k \equiv a_{kl}^{-1}$, and the inhomogeneities are $h_i = \delta_{il}$. Then Cramer's rule gives

$$a_{kl}^{-1} = (-)^{k+l} |\mathbf{a}|^{lk}/|\mathbf{a}|. \tag{9.76}$$

This could have been anticipated from comparing the first Laplace sum (9.67) with

$$(\mathbf{a}^{-1}\mathbf{a})_{lm} = \sum_k a_{lk}^{-1} a_{km} = \delta_{lm}.$$

Using the result (9.76), the solutions (9.75) of the general linear equations (9.73) may be rewritten

$$x_k = \sum_i a_{ki}^{-1} h_i, \tag{9.77}$$

and this also follows from the matrix equations $\mathbf{ar} = \mathbf{h} \to \mathbf{r} = \mathbf{a}^{-1}\mathbf{h}$, with $\mathbf{h}$ a column *vector* having the components $h_1, h_2, \ldots, h_n$.

The result (9.76) makes it plain that not every matrix can be inverted, for the inverse elements may not exist when $|\mathbf{a}| = 0$. Matrices $\mathbf{a}$ with the property $|\mathbf{a}| = 0$ are called "singular.'

9.4 TENSORS AND COVARIANCES

Further points of behavior in frame rotations will now be explored for the insight they give into quantities needed for describing the dynamics of rigid bodies.

When the definitions of the scalar and vector products of two three-dimensional vectors were introduced in §1.1, the question arose whether those can be said to exhaust the ways in which the products of two vectors could be defined. The question may be considered from this viewpoint: two arbitrary vectors, say **r** and **p**, each have three arbitrary components and so nine independent products, $x_k p_l$, may be formed from them. Giving the value of the scalar, or "inner," product, $\mathbf{r} \cdot \mathbf{p}$, and of each of the three components of the vector product, $\mathbf{r} \times \mathbf{p}$, serves to specify only four of the nine independent numbers. To specify the remaining five, a new type of product, having five components, must be introduced; this is a type known as a "reduced tensor" product. Its definition is best approached by considering first the tensor "outer product", **T**, defined as having the nine components $T_{kl} = x_k p_l$.

Tensor Covariances

Like scalars and vectors, tensors are defined through their having a characteristic behavior in frame rotations. A scalar is supposed to remain unchanged in value, to be *in*variant to frame transformations. A vector must have a specific type of *co*variance, like that of the "prototype" vector **r** itself. Thus **p** is a vector if

$$p'_j = a_{jl} p_l \qquad \text{when} \qquad x'_i = a_{ik} x_k, \tag{9.78}$$

in a rotation described by the orthonormal matrix $\mathbf{a} = \tilde{\mathbf{a}}^{-1}$. It is significant that the scalar character of the product $\mathbf{r} \cdot \mathbf{p}$ is dictated by the behavior of its factors: $\mathbf{r}' \cdot \mathbf{p}' = a_{ik} x_l \cdot a_{il} p_l = \mathbf{r} \cdot \mathbf{p}$ because $\tilde{\mathbf{a}}\mathbf{a} = \mathbf{1}$.

The covariance defining a tensor is like that of the prototype $T_{kl} = x_k p_l$. The tensor component $T'_{ij} = x'_i p'_j$ formed in the rotated frame is a linear combination,

$$T'_{ij} = a_{ik} a_{jl} x_k p_l = a_{ik} a_{jl} T_{kl}, \tag{9.79}$$

of its components on the original frame. Any nine-component quantity with such a behavior in rotations is called a tensor; examples will be seen in the next chapter.

The word "tensor" has also been given a more general meaning. An

n-dimensional tensor of the "rth rank" is defined as an entity having n^r components with the transformation property

$$T'_{i_1 i_2 \cdots i_r} = a_{i_1 j_1} a_{i_2 j_2} \cdots a_{i_r j_r} T_{j_1 j_2 \cdots j_r}, \tag{9.80}$$

each of the indices $i_1, i_2, \ldots, j_1, j_2, \ldots$ being one of the integers $1, 2, \ldots, n$. In this scheme, a vector is a tensor of rank $r = 1$, and a scalar has rank zero.

A tensor of rank $r - 2$ can always be formed from a tensor of rank $r \geq 2$ by a process called "contraction," amounting to summation over any pair of equal indices. For example,

$$\begin{aligned} T'_{iii_3 \cdots i_r} \left(\equiv \sum_i T'_{iii_3 \cdots i_r} \right) &= a_{ij_1} a_{ij_2} a_{i_3 j_3} \cdots a_{i_r j_r} T_{j_1 j_2 j_3 \cdots j_r} \\ &= a_{i_3 j_3} \cdots a_{i_r j_r} T_{j_1 j_1 j_3 \cdots j_r}, \end{aligned} \tag{9.81}$$

because $a_{ij_1} a_{ij_2} = (\tilde{\mathbf{a}}\mathbf{a})_{j_1 j_2} = \delta_{j_1 j_2}$. A contraction of just a second-rank tensor, T_{ij}, has a unique result—the scalar

$$T'_{it} = T_{jj} \equiv T_{11} + T_{22} + \cdots + T_{nn}. \tag{9.82a}$$

This is known variously as the "trace," the "spur," or the "diagonal sum" of the tensor, and it is symbolized by

$$\operatorname{tr}(\mathbf{T}) \equiv \operatorname{Sp}(\mathbf{T}) \equiv T_{jj}. \tag{9.82b}$$

For $T_{ij} = x_i p_j$, the trace is $\operatorname{tr}(\mathbf{T}) = \mathbf{r} \cdot \mathbf{p}$.

The Reduction of a Tensor

Any tensor T_{ij} may be decomposed into a symmetrical and an antisymmetrical tensor—that is, expressed as a sum

$$T_{ij} = S_{ij} + A_{ij}, \tag{9.83a}$$

with

$$S_{ij} \equiv \tfrac{1}{2}(T_{ij} + T_{ji}) = S_{ji}, \qquad A_{ij} \equiv \tfrac{1}{2}(T_{ij} - T_{ji}) = -A_{ji}. \tag{9.83b}$$

Each of these two tensors has $n^2 - n$ off-diagonal ($i \neq j$) components, only half of which are independent numbers. The antisymmetrical tensor has only zeros on the diagonal ($A_{ii} = 0$), and hence has just $\frac{1}{2}n(n-1)$ independent components. The symmetrical part, S_{ij}, contains the remaining $\frac{1}{2}n(n+1)$ pieces of information given by the n^2 numbers forming the original tensor T_{ij}.

In the special case of the three-dimensional tensor $T_{ij} = x_i p_j$, the antisymmetric tensor's components may be displayed as

$$\mathbf{A} = \tfrac{1}{2} \begin{pmatrix} 0 & (\mathbf{r} \times \mathbf{p})_3 & -(\mathbf{r} \times \mathbf{p})_2 \\ -(\mathbf{r} \times \mathbf{p})_3 & 0 & (\mathbf{r} \times \mathbf{p})_1 \\ (\mathbf{r} \times \mathbf{p})_2 & -(\mathbf{r} \times \mathbf{p})_1 & 0 \end{pmatrix}. \tag{9.84a}$$

Thus the vector product $\mathbf{r} \times \mathbf{p}$ fully represents all the information contained in the antisymmetric part of $T_{ij} = x_i p_j$. A general one of the components A_{ij} here may be expressed with the help of what is called a "Levi–Civita symbol," δ_{ijk}. This is so defined that it vanishes if any pair of its indices i, j, k are equal. The only nonvanishing components are defined to be $\delta_{123} = \delta_{231} = \delta_{312} = +1$ and $\delta_{132} = \delta_{321} = \delta_{213} = -1$; the positive components have indices that are even permutations of 1, 2, 3, and the negative units have odd permutations as indices. Now

$$A_{ij} = \tfrac{1}{2}\,\delta_{ijk}(\mathbf{r} \times \mathbf{p})_k, \qquad (9.84b)$$

with summation over $k = 1, 2, 3$ to be understood; it actually results in just one component of $(\mathbf{r} \times \mathbf{p})$ for each $A_{ij} \neq 0$.

The trace of the symmetric part of $T_{ij} = x_i p_j$ is just the scalar product $S_{ii} = T_{ii} = \mathbf{r} \cdot \mathbf{p}$. This may be separated off from the full symmetric tensor $S_{ij} = \frac{1}{2}(x_i p_j + x_j p_i)$ to define the *reduced* tensor, of zero trace:

$$s_{ij} \equiv \tfrac{1}{2}(x_i p_j + x_j p_i) - \tfrac{1}{3}\delta_{ij}(\mathbf{r} \cdot \mathbf{p}). \qquad (9.85)$$

This reduced tensor product of the vectors $\mathbf{r}$, $\mathbf{p}$ has just five independent components: three off-diagonal ones $s_{ij} = S_{ij} = s_{ji} (i \neq j)$, and three diagonal ones like $s_{11} = \frac{1}{3}(2x_1 p_1 - x_2 p_2 - x_3 p_3)$ subject to $s_{ii} = 0$. It can now be seen that the nine products $T_{ij} = x_i p_j$ may be analyzed into a reduced tensor, a vector and scalar:

$$T_{ij} = s_{ij} + \tfrac{1}{2}\delta_{ijk}(\mathbf{r} \times \mathbf{p})_k + \tfrac{1}{3}\,\delta_{ij}(\mathbf{r} \cdot \mathbf{p}). \qquad (9.86)$$

Such so-called "reductions" prove useful in elasticity theory and in "multipole analysis."

Sometimes the tensor products are written in terms of the vector symbols, like $\mathbf{r}$ and $\mathbf{p}$, for their factors. Thus

$$\mathbf{T} = \mathbf{rp} \neq \mathbf{pr}, \qquad (9.87)$$

with omission of dot and cross, is to be understood as a tensor with the nine $T_{ij} = x_i p_j$. Such symbols are sometimes called "dyadics." In terms of the column and row vector notations introduced in the preceding section, $\mathbf{T} \equiv \mathbf{r}\tilde{\mathbf{p}}$ (9.56), to be distinguished from the single number $\tilde{\mathbf{p}}\mathbf{r} = \tilde{\mathbf{r}}\mathbf{p} \equiv \mathbf{r} \cdot \mathbf{p}$ (9.55). In the "dyadic" notation of (9.87), the reduced tensor (9.85) is written

$$\mathbf{s} = \tfrac{1}{2}(\mathbf{rp} + \mathbf{pr}) - \tfrac{1}{3}(\mathbf{r} \cdot \mathbf{p})\mathbf{1}, \qquad (9.88)$$

where $\mathbf{1}$ stands for the unit tensor with components $(\mathbf{1})_{ij} = \delta_{ij}$. Its right to be called a tensor follows from $a_{ik} a_{jl}\,\delta_{kl} = a_{ik} a_{jk} = \delta_{ij}$.

Checks of Transformation Properties

Two quantities, the gradient operator ∇ and the vector product $\mathbf{r} \times \mathbf{p}$, have been introduced and treated as vectors, but ought to have the propriety of their classification as such checked. They are constructed from other vectors and so their behavior will be dictated by the nature of their constructions.

The propriety of treating the gradient operation as a vector is easy to show even for the n-dimensional generalization, which is often denoted $\partial_k \equiv \partial/\partial x_k$. Clearly

$$\partial'_k \equiv \partial/\partial x'_k = (\partial x_l/\partial x'_k)\, \partial_l ,$$

and because $x_l = a^{-1}_{lk}\, x'_k = a_{kl} x'_k$ according to (9.57) and (9.59),

$$\partial'_k = a_{kl}\, \partial_l , \tag{9.89}$$

which is the correct behavior for a vector (9.78).

The propriety of treating products like $\mathbf{L} = \mathbf{r} \times \mathbf{p}$ as vectors is not quite so obvious. Consider the first component of the product; its factors transform according to

$$\begin{aligned} L'_1 &\equiv x'_2\, p'_3 - x'_3\, p'_2 = a_{2k}\, a_{3l}(x_k\, p_l - x_l\, p_k) \\ &= (a_{22}\, a_{33} - a_{23}\, a_{32})L_1 + (a_{23}\, a_{31} - a_{21} a_{33})L_2 + (a_{21} a_{32} - a_{22}\, a_{31})L_3 . \end{aligned}$$

The coefficients of $L_{1,\,2,\,3}$ here should be recognized as just the 2×2 minor determinants, $|\mathbf{a}|^{11}$, $-|\mathbf{a}|^{12}$, $|\mathbf{a}|^{13}$, of the 3×3 determinant $|\mathbf{a}|$, so that

$$L'_1 = |\mathbf{a}|^{11} L_1 - |\mathbf{a}|^{12} L_2 + |\mathbf{a}|^{13} L_3 = (-)^{1+i}\, |\mathbf{a}|^{1i} L_i .$$

For an arbitrary one of the three components of $\mathbf{L}'$, the expression will be

$$L'_k = (-)^{k+i}\, |\mathbf{a}|^{ki} L_i = |\mathbf{a}|\, a^{-1}_{ik} L_i ,$$

with the last result following from the connection (9.76) between the minors and the inverse matrix elements. The rotational transformation is orthonormal, $\mathbf{a}^{-1} = \tilde{\mathbf{a}}$, and hence

$$L'_k = |\mathbf{a}|\, a_{ki}\, L_i . \tag{9.90}$$

This becomes the behavior expected of a vector, $(\mathbf{L})' = \mathbf{a}\mathbf{L}$, in "proper" transformations (ones having $|\mathbf{a}| = +1$).

The purely rotational transformations are indeed proper. It was seen in (9.72) that any orthonormal matrix has $|\mathbf{a}| = +1$ *or* $|\mathbf{a}| = -1$. This is valid irrespective of the size of the rotation angles, and so $|\lim \mathbf{a}| = |\mathbf{a}|$ in the limit of zero rotation angles. The rotational transformation matrices go over continuously into the unit matrix, $\mathbf{1}$, as is evident from $\lim(x'_k = a_{kl}x_l) \equiv x'_k = x_k \equiv \delta_{kl}x_l$. Thus $|\mathbf{a}| = |\lim \mathbf{a}| = |\mathbf{1}| = +1$.

Vector products can safely be treated as ordinary vectors in comparing results as judged from relatively rotated reference frames. However, it is also possible to use frames—even orthogonal ones with a common origin, which are not connected by a continuous rotation from one to the other. This fosters distinctions which it is sometimes important to recognize and which will be considered next.

Space Inversions

In three dimensions, an unambiguous distinction can be made between *right*- and *left*-handed orthogonal frames, respectively having $\mathbf{i} \times \mathbf{j} = \pm\mathbf{k}$ as relations among their axis directions. A frame of neither type can be obtained from one of the opposite type by continuous rotation alone. The transformation would have to include a discontinuous reversal of one of the axes *or* a reversal of all three. The latter transformation, by itself, is called a "space inversion." This involves changing from a frame in which the "polar" vector $\mathbf{r}$ has components x, y, z, to one in which the components of the same vector are $x' = -x$, $y' = -y$, $z' = -z$, as indicated in Figure 9.7(a). The transformation matrix then needed for $(\mathbf{r}') = \mathbf{ar}$ is plainly just

$$\mathbf{a} = \begin{pmatrix} -1 & 0 & 0 \\ 0 & -1 & 0 \\ 0 & 0 & -1 \end{pmatrix}. \tag{9.91}$$

That is, $\mathbf{a} \equiv -\mathbf{1}$ and $|\mathbf{a}| = -1$. The three-dimensional rotation matrix $\mathbf{a}(\varphi, \vartheta, \psi)$ of (9.49) cannot be reduced to $\mathbf{a} = -\mathbf{1}$ for any choice of angles; it was seen in the preceding subsection that $|\mathbf{a}(\varphi, \vartheta, \psi)| = +1$ independently of any values that may be given to the angles. Unlike rotations, the space inversions are discontinuous with the identity transformation $\mathbf{a} = +\mathbf{1}$; there is a "barrier" to be jumped* between right- and left-handedness. The contrast between rotations and space inversions may be illustrated by comparing the rotation of a right-handed glove, during which it stays right-handed, with turning it inside out, when it becomes left-handed.

The general transformation between right- and left-handed orthogonal frames may obviously be represented by $\mathbf{a} \equiv -\mathbf{a}(\varphi, \vartheta, \psi)$, or space-inversion conjoined with a general proper rotation. The resultant matrix is still orthonormal, and the results obtained for orthonormal matrices in

*At least within the given space.

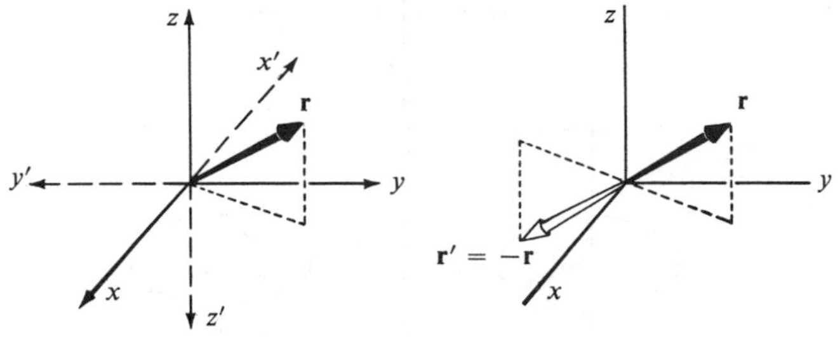

(a) "Passive" Space-Inversion (b) "Active" Space-Inversion

Figure 9.7

general still apply; only a change to $|\mathbf{a}| = -1$ is necessary. The general "improper rotation" includes mirrorings in any plane, such as the reflection in the (y, z)-plane, $x' = -x$, $y' = y$, $z' = z$, represented by $\mathbf{a} \equiv -\mathbf{a}(0, \pi, 0)$.

It may be remarked that for any *even* number of dimensions, $|\mathbf{a}| = +1$ when $\mathbf{a} \equiv -1$. This fits with the fact that the axis reversal of a two-dimensional frame is equivalent to a proper rotation through 180°; that is, $\varphi = \pi$ in $\mathbf{a}(\varphi)$ of (9.25). On the other hand, a reflection like $x' = +x, y' = -y$, represented by

$$\mathbf{a} \equiv \begin{pmatrix} 1 & 0 \\ 0 & -1 \end{pmatrix} \qquad \text{with} \quad |\mathbf{a}| = -1, \tag{9.92}$$

is *not* equivalent to any rotation. In any number of dimensions, it is an odd number of axes that must be reversed to get a transformation having $|\mathbf{a}| = -1$ and so unobtainable by a proper rotation.

Axial Vectors and Pseudoscalars

The improper transformations with $|\mathbf{a}| = -1$ provide a basis for a distinction between *polar* vectors like $\mathbf{r}$ and $\mathbf{p}$ of (9.78), and *axial* vectors like $\mathbf{L} = \mathbf{r} \times \mathbf{p}$ of (9.90). A relative sign difference in behavior arises because $|\mathbf{a}| = -1$ occurs as a factor in (9.90).

The distinction can be discussed as a physical one when space inversion, instead of being considered merely an adoption of a new frame from which to view a system, is regarded as a turning of the system inside out. The system itself is to be reconstructed, with every position vector $\mathbf{r}$ replaced by $-\mathbf{r}$ as indicated in Figure 9.7(b). Now $x' = -x$, $y' = -y$, $z' = -z$ because x', y', z' are components of a new vector $\mathbf{r}' = -\mathbf{r}$, rather than because they are projections of the original vector $\mathbf{r}$ on a new frame.

With $\mathbf{r} \to \mathbf{r}' = -\mathbf{r}$ in space inversions, the behavior characteristic of a polar vector is simple reversal, a change of sign in space inversions. That

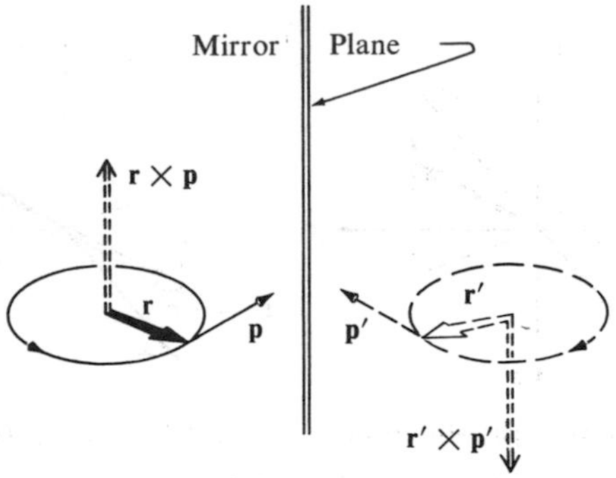

Figure 9.8

makes velocity ($\mathbf{v} = d\mathbf{r}/dt \to \mathbf{v}' = -\mathbf{v}$ because $d\mathbf{r}' = -d\mathbf{r}$) a polar vector as well, and consequently also momentum: $\mathbf{p} = m\mathbf{v} \to \mathbf{p}' = m\mathbf{v}' = -\mathbf{p}$. On the other hand, in the same process, an axial vector transforms like

$$\mathbf{L} = \mathbf{r} \times \mathbf{p} \to \mathbf{L}' = (-\mathbf{r}) \times (-\mathbf{p}) = +\mathbf{L}. \tag{9.93}$$

There is the characteristic sign difference from polar vector behavior, which has been represented in (9.90) by $|\mathbf{a}| = -1$. The distinction may be illustrated as in Figure 9.8, where a mirroring in the median plane is indicated; such a reflection differs from a space inversion merely by a 180° reorientation. The reflections $\mathbf{r}'$ and $\mathbf{p}'$ of the polar vectors might actually be seen, in a real mirror, as a solid rod and a direction of motion for example. The vector $\mathbf{r}' \times \mathbf{p}'$ is not representable by the mirror image of a directed axis, $\mathbf{r} \times \mathbf{p}$; such axial vectors are more purely conventional representations of physical properties. Another illustration could be afforded by the reflected image of a right-handed screw; a real screw constructed like the mirror image would be a left-handed one.

Another example of an axial vector is provided by angular velocity $\boldsymbol{\omega}$, such as leads to the linear velocity $\mathbf{v} = \boldsymbol{\omega} \times \mathbf{r}$. Upon space inversion,

$$\mathbf{v} = \boldsymbol{\omega} \times \mathbf{r} \to \mathbf{v}' = \boldsymbol{\omega}' \times \mathbf{r}' \equiv -\mathbf{v} = \boldsymbol{\omega}' \times (-\mathbf{r}),$$

and hence

$$\boldsymbol{\omega} \to \boldsymbol{\omega}' = +\boldsymbol{\omega}, \tag{9.94}$$

in contrast to the polar vector behavior $\mathbf{r} \to \mathbf{r}' = -\mathbf{r}$.

Such products as $\boldsymbol{\omega} \cdot \mathbf{r}$ cannot be true scalars, for merely by viewing them from another frame such as a space-inverted one:

$$\boldsymbol{\omega} \cdot \mathbf{r} \to \boldsymbol{\omega}' \cdot \mathbf{r}' = (+\boldsymbol{\omega}) \cdot (-\mathbf{r}) = -\boldsymbol{\omega} \cdot \mathbf{r}, \tag{9.95}$$

and the sign of the quantity is changed. On the other hand, in a proper rotation, $\boldsymbol{\omega}' \cdot \mathbf{r}' = \boldsymbol{\omega} \cdot \mathbf{r}$ still, and so the product is called a *pseudoscalar*. Axial vectors are also called *pseudovectors* occasionally.

9.5 OPERATOR MATRICES

The foregoing has made it plain that the information needed to locate body axes can be contained in a rotational transformation matrix, with elements consisting of the direction cosines (9.17), or expressed in terms of the Euler angles as in $\mathbf{a}(\varphi, \vartheta, \psi)$ of (9.49). However, the explicit connections of that description to rotational displacements, either the infinitesimal ones $\mathbf{dR} = \boldsymbol{\Omega}\, dt \times \mathbf{R}$ or the finite ones $\mathbf{R}' - \mathbf{R}$, have not yet been explored. The vector $\mathbf{R}$ gives the position of a point fixed to the body-frame, and its displacement relative to space-axes of fixed orientation is desired. A description is provided by the orthonormal rotation matrices when these are given the role of "operators."

The same linear relations $x_k' = a_{kl} x_l$, which give the components of a fixed vector on a rotating frame, may be reinterpreted as giving components of a rotating vector on a fixed frame. The new components x_k' are now taken to belong to a vector $\mathbf{r}'$ different from the original vector $\mathbf{r}$ and decomposed on the same basis as that used to get $x_l = \mathbf{e}_l \cdot \mathbf{r}$:

$$\mathbf{r}' = \mathbf{e}_l x_l' \neq \mathbf{r} = \mathbf{e}_l x_l, \tag{9.96}$$

in contrast to (9.20). No new basis vectors $\mathbf{e}_k'$ need to be introduced at all in this reinterpretation.

All the columns of the expression (9.51) are now taken to be displaying vector components on the same basis, $\{\mathbf{e}_l\} \equiv \mathbf{e}_1, \ldots, \mathbf{e}_n$, and it becomes appropriate to write the equivalent matrix equation as

$$\mathbf{r}' = \mathbf{ar}, \tag{9.97}$$

rather than $(\mathbf{r})' = \mathbf{ar}$ as in (9.52). Now the matrix $\mathbf{a}$ is functioning as a rotation *operator*, producing a new vector $\mathbf{r}'$ from an old one $\mathbf{r}$, rather than as frame-transformation matrix. The new vector $\mathbf{r}'$ bears the same relation to the axes $\{\mathbf{e}_l\}$ as did the original vector $(\mathbf{r})'$ to the rotated basis $\{\mathbf{e}_k'\}$ of the former interpretation. The reinterpretation replaces the frame rotation with a "back-rotation," $\mathbf{r} \to \mathbf{r}'$, of the vector itself. Thus, corresponding to the two-dimensional counterclockwise frame rotation of Figure 9.5 is the clockwise rotation of the position vector in Figure 9.9. The new interpretation of the relations $x_k' = a_{kl} x_l$ is sometimes called "active," as against the "passive" role played by the vector $\mathbf{r}$ when it was interpreted as remaining undisturbed while the frame was rotated.

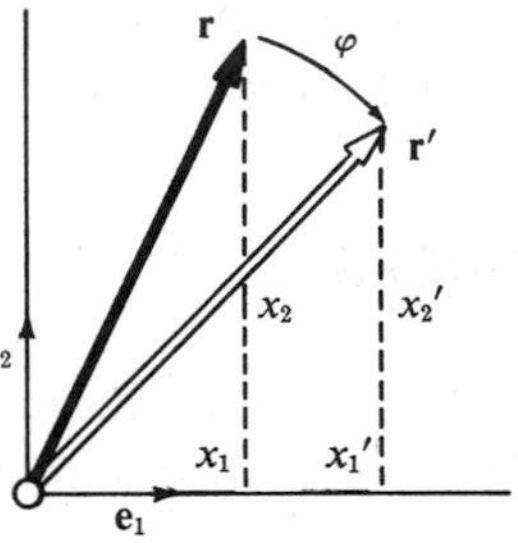

Figure 9.9

In the active interpretation, the result of applying the inverse $\mathbf{a}^{-1}$ to (9.97),

$$\mathbf{a}^{-1}\mathbf{r}' = \mathbf{r}, \tag{9.98}$$

rotates the vector $\mathbf{r}'$ back to $\mathbf{r}$. Since the operator $\mathbf{a}(\varphi)$ of (9.25) performs the clockwise rotation $\mathbf{r} \rightarrow \mathbf{r}'$ of Figure 9.9 (when $\varphi > 0$), it must be expected that $\mathbf{a}^{-1}(\varphi) = \mathbf{a}(-\varphi)$. This checks with the orthonormality property of the rotation matrix, $\mathbf{a}^{-1} = \tilde{\mathbf{a}}$, as inspection of (9.25) bears out. Corresponding results hold for the three-dimensional rotation (9.49):

$$\mathbf{a}^{-1}(\varphi, \vartheta, \psi) = \mathbf{a}(-\psi, -\vartheta, -\varphi) = \tilde{\mathbf{a}}(\varphi, \vartheta, \psi).$$

Notice the inversion in order of the negative angles.

Rotation Operators

The rotated position $\mathbf{R}'$ of a point that starts at $\mathbf{R}$ should be representable as the result of an operation

$$\mathbf{R}' = \mathbf{A}\mathbf{R}, \tag{9.99}$$

where $\mathbf{A}$ must be orthonormal, $\tilde{\mathbf{A}}\mathbf{A} = \mathbf{1}$, if the displacement is to be merely a rotation: $\tilde{\mathbf{R}}'\mathbf{R}' = \tilde{\mathbf{R}}\mathbf{R}$. The matrix representative of the operator $\mathbf{A}$ depends on the way the rotation is specified. If its axis $\hat{\mathbf{\Omega}}$ and the angle of rotation α are given, the z-axis may be chosen to lie along $\hat{\mathbf{\Omega}}$ and then $\mathbf{A} = \mathbf{a}(-\alpha) = \mathbf{a}^{-1}(\alpha) = \tilde{\mathbf{a}}(\alpha)$, where $\mathbf{a}(\alpha)$ has the form (9.46). It is then being supposed that $\alpha > 0$ for a *right*-handed rotation about $\hat{\mathbf{\Omega}}$, so that $\mathbf{a}(\alpha)$ itself is a left-handed rotation operator, as in Figure 9.9.

Sometimes the choice of z-axis is dictated by other considerations, so that the unit vector $\hat{\mathbf{\Omega}}$ cannot lie along it, and may have more than one cartesian component on the frame actually used. Then the matrix $\mathbf{A}(\hat{\mathbf{\Omega}}, \alpha)$ may be found from the expression (9.5) for the rotational displacement:

$$R_k' = [\delta_{kl} \cos \alpha - \delta_{klm} \hat{\Omega}_m \sin \alpha + \hat{\Omega}_k \hat{\Omega}_l (1 - \cos \alpha)] R_l, \tag{9.100}$$

where δ_{klm} is the antisymmetrical tensor defined in connection with (9.84). The square bracket here obviously specifies the element A_{kl} of the requisite rotation matrix, $\mathbf{A}(\hat{\mathbf{\Omega}}, \alpha)$. It is easy to see that the form for $\mathbf{A}$ here properly reduces to $\mathbf{A} \equiv \mathbf{a}(-\alpha)$ when the rotation is around the z-axis (that is, $\hat{\Omega}_k = \delta_{k3}$).

Examples are provided by the two rotations of 90°, around the x- and y-axes respectively, in Figure 9.3. In these examples, the general matrix element of (9.100) reduces to $A_{kl} = -\delta_{klm}\hat{\Omega}_m + \hat{\Omega}_k\hat{\Omega}_l$, with $\hat{\Omega}_k = \delta_{k1}$ and δ_{k2} in the respective cases:

$$\mathbf{A}(\mathbf{w}_1 = 2\mathbf{i}) = \begin{pmatrix} 1 & 0 & 0 \\ 0 & 0 & -1 \\ 0 & 1 & 0 \end{pmatrix}, \qquad \mathbf{A}(\mathbf{w}_2 = 2\mathbf{j}) = \begin{pmatrix} 0 & 0 & 1 \\ 0 & 1 & 0 \\ -1 & 0 & 0 \end{pmatrix}.$$

Now, for example, the passages from the orientations (a) to those of (b), in Figure 9.3, of the two initial position vectors $\mathbf{R} \equiv \mathbf{I} = \mathbf{i}$ and $\mathbf{J} = \mathbf{j}$, are respectively represented by

$$\mathbf{A}(\mathbf{w}_1 = 2\mathbf{i}) \begin{pmatrix} 1 \\ 0 \\ 0 \end{pmatrix} = \begin{pmatrix} 1 \\ 0 \\ 0 \end{pmatrix} = \mathbf{i}, \qquad \mathbf{A}(\mathbf{w}_1 = 2\mathbf{i}) \begin{pmatrix} 0 \\ 1 \\ 0 \end{pmatrix} = \begin{pmatrix} 0 \\ 0 \\ 1 \end{pmatrix} = \mathbf{k}.$$

The product rotation that takes the orientation (a) into the final one (c), as applied to $\mathbf{I} = \mathbf{i}$ of (a), is:

$$\mathbf{A}(\mathbf{w}_2 = 2\mathbf{j})\mathbf{A}(\mathbf{w}_1 = 2\mathbf{i})\mathbf{i} = \begin{pmatrix} 0 & 1 & 0 \\ 0 & 0 & -1 \\ -1 & 0 & 0 \end{pmatrix} \begin{pmatrix} 1 \\ 0 \\ 0 \end{pmatrix} = \begin{pmatrix} 0 \\ 0 \\ -1 \end{pmatrix} = -\mathbf{k}. \qquad (9.101)$$

Comparison* of the elements of the product rotation matrix here to the general ones of (9.100) shows that it represents a resultant rotation by 120° around $\hat{\mathbf{\Omega}} = 3^{-1/2}(\mathbf{i} + \mathbf{j} - \mathbf{k})$, just as found in connection with Figure 9.3. The example here illustrates the relationships between the representation of rotations by matrices and by the finite rotation vectors, $\mathbf{w}$.

Operator Matrix Elements

The rotation operations like (9.97) or (9.99) are special examples of a linear operator formalism that has a wide range of uses in physical descriptions. The discussion will therefore be continued for a more general linear operation, initially defined by *any* n^2 coefficients Q_{kl}. These can

---

* See page 309 for a more elegant and powerful procedure.

form a linear operator in that they can be applied to any set of n numbers, $u_1, u_2, \ldots, u_n$ to produce a new set $v_1, v_2, \ldots, v_n$, according to

$$v_k = Q_{kl} u_l . \tag{9.102a}$$

Such relations can always be put into a vector form simply by defining an n-dimensional vector space, and an orthonormal basis $\mathbf{e}_1, \ldots, \mathbf{e}_n$ in it, and then defining the column vectors

$$\mathbf{u} \equiv \mathbf{e}_l u_l \qquad \text{and} \qquad \mathbf{v} = \mathbf{e}_k v_k . \tag{9.103}$$

In this language, the given coefficients Q_{kl} will form an operator matrix $\mathbf{Q}$ in the matrix equation version

$$\mathbf{v} = \mathbf{Q}\mathbf{u} \tag{9.102b}$$

of (9.102*a*). Of course, an arbitrary set of coefficients Q_{kl} will not in general form a matrix with any simply grasped properties, such as the orthonormality possessed by the rotation operations; any properties $\mathbf{Q}$ has may depend on the way the various numbers involved have been ordered.

The adoption of the basis $\mathbf{e}_1, \ldots, \mathbf{e}_n$ as above has implications for the relation of the matrix elements Q_{kl} to it. The resolutions on it of the original vector $\mathbf{u}$ and of the resultant vector $\mathbf{Qu}(\equiv \mathbf{v})$ may be compared as in

$$\begin{aligned} \mathbf{u} &= \mathbf{e}_l u_l , \qquad \text{where} \qquad u_l = \tilde{\mathbf{e}}_l \mathbf{u} \equiv \mathbf{e}_l \cdot \mathbf{u}, \\ \mathbf{Qu} &= \mathbf{e}_l (\mathbf{Qu})_l , \qquad \text{where} \qquad (\mathbf{Qu})_k = \tilde{\mathbf{e}}_k \mathbf{Qu}, \end{aligned} \tag{9.104}$$

with the components following from the orthonormality $\tilde{\mathbf{e}}_k \mathbf{e}_l \equiv \mathbf{e}_k \cdot \mathbf{e}_l = \delta_{kl}$. The components $(\mathbf{Qu})_k$ are related to the matrix elements Q_{kl} as in (9.102*a*), and hence

$$(\mathbf{Qu})_k = (Q_{kl}) u_l = \tilde{\mathbf{e}}_k \mathbf{Qu} = (\tilde{\mathbf{e}}_k \mathbf{Q} \mathbf{e}_l) u_l . \tag{9.105}$$

This is to be useful for any vector components u_l to which the operation may be applied, and so the individual elements of the operator matrix are given representations

$$Q_{kl} = \tilde{\mathbf{e}}_k \mathbf{Q} \mathbf{e}_l . \tag{9.106}$$

That is, it becomes a projection $\mathbf{e}_k \cdot (\mathbf{Q}\mathbf{e}_l)$ on the kth axis of the adopted basis, of the vector resultant of applying the operation to the lth basis vector.

The representability of operator matrix elements as projections like (9.106) on any orthogonal basis, as well as various ramifications of it, becomes transparent when the columns representing the basis vectors themselves are made explicit:

$$\mathbf{e}_1 = \begin{pmatrix} 1 \\ 0 \\ \vdots \\ 0 \end{pmatrix}, \qquad \mathbf{e}_2 = \begin{pmatrix} 0 \\ 1 \\ \vdots \\ 0 \end{pmatrix}, \qquad \ldots, \qquad \mathbf{e}_n = \begin{pmatrix} 0 \\ 0 \\ \vdots \\ 1 \end{pmatrix}, \tag{9.107}$$

in consistency with the resolutions $\mathbf{e}_k = \sum_l \mathbf{e}_l(\mathbf{e}_l \cdot \mathbf{e}_k) = \sum_l \mathbf{e}_l \delta_{lk}$. In terms of these, a column vector like $\mathbf{u}$ of (9.103) can be displayed as

$$\mathbf{u} \equiv \begin{pmatrix} u_1 \\ u_2 \\ \vdots \\ u_n \end{pmatrix} = u_1 \begin{pmatrix} 1 \\ 0 \\ \vdots \\ 0 \end{pmatrix} + u_2 \begin{pmatrix} 0 \\ 1 \\ \vdots \\ 0 \end{pmatrix} + \cdots + u_n \begin{pmatrix} 0 \\ 0 \\ \vdots \\ 1 \end{pmatrix} \equiv u_l \mathbf{e}_l. \tag{9.108}$$

For the square matrix $\mathbf{Q}$,

$$\tilde{\mathbf{e}}_1 \mathbf{Q} \mathbf{e}_2 = \tilde{\mathbf{e}}_1 \begin{pmatrix} Q_{11} & Q_{12} & \cdots & Q_{1n} \\ Q_{21} & Q_{22} & \cdots & Q_{2n} \\ \vdots & \vdots & & \vdots \\ Q_{n1} & Q_{n2} & \cdots & Q_{nn} \end{pmatrix} \begin{pmatrix} 0 \\ 1 \\ \vdots \\ 0 \end{pmatrix} = (1, 0, \ldots, 0) \begin{pmatrix} Q_{12} \\ Q_{22} \\ \vdots \\ Q_{n2} \end{pmatrix} = Q_{12},$$

showing in detail how the units of the basis vectors "pick out" a matrix element. The general element is

$$\tilde{\mathbf{e}}_k \mathbf{Q} \mathbf{e}_l = (\mathbf{e}_k)_i \, Q_{ij} (\mathbf{e}_l)_j = \delta_{ki} \, Q_{ij} \, \delta_{jl} = Q_{kl}.$$

All these processes of resolving matrices onto a basis stem from properties of the basis like

$$\mathbf{e}_2 \, \tilde{\mathbf{e}}_2 = \begin{pmatrix} 0 \\ 1 \\ 0 \\ \vdots \\ 0 \end{pmatrix} (0, 1, 0 \ldots 0) = \begin{pmatrix} 0 & 0 & 0 & \cdots & 0 \\ 0 & 1 & 0 & \cdots & 0 \\ 0 & 0 & 0 & \cdots & 0 \\ \vdots & \vdots & \vdots & & \\ 0 & 0 & 0 & \cdots & 0 \end{pmatrix},$$

and $\mathbf{e}_l \tilde{\mathbf{e}}_k \equiv 0$, a "null matrix," whenever $l \neq k$. Plainly, the sum over the complete basis is

$$\left(\sum_{l=1}^{n} \right) \mathbf{e}_l \tilde{\mathbf{e}}_l = \mathbf{1}, \tag{9.109}$$

the unit matrix. Now the resolutions may be regarded as results of inserting a unit matrix in the form (9.109). For example, the resolution of the column vector $\mathbf{u}$ shown in (9.104) can be understood as

$$\mathbf{u} = \mathbf{1}\mathbf{u} = \mathbf{e}_l \tilde{\mathbf{e}}_l \mathbf{u} = \mathbf{e}_l u_l. \tag{9.110}$$

In a resolution of a square matrix, the unit is inserted in two places, as in

$$\mathbf{Qu} = \mathbf{1Q1u} = \mathbf{e}_k \tilde{\mathbf{e}}_k \mathbf{Q} \mathbf{e}_l \tilde{\mathbf{e}}_l \mathbf{u} = \mathbf{e}_k Q_{kl} u_l, \tag{9.111}$$

of (9.105). Because of these properties, the unit matrix $\mathbf{e}_l \tilde{\mathbf{e}}_l$ is called a "projection operator."

Transformations of the Basis

The basis $\{\mathbf{e}_l\} \equiv \mathbf{e}_1, \ldots, \mathbf{e}_n$ used in the preceding subsection was defined in such a way that the initially given operator coefficients Q_{kl} became projections $\tilde{\mathbf{e}}_k \mathbf{Q} \mathbf{e}_l$ on it, of an operator matrix $\mathbf{Q}$. In the space so defined, the essential property of the operation becomes the conversion of any vector $\mathbf{u} = \mathbf{e}_l u_l$ in the space into a resultant vector $\mathbf{v} \equiv \mathbf{Qu} = \mathbf{e}_k(Q_{kl} u_l)$, with the original coefficients Q_{kl} now seen as only special representatives of the operation, as resolved on the special frame $\{\mathbf{e}_l\}$. The connections to resolutions on some relatively rotated alternative frame, $\{\mathbf{e}_i'\} \equiv \mathbf{e}_1', \ldots, \mathbf{e}_n'$, will now be considered.

In the terms introduced at the end of the preceding section, projections by the operator $\mathbf{e}_l \tilde{\mathbf{e}}_l = \mathbf{1}$ are to be replaced by $\mathbf{e}_i' \tilde{\mathbf{e}}_i' = \mathbf{1}$. Thus the analyses of vectors like $\mathbf{u}$ into the components u_l of (9.110) are to be replaced by analyses into new components, u_i', as in

$$\mathbf{u} = \mathbf{e}_i'(\tilde{\mathbf{e}}_i' \mathbf{u}) = \mathbf{e}_i' u_i'. \tag{9.112}$$

Similarly the resolution of the operation $\mathbf{Qu}$ of (9.111) is to be replaced by

$$\mathbf{Qu} = \mathbf{e}_i'(\tilde{\mathbf{e}}_i' \mathbf{Q} \mathbf{e}_j')\tilde{\mathbf{e}}_j' \mathbf{u} = \mathbf{e}_i' Q_{ij}' u_j', \tag{9.113}$$

where the $Q_{ij}' \equiv \tilde{\mathbf{e}}_i' \mathbf{Q} \mathbf{e}_j'$ are *new matrix elements* of the operator matrix $\mathbf{Q}$ as resolved on the basis $\mathbf{e}_i'$. A new matrix of n^2 numbers has replaced the originally given Q_{kl} in defining the same operation $\mathbf{u} \to \mathbf{v} \equiv \mathbf{Qu}$.

The connections of the new to the old resolutions can be found from judicious insertions of the projection operators again:

$$u_i' = \tilde{\mathbf{e}}_i' \mathbf{u} = \tilde{\mathbf{e}}_i' \mathbf{e}_l \tilde{\mathbf{e}}_l \mathbf{u} = (\mathbf{e}_i' \cdot \mathbf{e}_l) u_l.$$

The cosine coefficients are just the elements $a_{il} = \mathbf{e}_i' \cdot \mathbf{e}_l$ of the orthonormal rotational transformation matrix $\mathbf{a} = \tilde{\mathbf{a}}^{-1}$ much discussed in the preceding section and, as there, the above transformation relations are components of a column matrix equation that may be written

$$(\mathbf{u})' = \mathbf{au}. \tag{9.114}$$

For the operator matrix elements,

$$\begin{aligned} Q'_{ij} &\equiv \tilde{\mathbf{e}}'_i \mathbf{Q} \mathbf{e}'_j = \tilde{\mathbf{e}}'_i \mathbf{e}_k \tilde{\mathbf{e}}_k \mathbf{Q} \mathbf{e}_l \tilde{\mathbf{e}}_l \mathbf{e}'_j \\ &= (\tilde{\mathbf{e}}'_i \mathbf{e}_k) Q_{kl} (\mathbf{e}'_j \cdot \mathbf{e}_l) = a_{ik} a_{jl} Q_{kl} . \end{aligned} \tag{9.115a}$$

These are components of a matrix equation

$$(\mathbf{Q})' = \mathbf{a}\mathbf{Q}\tilde{\mathbf{a}} = \mathbf{a}\mathbf{Q}\mathbf{a}^{-1}. \tag{9.115b}$$

This could have been obtained more directly by making use of the fact that $\mathbf{a}^{-1}\mathbf{a} = \mathbf{1}$ in

$$(\mathbf{v})^1 = \mathbf{a}\mathbf{v} \equiv (\mathbf{Q})^1 (\mathbf{u})^1 = \mathbf{a}\mathbf{Q}\mathbf{u} = (\mathbf{a}\mathbf{Q}\mathbf{a}^{-1})(\mathbf{a}\mathbf{u}). \tag{9.116}$$

It can be seen that matrix elements of an operator as resolved on orthogonal bases behave like components of a second-rank tensor (compare (9.115a) to (9.79)).

As an example, it may be checked that when $\mathbf{Q} = \mathbf{b}(\vartheta)$ of (9.47), an operator that rotates vectors through an angle $-\vartheta$ about the x-axis, then the transformation

$$\mathbf{a} = \begin{pmatrix} 0 & 0 & -1 \\ 0 & 1 & 0 \\ 1 & 0 & 0 \end{pmatrix},$$

which converts the x-axis into a z-axis, makes $(\mathbf{b}(\vartheta))' \equiv \mathbf{a}(\vartheta)$ of (9.46) a rotation by $-\vartheta$ around a z-axis.

On the new basis, it is the unit vectors $\mathbf{e}'_i = \sum_j \mathbf{e}'_j \, \delta_{ij}$ that acquire the basic column representations (9.107). The original basis vectors $\mathbf{e}_l$ become columns like

$$(\mathbf{e}_l)' = \mathbf{e}'_i(\tilde{\mathbf{e}}'_i \mathbf{e}_l) = \mathbf{e}'_i a_{il} = \begin{pmatrix} a_{1l} \\ a_{2l} \\ \vdots \\ a_{nl} \end{pmatrix} \equiv \mathbf{a}\mathbf{e}'_l . \tag{9.117}$$

That the columns of the transformation matrix are just the components of the old basis vectors on the new axes was pointed out in (9.27). Conversely, when the new basis vectors $\mathbf{e}'_i$ are resolved on the initial frame,

$$\mathbf{e}'_i \; [\neq (\mathbf{e}'_i)'] = \mathbf{e}_l \tilde{\mathbf{e}}_l \mathbf{e}'_i = \mathbf{e}_l \tilde{a}_{li} \equiv \mathbf{a}^{-1} \mathbf{e}_i , \tag{9.118}$$

each a *row* of $\mathbf{a}$. The last equivalence emphasizes that the "passive" frame transformation by $\mathbf{a}$ is the same as an "active" rotation by $\mathbf{a}^{-1}$.

Similarity Transformations

Sometimes it is necessary to consider more general types of transformations than the orthonormal ones, $\mathbf{a} = \tilde{\mathbf{a}}^{-1}$, corresponding to mere rotations of axes. Any nonsingular matrix of coefficients, $\mathbf{S}$, may be used to give new representations

$$(\mathbf{u})' = \mathbf{S}\mathbf{u} \qquad \text{and} \qquad (\mathbf{Q})' = \mathbf{S}\mathbf{Q}\mathbf{S}^{-1} \tag{9.119}$$

of the vectors and matrices. The changed representatives still correspond to the same operation as $\mathbf{u} \to \mathbf{v} \equiv \mathbf{Q}\mathbf{u}$, since

$$(\mathbf{Q}\mathbf{u})' = \mathbf{S}(\mathbf{Q}\mathbf{u}) = \mathbf{S}\mathbf{Q}\mathbf{S}^{-1}\mathbf{S}\mathbf{u} = (\mathbf{S}\mathbf{Q}\mathbf{S}^{-1})(\mathbf{u})',$$

and so are said to be "equivalent" to $\mathbf{u}$ and $\mathbf{Q}$. These "equivalence" transformations are more generally called "similarity transformations."

The similarity transformations form a wide class which includes the orthonormal ones. They can be said to convert initially orthogonal bases, with $\mathbf{e}_k \cdot \mathbf{e}_l = \delta_{kl}$, into new basis vectors with components on the initial basis as given by

$$\mathbf{u}_k = \mathbf{S}^{-1}\mathbf{e}_k = \mathbf{e}_l \tilde{\mathbf{e}}_l \mathbf{S}^{-1}\mathbf{e}_k = \mathbf{e}_l S_{lk}^{-1} \tag{9.120}$$

in place of (9.118). These are not necessarily an orthogonal set, for

$$\tilde{\mathbf{u}}_k \mathbf{u}_l = \tilde{\mathbf{e}}_k \tilde{\mathbf{S}}^{-1}\mathbf{S}^{-1}\mathbf{e}_l \tag{9.121}$$

does not reduce to $\tilde{\mathbf{e}}_k \mathbf{e}_l = \delta_{kl}$ unless $\tilde{\mathbf{S}}^{-1} = \mathbf{S}$ is an orthonormal matrix.

An important property of all similarity transformations, whether orthonormal or not, is that they leave invariant the trace (9.82) of every matrix:

$$Q'_{ii} = S_{ik} Q_{kl} S_{li}^{-1} = (\mathbf{S}^{-1}\mathbf{S})_{lk} Q_{kl} = Q_{ll}\,. \tag{9.122}$$

This property has already been exhibited, as a scalar behavior of traces in orthogonal transformations, for the tensor (9.82). Examples of invariant traces will be encountered in the next chapter.

9.6 INFINITESIMAL ROTATIONS

The foregoing discussion has shown that finite rotational displacements can be represented by $\mathbf{R}' - \mathbf{R} = (\mathbf{A} - \mathbf{1})\mathbf{R}$, with the help of the orthonormal rotation matrices, $\mathbf{A}$. How these displacements reduce to the infinitesimal displacements $\boldsymbol{\delta}\mathbf{R} = \boldsymbol{\Omega}\,\delta t \times \mathbf{R}$ can be seen from a consideration of the rotation matrix elements of (9.100) in the limit $\alpha \equiv \delta\alpha \to 0$:

$$A_{kl} \to \delta_{kl} - \delta_{klm}\hat{\boldsymbol{\Omega}}_m\,\delta\alpha. \tag{9.123}$$

If $\delta\alpha = \dot{\alpha}\,\delta t$, then $\hat{\boldsymbol{\Omega}}\dot{\alpha} \equiv \boldsymbol{\Omega}$ is an angular velocity, and the displacement reduces to

$$(\mathbf{R}' - \mathbf{R})_k \to \delta R_k = -\delta_{klm}(\Omega_m\,\delta t)R_l, \tag{9.124}$$

which is just a component of $\boldsymbol{\delta}\mathbf{R} = \boldsymbol{\Omega} \times \mathbf{R}\,\delta t$, according to the definition of δ_{klm} given in connection with (9.84).

The equation $\dot{\mathbf{R}} = \boldsymbol{\Omega} \times \mathbf{R}$ can be written as a column vector equation with the help of an angular velocity *tensor* having the components

$$\boldsymbol{\Omega}_{kl} = -\delta_{klm}\,\Omega_m. \tag{9.125}$$

The tensor is an antisymmetric one, $\tilde{\boldsymbol{\Omega}} = -\boldsymbol{\Omega}$, and can be displayed as in

$$\dot{\mathbf{R}} = \boldsymbol{\Omega}\mathbf{R} \equiv \begin{pmatrix} \dot{X} \\ \dot{Y} \\ \dot{Z} \end{pmatrix} = \begin{pmatrix} 0 & -\Omega_Z & \Omega_Y \\ \Omega_Z & 0 & -\Omega_X \\ -\Omega_Y & \Omega_X & 0 \end{pmatrix} \begin{pmatrix} X \\ Y \\ Z \end{pmatrix}. \tag{9.126}$$

Thus found is a matrix representation of infinitesimal rotational displacements $\boldsymbol{\delta}\mathbf{R} = \dot{\mathbf{R}}\,\delta t$.

Especially in spaces of higher dimensionality (for example, the four-dimensional space-time of relativistic theories) it is simpler to begin with infinitesimal rotations, and to build up finite rotations from those. The type of procedure used will be illustrated in two and three dimensions here, for it can then be connected to already known results. Moreover, conceptions which are also valuable in the lower dimensions will thereby be introduced.

Rotations in a Plane

Consider an infinitesimal rotational displacement $\mathbf{R} \to \mathbf{R}' = \mathbf{R} + \boldsymbol{\delta}\mathbf{R}$, of a vector confined to the x, y plane. If the rotation angle is $\delta\varphi \to 0$,

$$\begin{aligned} X' &= X\cos\delta\varphi - Y\sin\delta\varphi \approx X - Y\,\delta\varphi, \\ Y' &= X\sin\delta\varphi + Y\cos\delta\varphi \approx Y + X\,\delta\varphi. \end{aligned} \tag{9.127}$$

This is to be represented as the result of an operation $\mathbf{R}' = \mathbf{A}(\delta\varphi)\mathbf{R} = \mathbf{R} + \boldsymbol{\Delta}\mathbf{R}$, where

$$\mathbf{A} = \mathbf{1} + \boldsymbol{\Delta}, \tag{9.128}$$

with the $\boldsymbol{\Delta}$ an "infinitesimal operator."

One way to represent the rotation operation was learned in §3.3. The vectors **R** and **R**′ are put into correspondence with complex numbers $R = X + iY$ and $R' = X' + iY'$. Then $R' = (1 + i\,\delta\varphi)R$ reproduces (9.127) exactly, so that the imaginary number $\Delta = i\,\delta\varphi$ is the infinitesimal operator in this representation. That is of course to be expected when $A = \exp\colon i\varphi$ is the finite rotation operator as found in §3.3, since $\exp\colon i\,\delta\varphi \approx 1 + i\,\delta\varphi$.

When the equations (9.127) are treated as components of a column vector, the rotation operator acquires the matrix representative

$$\mathbf{A} = \mathbf{1} - i\,\delta\varphi\sigma_2, \qquad \text{with} \quad \sigma_2 = \begin{pmatrix} 0 & -i \\ i & 0 \end{pmatrix}, \tag{9.129}$$

as already found for $\mathbf{a}(\varphi) = \tilde{\mathbf{A}}(\varphi)$ in (9.45). One reason for defining the matrix σ_2 in this way is that it then has the property $\sigma_2^2 = +\mathbf{1}$. In this representation $\boldsymbol{\Delta} = -i\,\delta\varphi\sigma_2$ is a matrix; its connection to the complex plane representation will be found in §10.5.

It should now be possible to build up the finite rotation by φ in infinitesimal steps $\delta\varphi = \varphi/N$, with $N \to \infty$. It should be the result of applying $\mathbf{A} = \mathbf{1} + \boldsymbol{\Delta}$, $N \to \infty$ times;

$$\mathbf{A}(\varphi) = \lim_{N\to\infty} (1 + \boldsymbol{\Delta})^N. \tag{9.130}$$

According to the binomial theorem (proved by Newton in his crib),

$$(1 + \boldsymbol{\Delta})^N = \sum_{s=0}^{N} \frac{N!}{s!(N-s)!} \boldsymbol{\Delta}^s, \tag{9.131}$$

where $s! = 1, 2, 3 \ldots s$, etc. Now $\boldsymbol{\Delta}$ is inversely proportional to N because it is proportional to $\delta\varphi = \varphi/N$; put $\boldsymbol{\Delta} = \mathbf{C}/N$ where $C = i\varphi$ in the complex plane representation above, and $\mathbf{C} = -i\varphi\sigma_2$ in the matrix representation. Then

$$\mathbf{A} = \lim_{N\to\infty} \sum_{s=0}^{N} \frac{N(N-1)\cdots(N-s+1)}{s!} \left(\frac{\mathbf{C}}{N}\right)^s = \sum_{s=0}^{\infty} \frac{\mathbf{C}^s}{s!} = \exp \mathbf{C}, \tag{9.132}$$

the last equality following from the well-known Taylor series expansion (as in (3.7)): $e^x = \sum x^s/s!$. The expected result $A = \exp\colon i\varphi$ thus emerges in the complex number representation, and

$$\mathbf{A}(\varphi) = e^{-i\varphi\sigma_2} \tag{9.133}$$

in the matrix representation. The fact that **C** is a matrix in this case does not interfere with the validity of the conclusion (9.132), since **C** commutes with itself. Because $\sigma_2 = -\tilde{\sigma}_2$ is antisymmetric, $\tilde{\mathbf{A}} = \exp\colon +i\varphi\sigma_2$ and $\tilde{\mathbf{A}}\mathbf{A} = \mathbf{1}$, as is proper for a rotation matrix.

The matrix (9.133) should be identifiable with $\tilde{\mathbf{a}}(\varphi)$ of (9.25). That it is can be seen by taking advantage of the property $\sigma_2^2 = +1$ to linearize the expression in σ_2. The first few terms of the power series (9.132), with $\mathbf{C} = -i\varphi\sigma_2$ and $\mathbf{C}^2 = -\varphi^2$, yield

$$\mathbf{A} = \mathbf{1}\left[1 - \frac{1}{2!}\varphi^2 + \frac{1}{4!}\varphi^4 - \cdots\right] - i\sigma_2\left[\varphi - \frac{1}{3!}\varphi^3 + \frac{1}{5!}\varphi^5 - \cdots\right]$$

$$= \mathbf{1}\cos\varphi - i\sigma_2 \sin\varphi = \begin{pmatrix} \cos\varphi & -\sin\varphi \\ \sin\varphi & \cos\varphi \end{pmatrix} \equiv \tilde{\mathbf{a}}(\varphi). \tag{9.134}$$

For this, it was necessary to recognize that the square brackets are just the Taylor expansions of $\cos\varphi$ and $\sin\varphi$; these identifications also follow from comparing $e^{i\varphi} = \cos\varphi + i\sin\varphi$ against the Taylor expansion of the exponential.

Three-Dimensional Rotations

In rotations of three-dimensional vectors about the z-axis, the relations (9.127) are supplemented with $Z' = Z$. Then the matrix representation of the operation $\mathbf{R}' = \mathbf{A}(\delta\varphi)\mathbf{R}$ requires

$$\mathbf{A}(\delta\varphi) = \begin{pmatrix} 1 & -\delta\varphi & 0 \\ \delta\varphi & 1 & 0 \\ 0 & 0 & 1 \end{pmatrix} = \mathbf{1} - i\,\delta\varphi \mathsf{S}_z, \tag{9.135}$$

where the matrix

$$\mathsf{S}_z = \begin{pmatrix} 0 & -i & 0 \\ +i & 0 & 0 \\ 0 & 0 & 0 \end{pmatrix} \tag{9.136a}$$

replaces the σ_2 of the two-dimensional case (9.129). Whereas $\sigma_2^2 = 1$,

$$\mathsf{S}_z^2 = \begin{pmatrix} 1 & 0 & 0 \\ 0 & 1 & 0 \\ 0 & 0 & 0 \end{pmatrix} \quad \text{and} \quad \mathsf{S}_z^3 = \mathsf{S}_z. \tag{9.136b}$$

The last relation does *not* imply that S_z^2 is equal to the unit matrix, because S_z has no inverse.

The infinitesimal matrix here is $\mathbf{\Delta} \equiv \mathbf{A}(\delta\varphi) - \mathbf{1} = -i\,\delta\varphi \mathsf{S}_z$. The rotation through the finite angle $\varphi = \lim(N\,\delta\varphi)$ may again be given the exponential form (9.132):

$$\mathbf{A}(\varphi) = e^{-i\varphi \mathsf{S}_z}. \tag{9.137a}$$

Because of the properties (9.136) of S_z, the power series expansion yields

$$\mathbf{A}(\varphi) = \mathbf{1} - iS_z\left[\varphi - \frac{\varphi^3}{3!} + \frac{\varphi^5}{5!} - \cdots\right] - S_z^2\left[\frac{\varphi^2}{2!} - \frac{\varphi^4}{4!} + \frac{\varphi^6}{6!} - \cdots\right]$$

$$= \mathbf{1} - iS_z \sin\varphi - S_z^2(1 - \cos\varphi) \tag{9.137b}$$

$$= \begin{pmatrix} 1 & -\sin\varphi & 0 \\ \sin\varphi & 1 & 0 \\ 0 & 0 & 1 \end{pmatrix} - \begin{pmatrix} 1-\cos\varphi & 0 & 0 \\ 0 & 1-\cos\varphi & 0 \\ 0 & 0 & 0 \end{pmatrix},$$

and this properly adds up to the form $\tilde{\mathbf{a}}(\varphi)$ of (9.46).

Reference to the rotational transformation $\mathbf{b}(\vartheta)$ of (9.47), when the rotation is about the x-axis, should now make it plain that, just as $\mathbf{a}(\varphi) = \exp\colon +i\varphi S_z$, so

$$\mathbf{b}(\vartheta) = e^{+i\vartheta S_x}, \tag{9.138}$$

where

$$S_x = \begin{pmatrix} 0 & 0 & 0 \\ 0 & 0 & -i \\ 0 & +i & 0 \end{pmatrix}, \qquad S_x^2 = \begin{pmatrix} 0 & 0 & 0 \\ 0 & 1 & 0 \\ 0 & 0 & 1 \end{pmatrix} \tag{9.139}$$

and $S_x^3 = S_x$. It follows that

$$\mathbf{b}(\vartheta) = \mathbf{1} + iS_x \sin\vartheta - S_x^2(1 - \cos\vartheta). \tag{9.140}$$

Finally, the matrix, $-S_y$, with the properties

$$-S_y = \begin{pmatrix} 0 & 0 & -i \\ 0 & 0 & 0 \\ +i & 0 & 0 \end{pmatrix}, \qquad S_y^2 = \begin{pmatrix} 1 & 0 & 0 \\ 0 & 0 & 0 \\ 0 & 0 & 1 \end{pmatrix}, \tag{9.141}$$

and $S_y^3 = S_y$, will play the same role in rotations about the y-axis as do S_x and S_z in those about the x- and z-axes, respectively. The reason for the choice of sign in naming S_y will appear below.

A significant new expression for the general Euler angle transformation (9.48) is now possible:

$$\mathbf{a}(\varphi, \vartheta, \psi) = \mathbf{a}(\psi)\mathbf{b}(\vartheta)\mathbf{a}(\varphi) = e^{i\psi S_z} e^{i\vartheta S_x} e^{i\varphi S_z} \tag{9.142}$$

The exponents cannot simply be added to produce a single exponential because S_x and S_z do not commute. Indeed,

$$S_z S_x = \begin{pmatrix} 0 & 0 & -1 \\ 0 & 0 & 0 \\ 0 & 0 & 0 \end{pmatrix}, \qquad S_x S_z = \begin{pmatrix} 0 & 0 & 0 \\ 0 & 0 & 0 \\ -1 & 0 & 0 \end{pmatrix}.$$

The difference can be readily seen to be $\mathsf{S}_z \mathsf{S}_x - \mathsf{S}_x \mathsf{S}_z = i\mathsf{S}_y$. Such differences are called "commutators" and are denoted

$$\mathsf{S}_z \mathsf{S}_x - \mathsf{S}_x \mathsf{S}_z \equiv [\mathsf{S}_z, \mathsf{S}_x]_- . \tag{9.143}$$

The commutator bracket symbol here should not be confused with the Poisson bracket symbol (8.36), which applies to differentiable functions. It can then be readily found that

$$[\mathsf{S}_x, \mathsf{S}_y]_- = i\mathsf{S}_z, \qquad [\mathsf{S}_y, \mathsf{S}_z]_- = i\mathsf{S}_x, \qquad [\mathsf{S}_z, \mathsf{S}_x]_- = i\mathsf{S}_y. \tag{9.144a}$$

The sign choices as above lead to the uniformity of sign here, under cyclic permutations of the labels x, y, z. Moreover,

$$\mathsf{S}_x^2 + \mathsf{S}_y^2 + \mathsf{S}_z^2 = \begin{pmatrix} 2 & 0 & 0 \\ 0 & 2 & 0 \\ 0 & 0 & 2 \end{pmatrix} = \mathbf{2}. \tag{9.144b}$$

Any operators with these properties are called "unit spin" operators, It is easy to appreciate that the noncommutativity of these operators is at bottom the result of the noncommutativity of rotations about different axes, as in (9.142).

On the basis here, the "unit spin" matrices are all antisymmetric. This is necessary if they are to be proportional to the infinitesimal matrices $\Delta \approx \mathbf{A} - \mathbf{1}$ of an orthonormal operation $\mathbf{A}$, as in $\Delta = -i\,\delta\varphi \mathsf{S}_z$, for example. The orthonormality property requires

$$\tilde{\mathbf{A}}\mathbf{A} = \mathbf{1} \equiv (\mathbf{1} + \tilde{\Delta})(\mathbf{1} + \Delta) \approx \mathbf{1} + (\Delta + \tilde{\Delta}), \tag{9.145}$$

and hence $\tilde{\Delta} = -\Delta$. This also leads correctly to such properties as

$$\tilde{\mathbf{a}}(\varphi, \vartheta, \psi) = \mathbf{a}^{-1}(\varphi, \vartheta, \psi) = \mathbf{a}(-\psi, -\vartheta, -\varphi)$$

in the form (9.142) (see the discussion following (9.98)).

The general rotation operator $\mathbf{A}(\hat{\mathbf{\Omega}}, \alpha)$ of (9.100), about an axis $\hat{\mathbf{\Omega}}$ which has an arbitrary orientation relative to the x, y, z axes, is made up of infinitesimal steps which were described in (9.123) to (9.126) with the help of an angular velocity matrix, $\mathbf{\Omega}$. This matrix may be analyzed as

$$\begin{aligned} \mathbf{\Omega} &= -i\Omega_x \begin{pmatrix} 0 & 0 & 0 \\ 0 & 0 & -i \\ 0 & +i & 0 \end{pmatrix} - i\Omega_y \begin{pmatrix} 0 & 0 & +i \\ 0 & 0 & 0 \\ -i & 0 & 0 \end{pmatrix} - i\Omega_z \begin{pmatrix} 0 & -i & 0 \\ +i & 0 & 0 \\ 0 & 0 & 0 \end{pmatrix}, \\ &= -i(\Omega_x \mathsf{S}_x + \Omega_y \mathsf{S}_y + \Omega_z \mathsf{S}_z) \equiv -i\mathbf{\Omega} \cdot \mathbf{S}. \end{aligned} \tag{9.146}$$

This outcome is actually sufficient to show that the three matrices S_x, S_y, S_z should be treated as components of an axial vector like $\mathbf{\Omega}$, in frame rotations. The relative sign choices above were directed toward just this result. The commutation relations (9.144*a*) can now be written in the economical form

$$\mathbf{S} \times \mathbf{S} = i\mathbf{S}. \tag{9.147}$$

This kind of relation could be satisfied only by an *axial* vector, since the cross-product of two vectors is an axial vector when the factors are either both axial or both polar.

The infinitesimal operator $\Delta = \mathbf{A}(\hat{\mathbf{\Omega}}, \delta\alpha) - \mathbf{1}$, in which $\mathbf{\Omega}\,\delta t = \hat{\mathbf{\Omega}}\,\delta\alpha$, is just $\Delta = \mathbf{\Omega}\,\delta t = -i(\hat{\mathbf{\Omega}} \cdot \mathrm{S})\,\delta\alpha$, and so the finite rotation operator can be given the forms

$$\begin{aligned}\mathbf{A}(\hat{\mathbf{\Omega}}, \alpha) &= \exp - i\alpha\hat{\mathbf{\Omega}} \cdot \mathbf{S} \\ &= \mathbf{1} - i\hat{\mathbf{\Omega}} \cdot \mathbf{S} \sin\alpha - (\hat{\mathbf{\Omega}} \cdot \mathbf{S})^2(1 - \cos\alpha),\end{aligned} \tag{9.148}$$

the final one resting on the fact that $(\hat{\mathbf{\Omega}} \cdot \mathbf{S})^3 = \hat{\mathbf{\Omega}} \cdot \mathbf{S}$ for any projection of the unit-spin vector. The result has just the matrix elements exhibited in (9.100).

The procedures for obtaining this are valid because $\hat{\mathbf{\Omega}} \cdot \mathbf{S}$ can be written as a single matrix. Note that

$$e^{-i\alpha\hat{\Omega} \cdot \mathrm{S}} \neq e^{-i\alpha\hat{\Omega}_x \mathrm{S}_x} e^{-i\alpha\hat{\Omega}_y \mathrm{S}_y} e^{-i\alpha\hat{\Omega}_z \mathrm{S}_z}$$

because S_x, S_y, S_z do not commute with each other.

Relations to the Canonical Generators

The means developed in this chapter for generating rotations should be related to the canonical transformation procedures already developed in (8.17), (8.24), (8.29), (8.38), and (8.45). The active rotations $\mathbf{r}' = \mathbf{A}\mathbf{r}$ and $\mathbf{p}' = \mathbf{A}\mathbf{p}$ may be generated out of the phase functions $S(\mathbf{r}, \mathbf{p}') = \tilde{\mathbf{p}}'\mathbf{A}\mathbf{r}$:

$$x_i' = \partial S/\partial p_i' = (\partial/\partial p_i')[p_k' A_{kj} x_j] = A_{ij} x_j, \tag{9.149a}$$

and

$$p_i = \partial S/\partial x_i = p_k' A_{ki} = \tilde{A}_{ik} p_k' = (\mathbf{A}^{-1}\mathbf{p}')_i, \tag{9.149b}$$

because $\tilde{\mathbf{A}} = \mathbf{A}^{-1}$.

For infinitesimal transformations (8.29), $S = S_1(\mathbf{r}, \mathbf{p}') + \varepsilon G(\mathbf{r}, \mathbf{p})$, the part $S_1 = \mathbf{p}' \cdot \mathbf{r}$ yields the identity transformation. The corresponding analysis of the matrix operator is $\mathbf{A} = \mathbf{1} + \Delta$, and hence

$$\varepsilon G = \tilde{\mathbf{p}}\,\Delta\mathbf{r} = \mathbf{p} \cdot \delta\mathbf{r} \tag{9.150}$$

furnishes the connection between the infinitesimal generator and the infinitesimal matrix, Δ.

The matrix may be given the form $\mathbf{\Delta} = \mathbf{A}(\hat{\mathbf{\Omega}}, \delta\alpha) - \mathbf{1} = -i\,\delta\alpha\hat{\mathbf{\Omega}}\cdot\mathbf{S}$ as noted just above (9.148). Then $\varepsilon = \delta\alpha$ and

$$G = -ip_k(\hat{\mathbf{\Omega}}\cdot\mathbf{S})_{kl}x_l \equiv -i\hat{\Omega}_m(\tilde{\mathbf{p}}\mathsf{S}_m\mathbf{r}). \tag{9.151}$$

A factor in this is

$$-i(\tilde{\mathbf{p}}\mathsf{S}_x\mathbf{r}) = (p_x, p_y, p_z)\begin{pmatrix}0 & 0 & 0\\ 0 & 0 & -1\\ 0 & 1 & 0\end{pmatrix}\begin{pmatrix}x\\ y\\ z\end{pmatrix} = -p_y z + p_z y,$$

and

$$\mathbf{L} = -\,ip_k(\mathbf{S})_{kl}x_l \equiv ix_l(\mathbf{S})_{lk}p_k \tag{9.152}$$

is just the angular momentum, already seen in (8.29) to generate active rotations $\boldsymbol{\delta}\mathbf{r} = \delta\alpha\hat{\mathbf{\Omega}}\times\mathbf{r}$ through $\varepsilon G = \delta\alpha\hat{\mathbf{\Omega}}\cdot\mathbf{L}$.

The Poisson bracket prescription (8.45) may be transcribed to $\mathbf{\Delta r} = -i\,\delta\alpha\hat{\Omega}_m[\mathbf{r}, \tilde{\mathbf{p}}\mathsf{S}_m\mathbf{r}] = [\mathbf{r}, \tilde{\mathbf{p}}\,\mathbf{\Delta r}]$, compatible with the property (8.38) of the Poisson brackets. More interesting are transcriptions of the Poisson brackets (8.49) among the angular momenta:

$$[L_k, L_l] = L_m \equiv -[\tilde{\mathbf{p}}\mathsf{S}_k\mathbf{r}, \tilde{\mathbf{p}}\mathsf{S}_l\mathbf{r}] = -\,i\tilde{\mathbf{p}}\mathsf{S}_m\mathbf{r}, \tag{9.153}$$

where klm may be xyz or yzx or zxy. When the derivatives implicit in the Poisson bracket definition (8.36) are carried out, the last line reduces to

$$\tilde{\mathbf{p}}(\mathsf{S}_k\mathsf{S}_l - \mathsf{S}_l\mathsf{S}_k)\mathbf{r} = i\tilde{\mathbf{p}}\mathsf{S}_m\mathbf{r}, \tag{9.154}$$

showing the correspondence between the Poisson brackets and the commutators (9.144):

$$\tilde{\mathbf{p}}[\mathsf{S}_k, \mathsf{S}_l]_-\mathbf{r} = [\tilde{\mathbf{p}}\mathsf{S}_k\mathbf{r}, \tilde{\mathbf{p}}\mathsf{S}_l\mathbf{r}].$$

Such results are important for establishing the correspondence between classical dynamical variables and "quantum-mechanical operators."

EXERCISES

9.1. Carry out in detail the indicated derivations of equations (9.4), (9.5), and (9.6) from (9.3).

9.2. Derive $\mathbf{w}_3$ of equation (9.7) as indicated.

9.3. Describe the rotation of a vector $\mathbf{R}$ to the direction of $\hat{\mathbf{\Omega}}\times\mathbf{R}$, and use this to abbreviate the calculations indicated in the text for Figure 9.3.

9.4. Show that the rule of combination (9.7) properly leads to a cancellation of $\mathbf{w}_2$ when $\mathbf{w}_3$ is followed by $-\mathbf{w}_2$, but that $\mathbf{w}_1$ is not canceled out when $\mathbf{w}_3$ is followed by $-\mathbf{w}_1$. Test your result in the latter case with the examples $\mathbf{w}_1 = 2\mathbf{i}$ and $\mathbf{w}_2 = 2\mathbf{j}$, of Figure 9.3.

9.5. Derive the results (9.16).

9.6. For the matrix $a(\varphi)$ of (9.25), show that

$$a^n(\varphi) = \begin{pmatrix} \cos n\varphi & \sin n\varphi \\ -\sin n\varphi & \cos n\varphi \end{pmatrix},$$

with any n for which you find this possible to prove.

9.7. With the help of the transformation matrix (9.46), the relations of Exercise 8.10 may be written as column-vector equations:

$$\mathbf{R} = \mathbf{a}(\omega t)\mathbf{r} = \nabla_P S, \qquad \nabla S(\mathbf{r}, \mathbf{P}) = \mathbf{p} = \mathbf{a}^{-1}\mathbf{P},$$

where $(\nabla_P)_X \equiv \partial/\partial P_X$, and so on.

(a) Verify this, as well as $S = \tilde{\mathbf{r}}\mathbf{a}^{-1}\mathbf{P} = \tilde{\mathbf{R}}\mathbf{P} = \tilde{\mathbf{r}}\mathbf{p}$.

(b) Show that $\mathbf{V} = \mathbf{a}\mathbf{v} + \dot{\mathbf{a}}\mathbf{r}$, where $\dot{\mathbf{a}}$ is a matrix having $\dot{a}_{kl}$ as elements.

9.8. Show by explicit evaluation that if a 3×3 determinant $|\mathsf{A}|$ has its elements changed by adding to one of its columns the corresponding elements of another column, multiplied by a constant ζ, then the value of the determinant is not changed. What happens if the corresponding elements of the two original columns differ by just the factor $-\zeta^{-1}$?

9.9. Check that (9.74) is indeed the solution of the equations (9.73) for the 2×2 case, by direct substitution into the equations.

9.10. Invert the matrix $\begin{pmatrix} m & \lambda \\ \lambda & m \end{pmatrix}$, having $\lambda \neq m$.

9.11. Show that (9.100) is indeed the kth component of (9.5).

9.12. Demonstrate the invariance of the volume elements $dx_1'\, dx_2'\, dx_3' = dx_1\, dx_2\, dx_3$ under the rotations $(\mathbf{r})' = \mathbf{ar}$.

9.13. Construct matrix operators $\mathsf{A}^{\pm 1}_{x,y}$ that will rotate any three-dimensional vector by $\pm 90°$ around the x and y axes respectively. Then find a simple product of $\pm 90°$ rotations, about x and y axes only, which will produce a $90°$ rotation about the z-axis.

9.14. Show that the determinant $|\mathsf{Q}|$, of the matrix in (9.119), is invariant to the similarity transformations.

9.15. Verify the "Jacobi identity" (compare Exercises 1.7 and 8.12),

$$[A, [B, C]_-]_- + [B, [C, A]_-]_- + [C, [A, B]_-]_- = 0,$$

for the *commutators* of matrices.

CHAPTER 10

RIGID BODY DYNAMICS AND EIGENVECTOR PROBLEMS

Descriptions of the motions of extended bodies may be developed from the Mechanics of Particles by treating the bodies as many-particle systems. From what was seen about the latter in Chapter 6, it is the treatment of the motions relative to a mass-center that requires further development.

The simplest extended-body model to treat is a rigid structure, one in which distances $|\mathbf{r}_i - \mathbf{r}_j|$, between points with which mass is associated, are held fixed. The model is adequate for solid bodies in which no relative displacement of parts is observed, or is negligible for the purposes of the description. The geometrical constraint conditions, $|\mathbf{r}_i - \mathbf{r}_j| = r_{ij}$ constant, replace specifications of interparticle forces.

A general rigid body will have just six degrees of freedom: three of translation and three of rotation. That should be seen to follow from the many-particle picture and the constraint conditions.

The three translational degrees of freedom may be associated with some one point, $\mathbf{r}_1$, of the rigid structure, and this point may thereafter be treated as a rotation center (it can be the mass-center but that is not essential). Once $\mathbf{r}_1$ is fixed, only rotation about it is permitted any second point, $\mathbf{r}_2$, of the structure, since $\mathbf{r}_2$ must remain at a fixed distance r_{12} from $\mathbf{r}_1$; there are thus added only two, orientational, degrees of freedom (for example, Θ, Φ). Once $\mathbf{r}_1$ and $\mathbf{r}_2$ are fixed, no additional point lying on their line is free to move (hence a "linear" rigid body, an ideal rod, has only five degrees of freedom). Any point, $\mathbf{r}_3$, off the line, lies in some plane through the line, and the plane's orientation, specifiable by a single azimuthal angle, provides the sixth degree of freedom. Once $\mathbf{r}_1$, $\mathbf{r}_2$, and $\mathbf{r}_3$ are fixed, no additional point of the rigid structure is free to move; there are then added three constraint conditions, $|\mathbf{r}_i - \mathbf{r}_{1,2,3}|$ constant, on the three components of any $\mathbf{r}_{i\neq 1,2,3}$ that is added.

Since only the direction, and not the distance, from $\mathbf{r}_1$ to $\mathbf{r}_2$ is pertinent, it may be specified by a unit vector, $\mathbf{I}$. The plane passing through $\mathbf{I}$ and $\mathbf{r}_3$ may be specified by giving another unit vector $\mathbf{J}$, fixed in that plane, and it may be taken orthogonal to $\mathbf{I}$. These two unit vectors, together with $\mathbf{K} = \mathbf{I} \times \mathbf{J}$, define a cartesian reference frame with origin at the rotation center. This is a frame that is attached rigidly to the structure and constitutes what was called a "body frame" in the preceding chapter. No point of the rigid body can move relative to it, and so the only motions are ones "imparted" through attachment to the moving frame. It is the rotational motions which most essentially characterize rigid body dynamics.

10.1 DYNAMICAL PROPERTIES OF RIGID BODIES

The observations on extended bodies that are found subject to classical mechanical principles, and the pertinent measurements on them, treat the bodies as continuous. Such a treatment consists in choosing the "mass-particles" making up the system to be a continuous infinity of "mass-elements," $\rho\, dV$, with dV a volume element and ρ a mass-density. Now integration replaces summation, as in evaluations of the total mass:

$$M = \sum_i m_i \to M = \oint \varrho(\mathbf{R})\, dV. \tag{10.1}$$

The integration symbol $\oint$ used here is meant to indicate that the integration is to be complete—that is, extend over all volume elements of the body. The possibility is allowed for that the mass-distribution is not uniform and the density is a function of the position $\mathbf{R}$ in the body. The distribution $\rho(\mathbf{R})$ will be independent of the motion if $\mathbf{R}$ denotes position relative to body axes and the body remains rigid.

The location of the body's mass-center is important for descriptions of the dynamical behavior, and, being fixed to the body, it may be regarded a "property" belonging to the body. It may be computed when the geometrical shape of, and density distribution in, the body are given, from

$$\overline{\mathbf{R}} = \sum_i \frac{m_i \mathbf{R}_i}{M} \rightarrow \overline{\mathbf{R}} = M^{-1} \oint \varrho \mathbf{R} \, dV. \tag{10.2}$$

Such computations are standard exercises in the calculus and will be given no further attention here. Of course, in a body frame having its origin at the mass-center, $\overline{\mathbf{R}} = 0$, but the origin will have to be chosen otherwise until after the location of the mass-center is known.

The Angular Momentum

In describing the rotation of a rigid body, its angular momentum $\mathbf{L}$ will be important. This is defined by

$$\mathbf{L} = \sum_i \mathbf{R}_i \times m_i \dot{\mathbf{R}}_i \rightarrow \mathbf{L} = \oint dV \varrho \mathbf{R} \times \dot{\mathbf{R}} \tag{10.3}$$

about a rotation center $R = 0$ fixed in the body. At any instant there will be an angular velocity $\mathbf{\Omega}(t)$ of the body as a whole, and this imparts the velocity $\dot{\mathbf{R}} = \mathbf{\Omega} \times \mathbf{R}$ (9.1) to the mass-element at $\mathbf{R}$. Thus

$$\begin{aligned} \mathbf{L} &= \oint dV \varrho \mathbf{R} \times (\mathbf{\Omega} \times \mathbf{R}) = \oint dV \varrho [R^2 \mathbf{\Omega} - \mathbf{R}(\mathbf{R} \cdot \mathbf{\Omega})], \\ &= \left(\oint dV \varrho R^2\right) \mathbf{\Omega} - \left(\oint dV \varrho \mathbf{R} X_l\right) \Omega_l, \end{aligned} \tag{10.4}$$

with summation over l, enumerating components of the vectors, to be understood. It is now clear that any one component of the angular momentum vector may be written

$$L_k = \left[\oint dV \varrho (R^2 \, \delta_{kl} - X_k X_l)\right] \Omega_l \equiv \mathsf{I}_{kl} \Omega_l, \tag{10.5}$$

in terms of the so-called "inertia tensor,"

$$\mathsf{I}_{kl} \equiv \oint dV \varrho (R^2 \, \delta_{kl} - X_k X_l) = \mathsf{I}_{lk}, \tag{10.6}$$

characteristic of the body. That the contraction $(\sum_l) \mathsf{I}_{kl} \Omega_l$ of a tensor with a vector yields a vector is shown by the transformation properties (9.79):

$$\mathsf{I}'_{kl} \Omega'_l = a_{ka} a_{lb} a_{lc} \mathsf{I}_{ab} \Omega_c = a_{ka} \delta_{bc} \mathsf{I}_{ab} \Omega_c = a_{ka} (\mathsf{I}_{ab} \Omega_b).$$

Because the vector $\mathbf{\Omega}$ here is an axial vector, so is $\mathbf{L}$.

In the dyadic notation like that of (9.88), the inertia tensor is

$$\mathsf{I} = \oint dV\rho[R^2\mathbf{1} - \mathbf{RR}] \tag{10.7}$$

and the vector resultant of the contraction of the tensor with the angular momentum vector is written

$$\mathbf{L} = \mathsf{I} \cdot \mathbf{\Omega}. \tag{10.8}$$

In this special case of a symmetric tensor, $\mathsf{I} = \tilde{\mathsf{I}}$, indicated in (10.6), the "right" and "left" products are equal: $(\mathbf{\Omega} \cdot \mathsf{I})_k = \Omega_l \mathsf{I}_{lk} = \mathsf{I}_{kl}\Omega_l = (\mathsf{I} \cdot \mathbf{\Omega})_k$. Notice that the unit tensor term, $\sim \mathbf{1}$ of (10.7), will lead to $(\mathbf{1} \cdot \mathbf{\Omega})_k = \delta_{kl}\Omega_l = \Omega_k$.

When the rotation is about a fixed axis in the body, this may be chosen as the Z-axis (that is, $\Omega = \Omega_Z$), and only the inertia coefficients I_{XZ}, I_{YZ}, I_{ZZ} are needed for expressing the angular momentum: $L_k = \mathsf{I}_{kZ}\Omega$. If, moreover, the body has axial symmetry about the rotation axis,

$$\mathsf{I}_{XZ} \equiv -\oint dV\varrho XZ = -\oint dV\varrho(-X)Z = 0, \tag{10.9}$$

and also $\mathsf{I}_{YZ} = 0$. These results follow from the fact that the replacements $X \to -X$, $Y \to -Y$ should make no difference to a body symmetrical about the Z-axis. The "moment of inertia" defined by

$$\mathsf{I}_{ZZ} = \oint dV\varrho[R^2 - Z^2] = \oint dV\varrho(X^2 + Y^2) \equiv I \tag{10.10}$$

survives, and the angular momentum is

$$\mathbf{L} = I\mathbf{\Omega}, \tag{10.11}$$

parallel to the rotation axis. This is an expression familiar in elementary physics, and the above outlines the restricted conditions under which it applies. In general, the angular momentum is *not* parallel to the angular velocity because the mass-distribution about the rotation axis may be "lopsided," thus throwing the momentum off from the rotation direction.

The Kinetic Energy

The kinetic energy of the rotational motion is given by $\tau = \oint dV\rho\frac{1}{2}\dot{R}^2$ with $\dot{\mathbf{R}} = \mathbf{\Omega} \times \mathbf{R}$. The substitution gives

$$\tau = \tfrac{1}{2}\oint dV\varrho(\mathbf{\Omega} \times \mathbf{R}) \cdot (\mathbf{\Omega} \times \mathbf{R}) = \tfrac{1}{2}\mathbf{\Omega} \cdot \oint dV\varrho[\mathbf{R} \times (\mathbf{\Omega} \times \mathbf{R})],$$

after interchanging dot and cross. Thus

$$\tau = \tfrac{1}{2}\boldsymbol{\Omega} \cdot \mathbf{L} = \tfrac{1}{2}\boldsymbol{\Omega} \cdot \mathbf{I} \cdot \boldsymbol{\Omega}. \tag{10.12}$$

The last of these expressions is equivalent to $\frac{1}{2}\Omega_k \mathsf{I}_{kl} \Omega_l$, or

$$\tau = \tfrac{1}{2}(\mathsf{I}_{XX}\Omega_X^2 + \mathsf{I}_{YY}\Omega_Y^2 + \mathsf{I}_{ZZ}\Omega_Z^2) + \mathsf{I}_{XY}\Omega_X\Omega_Y + \mathsf{I}_{YZ}\Omega_Y\Omega_Z + \mathsf{I}_{ZX}\Omega_Z\Omega_X. \tag{10.13}$$

If this were plotted on cartesian $\Omega_{X,Y,Z}$ axes, for a given value of τ, it would yield an ellipsoidal surface known as the "ellipsoid of inertia." (This may be shown with the help of Exercises 4.1(c), 10.2(a).)

The kinetic energy expression reduces to the elementary one

$$\tau = \tfrac{1}{2}I\Omega^2 \tag{10.14}$$

if the rotation is about a *fixed* axis in the body, so that the angular velocity can be taken to have only one component. The moment of inertia I may be expressed as in (10.10) if the fixed rotation axis is chosen to be the Z-axis of the body frame. Unlike (10.11), the elementary expression for the kinetic energy holds even for an unsymmetrical body, as long as it continues to rotate about the fixed axis.

Moments and Products of Inertia

Like the total mass (10.1), and the mass-center position (10.2), the inertia tensor (10.7) is a property of the rigid body that may be measured or computed once and for all. It is a measure of the mass-distribution which gives greater weight to mass farther from the rotation center, in a way that is just appropriate for evaluating angular momentum. The particular numbers used to specify the components of an inertia tensor, like those used to specify the components of the mass-center's position-vector, will depend on the choice of body axes.

The diagonal components of the inertia tensor, I_{XX}, I_{YY}, I_{ZZ}, are called the "moments of inertia" about the X, Y and Z axes, respectively. The three independent off-diagonal components, $\mathsf{I}_{XY} = \mathsf{I}_{YX}$, I_{YZ}, I_{ZX}, are called "products of inertia."

A simple example is provided by the uniform triangular sheet of Figure 10.1, having mass-density per unit area $M/\frac{1}{2}l^2$. For this case,

$$\mathsf{I}_{XX} = (2M/l^2)\int_0^l dX \int_0^{l-X} dY\, Y^2 = \tfrac{1}{6}Ml^2,$$

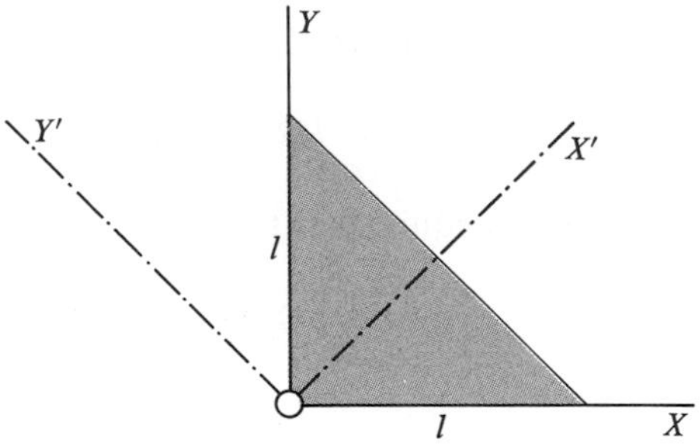

Figure 10.1

and I_{YY} is obviously the same. Then (see Exercise 10.2)

$$\mathsf{I}_{ZZ} = (2M/l^2) \iint dX\, dY(X^2 + Y^2) = \mathsf{I}_{XX} + \mathsf{I}_{YY} = \tfrac{1}{3}Ml^2.$$

The products of inertia I_{YZ}, I_{ZX} vanish because $Z = 0$ for all the mass-elements, but

$$\mathsf{I}_{XY} = -(2M/l^2) \int dXX \int dYY = -Ml^2/12.$$

The result is the inertia tensor

$$\mathbf{I} = \frac{Ml^2}{12}\begin{pmatrix} 2 & -1 & 0 \\ -1 & 2 & 0 \\ 0 & 0 & 4 \end{pmatrix}, \tag{10.15a}$$

on the basis in which the X, Y axes lie along the equal edges of the right triangle. On axes rotated through 45° as shown in Figure 10.1, $\mathsf{I}_{X'Y'} = 0$ and the inertia tensor becomes diagonal:

$$\mathbf{I}' = \frac{Ml^2}{12}\begin{pmatrix} 1 & 0 & 0 \\ 0 & 3 & 0 \\ 0 & 0 & 4 \end{pmatrix}. \tag{10.15b}$$

Notice that the trace of this tensor is the same as the trace of the preceding one, in conformity to the scalar character of traces shown in (9.82).

When the center of rotation is shifted from the apex of the triangle to its mass-center, at $\bar{X} = \bar{Y} = \tfrac{1}{3}l$ ($\bar{X}' = 2^{1/2}l/3$, $\bar{Y}' = 0$), the inertia tensors become

$$\mathbf{I}_0 = \frac{Ml^2}{36}\begin{pmatrix} 2 & 1 & 0 \\ 1 & 2 & 0 \\ 0 & 0 & 4 \end{pmatrix}, \qquad \mathbf{I}'_0 = \frac{Ml^2}{36}\begin{pmatrix} 3 & 0 & 0 \\ 0 & 1 & 0 \\ 0 & 0 & 4 \end{pmatrix}, \tag{10.15c}$$

on axes parallel to those of $\mathbf{I}$ and $\mathbf{I}'$, respectively. The last results may be obtained most easily from the preceding ones by using a general theorem giving the inertia tensor about the mass-center in terms of the inertia tensor on axes centered at an origin relative to which the mass-center has the location $\overline{\mathbf{R}}$:

$$\mathbf{I}_0 = \mathbf{I} - M[\bar{R}^2 \cdot \mathbf{1} - \overline{\mathbf{R}}\,\overline{\mathbf{R}}]. \tag{10.16}$$

This is proved very easily by substituting $\mathbf{R} = \overline{\mathbf{R}} + \mathbf{r}$ into the expression (10.7) for the inertia tensor.

Another example will be provided by the solid tetrahedron of Figure 10.2. The inertia tensor is

$$\mathbf{I} = \frac{Ml^2}{20}\begin{pmatrix} 4 & -1 & -1 \\ -1 & 4 & -1 \\ -1 & -1 & 4 \end{pmatrix}, \tag{10.17}$$

on the axes indicated. The tensor becomes diagonal if one of the axes is chosen to lie along the perpendicular dropped from the origin to the equilateral triangle face opposite the origin. Moreover, the moments of inertia about the two axes which then remain are equal however those two axes are oriented (orthogonally to the first of the new axes, of course). These facts may be checked by direct evaluation of the pertinent integrals, but a far more powerful procedure will be developed in the next section.

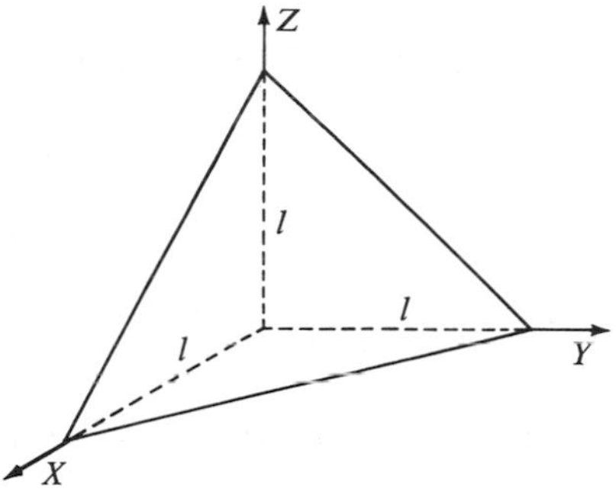

Figure 10.2

The Principal Axes

The preceding examples have indicated the possibility of finding body axes so oriented that the inertia tensor has only diagonal components, with all products of inertia vanishing, when referred to them. The same possibility is also indicated by the "ellipsoid of inertia" (10.13). It is well known that when the cartesian basis is chosen so that its axes are along the principal radii of the ellipsoid, then the quadratic form (10.13) can be reduced to one like

$$\frac{\Omega_1^2}{(2\tau/\mathsf{I}_{11})} + \frac{\Omega_2^2}{(2\tau/\mathsf{I}_{22})} + \frac{\Omega_3^2}{(2\tau/\mathsf{I}_{33})} = 1. \tag{10.18}$$

The orientation of the "principal axes," numbered 1, 2, 3, is such that the products of inertia vanish: $\mathsf{I}_{12} = \mathsf{I}_{23} = \mathsf{I}_{31} = 0$. Consequently the process of finding the body-frame orientation in which any symmetrical tensor is diagonalized is frequently referred to as "setting the ellipsoid on its principal axes."

There now have emerged motives for choosing the body frame in a very specific way. Because an arbitrary motion of a rigid body can be analyzed into a translation of its mass-center and a rotation about it, it becomes appropriate to put the origin of the frame at the mass-center and to orient the frame so that the inertia tensor is diagonal. The choice is modified for the special case in which some point of the rigid body, not the mass-center, is held fixed. Then there is only rotational motion about the fixed center and it becomes most convenient to choose it as the frame origin. The frame orientation may still be chosen so as to diagonalize the inertia tensor.

The main advantage of using principal axes is that this simplifies the representations. Using the notations $\mathsf{I}_{11} \equiv I_1$, $\mathsf{I}_{22} \equiv I_2$, $\mathsf{I}_{33} \equiv I_3$ for the only surviving inertia tensor components reduces the angular momentum expressions (10.5) from three, such as

$$L_X = \mathsf{I}_{XX}\,\Omega_X + \mathsf{I}_{XY}\,\Omega_Y + \mathsf{I}_{XZ}\,\Omega_Z, \tag{10.19}$$

to the three simpler ones

$$L_1 = I_1\Omega_1, \qquad L_2 = I_2\,\Omega_2, \qquad L_3 = I_3\,\Omega_3. \tag{10.20}$$

As already seen in (10.18), the kinetic energy expression reduces to

$$\tau = \tfrac{1}{2}(I_1\,\Omega_1^2 + I_2\,\Omega_2^2 + I_3\,\Omega_3^2). \tag{10.21}$$

Rotation about a principal axis, when only one of the three components of angular velocity $\Omega_{1,2,3}$ survives, makes this particularly simple. The three inertia coefficients $I_{1,2,3}$ that survive on principal axes are called the "principal moments of inertia."

The problem of diagonalizing the inertia tensor is an example of a frequently occurring type known as an "eigenvalue problem." The tensor components may be treated as elements of a 3×3 matrix, and then the problem becomes a special case of finding the "eigenvectors" of symmetric "operator matrices," to be discussed in the next section.

10.2 THE MATRIX EIGENVALUE PROBLEM

The linear relations for the components of the angular momentum $\mathbf{L} = \mathsf{I} \cdot \boldsymbol{\Omega}$ (10.8) may be treated as results of operation by a *matrix* I having the inertia tensor components as elements,

$$\mathbf{L} = \mathsf{I}\boldsymbol{\Omega} \equiv \begin{pmatrix} L_X \\ L_Y \\ L_Z \end{pmatrix} = \begin{pmatrix} \mathsf{I}_{XX} & \mathsf{I}_{XY} & \mathsf{I}_{XZ} \\ \mathsf{I}_{YX} & \mathsf{I}_{YY} & \mathsf{I}_{YZ} \\ \mathsf{I}_{ZX} & \mathsf{I}_{ZY} & \mathsf{I}_{ZZ} \end{pmatrix} \begin{pmatrix} \Omega_X \\ \Omega_Y \\ \Omega_Z \end{pmatrix}, \tag{10.22}$$

an example of the considerations leading to the general matrix Q of (9.102). The matrix elements displayed here are projections like (9.106),

$$\mathsf{I}_{kl} = \tilde{\mathbf{e}}_k \, \mathsf{I} \mathbf{e}_l , \tag{10.23}$$

with $\mathbf{e}_1, \mathbf{e}_2, \mathbf{e}_3 \equiv \mathbf{I}, \mathbf{J}, \mathbf{K}$, the initially chosen body axes. Thus $\tilde{\mathbf{e}}_1 \mathsf{I} \mathbf{e}_2 \equiv \mathbf{I} \cdot (\mathsf{I}\mathbf{J})$ is just a component of the inertia tensor (10.7):

$$\oint dV\varrho[R^2\mathbf{I} \cdot \mathbf{J} - (\mathbf{I} \cdot \mathbf{R})(\mathbf{R} \cdot \mathbf{J})] = -\oint dV\varrho XY = \mathsf{I}_{XY} .$$

With the initial frame chosen before the principal axes of the body are known, the off-diagonal elements will generally be nonvanishing; that is, products of inertia will exist in it. The problem is to find, from a knowledge of the projections (10.23) on the initial frame, a transformation to a new body frame in which the matrix becomes diagonal.

Matrix Diagonalization

Because of its many future applications, the problem will be discussed for a general matrix operator Q. Then the change of basis needed for the diagonalization may have to be a nonorthogonal one, and a more general similarity transformation, as by S of (9.119), should be considered. If the appropriate transformation exists, the matrix representation of the operator on the new basis will take the form

$$(\mathsf{Q})' = \mathsf{S}\mathsf{Q}\mathsf{S}^{-1} = \begin{pmatrix} \lambda_1 & 0 & \dots & 0 \\ 0 & \lambda_2 & \dots & 0 \\ \vdots & \vdots & & \vdots \\ 0 & 0 & \dots & \lambda_n \end{pmatrix} \equiv \boldsymbol{\lambda}, \tag{10.24}$$

with some diagonal elements $\lambda_1 \dots \lambda_n$, which are called the *eigenvalues* of the matrix operator. In the case of the inertia tensor, such diagonal elements are just the principal moments of inertia, denoted I_1, I_2, I_3 in (10.20), rather than

$\lambda_{1,2,3}$. In the example of the triangular sheet (10.15), the eigenvalues were 1, 3, and 4 in units of $Ml^2/12$; there the appropriate new basis (the principal axes) was stumbled upon when the frame was chosen with an eye for the symmetries of the body. It may be noted that, quite generally, the diagonal sum $\sum_m \lambda_m$ equals the trace Q_{kk} of the untransformed matrix, because of the invariance (9.122) in any similarity tranformation.

After multiplication by $\mathbf{S}^{-1}$ from the left, the matrix equation (10.24) becomes

$$\mathbf{QS}^{-1} = \mathbf{S}^{-1}\boldsymbol{\lambda}, \tag{10.25}$$

and an element of this result may be written*

$$Q_{kl}\, S_{l\mu}^{-1} = S_{k\nu}^{-1}\lambda_{\nu\mu} = S_{k\nu}^{-1}\,\delta_{\nu\mu}\,\lambda_{\mu} = S_{k\mu}^{-1}\lambda_{\mu}. \tag{10.26}$$

Greek designations are used for some of the indices here simply to emphasize that one index of any transformation matrix element refers to one frame, the other to another, just as in the orthogonal transformation $a_{kl} = \mathbf{e}_k' \cdot \mathbf{e}_l$ of (9.22). Thus, the elements Q_{kl} may be projections, like (10.23), on an initially chosen frame in which they are known, and the unknown elements λ_μ refer to the new basis, to be reached in the transformation by $\mathbf{S}$.

The equations (10.26), written sans the summation convention, are

$$\sum_l (Q_{kl} - \delta_{kl}\,\lambda_\mu)S_{l\mu}^{-1} = 0. \tag{10.27}$$

They are plainly a sequence of n simultaneous linear equations (one for each $k = 1, \ldots, n$) in the n unknowns $S_{1\mu}^{-1}, \ldots, S_{n\mu}^{-1}$ (that is, for a given μ and λ_μ) They are called the "secular equations" of the eigenvalue problem. The coefficients Q_{kl} are presumed given but λ_μ is also unknown. Homogeneous equations can at best determine the unknowns $S_{l\mu}^{-1}$ only within a constant factor, since multiplying any solutions by a common factor will not interfere with the satisfaction of the equations. Thus, the n equations determine only $(n-1)$ *ratios* among $S_{1\mu}^{-1}, \ldots, S_{n\mu}^{-1}$, and the nth unknown they can determine is the corresponding eigenvalue λ_μ. Discussed in terms of the general solution (9.74) of simultaneous linear equations, the numerator determinant $|\mathbf{h}|$ vanishes for the homogeneous equations here, since the whole column of it, $h_1 = h_2 = \cdots = h_n = 0$, vanishes. Thus the homogeneous equations here would only have the trivial solutions $S_{1\mu}^{-1} = S_{2\mu}^{-1} = \cdots = S_{n\mu}^{-1} = 0$ unless the determinant of the coefficients, forming the denominator of the solution

* Notice that the index μ in the last equality of (10.26) is *not* to be summed on, since it appears unrepeated, as a "free" (chosen) index, on the left side of the equation.

(9.74), also vanishes. Thus, the desired transformation **S** will exist only for eigenvalues λ_μ such that

$$\boxed{|Q_{kl} - \delta_{kl}\lambda_\mu| = 0.} \tag{10.28}$$

Reference to the definition (9.64) of a determinantal value shows that this amounts to an algebraic equation of the nth degree for the unknown λ_μ. There will generally be n roots, $\lambda_1, \ldots, \lambda_n$, and hence a determination of enough elements for the eigenvalue matrix $\boldsymbol{\lambda}$. The corresponding n sets of solutions $S_{1\mu}^{-1}, \ldots, S_{n\mu}^{-1}$, one set for each $\lambda_\mu = \lambda_1, \ldots, \lambda_n$ that is put into the secular equations (10.27), will constitute the n^2 elements of the desired transformation matrix.

Depending on the given coefficients Q_{kl}, the solutions $S_{l\mu}^{-1}$ of the secular equations may or may not turn out to form an orthonormal matrix; that is, $\mathbf{S} \to \mathbf{a}$ with $a_{l\mu}^{-1} = \tilde{a}_{l\mu} = a_{\mu l}$. It is easy to see what conditions a given matrix $\mathbf{Q}$ must fulfill if it is to be diagonalizable by an orthonormal transformation, $\mathbf{a} = \tilde{\mathbf{a}}^{-1}$. According to the diagonal form (10.24), the transpose of the transformed matrix must equal itself; now

$$(\widetilde{\mathbf{a}\mathbf{Q}\mathbf{a}^{-1}}) = (\widetilde{\mathbf{a}\mathbf{Q}\tilde{\mathbf{a}}}) = \mathbf{a}\tilde{\mathbf{Q}}\tilde{\mathbf{a}} = \mathbf{a}\tilde{\mathbf{Q}}\mathbf{a}^{-1}$$

and this equals $\mathbf{a}\mathbf{Q}\mathbf{a}^{-1}$ itself if

$$\mathbf{a}(\mathbf{Q} - \tilde{\mathbf{Q}})\mathbf{a}^{-1} = 0 \text{ or } \mathbf{Q} - \tilde{\mathbf{Q}} = \mathbf{0}. \tag{10.29}$$

Thus, an orthonormal transformation **a**, a mere rotation of the initial frame, is sufficient to diagonalize a symmetric matrix, with $\tilde{\mathbf{Q}} = \mathbf{Q}$. It was found that the inertia tensor (10.7) of any rigid body is symmetric, $\tilde{\mathbf{I}} = \mathbf{I}$, as resolved on any initial orthogonal frame. Thus an important property emerges. The principal axes of a rigid body, on which the inertia tensor has only the principal moments of inertia and no products of inertia, may always constitute an orthogonal body frame, merely rotated from any initial orthogonal body axes, no matter how irregular the body.

Before specific examples of the eigenvalue problem are considered, attention should be called to another view of it.

Eigenvectors

The n transformation matrix elements $S_{1\mu}^{-1}, \ldots, S_{n\mu}^{-1}$ for a given μ (a column of $\mathbf{S}^{-1}$) may be defined to be components of a vector $\mathbf{u}_\mu$ (9.120) resolved on the initial basis $\mathbf{e}_1, \ldots, \mathbf{e}_n$. Then

$$\mathbf{u}_\mu = \mathbf{e}_l S_{l\mu}^{-1} \qquad \text{and} \qquad S_{l\mu}^{-1} = \tilde{\mathbf{e}}_l \mathbf{u}_\mu, \tag{10.30}$$

and the diagonalization condition (10.26) becomes

$$\tilde{\mathbf{e}}_k\,\mathbf{Q}\mathbf{e}_l\,S_{l\mu}^{-1} = \tilde{\mathbf{e}}_k\,\mathbf{Q}\mathbf{e}_l\,\tilde{\mathbf{e}}_l\,\mathbf{u}_\mu = \tilde{\mathbf{e}}_k\,\mathbf{Q}\mathbf{u}_\mu = \tilde{\mathbf{e}}_k\,\mathbf{u}_\mu\,\lambda_\mu,$$

since $\mathbf{e}_l\tilde{\mathbf{e}}_l = \mathbf{1}$ is just the unit projection operator (9.109). This result is the kth component of the vector equation

$$\boxed{\mathbf{Q}\mathbf{u}_\mu = \lambda_\mu\,\mathbf{u}_\mu.} \tag{10.31}$$

The matrix diagonalization problem is thus made equivalent to the finding of vectors $\mathbf{u}_\mu$ that are unchanged in direction when operated on by $\mathbf{Q}$. They are called *eigenvectors* of the operator. The operation on an eigenvector becomes equivalent to multiplying it by a number λ_μ, now called the eigenvalue of the operator $\mathbf{Q}$ belonging to its eigenvector $\mathbf{u}_\mu$.

The satisfaction of the eigenvector equations (10.31) would obviously be unimpaired if any given eigenvector solution were multiplied by an arbitrary constant. Only the directional property of an eigenvector is significant; hence it is usually standardized as a unit vector: $\mathbf{u}_\mu \cdot \mathbf{u}_\mu = 1$. Whenever some procedure happens to yield a solution $\mathbf{u}'_\mu$ such that $\mathbf{u}'_\mu \cdot \mathbf{u}'_\mu = l_\mu^2 \neq 1$, then $\mathbf{u}_\mu = \mathbf{u}'_\mu/l_\mu$ may always be taken to be the equivalent eigenvector, a process called "normalization to unity."

The special properties of a symmetric operator, $\mathbf{Q} = \tilde{\mathbf{Q}}$, may now be discussed in terms of its eigenvectors. Suppose it has several eigenvectors and let $\mathbf{u}_\nu$ be one additional to $\mathbf{u}_\mu$ of (10.31), with $\mathbf{Q}\mathbf{u}_\nu = \lambda_\nu\mathbf{u}_\nu$. Now subtract the projection of this vector equation on $\mathbf{u}_\mu$ from a projection on $\mathbf{u}_\nu$ of the equation (10.31):

$$\tilde{\mathbf{u}}_\nu\,\mathbf{Q}\mathbf{u}_\mu - \tilde{\mathbf{u}}_\mu\,\mathbf{Q}\mathbf{u}_\nu = (\lambda_\mu - \lambda_\nu)\mathbf{u}_\mu \cdot \mathbf{u}_\nu,$$

since any scalar product like $\tilde{\mathbf{u}}_\nu\mathbf{u}_\mu \equiv \mathbf{u}_\nu \cdot \mathbf{u}_\mu$ is symmetric (9.55). For the same reason the first term may be replaced by $(\widetilde{\mathbf{Q}\mathbf{u}_\mu})\mathbf{u}_\nu = \tilde{\mathbf{u}}_\mu\,\tilde{\mathbf{Q}}\mathbf{u}_\nu$, and so

$$\tilde{\mathbf{u}}_\mu(\tilde{\mathbf{Q}} - \mathbf{Q})\mathbf{u}_\nu = (\lambda_\mu - \lambda_\nu)\mathbf{u}_\mu \cdot \mathbf{u}_\nu. \tag{10.32}$$

Thus, for a symmetric operator, $\mathbf{u}_\mu \cdot \mathbf{u}_\nu$ must vanish unless $\mu = \nu$. Any several eigenvectors of a symmetric matrix are orthonormal, $\mathbf{u}_\mu \cdot \mathbf{u}_\nu = \delta_{\mu\nu}$, after normalization to unity. They may then be used as directions of an orthogonal frame. This obviously corresponds to the fact that the transformation needed to diagonalize a symmetric matrix is an orthonormal one, $\mathbf{S} \to \tilde{\mathbf{a}} = \tilde{\mathbf{a}}^{-1}$, as found in connection with (10.29), and hence the vectors (10.30),

$$\mathbf{u}_\mu = \mathbf{e}_l\,a_{l\mu}^{-1} = \mathbf{e}_l\,\tilde{a}_{l\mu} \equiv \mathbf{e}'_\mu = \mathbf{a}^{-1}\mathbf{e}_\mu, \tag{10.33}$$

according to (9.118), are merely rotated from the initial orthogonal basis, $\mathbf{e}_\mu$.

In terms of the vector components $(\mathbf{u}_\mu)_l = u_{\mu l} \equiv S^{-1}_{l\mu}$, the secular equations (10.27) have the appearance

$$\sum_l (Q_{kl} - \delta_{kl}\lambda_\mu)u_{\mu l} = 0, \tag{10.34}$$

and follow directly from projecting the eigenvector equation (10.31) on $\tilde{\mathbf{e}}_k$, as seen in the equation preceding it. This is the form in which they will be used in the following.

Diagonalizing Symmetric Tensors

For inertia matrices $\mathbf{I}$, the secular equations (10.34) have the form

$$\sum_l (\mathbf{I}_{kl} - \delta_{kl} I_\mu)u_{\mu l} = 0, \tag{10.35}$$

where the Latin indices refer to the X, Y, Z axes of (10.22) and $I_{1,2,3}$ are the principal moments of inertia. For the symmetric inertia tensors, the eigenvectors $\mathbf{u}_\mu \equiv \mathbf{u}_{1,2,3}$ will form an orthogonal set, $\mathbf{u}_{1,2,3} \equiv \mathbf{e}'_{1,2,3}$ (10.33), and are just the principal axes, on which products of inertia,

$$\mathbf{I}_{\mu\nu} = \tilde{\mathbf{u}}_\mu \mathbf{I}\mathbf{u}_\nu = I_\nu \delta_{\mu\nu},$$

vanish because the $\mathbf{u}$'s are eigenvectors: $\mathbf{I}\mathbf{u}_\nu = I_\nu \mathbf{u}_\nu$.

The procedure for solving the eigenvector problem is well illustrated by the example of the inertia matrix $\mathbf{I}$ of (10.15*a*). The secular equations in this case are

$$\begin{aligned}
\sum_l (\mathbf{I}_{Xl} - \delta_{Xl} I_\mu)u_{\mu l} &= (Ml^2/12)[(2 - I'_\mu)u_{\mu X} - u_{\mu Y}] = 0,\\
\sum_l (\mathbf{I}_{Yl} - \delta_{Yl} I_\mu)u_{\mu l} &= (Ml^2/12)[-u_{\mu X} + (2 - I'_\mu)u_{\mu Y}] = 0, \qquad (10.36a)\\
\sum_l (\mathbf{I}_{Zl} - \delta_{Zl} I_\mu)u_{\mu l} &= (Ml^2/12)[(4 - I'_\mu)u_{\mu Z}] = 0,
\end{aligned}$$

with $I'_\mu = I_\mu/(Ml^2/12)$ the eigenvalue in units of $(Ml^2/12)$. The first two equations give alternative results for the ratio

$$\frac{u_{\mu Y}}{u_{\mu X}} = 2 - I'_\mu = \frac{1}{2 - I'_\mu},$$

and therefore can be satisfied only by $u_{\mu X} = u_{\mu Y} = 0$, *or* if the eigenvalue is one of the two roots of the quadratic equation for it, implicit here: $I'_1 = 1$ or $I'_2 = 3$. For either of these values, the last of the secular equations requires

$u_{\mu Z} = 0$, but $I'_3 = 4$ is possible when $u_{3X} = u_{3Y} = 0$. Thus, the three-dimensional matrix possesses three eigenvalues, $I'_{1,2,3}$, corresponding to three eigenvectors $\mathbf{u}_{1,2,3}$, with

$$\frac{u_{1Y}}{u_{1X}} = +1, \quad u_{1Z} = 0, \qquad \text{for} \quad I'_1 = 1,$$

$$\frac{u_{2Y}}{u_{2X}} = -1, \quad u_{2Z} = 0, \qquad \text{for} \quad I'_2 = 3,$$

$$u_{3X} = u_{3Y} = 0, \quad u_{3Z} \text{ arbitrary}, \qquad \text{for} \quad I'_3 = 4.$$

A normalized eigenvector $\mathbf{u}_3$ will just be $\mathbf{u}_3 = \mathbf{K}$, the unit vector along the Z-axis. The other eigenvectors, $\mathbf{u}_1$ and $\mathbf{u}_2$, lie in the X, Y plane and make 45° angles with the X and Y axes; they have the directions of the X' and Y' axes drawn in Figure 10.1. After each is normalized,

$$\mathbf{u}_1 = 2^{-1/2}(\mathbf{I} + \mathbf{J}), \qquad \mathbf{u}_2 = -2^{-1/2}(\mathbf{I} - \mathbf{J}), \qquad \mathbf{u}_3 = \mathbf{K}, \tag{10.36b}$$

forming a complete, orthonormal basis and giving the directions of the principal axes. The minus sign is chosen for $\mathbf{u}_2$ in order that the frame be right-handed: $\mathbf{u}_1 \times \mathbf{u}_2 = +\mathbf{K}$. The results check with the diagonal matrix $\mathbf{I}'$ of (10.15*b*).

The secular determinant (10.28) was not explicitly employed here but might have been. It amounts to

$$\begin{vmatrix} 2 - I'_\mu & -1 & 0 \\ -1 & 2 - I'_\mu & 0 \\ 0 & 0 & 4 - I'_\mu \end{vmatrix} = 0. \tag{10.36c}$$

Expanded as in (9.66), it forms the cubic in I'_μ:

$$(4 - I'_\mu)[(2 - I'_\mu)^2 - 1] = 0,$$

having just the roots $I'_{1,2,3}$ found otherwise above.

The identification $u_{\mu l} = a^{-1}_{l\mu}$ noted in (10.33), for example, leads to the orthonormal transformation matrix

$$\mathbf{a} = 2^{-1/2}\begin{pmatrix} 1 & 1 & 0 \\ -1 & 1 & 0 \\ 0 & 0 & 2^{1/2} \end{pmatrix}, \tag{10.36d}$$

which is a special case of the rotation matrix $\mathbf{a}(\varphi, \vartheta, \psi)$ of (9.49), with $\vartheta = 0$ and $\varphi + \psi = \pi/4$. It is a proper rotation, with $|\mathbf{a}| = +1$, as a consequence of choosing the principal axis directions (10.36*b*) to be right-handed.

Degeneracy

A contingency may arise which is well illustrated by the example of the highly symmetrical inertia matrix of (10.17) for the tetrahedron of Figure 10.2. The secular determinant in this case is

$$\begin{vmatrix} 4 - I_\mu & -1 & -1 \\ -1 & 4 - I_\mu & -1 \\ -1 & -1 & 4 - I_\mu \end{vmatrix} = 0,$$

if I_μ is an eigenvalue in units of $Ml^2/20$. The three roots of the cubic implicit here may be obtained by making use of the various properties of determinants outlined after (9.68). Thus the last two columns may be added to the first to give:

$$\begin{vmatrix} 2 - I_\mu & -1 & -1 \\ 2 - I_\mu & 4 - I_\mu & -1 \\ 2 - I_\mu & -1 & 4 - I_\mu \end{vmatrix} = (2 - I_\mu) \begin{vmatrix} 1 & -1 & -1 \\ 1 & 4 - I_\mu & -1 \\ 1 & -1 & 4 - I_\mu \end{vmatrix} = 0.$$

Adding the first column to each of the others in the last determinant gives

$$(2 - I_\mu) \begin{vmatrix} 1 & 0 & 0 \\ 1 & 5 - I_\mu & 0 \\ 1 & 0 & 5 - I_\mu \end{vmatrix} = (2 - I_\mu)(5 - I_\mu)^2 = 0. \qquad (10.37a)$$

Thus two of the eigenvalues are identical, $I_1 = I_2 = 5$, and $I_3 = 2$ (in units of $Ml^2/20$). The falling together of the two eigenvalues $I_1 = I_2 = 5$ is called a "degeneracy," and its consequences are of chief interest here. Disposing first of $I_3 = 2$, it leads to the secular equations for $\mathbf{u}_3$:

$$2u_{3X} - u_{3Y} - u_{3Z} = 0,$$

$$-u_{3X} + 2u_{3Y} - u_{3Z} = 0,$$

$$-u_{3X} - u_{3Y} + 2u_{3Z} = 0,$$

having the solution $u_{3X} = u_{3Y} = u_{3Z}$; the normalized eigenvector with such components is

$$\mathbf{u}_3 = 3^{-1/2}(\mathbf{I} + \mathbf{J} + \mathbf{K}). \qquad (10.37b)$$

This is the principal axis of the tetrahedron, which is normal to the equililateral triangle face (Figure 10.2); the mass is distributed most closely to it, and so its having the smallest moment of inertia, $I_3 = 2(Ml^2/20)$, is to be expected. For the degenerate eigenvalues $I_1 = I_2 = 5$, each secular equation reduces to the same one:

$$u_{\mu X} + u_{\mu Y} + u_{\mu Z} = 0.$$

This is the only restriction on the eigenvectors $\mathbf{u}_1$ and $\mathbf{u}_2$, and so they may be any vectors of the form

$$\mathbf{u}_{1,2} = \alpha\mathbf{I} + \beta\mathbf{J} - (\alpha + \beta)\mathbf{K}, \tag{10.37c}$$

with arbitrary α and β. It is easy to check that this form is orthogonal to $\mathbf{u}_3$ whatever α and β are chosen; accordingly, any axis in the plane normal to $\mathbf{u}_3$ is a principal axis, with the moment of inertia $I_{1,2} = 5(Ml^2/20)$ about it.

The last result is typical of degeneracy. Whenever g orthogonal eigenvectors all yield the same eigenvalue λ_g (in a space of $n \geq g$ dimensions) —that is, $\mathbf{Q}\mathbf{u}_s = \lambda_g \mathbf{u}_s$ for $s = 1, \ldots, g$—then any linear combination $\sum_s \alpha_s \mathbf{u}_s$ is plainly also an eigenvector having the same eigenvalue:

$$\mathbf{Q}\left(\sum_s \alpha_s \mathbf{u}_s\right) = \lambda_g\left(\sum_s \alpha_s \mathbf{u}_s\right), \tag{10.38}$$

with arbitrary coefficients $\alpha_1, \alpha_2, \ldots, \alpha_g$. Any vector of this g-dimensional "subspace" is an eigenvector, just as any vector in the plane normal to $\mathbf{u}_3$, of the above example, is a principal axis.

When degeneracy occurs, the criterion that they be principal axes is no longer sufficient to determine a completely specific choice of body axes. Additional criteria sometimes arise in special problems. Suppose, for example, it becomes for some reason convenient to have one of the body axes, of the above problem, in the plane of an isosceles triangle face of the tetrahedron. Taking this face to be in the X, Y plane of Figure 10.2, the additional criterion requires choosing $(\alpha + \beta) = 0$ for one of the eigenvectors encompassed by (10.37c). This leads to the normalized choice

$$\mathbf{u}_1 = 2^{-1/2}(\mathbf{I} - \mathbf{J}) \tag{10.37d}$$

for one of the body axes. With it, $\mathbf{u}_2$ is completely determinable from

$$\mathbf{u}_2 = \mathbf{u}_3 \times \mathbf{u}_1 = 6^{-1/2}(\mathbf{I} + \mathbf{J} - 2\mathbf{K}). \tag{10.37e}$$

This corresponds to $\mathbf{u}_1 \cdot \mathbf{u}_2 = 2^{-1/2}(\alpha - \beta) = 0$ for the coefficients in the form (10.37c) and to $\mathbf{u}_2 \cdot \mathbf{u}_2 = 1$. Of course, an infinity of other choices for $\mathbf{u}_{1,2}$ are consistent with (10.37c).

10.3 THE EULER EQUATIONS OF RIGID BODY ROTATION

Any system must move in consistency with the "body" equations of motion,

$$\boxed{\dot{\mathbf{p}} = \mathbf{f}, \qquad \dot{\mathbf{l}} = \mathbf{n},} \tag{10.39}$$

according to the findings (6.5) and (6.12). The first of these represents the effect of any resultant external force **f** on the translation, $\dot{\bar{\mathbf{r}}} = \bar{\mathbf{p}}/M$, of the mass-center of the system. The second governs independent rotational motion about the mass-center if **l** is the angular momentum about it and **n** is the resultant of all external torques. For a general system, there must be additional equations for detailed motions relative to the mass-center, but a rigid body can only rotate about it. The six degrees of freedom possessed by such a body are just determinable from the six components of the above equations and no more need be considered. Moreover, the three translational degrees of freedom move like those of a single particle, and so need little further attention.

The rotation may instead be gauged about any point fixed in space, and then it is governed by the equation of motion found in (6.7),

$$\dot{\mathbf{L}} = \mathbf{N}, \tag{10.40}$$

with **L** the angular momentum and **N** the resultant external torque about the fixed point. This is the appropriate equation to use when one of the points of the rigid body, not necessarily its mass-center, is fixed in space so that the body can only rotate about this point. The rotation equation of (10.39) will hereafter also be written in the notation (10.40) so that both may be treated simultaneously. There is just the difference of interpretation to bear in mind.

Reference to Arbitrary Body Axes

The connection between the angular momentum **L**, occuring in the rotation equation $\dot{\mathbf{L}} = \mathbf{N}$, and the instantaneous angular velocity $\mathbf{\Omega}(t)$ of the body is $\mathbf{L} = \mathsf{I} \cdot \mathbf{\Omega}$, according to (10.8). The inertia tensor has components that are constant properties of a rigid body only if they refer to axes fixed to the body. For that reason, it becomes convenient to refer all the vectors **L**, $\mathbf{\Omega}$, **N** to some set of body axes, as in $\mathbf{L} = \mathbf{I}L_X + \mathbf{J}L_Y + \mathbf{K}L_Z$, when **I**, **J**, **K** are the unit vector directions of the chosen body axes. It was learned as early as in §1.4 that when a vector like **L** is referred to rotating axes, having angular velocity $\mathbf{\Omega}(t)$, its rate of change is the sum of $\partial\mathbf{L}/\partial t$, relative to those axes, and a rate $\mathbf{\Omega} \times \mathbf{L}$ imparted by the rotation of the axes. On the other hand, no rotation is thus imparted to the rotation axis itself and $\dot{\mathbf{\Omega}} = \partial\mathbf{\Omega}/\partial t$ only. As a consequence, the equation of rotational motion may be written as

$$\dot{\mathbf{L}} = \partial\mathbf{L}/\partial t + \mathbf{\Omega} \times \mathbf{L} = \mathsf{I} \cdot \dot{\mathbf{\Omega}} + \mathbf{\Omega} \times (\mathsf{I} \cdot \mathbf{\Omega}) = \mathbf{N}, \tag{10.41}$$

in terms of the inertia tensor.

The scalar multiplication of the angular velocity into the equation of motion yields

$$\mathbf{\Omega} \cdot \mathsf{I} \cdot \dot{\mathbf{\Omega}} = (d/dt)\tfrac{1}{2}\mathbf{\Omega} \cdot \mathsf{I} \cdot \mathbf{\Omega} = \mathbf{\Omega} \cdot \mathbf{N}. \tag{10.42}$$

This states that changes in the rotational kinetic energy (10.12) result from work by external torques, at the rate $\mathbf{N} \cdot \boldsymbol{\Omega}$.

The equation of rotational motion becomes the simple one of elementary physics, $I\dot{\Omega} = N$, when the motion is *constrained* to rotation about a fixed axis. If this is chosen to be the Z-axis, as in (10.10), the Z-component of $\dot{\mathbf{L}} = \mathbf{N}$ (10.40) reduces to just

$$\mathsf{I}_{ZZ}\dot{\Omega} = N_Z, \tag{10.43a}$$

since $\Omega \equiv \Omega_Z$ under the constraint, and $(\mathbf{I} \cdot \boldsymbol{\Omega})_k = \mathsf{I}_{kZ}\,\Omega$. The coefficient I_{ZZ} is the moment of inertia about the rotation axis, denoted I in the elementary formula. The remaining two components of (10.41) give

$$\begin{aligned} N_X &= \mathsf{I}_{XZ}\,\dot{\Omega} - \mathsf{I}_{YZ}\,\Omega^2 = (\mathsf{I}_{XZ}/\mathsf{I}_{ZZ})N_Z - \mathsf{I}_{YZ}\,\Omega^2, \\ N_Y &= \mathsf{I}_{YZ}\,\dot{\Omega} + \mathsf{I}_{XZ}\,\Omega^2 = (\mathsf{I}_{YZ}/\mathsf{I}_{ZZ})N_Z + \mathsf{I}_{XZ}\,\Omega^2, \end{aligned} \tag{10.43b}$$

torques which must be supplied to maintain the constraint. Notice that there would be no such stresses on the rotation axis if it were one of the principal axes of the body, so that $\mathsf{I}_{XZ} = \mathsf{I}_{YZ} = 0$. This fact is important for the "dynamical balancing" of flywheels and other such devices.

Reference to Principal Axes

The decomposition of the vector equation of rotation (10.41) is simplest when it refers to principal axes, as do the expressions (10.20) for the angular momentum components. Then, for example, $(\boldsymbol{\Omega} \times \mathbf{L})_1 = \Omega_2(I_3\,\Omega_3) - \Omega_3(I_2\,\Omega_2)$ and the equations of rotational motion become

$$\begin{aligned} I_1\dot{\Omega}_1 - (I_2 - I_3)\Omega_2\,\Omega_3 &= N_1, \\ I_2\,\dot{\Omega}_2 - (I_3 - I_1)\Omega_3\,\Omega_1 &= N_2, \\ I_3\,\dot{\Omega}_3 - (I_1 - I_2)\Omega_1\Omega_2 &= N_3. \end{aligned} \tag{10.44}$$

These are known as the Euler equations of motion. Notice that they contain terms of an "inertial force" type, intervening between the angular acceleration $\dot{\boldsymbol{\Omega}}$ and the torque $\mathbf{N}$.

When the three Euler equations are respectively multiplied by Ω_1, Ω_2, Ω_3 and added together, the result gives the rate of change of the rotational kinetic energy in the form

$$(d/dt)\tfrac{1}{2}(I_1\Omega_1^2 + I_2\,\Omega_2^2 + I_3\,\Omega_3^2) = \mathbf{N} \cdot \boldsymbol{\Omega}. \tag{10.45}$$

The expression for the kinetic energy here has already been seen in (10.21).

Free Rotation

The Euler equations as they stand are ostensibly to be solved for the angular velocity $\mathbf{\Omega}(t)$, to obtain a description of a motion. That can be done, without first replacing $\mathbf{\Omega}$ by time-derivatives of some kind of coordinates (angles), only when the torque $\mathbf{N}$ is independent of such coordinates. This is the case for free rotation, when there is no torque at all.

A prominent example of free rotation is provided by any freely falling body. Uniform gravitation accelerates the mass-center (making it follow a parabolic path (1.58), in general), but exerts no torque about the mass-center (6.15).

With no torque to change it, the angular momentum $\mathbf{L} \equiv \mathbf{l}$ remains conserved. However, the angular velocity $\mathbf{\Omega}$ may still change, both in direction and magnitude, as reference to Euler's equations shows. A freely falling body will generally "wobble." Only when the initial rotation happens to be started about a principal axis, $\Omega(0) = \Omega_1$ say, will it continue around that axis without wobbling. Then $\dot{\Omega}_1(0) = \dot{\Omega}_2(0) = \dot{\Omega}_3(0) = 0$ and $\mathbf{L} = I_1\mathbf{\Omega}$ remains conserved.

The general free rotation can be understood with the help of the conservation relations, for the angular momentum $\mathbf{L}$ and the kinetic energy $\tau = \frac{1}{2}\mathbf{\Omega} \cdot \mathbf{L}$. If the initial angular velocity is not around a principal axis, so that $\mathbf{\Omega}$ is not parallel to $\mathbf{L}$, then the situation may be represented as in Figure 10.3. The axis of rotation, $\mathbf{\Omega}$, "precesses" about the constant direction of $\mathbf{L}$, with its projection along $\mathbf{L}$ staying constant. More detailed descriptions have been worked out, but will not be given further attention here.

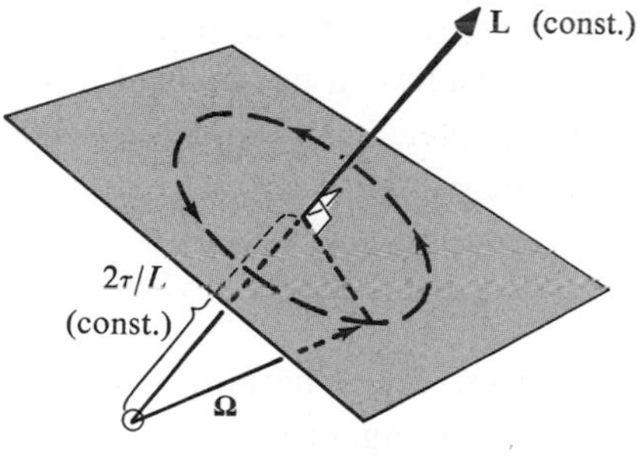

Figure 10.3

10.4 A SPINNING BODY IN A UNIFORM FIELD

Uniform gravitation does exert a torque about centers of a rigid body other than its mass-center. Its effect on a spinning symmetrical top, with its apex fixed, will be considered here (Figure 10.4). The symmetry is to be an

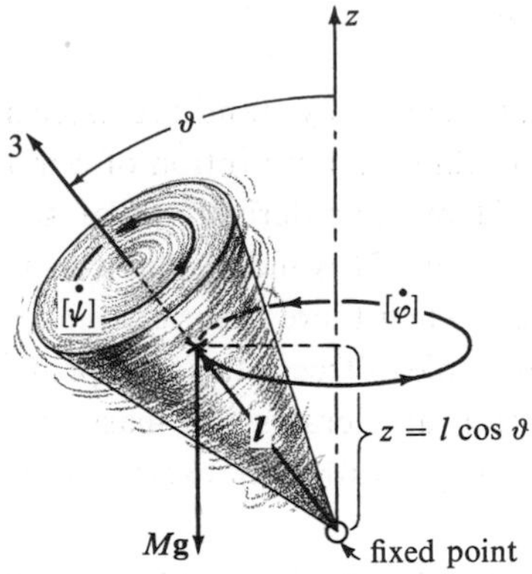

Figure 10.4

axial one, about the principal axis #3; the moments of inertia about all axes perpendicular to #3 are then degenerate, $I_1 = I_2 = I$. A position vector **l** along the symmetry axis will mark the mass-center, and then the gravitational torque is represented by $\mathbf{N} = \mathbf{l} \times M\mathbf{g}$ (6.14).

The Constants of the Motion

Because the torque $\mathbf{N} = \mathbf{l} \times M\mathbf{g}$ depends on the orientational position of the body, the Euler equations of motion (10.44), which are explicit in the angular velocity $\mathbf{\Omega}$ only, become inadequate as they stand. They may have the instantaneous orientation made explicit in them, through the relations (9.15) between $\mathbf{\Omega}$ and the Euler angles, for example, but this approach is unnecessarily laborious. The same results can be obtained through the more powerful lagrangian approach of Chapter 7. For this, the connections (9.15) need be used only to put the kinetic energy,

$$\tau = \tfrac{1}{2}I(\Omega_1^2 + \Omega_2^2) + \tfrac{1}{2}I_3\Omega_3^2, \tag{10.46a}$$

in terms of the Euler angles. Then, because the potential energy is simply $V = Mgz = Mgl\cos\vartheta$, the lagrangian $L = \tau - V$ becomes

$$L = \tfrac{1}{2}I(\dot{\vartheta}^2 + \dot{\varphi}^2\sin^2\vartheta) + \tfrac{1}{2}I_3(\dot{\psi} + \dot{\varphi}\cos\vartheta)^2 - Mgl\cos\vartheta. \tag{10.46b}$$

In this expression for $L(\vartheta, \dot{\varphi}, \dot{\vartheta}, \dot{\psi})$, the Euler angles φ, ϑ, ψ are serving as the generalized coordinates.

To be noticed immediately is that the variables φ, ψ are ignorable; hence the angular momenta p_φ and p_ψ, conjugate to them, are conserved (7.37). Moreover, the time variable is also ignorable, in correspondence to the energy conservation $\tau + V = E$, (7.39) and (7.41). Altogether, three conservation relations are immediately available:

$$p_\psi = \partial L/\partial\dot\psi = I_3(\dot\psi + \dot\varphi \cos \vartheta) \equiv I_3\,\Omega_3 = L_3\text{, a constant,} \tag{10.47}$$

$$p_\varphi = \partial L/\partial\dot\varphi = I\,\dot\varphi \sin^2 \vartheta + I_3\,\Omega_3 \cos \vartheta \equiv L_z\text{, a constant,} \tag{10.48}$$

$$E = \tfrac{1}{2}I(\dot\vartheta^2 + \dot\varphi^2 \sin^2 \vartheta) + \tfrac{1}{2}I_3\,\Omega_3^2 + Mgl \cos \vartheta\text{, a constant.} \tag{10.49}$$

All of these might have been anticipated:

$$\dot L_3 = N_3 = 0 \qquad \text{and} \qquad \dot L_z = N_z = 0$$

because the torque **N** has the direction of the nodal line, which lies in both the (1, 2) and (x, y) planes; from (10.45),

$$d\tau/dt = \mathbf{N}\cdot\mathbf{\Omega} = (Mgl \sin \vartheta)\Omega_a = -d(Mgl \cos \vartheta)/dt,$$

since the component of the angular velocity on the nodal line is just $\Omega_a = \dot\vartheta$ (9.12).

The three conservation relations are sufficient to replace the equations of motion in determining the motion: φ, ϑ, $\psi(t)$. The equation for the constant L_z, explicitly solved for the "precession velocity"

$$\dot\varphi = \frac{L_z - L_3 \cos \vartheta}{I \sin^2 \vartheta}, \tag{10.50}$$

can be used to eliminate all variables but ϑ from the energy equation:

$$E' \equiv E - \tfrac{1}{2}I_3\,\Omega_3^2 = \tfrac{1}{2}I\dot\vartheta^2 + \frac{(L_z - L_3 \cos \vartheta)^2}{2I \sin^2 \vartheta} + Mgl \cos \vartheta. \tag{10.51}$$

This allows expressing $t(\vartheta)$ as an integral from which $\vartheta(t)$ is in principle obtainable. Substitution of the result into (10.50) in principle allows integration for $\varphi(t)$. Then $\psi(t)$ is to be obtained from $\dot\psi(t) = \Omega_3 - \dot\varphi(t) \cos \vartheta(t)$, the relation implicit in the constancy of $L_3 = I_3\,\Omega_3$. More instructive than any expressions thus obtainable is a discussion of an "energy diagram" modeled on those used for the central force problems of Chapter 4 (Figures 4.3 and 4.4).

The Energy Diagram

The component of angular velocity along the top's symmetry axis, Ω_3, and so also the corresponding rotation energy, $\tfrac{1}{2}I_3\,\Omega_3^2$, stay constant in the motion (10.47). This portion of the rotational energy contains more than just the "spin energy," $\frac{1}{2}I_3\dot{\psi}^2$, since $\Omega_3 = \dot{\psi} + \dot{\varphi}\cos\vartheta$ includes the component of precession, $\dot{\varphi}$, along the symmetry axis. Exchanges can thus take place between the spin and the precession energies; this is not surprising in view of the fact that spin and precession become indistinguishable in the erect position of the top. Because $\frac{1}{2}I_3\Omega_3^2$ is conserved to itself, only the remainder of the energy, E' of (10.51), needs further discussion.

The conserved remainder, E', is shared among variable parts:
the potential energy, $V = Mgl\cos\vartheta$,
a "nutational" kinetic energy, $\tau_\vartheta \equiv \frac{1}{2}I\dot{\vartheta}^2$,
a precession energy fraction, $\tau'_\varphi \equiv \frac{1}{2}I\dot{\varphi}^2\sin^2\vartheta$.

The sharing can be studied as a function of $z = l\cos\vartheta$, the height of the mass-center, as in Figure 10.5. Plotted is $V = Mgz$ and

$$\tau'_\varphi(z) = I_3^2\Omega_3^2(\zeta - z)^2/2I(l^2 - z^2), \tag{10.52}$$

which follows from (10.50) and the definition $\zeta \equiv (L_z/L_3)l$. Then the difference $\tau_\vartheta(z) = E' - \tau'_\varphi(z) - Mgz$ is just the positive nutational energy, $\frac{1}{2}I\dot{\vartheta}^2$.

Figure 10.5 shows that the motion of the mass-center is confined between two horizontal planes at heights z_1 and z_2, the roots of $E' = \tau'_\varphi + V$. At each of those "turning points," where $\dot{z} = -(l\sin\vartheta)\dot{\vartheta} = 0$, the motion is turned back by an "inertial force" of precession, modified by gravity. The point marked ζ is a height at which the precession energy $\frac{1}{2}I\dot{\varphi}^2$ vanishes (10.52); while moving through this level, the mass-center has only a "nutational" component of velocity, along a (vertical) meridian of the sphere to which it is constrained.

In general, the spinning top will precess about the vertical, at a varying rate $\dot{\varphi}$; concurrently it may nutate ("nod") in an oscillatory fashion. Some special cases will now be examined in more detail.

A Pure Spin Start ($\dot{\vartheta}_0 = \dot{\varphi}_0 = 0$)

Suppose that the spinning ($\dot{\psi}_0 \neq 0$) top is started with its symmetry axis at rest, but inclined to the vertical ($\vartheta_0 > 0$). The various constants of the motion corresponding to such a start are

$$\begin{aligned} L_3 &= I_3\Omega_3 = I_3\dot{\psi}_0, \\ L_z &= I_3\Omega_3\cos\vartheta_0 = \frac{L_3 z_0}{l} \text{ and } \zeta = z_0, \\ E' &= E - \tfrac{1}{2}I_3\dot{\psi}_0^2 = Mgz_0. \end{aligned} \tag{10.53}$$

Initially, there is only "spin-energy," $\frac{1}{2}I_3\dot{\psi}_0^2$, and potential energy, Mgz_0.

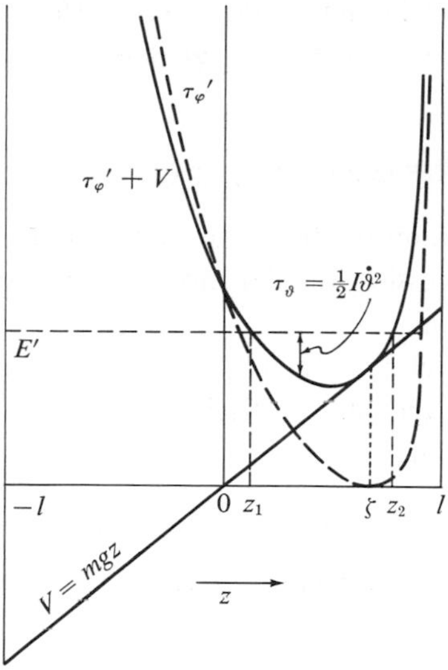

Figure 10.5

With the symmetry axis initially at rest, and "leaning over" by ϑ_0, the top might be expected to topple over, except that it must conserve the vertical component of angular momentum arising from the spin ($L_z = I_3 \dot{\psi}_0 \cos \vartheta_0$). The total vertical angular momentum component cannot be changed by the horizontally directed gravitational torque. The top may nevertheless *begin* to fall over ($\dot{\vartheta}_0 = 0$ but $\ddot{\vartheta}_0 \neq 0$) because it can compensate for a loss in the projection of its spin angular momentum ($I_3 \dot{\psi} \cos \vartheta$) by starting a precession ($\dot{\varphi} \neq 0$) about the vertical (10.48). Indeed, an initial "falling" must occur, and hence precession must be generated.

The necessity for an initial nutational acceleration ($\ddot{\vartheta}_0 \neq 0$) can be most directly seen from the third equation of motion, besides $\dot{p}_\varphi = \dot{p}_\psi = 0$, not explicitly used in the preceding development:

$$\dot{p}_\vartheta = (d/dt)(\partial L/\partial \dot{\vartheta}) = \partial L/\partial \vartheta, \tag{10.54a}$$

which is equivalent to

$$I\ddot{\vartheta} = Mgl \sin \vartheta + (I \dot{\varphi} \cos \vartheta - I_3 \Omega_3)\dot{\varphi} \sin \vartheta. \tag{10.54b}$$

(The torques on the right side are also derivable as the "negative gradients," $-(d/d\vartheta)(\tau_\varphi' + V)$, of the "effective potential" in the diagram; of course, L_z and L_3 must be held constant in taking the gradient of τ_φ', as expressed with the help of (10.50).) Because the motion is begun without precession ($\dot{\varphi}_0 = 0$), there is only uncompensated gravitational torque acting initially here, and that necessarily starts a falling motion, $\ddot{\vartheta}_0 = (Mgl/I) \sin \vartheta_0$, except from a vertical position. As the inclination ϑ grows, the precession following from (10.50) and (10.53),

$$\dot{\varphi} = I_3 \dot{\psi}_0(\cos \vartheta_0 - \cos \vartheta)/I \sin^2 \vartheta, \tag{10.55}$$

is generated. A precession of the same sign as that of the spin $\dot{\psi}_0$ is to be expected, since the precession angular momentum is to compensate for loss of spin angular momentum.

The lowest level $z_1 = l \cos \vartheta_1$, is to be obtained as a root of $E' = \tau'_\varphi + V$, according to the above discussion of the energy diagram. This involves the vanishing of

$$\begin{aligned} l^2 \sin^2 \vartheta(\tfrac{1}{2}I\dot{\vartheta}^2) = \tfrac{1}{2}I\dot{z}^2 &= (l^2 - z^2)[E' - \tau'_\varphi(z) - V(z)] \\ &= Mg(z_0 - z)[(l^2 - z^2) - \lambda(z_0 - z)], \end{aligned} \tag{10.56}$$

where

$$\lambda \equiv (I_3/I)(\tfrac{1}{2}I_3 \dot{\psi}_0^2/Mgl)l \tag{10.57}$$

is a length characteristic of the spin $\dot{\psi}_0$, under the initial conditions (10.53). The nutation comes to a stop ($\dot{\vartheta} = 0$ and $\dot{z} = 0$) at the turning points z_1, z_2; hence these are roots of the cubic resulting from putting (10.56) equal to zero:

$$z_2 = z_0, \qquad z_1 = \tfrac{1}{2}\lambda - [(\tfrac{1}{2}\lambda - l)^2 + \lambda(l - z_0)]^{1/2}. \tag{10.58}$$

The cubic has a third root, differing from z_1 by the sign of the radical, but it falls outside the domain $|z| < l$ being used in the description. That the upper bound z_2 should just be the starting level is only to be expected. It may be also checked that the formula for z_1 properly yields a height falling within the range $-l < z_1 < z_0$, by forming the expressions

$$l + z_1 = (l + \tfrac{1}{2}\lambda) - [(l + \tfrac{1}{2}\lambda)^2 - \lambda(l + z_0)]^{1/2} > 0,$$

$$z_0 - z_1 = (z_0 - \tfrac{1}{2}\lambda) + [(z_0 - \tfrac{1}{2}\lambda)^2 + (l^2 - z_0^2)]^{1/2} > 0.$$

For very fast spins ($\lambda \gg 2l$), the distance of drop tends to vanish,

$$z_0 - z_1 \approx (l^2 - z_0^2)/\lambda \to 0 \qquad \text{for } \lambda \gg 2l, \tag{10.59}$$

and $z_1 \to -l$ for $\lambda \to 0$ ($\dot{\psi}_0 \to 0$).

The drop from $z_2 = z_0$ to z_1 will, according to the energy diagram of Figure 10.5, be followed by recovery to z_0, and then the process will repeat itself. At the upper level, the precession (10.55) always disappears ($z_2 = z_0 = \zeta$), and at the lower one it takes on the finite value

$$\dot{\varphi}(z_1) = 2Mgl/I_3 \dot{\psi}_0. \tag{10.60}$$

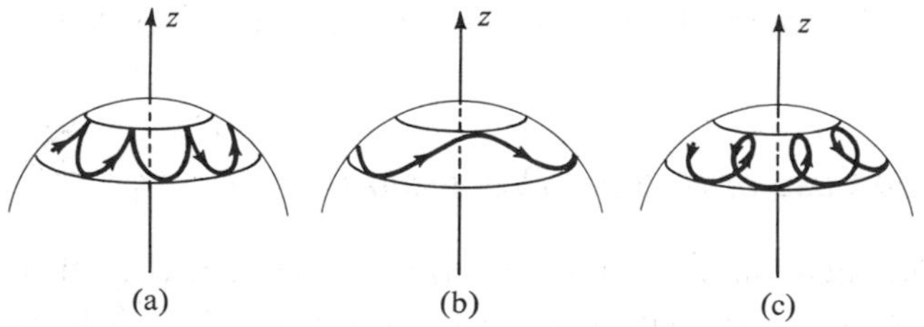

Figure 10.6

The mass-center will describe a curve on the sphere to which it is constrained, having an appearance as indicated in Figure 10.6(a). The cusps result from the fact that $\dot{\vartheta} = \dot{\varphi} = 0$ but $\ddot{\vartheta} \neq 0$ at $z = z_2 = z_0$. The curves in Figures 10.6(b) and (c) can be easily imagined to follow from starting conditions which make ζ, the level at which the precession $\dot{\varphi}$ vanishes, fall elsewhere than at $\zeta = z_2$. In (b), $\zeta > z_2$ (precession does not cease anywhere between z_1 and z_2 in (b)), and $z_1 < \zeta < z_2$ in (c).

The frequency of the cusped motion may be obtained from the nutational energy equation (10.56). The most easily interpretable results emerge in the "fast spin" case, when the maximum drop is $z_0 - z_1 \approx (l^2 - z_0^2)/\lambda \equiv \delta_m$. Then, in terms of intermediate distances $\delta = z_0 - z$, the equation may be approximated by

$$\dot{z}^2 \equiv \dot{\delta}^2 \approx (2Mg/I)\lambda\delta[\delta_m - \delta], \tag{10.61a}$$

with $\dot{\delta} = 0$ for both $\delta = 0$ and $\delta = \delta_m$. This is easily integrated to give

$$\delta = \tfrac{1}{2}\delta_m[1 - \cos(2Mg\lambda/I)^{1/2}t]. \tag{10.61b}$$

The period between the beginning of a drop and the subsequent recovery is then

$$T = 2\pi(I/2Mg\lambda)^{1/2} = (I/I_3)\,|2\pi/\dot{\psi}_0|, \tag{10.62}$$

longer or shorter than the spin period, $|2\pi/\dot{\psi}_0|$, according as $I \gtrless I_3$. As the spin rate is increased, the nutational frequency increases in proportion to it, but the amplitude δ_m becomes so small as to make the motion seem a regular, nutationless precession. When the corresponding precession velocity (10.55) is similarly approximated,

$$\dot{\varphi} \approx (I_3/I)\dot{\psi}_0\,\delta/l\sin^2\vartheta_0 = (2Mgl/I_3\,\dot{\psi}_0)\delta/\delta_m, \tag{10.63}$$

the result has the correct maximum (10.60) and vanishes at z_0 properly, as well. It oscillates in value with the same frequency as does the nutation, but it is the mean value,

$$\langle\dot{\varphi}\rangle \approx Mgl/I_3\,\dot{\psi}_0 \qquad (=\tfrac{1}{2}\dot{\varphi}(z_1)), \tag{10.64}$$

that is usually observed; the greater the spin rate, the slower the mean precession.

The Case of Regular Precession, $\dot{\vartheta} = 0$

The equation of motion (10.54) for ϑ makes it clear that an exactly regular precession, without nutation from $\vartheta = \vartheta_0 > 0$, can be started only when just enough precession, $\dot{\varphi}_0$, is imparted initially to cancel out the initial torque, $Mgl \sin \vartheta_0$. The precession will remain constant at

$$\dot{\varphi} = (L_z - L_3 \cos \vartheta_0)/I \sin^2 \vartheta_0 = \dot{\varphi}_0, \tag{10.65}$$

according to (10.50), and then, because $\Omega_3 = L_3/I_3$ is conserved, the spin will also remain constant at its initial value $\dot{\psi}_0 = \Omega_3 - \dot{\varphi}_0 \cos \vartheta_0$.

The balancing out of the torques in (10.54) determines the precession *needed* for a given spin $\dot{\psi}_0$ and height z_0:

$$Mgl = \dot{\varphi}_0[I_3 \dot{\psi}_0 - (I - I_3)\dot{\varphi}_0 z_0/l]. \tag{10.66a}$$

The same condition is deducible from the requirement that the energy E' just equal the minimum of $\tau'_\varphi + V$ in the energy diagram, this being needed for the nutation range to be reduced to zero ($z_1 = z_2 = z_0$). The equation for the required precession is quadratic, and therefore either of *two* values of regular precession may be started for a given height and spin:

$$\dot{\varphi}_0 = \frac{l}{2z_0} \cdot \frac{I_3 \dot{\psi}_0}{I - I_3} \{1 \pm [1 - 4Mgz_0(I - I_3)/I_3^2 \dot{\psi}_0^2]^{1/2}\}. \tag{10.66b}$$

The radical yields a lower limit for the spin which will allow a real regular precession to be started at all:

$$\tfrac{1}{2} I_3 \dot{\psi}_0^2 \geq 2Mgz_0[(I/I_3) - 1]. \tag{10.66c}$$

Any spin will do for a "squat" top with $I_3 > I$ (for example, a heavy ring of radius a and $l < 2^{-1/2}a$). Both the "fast" and "slow" regular precessions, as given by the upper and lower signs in (10.66*b*), have the same sign as the spin if $I_3 < I$. However, the fast precession, unlike the slow one, must be opposite to the spin in sign for $I_3 > I$. For a high enough spin, when the criterion (10.66*c*) is much exceeded, the slow precession becomes independent of the height: $\dot{\varphi}_0 \approx Mgl/I_3 \dot{\psi}_0$, just equal to the average (10.64).

The "Sleeping" Top

To be considered finally is the special case of the pure spin start, $\dot{\vartheta}_0 = 0$, from the erect position, $\vartheta_0 = 0$. In this position, spin and precession are indistinguishable, and this manifests itself in the expression (10.47), for the axial component L_3 of the angular momentum, through its reduction to $L_3 = I_3(\dot{\psi}_0 + \dot{\varphi}_0)$ when $\vartheta_0 = 0$. Only some value for the total $\Omega_0 \equiv \dot{\psi}_0 + \dot{\varphi}_0$ need be given initially, and then the conservation conditions (10.47) to (10.49) become

$$\begin{aligned} L_3 &= I_3\,\Omega_0 = I_3(\dot{\psi} + \dot{\varphi}\cos\vartheta), \\ L_z &= I_3\,\Omega_0 = I\dot{\varphi}\sin^2\vartheta + I_3\,\Omega_0\cos\vartheta, \\ E' &\equiv E - \tfrac{1}{2}I_3\,\Omega_0^2 = Mgl \\ &= \tfrac{1}{2}I(\dot{\vartheta}^2 + \dot{\varphi}^2\sin^2\vartheta) + Mgl\cos\vartheta, \end{aligned} \tag{10.67}$$

for a start with $\vartheta_0 = 0$, $\dot{\vartheta}_0 = 0$. These equations give all the possible motions $\vartheta(t)$, $\varphi(t)$, $\psi(t)$, which have $L_3 = L_z = I_3\,\Omega_0$ and $E' = Mgl$. $L_z = I_3\,\Omega_0$ is maintained even if a precession

$$\dot{\varphi} = \frac{I_3\,\Omega_0(1 - \cos\vartheta)}{I\sin^2\vartheta} = \frac{I_3\,\Omega_0\,l(l - z)}{I(l^2 - z^2)} \tag{10.68}$$

should develop, and then the conservation of $L_3 = I_3\,\Omega_0 = L_z$ will decrease the spin to $\dot{\psi} = \Omega_0 - \dot{\varphi}\cos\vartheta$. The energy conservation at $E' = Mgl$ allows nutation $\dot{\vartheta}$ only as given by

$$\tfrac{1}{2}I\dot{z}^2 = \tfrac{1}{2}I\dot{\vartheta}^2 l^2\sin^2\vartheta = Mg(l - z)^2(l + z - \lambda), \tag{10.69}$$

where

$$\lambda \equiv I_3^2\,\Omega_0^2/2MgI. \tag{10.70}$$

This parameter is actually identical to (10.57), since there is no need to discriminate between $\Omega_0 = \dot{\psi}_0 + \dot{\varphi}_0$ and $\Omega_0 = \dot{\psi}_0$ in a start from the erect position. The expression (10.69) could have been obtained directly from the condition (10.56), which is also valid for $\vartheta_0 = 0$, by specializing it to $z_0 = l\cos\vartheta_0 = l$. The considerations here were started afresh because it would seem that no nutation at all should really be expected, since there are obviously no torques operating on the erect position—neither a gravitational one, $Mgl\sin\vartheta$, nor such "inertial" ones as could generate $\ddot{\vartheta}_0$ according to (10.54*b*).

The possibilities offered by (10.69) may be explored by plotting $-\tfrac{1}{2}I\dot{z}^2 \equiv (l^2 - z^2)[\tau'_\varphi + V - E']$, as in Figure 10.7. The plot differs from an energy diagram like Figure 10.5 in that the energies are multiplied by $(l^2 - z^2) \equiv l^2\sin^2\vartheta$, thereby eliminating the singularities at $z = \pm l$ which arise from the inertial forces, and the zero of the total energy scale is shifted by Mgl.

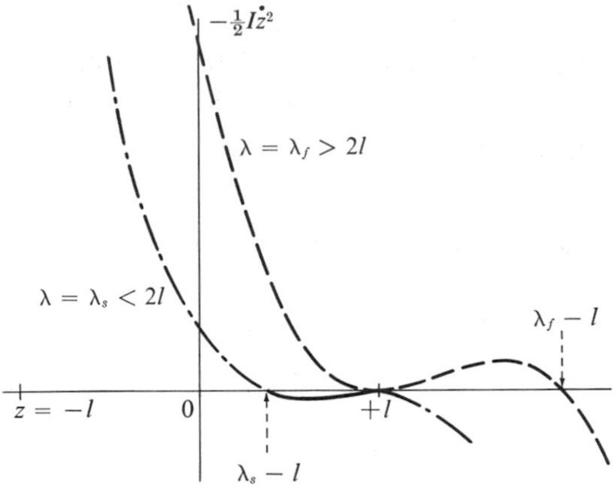

Figure 10.7

Two cases are plotted. One applies to a start with a "fast spin" $\Omega_0 > \Omega_c$, greater than a critical value with the spin energy

$$\tfrac{1}{2}I_3\Omega_c^2 = 2(I/I_3)Mgl, \tag{10.71}$$

corresponding to $\lambda = 2l$, and the other has $\lambda < 2l$. All cases yield a double root at $z = +l$ for the nutation, as (10.69) makes evident. The third root $z = \lambda - l$ lies within the domain of the description, $-l \le z \le l$, only in the slow spin cases with $\lambda < 2l$. Motion is possible only with $\frac{1}{2}I\dot{z}^2 \ge 0$, and hence the diagram shows that the top will "sleep" standing up, with $z = l$, while it has a "fast" spin, greater than the critical value Ω_c. On the other hand, if it has been given a spin energy less than the critical value, a nutation over the range $\lambda - l < z < l$ is open to it, in consistency with the conservation principles.

The reconciliation with the fact that there are no torques to start the nutation of the slow spin case is provided by the circumstance that $\ddot{\vartheta} = 0$ *at* $z = +l$. This means that the top will stay erect as long as the conservation principles are perfectly maintained; its nutation period is infinite! However, a perturbation with any $\ddot{\vartheta}_0 \ne 0$, no matter how small, will immediately start a slow "wobbling" down to an extreme $z = \lambda - l$. The erect spinning is *unstable* for spins less than the critical value for a given top. The instability is suggested by the curve in Figure 10.7 in a way comparable to that of the linear motion case in Figure 3.1. Since $-\frac{1}{2}I\dot{z}^2$ is proportional to the "effective potential," $V(z) + \tau_\varphi'(z)$, both diagrams show centers of instability as occurring at maxima in "potential energy" curves. Whereas $z = +l$ is the unstable point of a maximum in slow spin cases, it is the location of a stable minimum, at which the top may "sleep" in peace, in the fast spin cases.

10.5 EIGENVECTORS OF ROTATION OPERATORS

The language of "eigenvectors" introduced in this chapter permits expressing further important properties of the rotation operators introduced in the preceding chapter. To be considered now is the eigenvector problem,

$$\mathbf{A}\mathbf{u}_\mu = \lambda_\mu \mathbf{u}_\mu, \tag{10.72}$$

with $\mathbf{A} = \tilde{\mathbf{A}}^{-1}$ an *orthonormal* matrix, like the rotation operator in (9.99), rather than a symmetric one like the inertia operators $\mathbf{I} = \tilde{\mathbf{I}}$.

As in the general case (10.30) and (10.31), the problem is equivalent to finding a transformed basis, $\mathbf{u}_\mu = \mathbf{e}_l S_{l\mu}^{-1}$, on which the operator matrix becomes diagonal:

$$\mathbf{S}\mathbf{A}\mathbf{S}^{-1} = \boldsymbol{\lambda} \equiv (\delta_{\mu\nu}\lambda_\nu), \tag{10.73}$$

as in (10.24) and (10.27). Because $\mathbf{A}$ is not symmetric, $\mathbf{S}$ will no longer be a simple, real orthogonal transformation like the frame rotations $\mathbf{a}$ of (10.33) or (10.36*d*).

It is immediately clear that a rotation operator can be expected to possess only a single *real* eigenvector, the rotation axis $\hat{\boldsymbol{\Omega}}$ itself. Any other vector $\mathbf{R}$, not lying along the axis, is sure to have its direction changed ($\hat{\boldsymbol{\Omega}} \times \mathbf{R} \neq 0$ in (9.124)). Moreover, a unit eigenvalue must be expected for

$$\mathbf{A}\hat{\boldsymbol{\Omega}} = \hat{\boldsymbol{\Omega}}, \tag{10.74}$$

since it follows from $\tilde{\mathbf{A}}\mathbf{A} = 1$ and $\hat{\boldsymbol{\Omega}} \cdot \hat{\boldsymbol{\Omega}} = 1$ that $(\widetilde{\mathbf{A}\hat{\boldsymbol{\Omega}}})(\mathbf{A}\hat{\boldsymbol{\Omega}}) = 1$ (for a proper rotation, with $|\mathbf{A}| = +1$).

The special eigenvector problem (10.74) acquires some substance when the rotation axis is initially unknown, as for the product of the two rotations in the example (9.101). It is easy to find that

$$\begin{pmatrix} 0 & 1 & 0 \\ 0 & 0 & -1 \\ -1 & 0 & 0 \end{pmatrix} \begin{pmatrix} \hat{\Omega}_1 \\ \hat{\Omega}_2 \\ \hat{\Omega}_3 \end{pmatrix} = \hat{\boldsymbol{\Omega}} = 3^{-1/2} \begin{pmatrix} 1 \\ 1 \\ -1 \end{pmatrix}$$

follows from the three component equations $\hat{\Omega}_2 = \hat{\Omega}_1$, $-\hat{\Omega}_3 = \hat{\Omega}_2$, $-\hat{\Omega}_1 = \hat{\Omega}_3$ and the normalization $\hat{\boldsymbol{\Omega}} \cdot \hat{\boldsymbol{\Omega}} = 1$. The former representation of the same result was $\hat{\boldsymbol{\Omega}} = 3^{-1/2}(\mathbf{i} + \mathbf{j} - \mathbf{k})$.

Important new concepts emerge when other solutions of the eigenvector problem (10.72) are sought than just $\mathbf{u}_0 = \hat{\boldsymbol{\Omega}}$. It will henceforth be presumed that $\hat{\boldsymbol{\Omega}}$ has already been given in specifying the rotation operator, $\mathbf{A}(\hat{\boldsymbol{\Omega}}, \alpha)$.

The Complex Eigenvectors

The rotation eigenvector problem admits more solutions when the space of the vectors (9.19*a*) is generalized to include ones which may have complex numbers as components. This will be investigated specifically for a three-dimensional rotation $\mathbf{A}(\hat{\mathbf{\Omega}}, \alpha)$ about a given axis $\hat{\mathbf{\Omega}}$, through a given real angle α. There is no real loss of generality if the initial basis adopted, $\mathbf{e}_1, \mathbf{e}_2, \mathbf{e}_3$, is one having the rotation axis as its *z*-direction, $\mathbf{e}_3 \equiv \hat{\mathbf{\Omega}}$. When $\hat{\mathbf{\Omega}}$ has been given, a preliminary transformation to that basis may always be carried out. The interest is in possible eigenvectors which are normal to, and so independent of, $\hat{\mathbf{\Omega}}$, in the "subspace" of $\mathbf{e}_1$ and $\mathbf{e}_2$.

When $\hat{\mathbf{\Omega}} \equiv \mathbf{e}_3$,

$$\mathbf{A}(\hat{\mathbf{\Omega}}, \alpha) = \mathbf{a}(-\alpha) = \begin{pmatrix} \cos\alpha & -\sin\alpha & 0 \\ \sin\alpha & \cos\alpha & 0 \\ 0 & 0 & 1 \end{pmatrix}, \tag{10.75}$$

as discussed on page 267. Its eigenvalues λ are to be obtained as roots of the secular determinant (10.28),

$$\begin{vmatrix} \cos\alpha - \lambda_\mu & -\sin\alpha & 0 \\ \sin\alpha & \cos\alpha - \lambda_\mu & 0 \\ 0 & 0 & 1 - \lambda_\mu \end{vmatrix} = 0, \tag{10.76a}$$

which is equivalent to the cubic

$$(1 - \lambda_\mu)[1 - 2\lambda_\mu \cos\alpha + \lambda_\mu^2] = 0.$$

The eigenvalues are therefore

$$\lambda_\mu = +1 \qquad \text{and} \qquad \cos\alpha \pm i \sin\alpha = e^{\pm i\alpha}. \tag{10.76b}$$

Of these, $\lambda_0 = +1$ plainly belongs to the eigenvector $\mathbf{u}_0 = \mathbf{e}_3 \equiv \hat{\mathbf{\Omega}}$ already discussed in (10.74); formal confirmation follows from the secular equations (10.34) with $\lambda_0 = +1$.

The secular equations with $\lambda_\mu = \exp\colon \pm i\alpha$ yield the relations

$$\sin\alpha(\mp i u_{\mu 1} - u_{\mu 2}) = 0,$$

$$\sin\alpha(u_{\mu 1} \mp i u_{\mu 2}) = 0,$$

$$(1 - e^{\pm i\alpha}) u_{\mu 3} = 0.$$

The corresponding eigenvectors are both perpendicular to $\mathbf{e}_3 \equiv \hat{\mathbf{\Omega}}$, since their components on it vanish ($u_{\mu 3} = 0$); they are respectively proportional to

$$\mathbf{u}_\mu \sim (\mathbf{e}_1 \mp i\mathbf{e}_2).$$

Both have "null lengths" when the square of the length is defined in the usual way,

$$\mathbf{u}_\mu \cdot \mathbf{u}_\mu \sim (\mathbf{e}_1 \mp i\mathbf{e}_2)^2 = \mp 2i\mathbf{e}_1 \cdot \mathbf{e}_2 = 0.$$

Thus, normalization in the usual sense is impossible.

At this point certain conventional "normalizations" of the complex eigenvectors, and a system of labeling them, will be introduced:

$$\begin{aligned} \mathbf{u}_{+1} &= -2^{-1/2}(\mathbf{e}_1 + i\mathbf{e}_2) && \text{for } \lambda_{+1} = e^{-i\alpha}, \\ \mathbf{u}_{-1} &= +2^{-1/2}(\mathbf{e}_1 - i\mathbf{e}_2) && \text{for } \lambda_{-1} = e^{+i\alpha}, \\ \mathbf{u}_0 &= \mathbf{e}_3 && \text{for } \lambda_0 = +1. \end{aligned} \tag{10.77}$$

The appropriateness of this labeling will become evident below.

"Helicity Components" of Vectors

There are circumstances (for example, involving circularly polarized light) in which projections of a real, three-dimensional vector $\boldsymbol{\varepsilon}$ onto the rotation eigenvectors become useful. If the cartesian components of $\boldsymbol{\varepsilon}$ are $\varepsilon_x = \mathbf{e}_1 \cdot \boldsymbol{\varepsilon}$, $\varepsilon_y = \mathbf{e}_2 \cdot \boldsymbol{\varepsilon}$, $\varepsilon_z = \mathbf{e}_3 \cdot \boldsymbol{\varepsilon}$, then

$$\varepsilon_{\pm 1} \equiv \tilde{\mathbf{u}}_{\pm 1} \boldsymbol{\varepsilon} = \mp 2^{-1/2}(\varepsilon_x \pm i\varepsilon_y), \qquad \varepsilon_0 \equiv \tilde{\mathbf{u}}_0 \boldsymbol{\varepsilon} = \varepsilon_z. \tag{10.78}$$

These are often called the "spherical components" of the vector; the complex ones, $\varepsilon_{\pm 1}$, have lately come to be known as "helicity components."

The spherical components of an ordinary point-position vector **r**,

$$r_{\pm 1} = \mp 2^{-1/2}(x \pm iy) \qquad \text{and} \qquad r_0 = z, \tag{10.79}$$

are also known as "solid spherical harmonics of first order." Spherical harmonics of all orders, corresponding to projections of reduced tensors having any rank, will be encountered in §13.3.

Now it should be pointed out that

$$\mathbf{r} = \mathbf{e}_k x_k = r_\mu^* \mathbf{u}_\mu, \tag{10.80a}$$

rather than $r_\mu \mathbf{u}_\mu$, with the definitions $r_\mu \equiv \tilde{\mathbf{u}}_\mu \mathbf{r}$. This happens because $\tilde{\mathbf{u}}_{\pm 1}\mathbf{u}_{\pm 1} = 0$ rather than 1, as for $\mathbf{e}_k \mathbf{e}_k = 1$, and $\tilde{\mathbf{u}}_{\pm 1}^* \mathbf{u}_{\pm 1} = 1$ instead. The row vector $\tilde{\mathbf{u}}_\mu^*$, having as elements the complex conjugates of the column $\mathbf{u}_\mu$, is called the "hermitian conjugate" (h.c.) of $\mathbf{u}_\mu$ and is designated, with the help of a "dagger" symbol,

$$\mathbf{u}_\mu^\dagger \equiv \tilde{\mathbf{u}}_\mu^*. \tag{10.81}$$

Now, for the real vector **r**, the resolutions (10.80*a*) may be rewritten

$$\mathbf{r} = \mathbf{e}_k(\tilde{\mathbf{e}}_k \mathbf{r}) = \mathbf{u}_\mu(\mathbf{u}_\mu^\dagger \mathbf{r}). \tag{10.80b}$$

The projection operator

$$\left(\sum_\mu\right)\mathbf{u}_\mu \mathbf{u}_\mu^\dagger = \mathbf{1} \tag{10.82}$$

replaces $\mathbf{e}_k \tilde{\mathbf{e}}_k = \mathbf{1}$ of (9.109).

It can now been seen that the normalizations adopted in (10.77) may be represented in

$$\mathbf{u}_\mu^\dagger \mathbf{u}_\nu = \delta_{\mu\nu}. \tag{10.83}$$

This also shows the sense in which the complex eigenvectors form an orthonormal basis; the condition is a generalization of $\tilde{\mathbf{e}}_k \mathbf{e}_l = \delta_{kl}$. The important point is that projections on the different members of the basis $\mathbf{u}_0$, $\mathbf{u}_{\pm 1}$ are linearly independent, like cartesian components. No one component $\mathbf{u}_\mu^\dagger \mathbf{v}$ of an arbitrary vector **v** in the space can be represented as a linear combination of its other components, since $\mathbf{u}_\mu^\dagger \mathbf{v} = \mathbf{u}_\mu^\dagger(\mathbf{u}_\nu \mathbf{u}_\nu^\dagger)\mathbf{v} = \delta_{\mu\nu} \mathbf{u}_\nu^\dagger \mathbf{v}$.

Displaying the components $r_\mu^* = u_\mu^\dagger \mathbf{r}$ of the position vector $\mathbf{r} = \mathbf{u}_\mu r_\mu^*$ in a column on the rotation eigenvector basis,

$$\mathbf{r} = \begin{pmatrix} r_{+1}^* \\ r_0^* \\ r_{-1}^* \end{pmatrix} = 2^{-1/2}\begin{pmatrix} -(x_1 - ix_2) \\ 2^{1/2} x_3 \\ x_1 + ix_2 \end{pmatrix}, \tag{10.84}$$

makes it evident that the square of the length is represented by

$$\mathbf{r}^\dagger \mathbf{r} = \sum_\mu |r_\mu|^2 = \sum_k x_k^2. \tag{10.85}$$

Generalized to a vector with complex cartesian components, $\mathbf{v} = \mathbf{e}_k v_k$, this length definition gives

$$\mathbf{v}^\dagger \mathbf{v} = \tilde{\mathbf{e}}_k \mathbf{e}_l v_k^* v_l = \sum_k |v_k|^2. \tag{10.86}$$

The complex eigenvectors in (10.83) provide examples. Those orthonormality conditions also provide examples of more general "scalar products," like

$$\begin{aligned} \mathbf{v}^\dagger \mathbf{v}' &= \tilde{\mathbf{e}}_k \mathbf{e}_l v_k^* v_l' = \left(\sum_k\right) v_k^* v_k', \\ &\equiv \mathbf{u}_\mu^\dagger \mathbf{u}_\nu v_\mu (v_\nu')^* = \sum_\mu v_\mu (v_\mu')^*. \end{aligned} \tag{10.87}$$

It is such treatments of complex vectors that prove most useful; generalizations of them will be encountered in later chapters.

Unitary Transformations

Rotations in spaces containing complex vectors, in addition to the real ones, may be defined as transformations $\mathbf{U}\mathrm{v} = (\mathrm{v})'$ which preserve the absolute lengths of the vectors, i.e.,

$$\mathrm{v}^\dagger \mathrm{v} = (\mathrm{v})'^\dagger (\mathrm{v})' = \mathrm{v}^\dagger \mathbf{U}^\dagger \mathbf{U} \mathrm{v}. \tag{10.88}$$

Here the definition of hermitian conjugacy (10.81) has been extended to square matrices like $\mathbf{U}$:

$$\mathbf{U}^\dagger \equiv \widetilde{\mathbf{U}^*} \qquad \text{or} \qquad \mathbf{U}^\dagger_{kl} = \mathbf{U}^*_{lk}. \tag{10.88}$$

Then $(\mathbf{U}\mathrm{v})^\dagger = \mathrm{v}^\dagger \mathbf{U}^\dagger$ because hermitian conjugation differs from transposition (9.42) only by complex conjugation.

It is obvious that the invariance (10.88) requires

$$\mathbf{U}^\dagger \mathbf{U} = \mathbf{1} \qquad \text{or} \qquad \mathbf{U}^{-1} = \mathbf{U}^\dagger, \tag{10.90}$$

a generalization of the orthonormality $\tilde{\mathbf{a}}\mathbf{a} = \mathbf{1}$ occasioned by the possibility of complex matrix elements for $\mathbf{U}$. Matrices with the property (10.90) are called "unitary," and the transformations by them are sometimes called "unitary rotations." It is evident that they leave properly invariant all scalar products as defined in (10.87).

The similarity transformation $\mathbf{S}$, which was found to diagonalize the orthonormal rotation matrix in (10.73), can now be seen to have been a unitary one, preserving $\mathbf{r}^\dagger\mathbf{r} = r^2$ (10.85). While the ordinary, real rotations $\mathbf{a} = \tilde{\mathbf{a}}^{-1}$ are sufficient for diagonalizing symmetric operations only, the more general unitary rotations can also diagonalize the nonsymmetric, orthonormal rotation operators. The specific matrix found has an inverse $\mathbf{S}^{-1}$ with elements identifiable from (10.30), $S^{-1}_{l\mu} = \tilde{\mathbf{e}}_l \mathbf{u}_\mu$. Thus $\mathbf{S}^{-1}$ has columns each of which is one of the column vectors $\mathbf{u}_{\pm 1}$, $\mathbf{u}_0$ of (10.77):

$$\begin{aligned}\mathbf{S}^{-1} &= (\mathbf{u}_{+1}, \mathbf{u}_0, \mathbf{u}_{-1}) \\ &= 2^{-1/2}\begin{pmatrix} -1 & 0 & 1 \\ -i & 0 & -i \\ 0 & 2^{1/2} & 0 \end{pmatrix}.\end{aligned}$$

The hermitian conjugate of this is

$$2^{-1/2}\begin{pmatrix} -1 & i & 0 \\ 0 & 0 & 2^{1/2} \\ 1 & i & 0 \end{pmatrix} \equiv \mathbf{U}. \tag{10.91}$$

This is **S** itself since multiplying it with $\mathbf{S}^{-1}$ yields unity. **S** is therefore unitary and will henceforth, as indicated, be denoted **U** instead. The arrangement of the elements in the matrix has obviously depended on the fact that the ordering $\mathbf{e}_1, \mathbf{e}_2, \mathbf{e}_3$ was adopted for the initial basis, and the ordering $\mathbf{u}_{+1}, \mathbf{u}_0, \mathbf{u}_{-1}$ for the unitarily rotated basis. Other orderings would change the appearance of the matrix and of columns like (10.84), but not their essential interrelationships.

Now the rotation operator $(\mathbf{A})'$, which is to be applied to the column vector $(\mathbf{r})' \equiv \mathbf{r}$ of (10.84) in obtaining the vector $\mathbf{r}' \equiv (\mathbf{r}')'$ rotated from $\mathbf{r}$ by the angle α around the z-axis, is the diagonal one:

$$(\mathbf{A})' \equiv \mathbf{U}\mathbf{A}\mathbf{U}^\dagger = \begin{pmatrix} e^{-i\alpha} & 0 & 0 \\ 0 & 1 & 0 \\ 0 & 0 & e^{i\alpha} \end{pmatrix}, \tag{10.92}$$

with elements consisting of the eigenvalues found in (10.76). Notice that the trace has been properly preserved at the value $1 + 2\cos\alpha$.

Diagonalization of the Unit-Spin Matrix

The unitary rotation **U** of (10.91) which has diagonalized the orthonormal rotation matrix $\mathbf{A}(\mathbf{e}_3, \alpha)$ to (10.92) is independent of the size of the rotation angle, α. This means that exactly the same transformation diagonalizes the infinitesimal rotation operator (9.135),

$$\mathbf{A}(\mathbf{e}_3, \delta\alpha) = \mathbf{1} - i\delta\alpha \mathsf{S}_z, \tag{10.93}$$

and, since $\mathbf{1}$ is already diagonal, also the unit-spin matrix S_z of (9.136). Indeed, it is readily seen that

$$(\mathsf{S}_z)' \equiv \mathbf{U}\mathsf{S}_z\mathbf{U}^\dagger = \begin{pmatrix} 1 & 0 & 0 \\ 0 & 0 & 0 \\ 0 & 0 & -1 \end{pmatrix}. \tag{10.94}$$

The diagonal elements $1, 0, -1$ are naturally the eigenvalues of the operator S_z belonging to the eigenvectors $\mathbf{u}_{+1}, \mathbf{u}_0, \mathbf{u}_{-1}$. The result

$$\mathsf{S}_z \mathbf{u}_\mu = \mu \mathbf{u}_\mu \tag{10.95}$$

is a column matrix equation which should hold in any representation (resolved on any basis), and it can be checked, for instance, by applying the matrix displayed as in (9.136*a*), to the eigenvectors as decomposed on $\mathbf{e}_1, \mathbf{e}_2, \mathbf{e}_3$. It can now be seen that the labels chosen for the eigenvectors in (10.77) are just the eigenvalues of S_z.

It cannot be expected that the same transformation has simultaneously diagonalized the other components, $S_{x,y}$ (9.139) and (9.141), of the unit-spin vector. Any pair of diagonal matrices commutes, and rotations about different axes do not commute. Indeed, the nonvanishing commutators $\mathbf{S} \times \mathbf{S} = i\mathbf{S}$ (9.147) are expressed independently of frame, and hence must be expected to be preserved on any basis;

$$(\mathbf{S})' \times (\mathbf{S})' = \mathbf{USU}^{-1} \times \mathbf{USU}^{-1} = i\mathbf{USU}^{-1} = i(\mathbf{S})'$$

shows that they indeed are. The specific representations of $(S_x)'$ and $(S_y)'$, on the same basis as $(S_z)'$ of (10.94), are

$$(S_x)' = 2^{-1/2}\begin{pmatrix} 0 & 1 & 0 \\ 1 & 0 & 1 \\ 0 & 1 & 0 \end{pmatrix}, \qquad (S_y)' = 2^{-1/2}\begin{pmatrix} 0 & -i & 0 \\ i & 0 & -i \\ 0 & i & 0 \end{pmatrix}. \tag{10.96}$$

The vanishing traces of the original matrices $S_{x,y,z}$ are properly preserved in these transforms. Moreover, the "squared-length" (9.144*b*) of $\mathbf{S}$ as a vector is also properly preserved at the valuc $\mathbf{S} \cdot \mathbf{S} = 2$. The latter can also be written $\mathbf{S}^\dagger \cdot \mathbf{S} = 2$ because $\mathbf{S}^\dagger = \mathbf{S}$ for each cartesian component of $\mathbf{S}$.

Any matrix $\mathbf{H}$ with the property

$$\mathbf{H}^\dagger = \mathbf{H} \tag{10.97}$$

is called an *hermitian* matrix.* For spaces of complex vector components, the hermitian matrix is the appropriate generalization of the symmetry—like that ($\tilde{\mathbf{I}} = \mathbf{I}$) of the real inertia matrices—just as the unitarity $\mathbf{U}^\dagger = \mathbf{U}^{-1}$ proved to be the appropriate generalization of the real orthogonality $\tilde{\mathbf{a}} = \mathbf{a}^{-1}$. Hermitian matrices are considered appropriate generalizations because their eigenvalues are always real (Exercise 10.24), as for S_z in (10.94); hence they may be used to represent real measurable numbers, just as the eigenvalues of $\mathbf{I} = \tilde{\mathbf{I}}$ represent real moments of inertia. Deeper inquiry into the properties of hermitian operators will be reserved for the contexts in which they are put to use (§14.3).

The Two Dimensional Rotation Eigenvectors

Insight into what goes on in the diagonalization of rotation operators, as to (10.92), can be obtained from carrying out the same procedures for the rotations in a plane spanned by the two basis vectors, $\mathbf{e}_1$, $\mathbf{e}_2$. The two-

* It was to gain this property that $\mathbf{S}$ in (9.135) and $\boldsymbol{\sigma}$ in (9.45) were defined with a factor i separated off.

dimensional rotation operator, $\mathbf{A}(\alpha) = \mathbf{a}(-\alpha)$ of (9.25), is the same as the three-dimensional one (10.75) with the third column and third row eliminated, and so it can be expected to be diagonalized to (see 10.92))

$$(\mathbf{A})' = \mathbf{U}\mathbf{A}\mathbf{U}^{-1} = \begin{pmatrix} e^{-i\alpha} & 0 \\ 0 & e^{i\alpha} \end{pmatrix} \tag{10.98}$$

by the unitary transformation

$$\mathbf{U} = 2^{-1/2}\begin{pmatrix} -1 & i \\ 1 & i \end{pmatrix} = (\mathbf{U}^{-1})^{\dagger} \tag{10.99}$$

(see (10.91)) and this is readily verified by performing the operations in (10.98).

The same transformation gives the two-dimensional position vector the representation (see (10.84))

$$(\mathbf{r})' = \mathbf{U}\begin{pmatrix} x \\ y \end{pmatrix} = \begin{pmatrix} -2^{-1/2}(x - iy) \\ 2^{-1/2}(x + iy) \end{pmatrix} \tag{10.100}$$

on the basis of the eigenvectors (see (10.77))

$$\mathbf{u}_{\pm 1} = \mp 2^{-1/2}(\mathbf{e}_1 \pm i\mathbf{e}_2), \tag{10.101}$$

which follow from $\mathbf{U}^{-1}_{l\mu} \equiv u_{\mu l}$ (see (10.30)).

On the new basis, the rotation $\mathbf{r} \to \mathbf{r}' = \mathbf{A}\mathbf{r}$ becomes

$$\mathbf{r}' = 2^{-1/2}\begin{pmatrix} -x' + iy' \\ x' + iy' \end{pmatrix} = 2^{-1/2}\begin{pmatrix} e^{-i\alpha}(-x + iy) \\ e^{i\alpha}(\ \ x + iy) \end{pmatrix}. \tag{10.102}$$

The two components of this column vector equation yield just

$$x' + iy' = e^{i\alpha}(x + iy) \tag{10.103}$$

and its complex conjugate. Thus, on the new basis, the matrix representation is reduced to just the complex plane representation of a rotation, first encountered in §3.3 and considered also in connection with (9.127).

The infinitesimal rotation generator σ_2 of (9.129) is diagonalized to

$$\sigma_3 \equiv (\sigma_2)' = \mathbf{U}\sigma_2\mathbf{U}^{\dagger} = \begin{pmatrix} 1 & 0 \\ 0 & -1 \end{pmatrix}. \tag{10.104}$$

The two matrices σ_2, σ_3 and a third one defined by

$$\sigma_1 = \frac{1}{2i}[\sigma_2, \sigma_3]_- = \begin{pmatrix} 0 & 1 \\ 1 & 0 \end{pmatrix} \tag{10.105}$$

may be used to represent three operators on the same basis; as such they have become famous as the "Pauli spin matrices." They differ from each other by unitary rotations, as in (10.104), and obey commutator relations

$$\boldsymbol{\sigma} \times \boldsymbol{\sigma} = 2i\boldsymbol{\sigma}, \tag{10.106}$$

analogous to (9.147) for the unit-spin matrices, if $\sigma_1, \sigma_2, \sigma_3$ are treated as the cartesian components of a three-dimensional axial vector $\boldsymbol{\sigma}$. They can be put to use in classical problems, as in a so-called "Cayley-Klein parametrization" of gyroscopic motions, but their more fundamental uses appear in the quantum mechanics of spinning particles. (See Exercises 10.19, 10.22, and 10.23.)

EXERCISES

10.1. Homogeneous right-circular cones may be characterized by their heights and their base radii as parameters.

(a) Show that the mass-center position remains unchanged as the radius is varied.

(b) Show that the "radius of gyration" ($\sqrt{I/M}$) about the symmetry axis remains unchanged as the height is varied.

(c) Show that every body has properties corresponding to (a) and (b). [Think of independent scale factors for orthogonal dimensions.]

10.2. (a) Form the trace of the inertia tensor integral (10.7) and show how it implies that, among the three moments of inertia on a given basis, no one can exceed the sum of the other two in value.

(b) Express any one in terms of the other two for a *laminar* body occupying one of the planes of the basis.

10.3. Find the mass-center, and the principal moments of inertia about it, of a body formed out of a thin, uniform circular disk with radius a, by cutting out of it a circular section with radius $b < a$, *tangential* to the edge of the disk.

[About the axis in the plane of the disk normal to the line of centers:

$$I_2 = \tfrac{1}{4}M(a^2 + b^2) - M\left(\frac{1}{a} + \frac{1}{b}\right)^{-2}.$$

Consideration of why this does not vanish for $b = a$ will lead to the finding that the $b \to a$ limits of all the results here correspond to a linear mass distribution proportional to $(1 - \cos\varphi)$, on a ring of radius a.]

10.4. Investigation of the following very simple example will suffice to show that principal axes based on different rotation centers of the same body will generally *not* constitute parallel frames. Take the body to consist of three equal point-masses connected by relatively weightless rigid rods so that they form a right triangle with unequal legs.

(a) First, construct the inertia tensor on a basis which has the two legs as two of its axes.

(b) Do the same on axes parallel to those of (a) but centered at one end of the hypotenuse, instead.

(c) Show that principal axes at the two points *may* be parallel to each other when the legs are equal. Why?

10.5. Principal axes and nondegenerate moments of inertia, of a body rotating about its mass-center, are given. The center of rotation is then shifted by an arbitrary displacement **s**. Show that the principal axes at the new rotation center remain parallel to those at the mass-center only if **s** is a shift *along* one of the principal axes.

10.6. From the inertia tensor at the apex of a homogeneous right-circular cone with height h and base radius a, deduce the moment of inertia about a "generator" of the cone (an axis through its apex and along its surface). [The reader will acquire a proper respect for the value of knowing the transformation properties of inertia tensors if he checks his result by direct integration.]

10.7. Extend the investigation in the preceding exercise to finding the inertia tensor at a point on the edge of the base. One frame axis is to be a generator of the cone, and another is to be normal to the symmetry axis of the cone. Show that one product of inertia does not vanish, but has the magnitude

$$-\left(\frac{3}{20}\right) Mah(h^2 + 6a^2)/(h^2 + a^2).$$

[The reader may then want to diagonalize the tensor.]

10.8. A uniform rod is held vertically by a pivot point at its lower end. A very slight disturbance makes it swing down, around the pivot point, and it is released just as it reaches the downward vertical position. Describe the subsequent motion in quantitative detail [For example, show that the rod returns to a vertical position after falling distances $(n^2\pi^2/12)$ times its length, with $n = 1, 2, 3, \ldots$]

10.9. A uniform rod of length l can move smoothly with its ends constrained to a vertical circle of diameter $2a > l$. Find the length of an "equivalent" simple pendulum—one that has the same period as small oscillations of the rod from its stable equilibrium position. For what l is the equivalent pendulum length equal to a? [$l = \sqrt{3}\, a$.] Compare the behavior of a system consisting of three like rods forming a triangle "inscribed" in the circle.

10.10. Redo Exercise 7.7 with the two pendulums replaced by uniform rods. Show that now the two normal frequencies have the squares

$$3(1 \pm 2/\sqrt{7})g/l.$$

10.11. A particle of momentum $m\mathbf{v}_0$ impinges on a uniform rod, mass M and length l, at rest and subject to no other forces (this might be arranged by having all motions confined to a smooth horizontal plane). Suppose the collision is elastic and "smooth" (momentum transferred only normally to the rod) and that the particle may be incident at any distance $s < \frac{1}{2}l$ from the center, with any angle i to the normal.

(a) Show that the particle may impart three times as much rotational as translational (CM) energy, by impinging on a tip ($s = \frac{1}{2}l$). [Compare the system of Exercise 5.5 in this respect.]

(b) What velocity is generally imparted to the mass center?

$$[2mv_0 \cos i/\{M + m(1 + 12s^2/l^2)\}.]$$

(c) Find a point on the rod that receives no initial impulse, for a given s (a relative "center of percussion" is thus found).

10.12. The formula (10.48), for an angular momentum of an axially symmetric top, implies that its moment of inertia about an axis through the tip and making an angle ϑ with its symmetry axis must be

$$I \sin^2 \vartheta + I_3 \cos^2 \vartheta.$$

Verify this by rotating the basis of the inertia tensor from its principal axes.

10.13. A top consists of a heavy uniform circular disk and a relatively weightless pin as its symmetry axis. The point of the pin extends from the center by just half the disk radius. Show that the top can be made to precess uniformly at a rate independent of its orientation. Compare the *direction* of the precession, relative to the spin, with the precession induced by a rolling, without sliding, of the top's rim when it and the top's point are in contact with a level floor.

10.14. Suppose that the cone of Exercise 10.6 rolls, without sliding, on a plane surface, with an angular ("spin") velocity ω_0 about its symmetry axis.

(a) Express the kinetic energy in terms of the principal moments of inertia at the tip.

(b) Reexpress the kinetic energy in terms of the instantaneous angular velocity of the body about the line of contact with the plane, and show that the equation of this to the result (a) yields the same moment of inertia about the line of contact as was found in Exercise 10.6.

(c) In the result (a), transform the moments of inertia to the principal ones at the mass-center. Show how the part of the kinetic energy which becomes independent of the centroidal moments is to be interpreted.

(d) Use Euler's equations to show that the torque of reaction (about the tip) from a *level* surface is

$$(\tfrac{3}{4}Mgh + I_c\,\omega_0^2 \sin \alpha) \cos \alpha,$$

where $\alpha = \tan^{-1}(a/h)$ and I_c is the moment of inertia about the line of contact.

10.15. Analyze the rolling motion of the preceding exercise relative to the point of contact at the edge of the cone's base, as an instantaneous center of rotation.

(a) Find the variable angular momentum at the edge point. [The results of Exercise 10.7 may be found useful for this.]

(b) From (a), find the torque about the edge point which is needed to maintain the motion. [Refer to (1.46) or (10.41) for time-derivatives.]

(c) Find angular velocities at which the side of the cone will lose contact with the plane. [$\omega_0^2 > (Mgh/4I_c)(1 + 3\sin^2\alpha)/\sin\alpha\cos^2\alpha$ in terms used for Exercise 10.14(d). The findings in that exercise are themselves sufficient for obtaining the result here if the reader can persuade himself that the torque of reaction has the same magnitude at both edge and tip, and that just the part of this reaction needed to offset the inertial forces of rotation must not exceed the gravitational torque, if upset is to be avoided.]

10.16. Consider a thumbtack, consisting of a disk with radius a and a relatively weightless pin of length l, rolling without sliding, with the pin point and the rim in contact with an *inclined* plane. Show that if the rolling may start, at rest, from any position, then it is sure to continue without upset only if the angle of the plane with the horizontal does not exceed $\tan^{-1}(a/5l)$. [The procedure used in the preceding exercise may help.]

10.17. The rolling of a coin on a level plane may be investigated in the following steps (treating the coin as a sharp-edged uniform disk of radius a).

(a) Advantage may be taken of the degeneracy on principal axes at the center of the disk by using the frame of orthogonal unit-vectors $\mathbf{e}_{1,2,3}$ indicated in the diagram, which shows the disk edge-on. The axis $\mathbf{e}_2$ is to continue passing through the instantaneous point of contact, at $a\mathbf{e}_2$, so that the frame's angular velocity,

$$\boldsymbol{\Omega} = \mathbf{e}_1\dot{\vartheta} + \mathbf{e}_2(+\dot{\varphi}\cos\vartheta) + \mathbf{e}_3(-\dot{\varphi}\sin\vartheta),$$

does *not* include the "spin," ω. The body itself has the angular velocity $\boldsymbol{\Omega} + \omega\mathbf{e}_3$. Show that the center has the velocity

$$\bar{v} = \mathbf{e}_1 a(\omega - \dot{\varphi}\sin\vartheta) - \mathbf{e}_3 a\dot{\vartheta}$$

in these terms.

(b) The equation of motion for the angular momentum $\mathbf{L}_0$, about the disk center, will be $\dot{\mathbf{L}}_0 = a\mathbf{e}_2 \times \mathbf{C}$, where $\mathbf{C}$ is the force of reaction from the plane. Show that this unknown force may be eliminated by using the equation of motion for the mass-center velocity $\bar{\mathbf{v}}$, with the result

$$\frac{d}{dt}[\mathbf{L}_0 - a\mathbf{e}_2 \times M\bar{\mathbf{v}}] = \mathbf{e}_1 Mga\sin\vartheta.$$

Now show that the square bracket here is just the angular momentum about the point of contact,

$$\mathbf{L} = \mathbf{e}_1 I_1\dot{\vartheta} + \mathbf{e}_2 I_2(\dot{\varphi}\cos\vartheta) + \mathbf{e}_3 I_3(\omega - \dot{\varphi}\sin\vartheta),$$

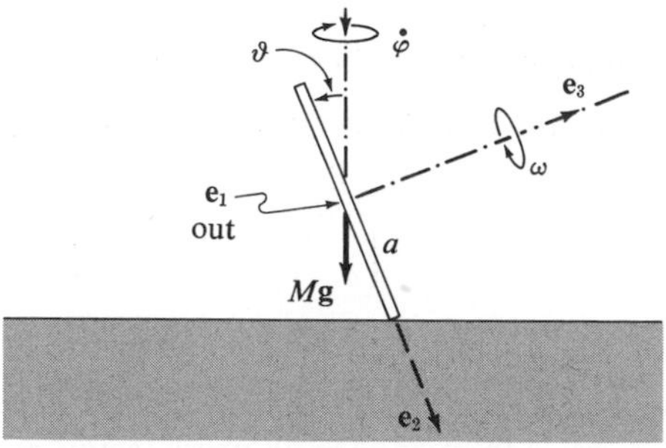

where $I_{1,2,3}$ are principal moments of inertia at the *edge* of the disk. (Thus, only the equation for $\dot{\mathbf{L}}$ would have been needed if the axes chosen had been centered at the point of contact.)

(c) Using $\dot{\mathbf{L}} = \partial\mathbf{L}/\partial t + \mathbf{\Omega} \times \mathbf{L}$ (1.46), three scalar equations of motion, for $\ddot{\vartheta}$, $\ddot{\varphi}$, and $\dot{\omega}$, can be found. Show that these are satisfied by a motion with $\vartheta = \vartheta_0$ constant, having a constant precession $\dot{\varphi}_0$ and a constant spin ω_0, if $\dot{\varphi}_0$ and ω_0 are related by

$$5a\dot{\varphi}_0^2 \sin\vartheta_0 \cos\vartheta_0 - 6a\omega_0\dot{\varphi}_0 \cos\vartheta_0 + 4g\sin\vartheta_0 = 0.$$

From this, it is easy to find that the coin will follow a circle of radius $r = (\bar{v}/\dot{\varphi}_0) + a\sin\vartheta_0$ which becomes as large as $r \to 3\bar{v}^2/2g\vartheta_0$ for $\vartheta_0 \to 0$, and approaches $r \approx a$ for $\vartheta_0 \to \frac{1}{2}\pi$.

(d) For $\vartheta_0 = 0$, the motion reduces to an erect rolling on a straight line. Its stability may be explored by writing the equations of motion for $\vartheta \ll 1$. Show that $\ddot{\vartheta}/\vartheta < 0$, and so any oscillations from $\vartheta = 0$ remain small, if only $\bar{v} > \frac{1}{3}\sqrt{ga}$. (The reader may also investigate the frequency with which the coin will "rattle" for $\vartheta_0 \to \pi/2$ and low speeds.)

10.18. Find helicity components, $v_{\pm 1}(t)$ and $r_{\pm 1}(t)$, of the expressions for the velocity, and for the position relative to an axis point of the orbit, in the motion of Exercise 7.12. Show that $r_{\pm 1} = \mp i v_{\pm 1}/\omega$ with an appropriate choice of axes.

10.19. For the so-called "Pauli spin-matrices" $\boldsymbol{\sigma}(\sigma_1, \sigma_2, \sigma_3)$ of (10.105), (9.129), and (10.104), show that

(a) $\boldsymbol{\sigma}^\dagger\boldsymbol{\sigma} = \boldsymbol{\sigma}\cdot\boldsymbol{\sigma} = 3$ [compare (9.144*b*)],

(b) $\sigma_i\sigma_j + \sigma_j\sigma_i = 2\delta_{ij}\mathbf{1}$ [the expression on the left is sometimes called an "anticommutator" and denoted $[\sigma_i, \sigma_j]_+$ or $\{\sigma_i, \sigma_j\}$],

(c) $\sigma_i\sigma_j = i\sigma_k$ if $i, j, k = 1, 2, 3$ or $2, 3, 1$ or $3, 1, 2$.

(d) Moreover, show that an arbitrary 2×2 matrix, $\begin{pmatrix} \xi & \eta \\ \zeta & \delta \end{pmatrix}$, may be represented on the basis of the Pauli spinors as

$$\rho_0\mathbf{1} + \boldsymbol{\rho}\cdot\boldsymbol{\sigma},$$

with appropriate definitions of ρ_0, $\rho_{1,2,3}$ in terms of ξ, η, ζ, δ.

(e) Finally, show that $(\boldsymbol{\sigma} \cdot \mathbf{a})(\boldsymbol{\sigma} \cdot \mathbf{b}) = \mathbf{a} \cdot \mathbf{b} + i\boldsymbol{\sigma} \cdot (\mathbf{a} \times \mathbf{b})$, when $a_{1,2,3}$, $b_{1,2,3}$ are numbers treated like vector components.

10.20. (a) Show that the general 2×2 *hermitian* matrix (10.97) may be given the form

$$\begin{pmatrix} a & c_1 + ic_2 \\ c_1 - ic_2 & b \end{pmatrix},$$

where a, b, $c_{1,2}$ are any four real quantities.

(b) Show that a general 2×2 *unitary* matrix depends on just three real parameters, aside from an over-all phase factor, and may be written

$$e^{+i\delta} \begin{pmatrix} e^{i\alpha} \cos \beta & e^{i\gamma} \sin \beta \\ -e^{-i\gamma} \sin \beta & e^{-i\alpha} \cos \beta \end{pmatrix}.$$

For $\delta = 0$ or $\frac{1}{2}\pi$, it becomes *unimodular* (determinant $= \pm 1$).

(c) Write the general 2×2 matrix which is *both* hermitian and unitary. [Note that the Pauli matrices of the preceding exercise all fit into this category. What is the trace and the determinant of each?]

10.21. The fact that 2×2 unitary matrices have three nontrivial parameters (Exercise 10.20(b)) makes it possible to represent *three*-dimensional rotations by *two*-dimensional matrices, as in the following.

(a) Show that the relations between x, y, z and x', y', z' in

$$\begin{pmatrix} z' & x' - iy' \\ x' + iy' & -z' \end{pmatrix} = \mathbf{U}_\vartheta \begin{pmatrix} z & x - iy \\ x + iy & -z \end{pmatrix} \mathbf{U}_\vartheta^{-1},$$

where

$$\mathbf{U}_\vartheta = \begin{pmatrix} \cos \vartheta/2 & i \sin \vartheta/2 \\ i \sin \vartheta/2 & \cos \vartheta/2 \end{pmatrix},$$

are just those characteristic of a frame rotation by ϑ around the x-axis. Note that the position vector is here represented by a *square* 2×2 hermitian matrix, $= \boldsymbol{\sigma} \cdot \mathbf{r}$ on a Pauli basis (Exercise 10.19(d)), instead of a column as in §9.3 or (10.84).

(b) Find a matrix U_φ which generates a frame rotation by φ around the z-axis in a similar way.

(c) Show that a general rotation, by Euler angles φ, ϑ, ψ, may be similarly represented, as generated by

$$\mathbf{U}_\psi \mathbf{U}_\vartheta \mathbf{U}_\varphi = \begin{pmatrix} e^{i(\varphi+\psi)/2} \cos \dfrac{\vartheta}{2} & ie^{-i(\varphi-\psi)/2} \sin \dfrac{\vartheta}{2} \\ ie^{i(\varphi-\psi)/2} \sin \dfrac{\vartheta}{2} & e^{-i(\varphi+\psi)/2} \cos \dfrac{\vartheta}{2} \end{pmatrix}.$$

[Thus found is a so-called "$SU(2)$ representation" of the "3-dimensional rotation group" (of transformations), as against an "$O(3)$ representation" provided by (9.49).]

10.22. (a) Express each of the matrices $\mathbf{U}_\vartheta$, $\mathbf{U}_\varphi$, $\mathbf{U}_\psi$ of the preceding exercise on a Pauli basis (Exercise 10.19(d)).

(b) Show that

$$\mathbf{U}_\psi \mathbf{U}_\vartheta \mathbf{U}_\varphi = e^{i\sigma_3\psi/2} e^{i\sigma_1\vartheta/2} e^{i\sigma_3\varphi/2}.$$

[Compare (9.142).]

10.23. Consider the abstract two-dimensional space of *column* vectors $\boldsymbol{\Psi} \equiv \begin{pmatrix} \psi_1 \\ \psi_2 \end{pmatrix}$ subject to operations by the 2×2 matrices like $\mathbf{U}_\vartheta$ of Exercise 10.21, or σ_i of Exercise 10.19.

(a) Show that the transform by $\mathbf{U} \equiv \mathbf{U}_\psi \mathbf{U}_\vartheta \mathbf{U}_\varphi$ of Exercise 10.21(c), $\mathbf{U}\boldsymbol{\Psi} \equiv (\boldsymbol{\Psi})'$, yields the column of components

$$\psi_1' = e^{i\psi/2}[\psi_1 e^{i\varphi/2} \cos \vartheta/2 + i\psi_2 e^{-i\varphi/2} \sin \vartheta/2],$$

$$\psi_2' = e^{-i\psi/2}[\psi_2 e^{-i\varphi/2} \cos \vartheta/2 + i\psi_1 e^{i\varphi/2} \sin \vartheta/2].$$

[Two-component quantities that are changed in this way, when evaluated in relatively rotated frames in ordinary space, are called "half-vectors" or "spinors." They have proved useful for describing many types of particles, said to be "spinning."]

(b) Find $\boldsymbol{\Psi}$'s which are "eigenspinors" of any one of the Pauli σ's; that is, $\sigma_i \boldsymbol{\Psi} = \lambda \boldsymbol{\Psi}$ with $\lambda = +1$ or -1.

10.24. Generalize the finding in (10.32) to the case of an *hermitian* operator (10.97), showing that its eigenvectors (for different eigenvalues) have the orthogonality (10.83). Notice that a specialization of the same equation to $\mu \equiv \nu$ tells you that all eigenvalues of an hermitian operator must be real.

CHAPTER 11

COUPLED OSCILLATIONS

To be considered now are descriptions that have interest in themselves and also prepare for such treatments of elastically deformable media as will be taken up in the succeeding chapters.

A body that is not quite rigid may be supposed to consist of particles that have definite relative positions, as do the particles of an ideally rigid body, only while in a state of motionless internal equilibrium. This equilibrium must be supposed vulnerable to disturbances in which the particles are displaced from their equilibrium positions. Whenever displacements do occur, restoring forces can be expected to arise, tending to return each particle to its equilibrium point. It is just the equilibrium among such forces which can be presumed responsible for determining the body's static configuration in the first place. As the force working on a displaced particle sends it back toward its equilibrium position, the particle gains kinetic energy. With the kind of energy conservation characteristic of an elastic medium (by definition), the equilibrium position will be overshot by the returning particle, and oscillations about the equilibrium configuration will be the result.

The restoring force on a particle might be taken to be proportional to its displacement, $\sim(\mathbf{r}_i - \mathbf{r}_i^o)$, on such a basis for Hooke's law as was discussed in connection with (3.7). However, with a force depending only on the particle's own displacement, the body would behave like an ensemble of independent oscillators. The restoring force on each particle should rather be expected to arise just from the presence of neighboring particles, and hence to depend quite essentially on their *relative* displacements. The body would be better represented as a system of *coupled* oscillators, a generalized version of the two-particle system treated in §3.7. Then the motions described for the various parts of the body will have some sort of coherence with each other, in better correspondence to real elastically deformable bodies.

11.1 THE EIGENFREQUENCY PROBLEM

To treat small departures from an equilibrium configuration, any coordinates $q_1, q_2, \ldots$ that are defined to vanish at equilibrium may be used —for example, $q_1 = x_1 - x_1^o$, $q_2 = y_1 - y_1^o$, $q_3 = z_1 - z_1^o$, $q_4 = x_2 - x_2^o, \ldots,$ where $\mathbf{r}_1^o, \mathbf{r}_2^o, \ldots$ are equilibrium positions. More generalized coordinates than these may be used instead; some of the q's might be angular displacements in the torsional twisting of wires, or even transient charges in electrical circuits.

If the various degrees of freedom interact only with each other, so that their energy as a system is conserved, the forces may be held derivable from some potential $V(q_1, q_2, \ldots) \equiv V(q)$. This is not the most general way to represent energy conservation, but it does encompass a large class of situations. Then for small enough displacements q_s, the potential may be approximated as in the "Taylor Series" development (3.15):

$$V(q) = V(0) + \sum_s q_s(\partial V/\partial q_s)_0 + \tfrac{1}{2}\sum_{r,s} q_r q_s(\partial^2 V/\partial q_r\, \partial q_s)_0 + \cdots. \tag{11.1}$$

A potential gauge with $V(0) = 0$ at equilibrium may always be adopted, and every generalized force component, $K_s = -\partial V/\partial q_s$ (7.22), must vanish in the equilibrium configuration. Thus the first effects of departures from equilibrium should arise from

$$V(q) \approx \tfrac{1}{2}\sum_{r,s} \kappa_{rs} q_r q_s, \qquad \text{with} \qquad \kappa_{rs} \equiv (\partial^2 V/\partial q_r\, \partial q_s)_0 = \kappa_{sr}. \tag{11.2}$$

When the displacements remain small enough for this to be a valid approximation, it is said that the "elastic limit" is not being passed.

The corresponding generalized force on the rth degree of freedom will be

$$K_r = -\partial V/\partial q_r \approx -\sum_s \kappa_{rs} q_s. \tag{11.3}$$

A coefficient κ_{rs} with $r \neq s$ measures the strength of a coupling between the degrees of freedom q_r and q_s, and $-\kappa_{rr}q_r$ is such a force as draws a particle back toward its equilibrium position even while all its fellows are undisplaced.

If q_r is simply a cartesian coordinate of a mass particle m_r, then the equation of motion for it is $m_r \ddot{q}_r = K_r$ and its kinetic energy is simply $\frac{1}{2}m_r\dot{q}_r^2$. However, for more generalized coordinates, the total kinetic energy may take the form (7.13)

$$\tau = \tfrac{1}{2}\sum_{r,s} \mu_{rs}\dot{q}_r\dot{q}_s, \tag{11.4}$$

perhaps containing "dynamic coupling" terms $r \neq s$. This was shown to be the most general form when explicit time dependence is avoided, as befits an energy conserving formulation. The mass coefficients $\mu_{rs} = \mu_{sr}$ may most generally be functions of the coordinates, but, in the approximation already adopted in (11.2), they should be treated as constants equal to their values at equilibrium. This may entail neglecting any "inertial effects," $\sim\partial\tau/\partial q_s$ (7.29), which might be present, but such secondary effects of motion should be negligible in small oscillations. With the mass coefficients treated as constants, the generalized momenta are simply $p_r = \sum_s \mu_{rs}\dot{q}_s$ as in (7.15), and the equations of motion (7.28) become

$$\dot{p}_r = \sum_s \mu_{rs}\ddot{q}_s = -\sum_s \kappa_{rs}q_s. \tag{11.5}$$

These are just Lagrange equations (7.33) with

$$L = \tfrac{1}{2}\sum_{r,s} [\mu_{rs}\dot{q}_r\dot{q}_s - \kappa_{rs}q_r q_s], \tag{11.6}$$

or canonical ones (8.5) with an hamiltonian equal to the conserved energy.

By defining a column vector $\mathbf{q}$, of components $q_1, q_2, q_3, \ldots$ on configuration-space axes, the equations of motion may be collected together as a matrix equation:

$$\boldsymbol{\mu}\ddot{\mathbf{q}} = -\boldsymbol{\kappa}\mathbf{q}. \tag{11.7}$$

Here $\boldsymbol{\mu}$ is a square matrix of elements $\mu_{rs} = \mu_{sr}$ and $\boldsymbol{\kappa}$ a square matrix of elements $\kappa_{rs} = \kappa_{sr}$, both matrices being symmetric ones: $\tilde{\boldsymbol{\mu}} = \boldsymbol{\mu}$ and $\tilde{\boldsymbol{\kappa}} = \boldsymbol{\kappa}$.

In the example (3.58) of two coupled degrees of freedom,

$$\boldsymbol{\mu} = \begin{pmatrix} m_1 & 0 \\ 0 & m_2 \end{pmatrix} \quad \text{and} \quad \boldsymbol{\kappa} = \begin{pmatrix} m_1\omega_1^2 + \kappa & -\kappa \\ -\kappa & m_2\omega_2^2 + \kappa \end{pmatrix}. \tag{11.8}$$

In the specialization to equal masses, $m_1 = m_2 = m$, operations by $\boldsymbol{\mu}$ are reduced to simple multiplications by the constant m.

The Spectrum of Frequencies

The form (11.7) for the equations of motion may suggest looking for a special solution describing a *simple* harmonic motion—that is, every degree of freedom oscillating simultaneously with the *same* frequency. That amounts to trying $q_s = A_s \exp\colon -i\omega t$ as a solution, a type of procedure already used in §3.4. Allowance is made for the possibility that the different degrees of freedom $q_1, q_2, q_3, \ldots$ may have to oscillate with different amplitudes $A_1, A_2, A_3, \ldots$. (After all, they may have different dimensions!)

Allowance will also be made for the possibility that several different such simple harmonic solutions may prove to exist, each with its own frequency $\omega_n = \omega_1$ or ω_2 or $\omega_3, \ldots$ and its own set of amplitudes $A_{n1}, A_{n2}, \ldots,$ $A_{ns}, \ldots$. Then $\omega_1, \omega_2, \omega_3, \ldots$ are said to form a *spectrum* of frequencies. Each of the distinct simple harmonic solutions may be represented by a "configuration vector"

$$\mathbf{q}_n = \mathrm{Re}\{\mathbf{A}_n e^{-i\omega_n t}\}, \tag{11.9a}$$

where $\mathbf{A}_n$ is a column with the elements $A_{n1}, A_{n2}, \ldots$. These may be complex, $A_{ns} \equiv |A_{ns}| \exp\colon i\alpha_{ns}$, corresponding to the different degrees of freedom being in different phases of oscillation, so that a component of $\mathbf{q}_n$ is

$$q_{ns}(t) = |A_{ns}| \cos(\omega_n t - \alpha_{ns}). \tag{11.9b}$$

It is to obtain such real expressions for the displacements that the real part of the trial form is taken in (11.9*a*). The imaginary part might be taken instead, as in §3.4, but that would correspond merely to changing the definition of α_{ns} by $\frac{1}{2}\pi$.

For the present, more attention will be given to the complex representative,

$$\mathbf{A}_n e^{-i\omega_n t}, \tag{11.10}$$

of each simple harmonic solution. Substituting the components of this column into the equations of motion (11.5) yields the column-vector equation

$$\boldsymbol{\mu}(-\omega_n^2 \mathbf{A}_n) = -\boldsymbol{\kappa}\mathbf{A}_n. \tag{11.11}$$

The mass matrix $\boldsymbol{\mu}$ can be expected to possess an inverse $\boldsymbol{\mu}^{-1}$ because only nonvanishing masses are taken into account, and so the equation may be rewritten as

$$(\boldsymbol{\mu}^{-1}\boldsymbol{\kappa})\mathbf{A}_n = \omega_n^2 \mathbf{A}_n. \tag{11.12}$$

This has the standard form (10.31) of an eigenvalue problem. The "eigen-frequencies" ω_n are then to be found as roots of a secular determinant like (10.28):

$$|(\boldsymbol{\mu}^{-1}\boldsymbol{\kappa}) - \omega_n^2 \cdot \mathbf{1}| = 0. \tag{11.13}$$

The degree of this algebraic equation for ω_n^2 is the same as the total number of degrees of freedom, f (equal to the order of the determinant); hence as many as f roots may be found, each an ω_r^2 for a possible simple harmonic motion. When f values $\omega^2 = \omega_1^2, \omega_2^2, \ldots, \omega_f^2$ are obtained, $2f$ values $\omega = \pm\omega_1, \pm\omega_2, \ldots, \pm\omega_f$ are found for the ω in the trial form $A_s \exp\colon -i\omega t$. This is taken into account when the real parts (11.9*a*) are used, since

$$\mathrm{Re}(A_s e^{-i\omega t}) = \tfrac{1}{2}[A_s e^{-i\omega t} + A_s^* e^{+i\omega t}].$$

The preliminary inversion of the mass matrix needed for (11.13) is an unnecessary labor (it is simple when $\boldsymbol{\mu}$ is diagonal, so that $(\boldsymbol{\mu}^{-1})_{rs} = m_r^{-1}\,\delta_{rs}$, as when cartesian coordinates are used for the displacements). This labor is not essential, since the components of (11.11) themselves form, for each n, a set of f linear homogeneous equations,

$$\sum_s [\kappa_{rs} - \mu_{rs}\omega_n^2]A_{ns} = 0, \tag{11.14}$$

one equation for each $r = 1, 2, \ldots, f$. For any homogeneous linear equations to possess solutions, the determinant of the coefficients must vanish,

$$|\boldsymbol{\kappa} - \omega_n^2\boldsymbol{\mu}| \equiv \begin{vmatrix} \kappa_{11} - \omega_n^2\mu_{11} & \kappa_{12} - \omega_n^2\mu_{12} & \cdots \\ \kappa_{21} - \omega_n^2\mu_{21} & \kappa_{22} - \omega_n^2\mu_{22} & \cdots \\ \vdots & \vdots & \end{vmatrix} = 0, \tag{11.15}$$

and this must furnish the same roots ω_n^2 as does the secular determinant (11.13), in which ω_n^2 appears only on the diagonal.

The matrix of coefficients $(\boldsymbol{\kappa} - \omega_n^2\boldsymbol{\mu})$ is symmetric whereas $(\boldsymbol{\mu}^{-1}\boldsymbol{\kappa})$ of (11.13) is not, in general, despite its factors $\boldsymbol{\mu}^{-1}$ and $\boldsymbol{\kappa}$ each being individually symmetric, since

$$\widetilde{(\boldsymbol{\mu}^{-1}\boldsymbol{\kappa})} = \tilde{\boldsymbol{\kappa}}\tilde{\boldsymbol{\mu}}^{-1} = \boldsymbol{\kappa}\boldsymbol{\mu}^{-1} \neq \boldsymbol{\mu}^{-1}\boldsymbol{\kappa} \tag{11.16a}$$

unless $\boldsymbol{\mu}^{-1}$ and $\boldsymbol{\kappa}$ commute. However, even in the simple case of a diagonal mass matrix

$$(\boldsymbol{\kappa}\boldsymbol{\mu}^{-1})_{rs} = \kappa_{rs} m_s^{-1} \neq (\boldsymbol{\mu}^{-1}\boldsymbol{\kappa})_{rs} = m_r^{-1}\kappa_{rs}, \tag{11.16b}$$

unless all the masses coupled together are equal, $m_r = m_s$.

In the example of (11.8), the determinant (11.15) becomes

$$\begin{vmatrix} m_1(\omega_1^2 - \omega_n^2) + \kappa & -\kappa \\ -\kappa & m_2(\omega_2^2 - \omega_n^2) + \kappa \end{vmatrix} = 0. \tag{11.17a}$$

For the case of identical coupled oscillators, $m_1 = m_2 = m$ and $\omega_1^2 = \omega_2^2 = \omega_0^2$, this yields exactly the eigenfrequencies already found in (3.60) and (3.61). The equations (11.14) then lead to

$$\kappa(A_{11} - A_{12}) = 0 \qquad \text{and} \qquad \kappa(A_{21} + A_{22}) = 0 \tag{11.17b}$$

for the eigenfrequencies ω_0 and $(\omega_0^2 + 2\kappa/m)^{1/2}$, respectively. Thus obtained are the special simple harmonic solutions, with $x_1 = x_2$ and $x_1 = -x_2$, illustrated in Figure 3.6.

Analysis into Simple Harmonic Modes

To be in a state of steady, simple harmonic motion, such as is represented by (11.9) or (11.10), the individual degrees of freedom are generally required to have a specific set of real amplitude ratios. These ratios are real because the linear homogeneous equations (11.14), which determine them, have real coefficients; A_{ns} and A_{ns}^* obey the same equations and so each will be the same unless the arbitrary constant factor left open by the equations, common to $A_{n1}, A_{n2}, \ldots$, is *chosen* to be complex. Each eigenvector $\mathbf{A}_1, \mathbf{A}_2, \ldots$ may be regarded as helping form a representative (11.10) of one completely defined "eigenstate" of simple harmonic motion, having some definite, standardized "unit of excitation." The standardization adopted may be a definite *real* scale factor for each $\mathbf{A}_n$ such that $\tilde{\mathbf{A}}_n \mathbf{A}_n \equiv \Sigma_s A_{ns}^2 = 1$, for example; each $\mathbf{A}_n$ would then be a real normalized eigenvector in the configuration space of $\mathbf{q}$.

Because any one of the simple harmonic modes, $\mathbf{A}_n \exp\!: -i\omega_n t$, has a characteristic set of component amplitudes, it can be excited by itself only if the motion is started with the displacements having the corresponding ratios. An *arbitrary* start will generally excite the entire spectrum of the eigenfrequencies, and the ensuing motion will consist of some superposition

$$\begin{aligned} \mathbf{q}(t) &= \tfrac{1}{2} \sum_n \mathbf{A}_n [C_n e^{-i\omega_n t} + c.c.] \\ &= \sum_n \mathbf{A}_n |C_n| \cos(\omega_n t + \gamma_n). \end{aligned} \tag{11.18}$$

Arbitrary constants $C_n = |C_n| \exp\!: -i\gamma_n$ must be introduced if the scale factors of the eigenvectors $\mathbf{A}_n$ have already been standardized in some such way as discussed. The expression (11.18) constitutes a general solution, adaptable to any start $q_s(0)$, $\dot{q}_s(0)$ which may be given the f degrees of freedom, if the superposition contains f (linearly) independent terms, each with two arbitrary real parameters, $|C_n|$ and γ_n. By linear independence is meant that no one term may be expressed as any linear combination of the others; clearly, adding such linear combinations would add nothing to the generality of the solution, for it would amount merely to putting a sum of arbitrary constants in the place of an already arbitrary constant.

Terms of the superposition having distinct frequencies *are* linearly independent of each other; no one, with its characteristic variation in time, can be matched by a linear combination of the others for all times. Problems arise when some of the roots ω_n^2 turn out to be identical (that is, when *degeneracy* occurs). Then, to have a total of f terms, the superposition will have to contain several with equal frequencies. Such a problem was resolved in the inertia tensor example of (10.37) by choosing any several eigenvectors of equal eigenvalue to be *orthogonal* to each other, and therefore linearly independent on that account. However, in the eigenfrequency problems, even eigenvectors $\mathbf{A}_n$, $\mathbf{A}_m$ corresponding to *different* frequencies cannot generally be orthogonal: $\tilde{\mathbf{A}}_n \mathbf{A}_m \neq 0$ even for $\omega_n \neq \omega_m$.

The failure of orthogonality is a consequence of the operator $(\boldsymbol{\mu}^{-1}\boldsymbol{\kappa})$ in (11.12) not being symmetric; that only symmetric operators have orthogonal eigenvectors was shown in (10.32). A formal remedy is simple. A "square-root matrix," $\boldsymbol{\mu}^{+1/2}$, and its inverse $\boldsymbol{\mu}^{-1/2}$, can be constructed in such a way* that $\boldsymbol{\mu}^{\pm 1/2}\boldsymbol{\mu}^{\pm 1/2} = \boldsymbol{\mu}^{\pm 1}$ while $\boldsymbol{\mu}^{\pm 1/2}\boldsymbol{\mu}^{\mp 1/2} = \mathbf{1}$. Each of the two matrices $\boldsymbol{\mu}^{\pm 1/2}$ cannot be other than symmetric since $\boldsymbol{\mu}$ is. Then $\boldsymbol{\mu}^{+1/2}$ may be used for a similarity transformation which makes $(\boldsymbol{\mu}^{-1}\boldsymbol{\kappa})$ symmetric; the result,

$$\boldsymbol{\mu}^{+1/2}(\boldsymbol{\mu}^{-1}\boldsymbol{\kappa})\boldsymbol{\mu}^{-1/2} = \boldsymbol{\mu}^{-1/2}\boldsymbol{\kappa}\boldsymbol{\mu}^{-1/2}, \tag{11.19}$$

cannot be other than symmetric, since each factor of the final symmetric product is. The eigenvector problem (11.12) is thus transformed into the equivalent one:

$$(\boldsymbol{\mu}^{-1/2}\boldsymbol{\kappa}\boldsymbol{\mu}^{-1/2})(\boldsymbol{\mu}^{+1/2}\mathbf{A}_n) = \omega_n^2(\boldsymbol{\mu}^{+1/2}\mathbf{A}_n). \tag{11.20}$$

Being eigenvectors of a symmetrical operator, the transformed ones,

$$\boldsymbol{\mu}^{+1/2}\mathbf{A}_n, \tag{11.21}$$

form an orthogonal set. Now the normalization in

$$(\tilde{\mathbf{A}}_n \boldsymbol{\mu}^{+1/2})(\boldsymbol{\mu}^{+1/2}\mathbf{A}_m) = \tilde{\mathbf{A}}_n \boldsymbol{\mu}\mathbf{A}_m = \delta_{nm} \tag{11.22}$$

---

* Even if not uniquely!

becomes more convenient than $\tilde{\mathbf{A}}_n \mathbf{A}_n = 1$. All these measures are unnecessary when dealing with a system of equal masses, for then $(\boldsymbol{\mu}^{-1}\boldsymbol{\kappa})$ is already symmetric and $\tilde{\mathbf{A}}_n \mathbf{A}_m = \delta_{nm}$ is obtainable.

To be representatives of each of the possible, independent, simple harmonic eigenmodes of motion, the column vectors (11.10) should be modified to

$$\mathbf{u}_n(t) \equiv \boldsymbol{\mu}^{+1/2} \mathbf{A}_n e^{-i\omega_n t}, \tag{11.23a}$$

forming a complete set of *orthogonal* complex unit vectors

$$\mathbf{u}_n^\dagger \mathbf{u}_m = \delta_{nm}, \tag{11.23b}$$

with $\mathbf{u}_n^\dagger$ an hermitian conjugate (10.83). Those members of the set that correspond to nondegenerate eigenfrequencies are fully defined by just the linear equations (11.14) and the normalization. Members of *degenerate* subsets are not, but then may be chosen in any way consistent with the equations (11.14) and the orthonormality condition (11.23*b*). The entire set will then be complete enough to yield for the general motion a superposition

$$\boldsymbol{\mu}^{+1/2} \mathbf{q}(t) = \mathrm{Re}\left\{\sum_n C_n \mathbf{u}_n\right\}, \tag{11.24}$$

which has the requisite $2f$ arbitrary constants in its f linearly independent terms. The right side amounts to a decomposition of an arbitrary vector in the q-space onto f orthogonal axes, $\mathbf{u}_n$, spanning the entire space. When $\mathbf{q}(t)$ is made explicit by operating on (11.24) with $\boldsymbol{\mu}^{-1/2}$, the result attains exactly the form (11.18) again.

The Normal Coordinates

The components of the general solution (11.18) may be written

$$q_s(t) = \sum_n A_{ns} Q_n(t) \tag{11.25a}$$

if

$$Q_n(t) \equiv |C_n| \cos(\omega_n t + \gamma_n). \tag{11.25b}$$

The first of these expressions may be regarded as a linear substitution of new coordinates Q_n for the original degrees of freedom q_s. The substitution has been of such a kind that each new coordinate has a simple harmonic variation, obeying an equation of motion $\ddot{Q}_n = -\omega_n^2 Q_n$. Coordinates with this property are called the "normal coordinates" of the system and the corresponding eigenfrequencies are also known as the "normal" frequencies. In the identical oscillators example (3.62), the normal coordinates were found to be $\bar{x} = \frac{1}{2}(x_1 + x_2)$ and $x = x_1 - x_2$, having the eigenfrequencies ω_0 and $(\omega_0^2 + 2\kappa/m)^{1/2}$, respectively.

The transformation (11.25) is being effected by a square matrix of elements A_{ns} such that

$$\mathbf{q} = \tilde{\mathbf{A}}\mathbf{Q},$$

and hence

$$\mathbf{Q} = (\tilde{\mathbf{A}})^{-1}\mathbf{q}.$$

Thus elements A_{ns}, which have been serving as components of the eigenvector column $\mathbf{A}_n$ in (11.12), here form the nth column of a square transformation matrix, $\tilde{\mathbf{A}}$ (and the nth *row* of $\mathbf{A}$ itself). The rth component of the eigenvector equation itself is

$$(\boldsymbol{\mu}^{-1}\boldsymbol{\kappa})_{rs} A_{ns} = \omega_n^2 A_{nr}.$$

Multiplying by $(\tilde{\mathbf{A}})^{-1}_{mr}$ and summing over r yields

$$(\tilde{\mathbf{A}})^{-1}_{mr}(\boldsymbol{\mu}^{-1}\boldsymbol{\kappa})_{rs}\tilde{A}_{sn} = \omega_n^2(\tilde{\mathbf{A}})^{-1}_{mr}\tilde{A}_{rn} = \omega_n^2\,\delta_{mn},$$

and this is just the (m, n)-element of a square matrix equation

$$\mathbf{S}(\boldsymbol{\mu}^{-1}\boldsymbol{\kappa})\mathbf{S}^{-1} = \boldsymbol{\omega}^2, \tag{11.26}$$

where

$$\mathbf{S} \equiv (\tilde{\mathbf{A}})^{-1} \tag{11.27}$$

and $\boldsymbol{\omega}^2$ is a diagonal matrix with eigenvalues, ω_n^2. Thus the transformation to normal coordinates,

$$\mathbf{Q} = (\tilde{\mathbf{A}})^{-1}\mathbf{q} \equiv \mathbf{S}\mathbf{q}, \tag{11.28}$$

is a similarity transformation, as in (9.119), which diagonalizes an operator matrix, $(\boldsymbol{\mu}^{-1}\boldsymbol{\kappa})$ of (11.26), as in (10.24). That any eigenvector problem like (11.12) is equivalent to a matrix diagonalization was shown in arriving at (10.31). Moreover, as pointed out in (10.30), the columns of the inverted transformation matrix $\mathbf{S}^{-1}$ are just the eigenvector columns, and that is also the case in (11.27),

$$S^{-1}_{ln} = \tilde{A}_{ln} = A_{nl} \equiv (\mathbf{A}_n)_l,$$

for the eigenvectors $\mathbf{A}_n$ here.

In the identical oscillators example mentioned in a preceding paragraph,

$$\mathrm{Q} = \mathbf{Sq} \equiv \begin{pmatrix} \bar{x} \\ x \end{pmatrix} = \begin{pmatrix} \frac{1}{2} & \frac{1}{2} \\ 1 & -1 \end{pmatrix} \begin{pmatrix} x_1 \\ x_2 \end{pmatrix}.$$

A trivial change of scales in defining the normal coordinates, to $Q_{1,2} \equiv \pm 2^{-1/2}(x_1 \pm x_2)$, makes the transformation an orthonormal one,

$$\mathbf{S} \to \mathbf{a} = 2^{-1/2}\begin{pmatrix} 1 & 1 \\ -1 & 1 \end{pmatrix},$$

with $\tilde{\mathbf{a}}\mathbf{a} = \mathbf{1}$, yielding a rotation $\mathbf{aq}$ by 45°. This possibility of obtaining normal coordinates by mere rotation in the q-space stems from the likeness of the coupled oscillators, and the consequent symmetry of the operator matrix $(\boldsymbol{\mu}^{-1}\boldsymbol{\kappa})$.

In general, the direct transformation to the normal coordinates, $\mathrm{Q} = \mathbf{Sq}$, cannot be orthonormal (that is, a mere rotation in the configuration space) because it must diagonalize a matrix $(\boldsymbol{\mu}^{-1}\boldsymbol{\kappa})$, in (11.26), which is generally not symmetric. That orthonormal transformations can diagonalize only symmetric matrices was shown in (10.29). On the other hand, the diagonalization required here can be equivalently expressed as

$$(\mathbf{S}\boldsymbol{\mu}^{-1/2})(\boldsymbol{\mu}^{-1/2}\boldsymbol{\kappa}\boldsymbol{\mu}^{-1/2})(\boldsymbol{\mu}^{+1/2}\mathbf{S}^{-1}) = \boldsymbol{\omega}^2,$$

with the final parenthesis just the inverse of the first one. Thus the matrix product

$$\mathbf{a} \equiv \mathbf{S}\boldsymbol{\mu}^{-1/2} \tag{11.29}$$

is here being required to diagonalize a *symmetric* matrix, $(\boldsymbol{\mu}^{-1/2}\boldsymbol{\kappa}\boldsymbol{\mu}^{-1/2})$, and can therefore be orthonormalized to $\tilde{\mathbf{a}}\mathbf{a} = \mathbf{a}\tilde{\mathbf{a}} = \mathbf{1}$. The upshot is that although $\mathrm{Q} = \mathbf{Sq}$ is not generally a mere rotation, it is yet of a restricted type:

$$\mathrm{Q} = \mathbf{a}(\boldsymbol{\mu}^{+1/2}\mathbf{q}), \tag{11.30}$$

a rotation by $\mathbf{a}$ after a symmetrization in mass performed by $\boldsymbol{\mu}^{+1/2}$.

An instructive consequence of the orthonormality of the matrix product (11.29) follows from

$$\mathbf{a}^{-1} = \boldsymbol{\mu}^{+1/2}\mathbf{S}^{-1} = \tilde{\mathbf{a}}, \qquad \widetilde{(\mathbf{a}^{-1})} = \widetilde{(\mathbf{S}^{-1})}\boldsymbol{\mu}^{+1/2} = \mathbf{a},$$

in which the symmetry of the mass matrix has been used. Now

$$\mathbf{a}\tilde{\mathbf{a}} = \mathbf{1} = \widetilde{(\mathbf{S}^{-1})}\boldsymbol{\mu}\mathbf{S}^{-1} \tag{11.31}$$

is a transformation in which the mass matrix is diagonalized to unity. Any transformation of the type $\tilde{\mathbf{F}}\boldsymbol{\mu}\mathbf{F}$ is called a "congruence transformation"; it reduces to an orthonormal one whenever $\tilde{\mathbf{F}} = \mathbf{F}^{-1}$.

The Energies

The total energy of the coupled motions described in (11.2) to (11.6) is conserved at some constant value

$$E = \tfrac{1}{2}\sum_{r,s}[\mu_{rs}\dot{q}_r\dot{q}_s + \kappa_{rs}q_r q_s], \tag{11.32}$$

but energy exchanges generally take place among the individual degrees of freedom. The kinetic and potential energies here may each be written as a scalar product of column vectors:

$$\tau = \tfrac{1}{2}(\tilde{\dot{\mathbf{q}}})\boldsymbol{\mu}\dot{\mathbf{q}} \qquad \text{and} \qquad V(q) = \tfrac{1}{2}\tilde{\mathbf{q}}\boldsymbol{\kappa}\mathbf{q}. \tag{11.33}$$

In the example of (11.8),

$$\tfrac{1}{2}\tilde{\mathbf{q}}\boldsymbol{\kappa}\mathbf{q} = \tfrac{1}{2}[m_1\omega_1^2 q_1^2 + m_2\omega_2^2 q_2^2 + \kappa(q_1 - q_2)^2], \tag{11.34}$$

in agreement with (3.59).

The energy may be expressed in terms of the normal coordinates by inserting $\mathbf{q} = \mathbf{S}^{-1}\mathbf{Q}$:

$$E = \tfrac{1}{2}(\tilde{\dot{\mathbf{Q}}})[(\widetilde{\mathbf{S}^{-1}})\boldsymbol{\mu}\mathbf{S}^{-1}]\dot{\mathbf{Q}} + \tfrac{1}{2}\tilde{\mathbf{Q}}[(\widetilde{\mathbf{S}^{-1}})\boldsymbol{\kappa}\mathbf{S}^{-1}]\mathbf{Q}. \tag{11.35}$$

The last square bracket may have $\boldsymbol{\mu}\boldsymbol{\mu}^{-1} = \mathbf{1}$ and $\mathbf{S}^{-1}\mathbf{S} = \mathbf{1}$ inserted into it to obtain

$$(\widetilde{\mathbf{S}^{-1}})\boldsymbol{\kappa}\mathbf{S}^{-1} = (\widetilde{\mathbf{S}^{-1}})\boldsymbol{\mu}\mathbf{S}^{-1}[\mathbf{S}(\boldsymbol{\mu}^{-1}\boldsymbol{\kappa})\mathbf{S}^{-1}] = [(\widetilde{\mathbf{S}^{-1}})\boldsymbol{\mu}\mathbf{S}^{-1}]\boldsymbol{\omega}^2,$$

as a consequence of the diagonalization (11.26) of $(\boldsymbol{\mu}^{-1}\boldsymbol{\kappa})$ to $\boldsymbol{\omega}^2$. Thus the kinetic and potential energies have in common the constant matrix $(\widetilde{\mathbf{S}^{-1}})\boldsymbol{\mu}\mathbf{S}^{-1}$, which becomes the unit matrix according to (11.31). As a result,

$$E = \tfrac{1}{2}[(\tilde{\dot{\mathbf{Q}}})\dot{\mathbf{Q}} + \tilde{\mathbf{Q}}\boldsymbol{\omega}^2\mathbf{Q}]. \tag{11.36}$$

The same congruence transformation which diagonalized the mass matrix to unity in (11.31) simultaneously diagonalizes the coupling matrix to $\boldsymbol{\omega}^2$, as comparison of (11.36) to (11.35) shows.

The energy expression (11.36) incorporates a sum of f *individually* conserved energies; that is, $E = \sum_n E_n$ with

$$E_n = \tfrac{1}{2}[\dot{Q}_n^2 + \omega_n^2 Q_n^2] = \tfrac{1}{2}\omega_n^2|C_n|^2. \tag{11.37}$$

The separate constancy, at a value determined by initial conditions giving C_n, follows most readily from the simple substitution of the harmonic variations (11.25*b*) for $Q_n(t)$. The analyses here, into the simple harmonic oscillations, might have been called "decoupling transformations," since they separate motions of interacting degrees of freedom into independent component modes that exchange no energy. An early example of such a result was seen in (3.64).

11.2 THE LINEAR TRIATOMIC MOLECULE

The simplest example that illustrates well the above formalism is provided by the linear three-particle system indicated in Figure 11.1. This differs from the two-particle system of Figure 3.6 in that here the particles are not to be coupled directly to the indicated equilibrium points but only to each other. Indeed, the only couplings to be taken into account are equal ones, of each of the equal outside masses to the one between them. The model is used to represent, in roughest approximation, such molecules as CO_2, during linear vibrations that lead to some of their most interesting manifestations.

The model to be used here is in the end defined by the equations of motion adopted:

$$\begin{aligned} m_1 \ddot{q}_1 &= -\kappa(q_1 - q_2), \\ m_2 \ddot{q}_2 &= +\kappa(q_1 - q_2) - \kappa(q_2 - q_3), \\ m_1 \ddot{q}_3 &= +\kappa(q_2 - q_3). \end{aligned} \tag{11.38}$$

Notice that the contributions to the acceleration of each particle are proportional to the negatives of its own displacement. The three equations are the components of a column-vector equation like (11.7), with mass and coupling matrices having the forms

$$\boldsymbol{\mu} = \begin{pmatrix} m_1 & 0 & 0 \\ 0 & m_2 & 0 \\ 0 & 0 & m_1 \end{pmatrix} \quad \text{and} \quad \boldsymbol{\kappa} = \begin{pmatrix} \kappa & -\kappa & 0 \\ -\kappa & 2\kappa & -\kappa \\ 0 & -\kappa & \kappa \end{pmatrix}. \tag{11.39}$$

The kinetic and potential energies (11.33) are

$$\begin{aligned} \tau &= \tfrac{1}{2}\tilde{\dot{\mathbf{q}}}\boldsymbol{\mu}\dot{\mathbf{q}} = \tfrac{1}{2}[m_1(\dot{q}_1^2 + \dot{q}_3^2) + m_2\dot{q}_2^2], \\ V &= \tfrac{1}{2}\tilde{\mathbf{q}}\boldsymbol{\kappa}\mathbf{q} = \tfrac{1}{2}\kappa[(q_1 - q_2)^2 + (q_2 - q_3)^2]. \end{aligned} \tag{11.40}$$

The latter contains only mutual couplings, corresponding to the mutual forces $F_s = -\partial V/\partial q_s$ exhibited in the three equations of motion.

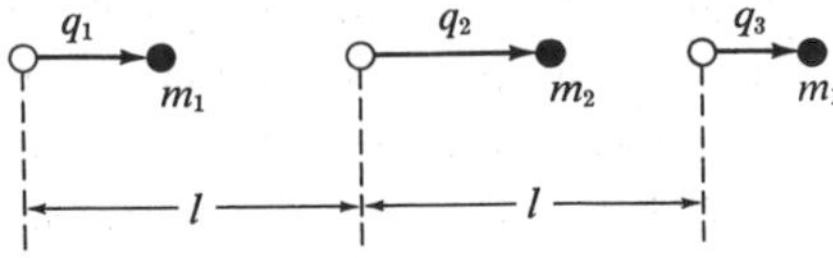

Figure 11.1

The trial forms $q_s = A_{ns} \exp\colon -i\omega_n t$ lead to linear homogeneous equations like (11.14) for the amplitudes A_{ns}, and these have nonvanishing solutions only when the determinant like (11.15) vanishes:

$$\begin{vmatrix} \kappa - m_1\omega_n^2 & -\kappa & 0 \\ -\kappa & 2\kappa - m_2\omega_n^2 & -\kappa \\ 0 & -\kappa & \kappa - m_1\omega_n^2 \end{vmatrix} = 0. \tag{11.41}$$

The secular determinant (11.13) differs from this only in that its first and third rows are divided by m_1, while the middle row is divided by m_2. Both determinants have the same roots:

$$\omega_1^2 = \frac{\kappa}{m_1}, \qquad \omega_2^2 = \frac{M\kappa}{m_1 m_2}, \qquad \omega_3^2 = 0, \tag{11.42}$$

with $M = 2m_1 + m_2$ the total mass of the system. It is easy to check that $\sum_n \omega_n^2 = \mathrm{Tr}(\boldsymbol{\mu}^{-1}\boldsymbol{\kappa})$, as is required by the trace invariance (9.122).

By substituting each of the three eigenfrequencies in turn into the linear equations (11.14) for the amplitudes, it is easy to find that

$$\begin{aligned} &\text{for } \omega_1, && A_{13} = -A_{11},\ A_{12} = 0, \\ &\text{for } \omega_2, && A_{22} = -2(m_1/m_2)A_{21},\ A_{23} = A_{21}, \\ &\text{for } \omega_3 = 0, && A_{31} = A_{32} = A_{33}. \end{aligned} \tag{11.43}$$

The character of each of the three independent motions can be seen most directly by using these results as in (11.9). Then,

$$\begin{aligned} &\text{for } \omega_1, && q_1 = -q_3 = |A_{11}|\cos(\omega_1 t - \alpha_{11}),\ q_2 = 0 \\ &\text{for } \omega_2, && q_1 = +q_3 = |A_{21}|\cos(\omega_2 t - \alpha_{21}) = -(m_2/2m_1)q_2 \\ &\text{for } \omega_3 = 0, && q_1 = q_2 = q_3 = |A_{31}|\cos\alpha_{31}, \end{aligned} \tag{11.44}$$

as indicated in Figure 11.2.

The first eigenfrequency, with $\omega_1^2 = \kappa/m_1$, is like that for a single mass m_1 coupled to a fixed center with force $-\kappa q$, because m_2 acts as a fixed center in the corresponding modes of motion. The frequency $\omega_2 = (M/m_2)^{1/2}\omega_1$ is higher because the motion of the middle mass relative to the others must work against the stiffness of both bonds together.

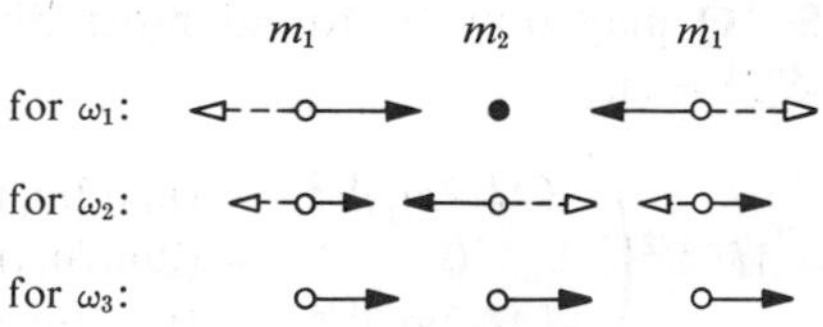

Figure 11.2

In the third of the modes, the couplings do no work at all. It corresponds merely to a steady bodily shift of the whole system, distorting no bond because no couplings to points fixed in space were introduced. The third of the solutions given in (11.44) is obviously not the most general one of the type $q_1 = q_2 = q_3$, since the two constants $|A_{31}|$, α_{31} combine to form a single arbitrary one in this case. More generally, the equations of motion $\ddot{q}_1 = \ddot{q}_2 = \ddot{q}_3 = 0$ yield a uniform translation with some constant velocity, just such a uniform center-of-mass translation as is to be expected of an isolated system. The modes with ω_1 and ω_2 are independent "internal" motions, relative to the mass-center

$$\bar{q} = (m_1 q_1 + m_2 q_2 + m_1 q_3)/M, \tag{11.45}$$

with $\bar{q} = 0$ for each of them.

The general motion arising from arbitrary excitation is most elegantly formulated in terms of normal coordinates $\mathbf{Q} = \mathbf{Sq} = (\tilde{\mathbf{A}})^{-1}\mathbf{q}$ (11.28). To find the transformation, the three eigenvectors $\mathbf{A}_{1,2,3}$ implicit in (11.43) are normalized according to the prescription $\tilde{\mathbf{A}}_n \boldsymbol{\mu} \mathbf{A}_n = 1$ (11.22) and presented as column vectors:

$$\mathbf{A}_1 = (2m_1)^{-1/2}\begin{pmatrix} 1 \\ 0 \\ -1 \end{pmatrix},$$

$$\mathbf{A}_2 = \left(\frac{m_2}{2Mm_1}\right)^{1/2}\begin{pmatrix} 1 \\ -2m_1/m_2 \\ 1 \end{pmatrix}, \qquad \mathbf{A}_3 = M^{-1/2}\begin{pmatrix} 1 \\ 1 \\ 1 \end{pmatrix}. \tag{11.46}$$

These are not orthonormal, but the representatives $\mathbf{u}_n$ (11.23) are:

$$\mathbf{u}_1 = 2^{-1/2}\begin{pmatrix} 1 \\ 0 \\ -1 \end{pmatrix} e^{-i\omega_1 t},$$

$$\mathbf{u}_2 = \left(\frac{m_2}{2M}\right)^{1/2}\begin{pmatrix} 1 \\ -2(m_1/m_2)^{1/2} \\ 1 \end{pmatrix} e^{-i\omega_2 t}, \tag{11.47}$$

$$\mathbf{u}_3 = \left(\frac{m_1}{M}\right)^{1/2}\begin{pmatrix} 1 \\ (m_2/m_1)^{1/2} \\ 1 \end{pmatrix}.$$

The relations $\mathbf{q} = \mathbf{S}^{-1}\mathbf{Q}$ may now be found most directly by forming the matrix of elements $S_{sn}^{-1} = A_{ns}$:

$$\mathbf{S}^{-1} = M^{-1/2}\begin{pmatrix} (M/2m_1)^{1/2} & (m_2/2m_1)^{1/2} & 1 \\ 0 & -(2m_1/m_2)^{1/2} & 1 \\ -(M/2m_1)^{1/2} & (m_2/2m_1)^{1/2} & 1 \end{pmatrix}. \qquad (11.48)$$

The components (11.25) of $\mathbf{q} = \mathbf{S}^{-1}\mathbf{Q} = \tilde{\mathbf{A}}\mathbf{q}$ are then

$$\begin{aligned} q_1 &= M^{-1/2}[(M/2m_1)^{1/2}Q_1 + (m_2/2m_1)^{1/2}Q_2 + Q_3], \\ q_2 &= M^{-1/2}[\qquad - (2m_1/m_2)^{1/2}Q_2 + Q_3], \\ q_3 &= M^{-1/2}[-(M/2m_1)^{1/2}Q_1 + (m_2/2m_1)^{1/2}Q_2 + Q_3]. \end{aligned} \qquad (11.49a)$$

With

$$Q_1 = |C_1| \cos(\omega_1 t + \gamma_1), \qquad Q_2 = |C_2| \cos(\omega_2 t + \gamma_2), \quad (11.49b)$$

and Q_3 any uniform translation, these constitute a general solution like (11.18).

The inversion to the relations $\mathbf{Q} = \mathbf{S}\mathbf{q}$ may be carried out by taking advantage of the orthonormality (11.31):

$$\mathbf{S} = (\widetilde{\mathbf{S}^{-1}})\boldsymbol{\mu} = M^{-1/2}\begin{pmatrix} (m_1 M/2)^{1/2} & 0 & -(m_1 M/2)^{1/2} \\ (m_1 m_2/2)^{1/2} & -(2m_1 m_2)^{1/2} & (m_1 m_2/2)^{1/2} \\ m_1 & m_2 & m_1 \end{pmatrix} \qquad (11.50)$$

Thus

$$\begin{aligned} Q_1 &= (m_1/2)^{1/2}(q_1 - q_3), \\ Q_2 &= (m_1 m_2/2M)^{1/2}(q_1 - 2q_2 + q_3), \\ Q_3 &= M^{-1/2}(m_1 q_1 + m_2 q_2 + m_1 q_3) \equiv M^{+1/2}\bar{q}. \end{aligned} \qquad (11.51)$$

It can be checked directly from the initial equations of motion (11.38) that each satisfies an equation $\ddot{Q}_n = -\omega_n^2 Q_n$, with just the eigenfrequencies (11.42). The third of the normal coordinates is just the center of mass position (11.45) rescaled.

The separately conserved modal energies (11.37) are

$$\begin{aligned} E_1 &= \tfrac{1}{4}[m_1(\dot{q}_1 - \dot{q}_3)^2 + \kappa(q_1 - q_3)^2], \\ E_2 &= \tfrac{1}{4}[(m_1 m_2/M)(\dot{q}_1 - 2\dot{q}_2 + \dot{q}_3)^2 + \kappa(q_1 - 2q_2 + q_3)^2], \\ E_3 &= \tfrac{1}{2}M\dot{\bar{q}}^2 = (m_1\dot{q}_1 + m_2\dot{q}_2 + m_1\dot{q}_3)^2/2M. \end{aligned} \qquad (11.52)$$

They add up properly to the total (11.40).

11.3 THE PARTICULATE STRING

To be considered next is a long string of identical particles. If the motions studied are confined to the straight line formed by the particles in their equilibrium configuration, then longitudinal vibrations in a uniform near-rigid rod may be approximated. When the displacements treated are ones transverse to the equilibrium line, that may serve as a crude model of a stretched string. The readier visibility of the transverse motions makes their consideration better for illustrative purposes, but practically no modifications in the treatment would be needed for it to apply to the longitudinal vibrations as well.

The Problem

The system indicated by Figure 11.3 will be considered. The total mass M of the string is shared among f particles of equal mass, $m = M/f$.

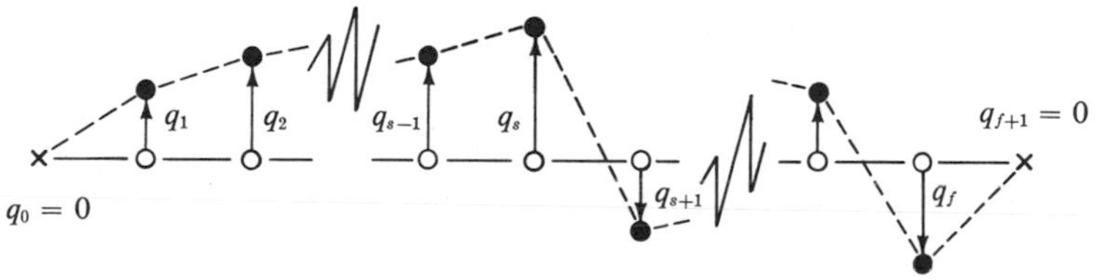

Figure 11.3

Each will be coupled just to its two neighbors, with all such couplings equal (per unit relative displacement). Couplings of the end particles to fixed points on the line will also be introduced, in order to avoid considering translations of the system as a whole (like the "third mode" of the preceding problem) and to restrict attention to internal motions. The model will then have correspondence to a stretched string held down at both ends. It will be convenient to introduce symbols $q_0 = 0$ and $q_{f+1} = 0$ to represent the fixed end points, in addition to the transverse displacements $q_1(t), q_2(t), \ldots, q_s(t), \ldots, q_f(t)$ of the f particles. For simplicity, these displacements will all be confined to the same plane; that will suffice for the most interesting starts that can be given to the motion. Moreover, there may in any case be no appreciable coupling between the transverse motions in one plane and ones in the plane normal to it; the two may be as separably independent as were the cartesian components of the spatial oscillations in §4.1.

The equations of motion in the circumstances just outlined may be written

$$m\ddot{q}_s = -\kappa(q_s - q_{s-1}) - \kappa(q_s - q_{s+1}) \qquad (11.53)$$

for $s = 1, 2, \ldots, f$. Under the convention that $q_0 = q_{f+1} \equiv 0$, the equations for the first and last particles become

$$m\ddot{q}_1 = -\kappa q_1 - \kappa(q_1 - q_2) \qquad \text{and} \qquad m\ddot{q}_f = -\kappa(q_f - q_{f-1}) - \kappa q_f .$$

The forces $-\kappa q_1$ and $-\kappa q_f$ in these equations represent the couplings to the fixed end points of the string.

An ordinary string possesses a straight-line equilibrium configuration only by virtue of a fixing of its end points under "tension." In that case, all the couplings should depend on the existence of the tension, and the coupling constant κ should be derivable from the force of tension T applied at the ends. It may be supposed that, because the particles of the string hold together, a force on any one particle is instantaneously transmitted to its neighbor. Then, under the tension, each particle may be treated as having two forces of magnitude T applied to it, respectively directed toward each neighbor, as indicated in Figure 11.4 for the sth particle. On this model, the transverse component of the force to be equated to $m\ddot{q}_s$ should be

$$-T(q_s - q_{s-1})/l - T(q_s - q_{s+1})/l,$$

in the approximation linear in the relative displacements. Here l may be taken to be a uniform equilibrium distance between neighbors such that $(f + 1)l = L$ is the total length of the string. The model thus yields

$$\kappa = T/l \tag{11.54}$$

as the relation between coupling coefficient and the applied force of tension; the restoring force per unit relative displacement becomes equal to the tension per unit of the particle separation. The equations (11.53) may also apply to cases in which the coupling has other origins.

The forces introduced into the equations of motion here are derivable, through $m\ddot{q}_s = F_s = -\partial V/\partial q_s$, from the potential

$$V(q) = \sum_{s=1}^{f+1} \tfrac{1}{2}\kappa(q_s - q_{s-1})^2 \equiv \sum_{s=0}^{f} \tfrac{1}{2}\kappa(q_{s+1} - q_s)^2, \tag{11.55}$$

with the convention $q_0 = q_{f+1} = 0$ to be observed. There are $f + 1$ terms because that is the number of bonds.

The equations of motion have the form (11.7), $\boldsymbol{\mu}\ddot{\mathbf{q}} = -\boldsymbol{\kappa}\mathbf{q}$, with a mass matrix $\boldsymbol{\mu}$ which is equivalent to mere multiplication by a constant m. It is then convenient to define a "frequency unit" ω_0 by way of

$$\omega_0^2 \equiv \kappa/m \qquad (= T/ml). \tag{11.56}$$

Accordingly, the $f \times f$ matrix $(\boldsymbol{\mu}^{-1}\boldsymbol{\kappa})$ of the eigenvector problem (11.12) is the

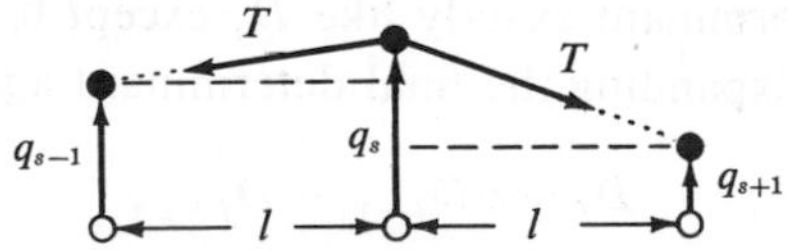

Figure 11.4

simple symmetric one

$$m^{-1}\boldsymbol{\kappa} = \begin{pmatrix} 2\omega_0^2 & -\omega_0^2 & 0 & 0 & \cdots \\ -\omega_0^2 & 2\omega_0^2 & -\omega_0^2 & 0 & \cdots \\ 0 & -\omega_0^2 & 2\omega_0^2 & -\omega_0^2 & \cdots \\ 0 & 0 & -\omega_0^2 & 2\omega_0^2 & \cdots \\ \vdots & \vdots & \vdots & \vdots & \cdots \end{pmatrix}. \tag{11.57}$$

It is this that must be diagonalized to find the possible eigenfrequencies. Being a symmetric operator, a mere rotation $\mathbf{Sq} \equiv \mathbf{aq} = \mathbf{Q}$ will suffice to change to the basis on which it is diagonal; the normal coordinates Q_n will be projections on axes that are merely rotated from the initial cartesian ones of the q-space. The eigenvectors $\mathbf{A}_n$ will form an orthonormal set, $\tilde{\mathbf{A}}_n \mathbf{A}_m = \delta_{nm}$, without any preliminary symmetrization in mass.

The Eigenfrequencies

The eigenvalues ω_n^2 of $(\boldsymbol{\mu}^{-1}\boldsymbol{\kappa})$ are to be obtained as roots of the secular determinant (11.13):

$$|(\boldsymbol{\mu}^{-1}\boldsymbol{\kappa}) - \omega_n^2 \cdot \mathbf{1}| = 0 \equiv (\omega_0^2)^f D_f. \tag{11.58}$$

If the determinant D_f is defined as indicated here, and its diagonal elements are designated

$$\alpha \equiv 2 - \omega_n^2/\omega_0^2, \tag{11.59}$$

it will have the appearance

$$D_f \equiv \begin{vmatrix} \alpha & -1 & 0 & 0 & \cdots \\ -1 & \alpha & -1 & 0 & \cdots \\ 0 & -1 & \alpha & -1 & \cdots \\ 0 & 0 & -1 & \alpha & \cdots \\ \vdots & \vdots & \vdots & \vdots & \cdots \end{vmatrix}. \tag{11.60a}$$

Expansion into minors (9.67) of the first column yields

$$D_f = \alpha D_{f-1} + \begin{vmatrix} -1 & 0 & 0 & \cdots \\ -1 & \alpha & -1 & \cdots \\ 0 & -1 & \alpha & \cdots \\ \vdots & \vdots & \vdots & \cdots \end{vmatrix},$$

where D_{f-1} is a determinant exactly like D_f except for having only $(f-1)$ rows and columns. Expanding the final determinant again leads to

$$D_f = \alpha D_{f-1} - D_{f-2}, \tag{11.60b}$$

a relation among determinants of the type D_f which differ only as to degree. It is easy to compute directly the values of D_1, D_2, and D_3, applying to $f = 1$, 2, and 3 degrees of freedom, respectively: $D_1 = \alpha$, $D_2 = \alpha^2 - 1$, $D_3 = \alpha^3 - 2\alpha$. These can be seen to fulfill the "recurrence" relation (11.60*b*) and will help give a start toward finding D_f for arbitrary f. It will also help to notice that the formal relation is satisfied by $D_0 = \alpha D_1 - D_2 = +1$ and $D_{-1} = \alpha D_0 - D_1 = 0$.

The equation (11.60*b*) is of a type known as a "difference equation," for D_f as a function of the discrete variable f, analogous to a differential equation in a continuous variable. The analogy can be made plainer by rewriting the difference equation as

$$(D_f - D_{f-1}) - (D_{f-1} - D_{f-2}) = -(2-\alpha)D_{f-1} = -(\omega_n/\omega_0)^2 D_{f-1},$$

and then *symbolizing* it with the help of $\Delta f \equiv 1$ as

$$\left(\frac{\Delta D}{\Delta f}\right)_{f-\frac{1}{2}} - \left(\frac{\Delta D}{\Delta f}\right)_{f-\frac{3}{2}} \equiv \frac{\Delta^2 D_{f-1}}{(\Delta f)^2} = -(\omega_n/\omega_0)^2 D_{f-1}.$$

This, together with the requirement for consistency with the difference equation represented as $D_{-1} = 0$, should suggest *trying* $D_{f-1} = C \sin f\theta$ as a form of solution, with C and θ to be determined under the expectation that they will be "constants" (independent of f). Substitution into the difference equation (11.60*b*) gives

$$\begin{aligned}\alpha C \sin f\theta &= C\,[\sin(f+1)\theta + \sin(f-1)\theta] \\ &= 2\,C \cos\theta \sin f\theta.\end{aligned}$$

This shows that $D_f = C \sin(f+1)\theta$ will indeed be a solution if θ is such that

$$2\cos\theta = \alpha \equiv 2 - (\omega_n/\omega_0)^2. \tag{11.61}$$

Moreover, for consistency with the known "boundary condition," $D_0 = 1$, the quantity C must be chosen so that $C \sin\theta = 1$. Thus obtained is

$$D_f = \frac{\sin(f+1)\theta}{\sin\theta}. \tag{11.62}$$

This result agrees with the expectations that $D_1 = \alpha$, $D_2 = \alpha^2 - 1$, $D_3 = \alpha^3 - 2\alpha$, ...

The eigenvalues ω_n^2 are to be obtained from setting $D_f = 0$ (11.58); the form (11.62) for D_f vanishes for $\theta = n\pi/(f+1)$, where n is any integer other than $n = 0, n = f + 1, \ldots$ (for $\theta = 0$, (11.58) gives $D_f = f + 1$; for $\theta = \pi$, $D_f = (-)^{(f+1)}$, and so on). The relation (11.61) of ω_n to θ then leads to

$$\omega_n^2 = 2\omega_0^2\left[1 - \cos\left(\frac{n\pi}{f+1}\right)\right] = 4\omega_0^2 \sin^2\left(\frac{n\pi}{2(f+1)}\right) \tag{11.63}$$

for the eigenvalues, each ω_n lying between 0 and $2\omega_0$. At most f distinct eigenvalues are to be expected and indeed, just the choices $n = 1, 2, \ldots, f$ are sufficient to yield all possible ones, as Figure 11.5, drawn for the case $f = 5$, may help make clear. The figure also makes clear that the sum of all the cosines in (11.63), for $n = 1, 2, \ldots, f$, vanishes. Thus $\sum_n \omega_n^2 = 2f\omega_0^2$, and this agrees with the trace of the matrix $(\boldsymbol{\mu}^{-1}\boldsymbol{\kappa})$ of (11.57), as it should according to the trace invariance (9.122).

In the case of $f = 1$, (11.63) gives the eigenvalue $\omega_1^2 = 2\kappa/m$. This is the case of a single particle oscillating between two fixed points, with the equation of motion $m\ddot{q}_1 = -2\kappa q_1$ which follows from (11.53) for this case. The $f = 2$ case corresponds to the identical coupled oscillators of §3.7 if $\omega_0^2 \equiv \kappa/m$ is chosen for them. The eigenvalues found in (3.60) and (3.61) were $\omega_1^2 = \omega_0^2$ and $\omega_2^2 = \omega_0^2 + 2\kappa/m = 3\omega_0^2$. The same ones follow from (11.63).

The Modes of Oscillation

The amplitude ratios characteristic of any one eigenmode are to be obtained from f linear equations like (11.14), evident from the equations of motion (11.53) with $\ddot{q}_s = -\omega_n^2 q_s$:

$$-m\omega_n^2 A_{ns} = \kappa(A_{n,s-1} - 2A_{ns} + A_{n,s+1}), \tag{11.64}$$

where A_{ns} is the amplitude of $q_s(t)$ in the eigenmode of frequency ω_n. Here again are difference equations very similar to (11.60*b*):

$$A_{n,s+1} = \alpha A_{ns} - A_{n,s-1}.$$

Since A_{n0} here refers to the "amplitude" of $q_0 \equiv 0$, this serves as a boundary condition in choosing the solution to be of the form $A_{ns} \sim \sin s\theta$, with θ to be related to ω_n by (11.61) again. Thus obtained is the proportionality

$$A_{ns} \sim \sin[sn\pi/(f+1)], \tag{11.65}$$

with the magnitude to be obtained by normalization. $A_{n0} = A_{n,f+1} = 0$, in proper agreement with $q_0 = q_{f+1} = 0$.

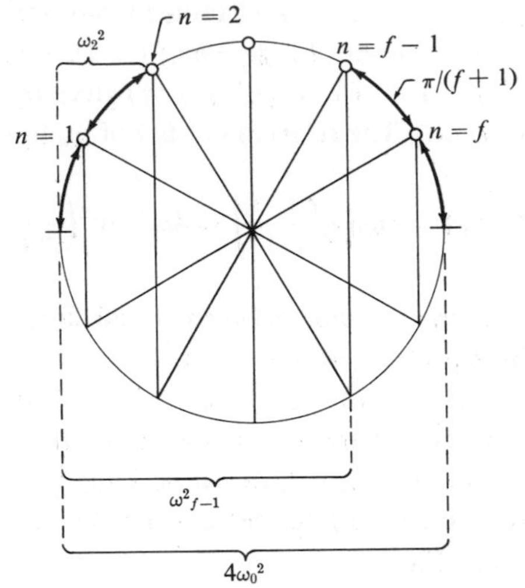

Figure 11.5

To find the proportionality constant for (11.65) which will normalize the eigenvector to $\tilde{\mathbf{A}}_n \mathbf{A}_n = 1$, the sum

$$\sum_{s=1}^{f} \sin^2 \frac{sn\pi}{f+1} = \tfrac{1}{2} \sum \left[1 - \cos s\left(\frac{2n\pi}{f+1}\right)\right]$$
$$= \tfrac{1}{2} f - \tfrac{1}{4} \sum [a^s + c.c.] \tag{11.66}$$

with $a \equiv e^{i2\pi n/(f+1)}$, must be evaluated. This may be done with the help of the "geometric" series with any ratio a:

$$\sum_{s=0}^{\infty} a^s = 1 + a \sum_{s=0}^{\infty} a^s = \frac{1}{1-a}$$
$$= 1 + a + a^2 + \cdots + a^N(1-a)^{-1} \tag{11.67}$$

with any N. Now the polynomial occuring in (11.66) may be evaluated as

$$\sum_{s=1}^{f} a^s = \sum_{s=1}^{\infty} a^s - \sum_{s=f+1}^{\infty} a^s = \frac{a}{1-a} - \frac{a^{f+1}}{1-a} = -1,$$

since $a^{f+1} = \exp\colon 2\pi i n = 1$ for any integer n. The final result for the sum (11.66) is $\frac{1}{2}(f+1)$ and so

$$u_{ns}(t) \equiv A_{ns}\, e^{-i\omega_n t} = \left(\frac{2}{f+1}\right)^{1/2} \sin\left(\frac{sn\pi}{f+1}\right) \cdot e^{-i\omega_n t} \tag{11.68}$$

are the components of an orthonormalized representative like (11.23), for an eigenmode of the f-particle string. Symmetrization in mass, by $\boldsymbol{\mu}^{+1/2}$ ($= m^{1/2}$ here), is unnecessary for a system of equal masses, as already remarked. The eigenvectors $\mathbf{A}_n$ themselves form an orthogonal set; this is expected from the

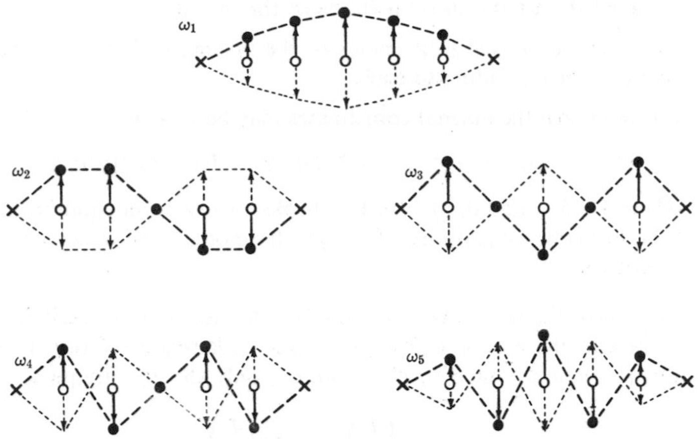

Figure 11.6

symmetry of the operator $(\boldsymbol{\mu}^{-1}\boldsymbol{\kappa}) = m^{-1}\boldsymbol{\kappa}$, and $\tilde{\mathbf{A}}_n \mathbf{A}_m = \delta_{nm}$ may also be checked directly in the same way that (11.66) was evaluated.

The motions in the individual simple harmonic modes of oscillation may be studied by forming real expressions like (11.9):

$$q_{ns} = \sin\left(\frac{sn\pi}{f+1}\right)\cos(\omega_n t + \gamma_n) \qquad (11.69)$$

when a "unit" amplitude is excited. Figure 11.6 indicates the five independent modes possible to a five-particle string, respectively having frequencies (11.63) with $\omega_1^2 = (2 - 3^{1/2})\omega_0^2$, $\omega_2^2 = \omega_0^2$, $\omega_3^2 = 2\omega_0^2$, $\omega_4^2 = 3\omega_0^2$, and $\omega_5^2 = (2 + 3^{1/2})\omega_0^2$. It is to be expected that the modes requiring the greatest relative displacements of neighbors will have the greatest frequencies.

An arbitrary superposition of the simple harmonic modes (11.69) will constitute the general solution like (11.18):

$$q_s(t) = \sum_{n=1}^{f} C_n \sin\left(\frac{sn\pi}{f+1}\right)\cos(\omega_n t + \gamma_n) \qquad (11.70)$$

for $q_1, q_2, \ldots, q_f$, with C_n and γ_n the requisite $2f$ arbitrary constants.

EXERCISES

11.1. A mass m is suspended (under uniform gravity) from an elastic cord of natural (unstretched) length l and a force constant $m\omega_0^2$ (as on page 153). A second, equal mass is suspended from the first one by a similar cord.

(a) Find the equilibrium positions of the masses.

(b) Find the normal frequencies of the system, and describe the vertical motions corresponding to each.

(c) Show that the normal coordinates may be written

$$Q_{1,2} = s^{-1/2}\{[2/(s \mp 1)]^{1/2}q_1 \mp [(s \mp 1)/2]^{1/2}q_2\},$$

where $s \equiv 5^{1/2}$ and q_1, q_2 are the displacements from equilibrium. Check that they obey equations of simple harmonic motion with the correct frequencies.

11.2. Show how the results (11.44) may be obtained more directly by working relative to the mass-center at rest—that is, letting $\bar{q} = 0$ in (11.45). When this restriction is used to eliminate q_2, only the eigenfrequency problem

$$\begin{pmatrix} \ddot{q}_1 \\ \ddot{q}_3 \end{pmatrix} = -\boldsymbol{\kappa}' \begin{pmatrix} q_1 \\ q_3 \end{pmatrix}$$

remains, with $\boldsymbol{\kappa}'$ a *symmetric* matrix.

11.3. (a) Argue that, of the nine degrees of freedom possessed by any three particles, five of the ones possessed by the system of §11.2 may be classed as "external" ones, varying only if there is translation and rotation "as a body" (as of a rigid rod joining the equilibrium points, in this case).

(b) Consider now only the "internal" motions—that is, relative to a frame in which $\bar{q} = 0$ and there is no net angular momentum about any point fixed in space. Argue that when the longitudinal vibrations already treated in Exercise 11.2 are also eliminated (the interparticle bonds are kept rigid, perhaps at some small angle to each other), then the two remaining degrees of freedom can only result in such rotations about the equilibrium line as are indicated by the figure, because of the symmetry

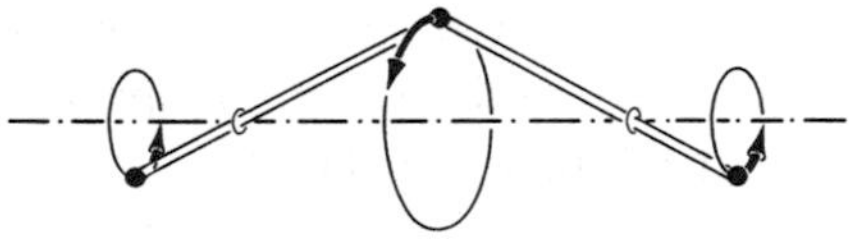

in the outer particles. (Thus, the general "internal" motion will consist of such rotations, conserving the internal angular momentum, superposed on the longitudinal vibrations.)

11.4. A system consists of six equal mass particles lying at the vertices of a regular hexagon when in equilibrium (molecular physicists use this as a model of a "benzene ring"). Investigate the part of the frequency spectrum that may be expected to arise from small displacements, $q_1, q_2, \ldots, q_6$, along the circle circumscribing the equilibrium hexagon. A lagrangian of the form

$$L = \sum_{s=1}^{6} \tfrac{1}{2}m[\dot{q}_s^2 - \omega_0^2(q_s - q_{s-1})^2],$$

with $q_0 \equiv q_6$, should be adequate

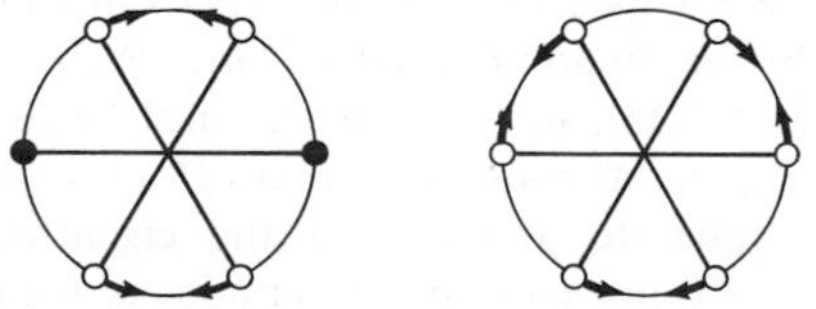

(a) If no further constraints are imposed, there should be six eigenfrequency magnitudes. Show that only four distinct ones arise, with the two highest ones associated with oscillations indicated by the figure. The first of these is two-fold degenerate.

(b) Show that the lowest nonvanishing eigenfrequency ($1.\ \omega_0$) is two-fold degenerate because it requires some superposition of the two orthogonal mass-center displacements proportional to

$$\bar{x} = q_1 - q_4 + \tfrac{1}{2}(q_2 - q_3 - q_5 + q_6),$$

$$\bar{y} = \tfrac{1}{2}\sqrt{3}(q_2 + q_3 - q_5 - q_6).$$

[Check that $\ddot{\bar{x}} = -\omega_0^2 \bar{x}$ and $\ddot{\bar{y}} = -\omega_0^2 \bar{y}$ follow from the above lagrangian.]

(c) Oscillations of $\bar{x}$, $\bar{y}$ must be regarded as "spurious" (see Exercise 7.16) for an isolated molecule, since then no restoring forces should arise from mass-center displacements. To get motions relative to the mass center only, impose the constraints $\bar{x} = \bar{y} = 0$. Show that now only the oscillations indicated in the figure arise.

11.5. Consider a system of degrees of freedom having the kinetic energy $T = \frac{1}{2} \sum m_s \dot{q}_s^2$, and a potential energy of the form (11.2). Suppose that the motion of each degree of freedom is damped by a force proportional to its momentum, $-\gamma m_s \dot{q}_s$, as in §3.2. Show that the motions can be found as in §11.1, with the real eigenfrequencies replaced by the complex ones,

$$\Omega_n = \pm(\omega_n^2 - \tfrac{1}{4}\gamma^2)^{1/2} - \tfrac{1}{2}i\gamma,$$

where the ω_n's are the real roots of the determinant (11.15). For small damping, $\Omega_n \approx \pm\omega_n - \frac{1}{2}i\gamma$ and the general solution has the form (11.18) multiplied by exp: $-\gamma t/2$. Thus every mode is *equally* damped in amplitude despite the differences of frequency.

11.6. Now suppose that the system of the preceding exercise is being excited by an "incident" disturbance which imposes a periodic force of frequency ω on each degree of freedom, as in

$$m_r(\ddot{q}_r + \gamma\dot{q}_r) + \textstyle\sum_s \kappa_{rs} q_s = \mathrm{Re}\,\{F_r e^{-i(\omega t + \delta_r)}\},$$

allowance being made for the possibility that the magnitude and phase of the external force are different at each degree of freedom.

(a) Show that these equations of motion have a special solution $q_r(\omega t) = \mathrm{Re}\{a_r e^{-i\omega t}\}$, and that any additional motions are only transient (compare §3.6).

(b) Use (9.74) to show that each excited amplitude can be expressed as a ratio of determinants, $a_r = D_r(\omega)/D(\omega)$, where

$$D(\omega) \equiv |\kappa_{rs} - \delta_{rs} m_r \Omega^2|,$$

with $\Omega^2 = \omega^2 + i\gamma\omega$, and $D_r(\omega)$ differs from $D(\omega)$ by having its rth column replaced by the "inhomogeneities" $F_1 e^{-i\delta_1}, F_2 e^{-i\delta_2}, \ldots$. The determinant $D(\omega)$ is just (11.15), and has the "natural" eigenfrequencies of the system, $\omega_1, \omega_2, \ldots, \omega_f$, as roots. However, $D(\omega) \neq 0$ here, even for an imposed frequency equal to any one of the eigenfrequencies, unless $\gamma = 0$; a damping $\gamma \neq 0$ prevents the amplitudes a_r from becoming infinite at the "resonances" $\omega = \omega_1$ or ω_2 or $\ldots$.

(c) Show that the above determinant must be expressible as

$$D(\omega) = m_1(\Omega^2 - \omega_1^2) \cdot m_2(\Omega^2 - \omega_2^2) \cdots m_f(\Omega^2 - \omega_f^2)$$
$$\equiv \left[\prod m_n(\Omega + \omega_n)\right]\left[\prod(\Omega - \omega_n)\right],$$

where the symbol $\prod$ stands for a product of f factors like the ones indicated ($n = 1, 2, 3, \ldots, f$ in the successive factors). Then, for small damping, $\Omega \approx \omega + \frac{1}{2}i\gamma$, show that

$$|a_r|^2 \approx \frac{|D_r(\omega)|^2\left[\prod m_n^{-2}(\omega + \omega_n)^{-2}\right]}{\prod\left[(\omega - \omega_n)^2 + \frac{1}{4}\gamma^2\right]}.$$

As a function of ω, this exhibits resonance peaks at $\omega = \omega_1, \omega_2, \ldots, \omega_f$. The width of each resonance (see (3.53)) is γ, insofar as the numerator can be taken to vary negligibly over spans $\Delta\omega \lesssim \gamma$.

CHAPTER 12

OSCILLATIONS IN A LINEAR CONTINUUM

A string or a rod is better represented as a linear *continuum* of mass elements dm than by any number of discrete particles, insofar as Classical Mechanics is valid for it. A continuum model has already been employed for rigid bodies in Chapter 10. That was simple to relate to the Mechanics of Particles because there is no need to describe internal motions within a rigid body. For a deformable body, it is just the internal motions which provide the characteristic concern. A continuum model of a *linear* deformable body may be constructed by letting the number of particles f, chosen to represent the string in the preceding section, tend to infinity.

12.1 A BOUNDED LINEAR CONTINUUM

A string of f point-masses, m, equally spaced at a distance l apart while in their straight-line equilibrium configuration, can be made to approach representation of a continuous string by letting $f \to \infty$ in such a way that $(f+1)l \to L$, a finite length of string. It is necessary to take $l \to 0$ such that $l = L/(f+1)$ with the same L at every stage of the limit process.

As the particles approach a continuous infinity in number, it becomes inconvenient to continue labeling them by the integers, s, as in the preceding chapter. It becomes appropriate to label each by a value of a continuous variable, such as its position x along the straight line of the equilibrium configuration. Then the displacements $q_s(t)$ are to be treated as forming a function $q(x, t)$, of the distance $0 < x < L$ as well as of time. The velocity $\dot{q}_s(t)$ goes over into the time rate of change of the displacement $q(x, t)$ of the particle *at* x, and hence into the *partial* derivative $(\partial/\partial t)q(x, t)$ for which x is held constant. Wherever the integer s had appeared otherwise than as just a label, the correspondence

$$x = sl = sL/(f + 1) \qquad \text{or} \qquad x/L = s/(f + 1) \tag{12.1}$$

must be recognized. Thus the integer label s, corresponding to a particle at a finite position x, must be made to approach infinity in such a way that, as $f \to \infty$, the ratio $s/(f + 1)$ remains equal to the finite fraction, x/L. Going from one particle to the next, formerly described as a change of the label s by $\Delta s = 1$, now corresponds to the change of x by $\Delta x = l\Delta s = l$, so that $l \to dx$ in the limit.

The discrete particles of the former treatment were each assigned a finite mass, m. Now a mass ρ per unit length will be assigned, such that $fm \to \rho L$ as $f \to \infty$. Thus $m \to 0$ in such a way that $m/l \equiv m/\Delta x \to \rho$.

Results for the Continuum Limit

The equations of motion (11.53) for the string of particles are now to be relabeled according to

$$\frac{\partial^2}{\partial t^2} q(x, t) = \frac{\kappa l}{m}\left\{\frac{q(x + l, t) - q(x, t)}{l} - \frac{q(x, t) - q(x - l, t)}{l}\right\}.$$

As $l \to 0$,

$$q(x \pm l, t) \to q(x, t) \pm l\left(\frac{\partial q}{\partial x}\right) + \tfrac{1}{2}l^2\left(\frac{\partial^2 q}{\partial x^2}\right),$$

and hence the equation approached in the limit is

$$\frac{\partial^2 q}{\partial t^2} = c^2 \frac{\partial^2 q}{\partial x^2}, \tag{12.2}$$

where

$$c^2 = \lim \frac{\kappa l^2}{m} = \lim \frac{\kappa l}{\rho} \quad (=T/\rho). \tag{12.3}$$

The parenthesis applies in the case (11.54), when the couplings arise from tension T applied to the string. Whatever the origin of the couplings, they must be representable by some finite limit, ρc^2, for $\kappa l = \kappa \Delta x$, if the equation (12.2) is to govern the motions of the continuum. This is a partial differential equation in the two continuous variables x, t, and its development here shows that it may be regarded as a continuous infinity of ordinary differential equations, "one for each x." It is an equation of a type that occurs frequently since it expresses first effects of departures from equilibrium, among a continuity of elements and in a "Hooke's Law" approximation.

The expression (11.55) for the potential energy should now be rewritten with $l \equiv \Delta x$, as

$$V = \sum_{s=0}^{f} \Delta x \tfrac{1}{2}(\kappa \Delta x)\left[\frac{q(x+\Delta x, t) - q(x, t)}{\Delta x}\right]^2,$$

having the limit

$$V = \int_0^L dx \tfrac{1}{2}\rho c^2 (\partial q/\partial x)^2. \tag{12.4}$$

For the kinetic energy,

$$\tau = \sum_{s=1}^{f} \tfrac{1}{2} m \dot{q}_s^2 = \sum \Delta x \, \tfrac{1}{2}\left(\frac{m}{\Delta x}\right)\left(\frac{\partial q}{\partial t}\right)^2,$$

and the limit is

$$\tau = \int_0^L dx \tfrac{1}{2}\rho (\partial q/\partial t)^2. \tag{12.5}$$

Each contribution to the energy here is positive-definite.

The eigenfrequencies which were found in (11.63), and expressed in terms of $\omega_0^2 \equiv \kappa/m$, can be reexpressed as

$$\omega_n = 2\left(\frac{\kappa}{m}\right)^{1/2} \sin \frac{n\pi}{2(f+1)} \to 2\left(\frac{\rho c^2/l}{\rho l}\right)^{1/2} \sin \frac{n\pi l}{2L}.$$

When $l \approx L/f \to 0$, the sine function becomes replaceable by its vanishingly small argument* and

$$\omega_n = 2(c/l)(n\pi l/2L) \to n\pi c/L. \tag{12.6}$$

* For instance, the Taylor expansion of the sine function gives $\sin \varepsilon \approx \varepsilon - \frac{1}{6}\varepsilon^3 + \cdots$ (see (9.134)).

The general motion (11.70) is now to be described by

$$q(x, t) = \sum_{n=1}^{\infty} C_n \sin\left(\frac{n\pi x}{L}\right)\cos\left(\frac{n\pi ct}{L} + \gamma_n\right), \tag{12.7}$$

as reference to (12.1) helps show.

Solution of the Partial Differential Equation

It is possible to obtain solutions of the partial differential equation of motion (12.2) directly, instead of starting with the discrete particle picture as the basis for (12.7). An approach to the solution analogous to trying $q_s = A_s \exp\colon -i\omega t$ for the particle displacement, as in (11.9), is to try $q(x, t) = A(x) \exp\colon -i\omega t$. Then $A(x)$ must be a solution of the ordinary differential equation

$$d^2A/dx^2 = -(\omega/c)^2 A(x), \tag{12.8}$$

which is also the limit of the secular equations (11.64).

The ratio ω/c occuring in the last equation is customarily designated

$$k \equiv \omega/c \tag{12.9}$$

and is called a "wave number" (for reasons to be seen). The general solution of (12.8) is

$$A(x) = C \sin(kx + \varphi), \tag{12.10}$$

with arbitrary amplitude C and phase φ. This describes a sinusoidal "wave" form which is repeated in value over each distance:

$$\lambda = 2\pi/k. \tag{12.11}$$

Then $k = 2\pi/\lambda$ is the number of wavelengths λ per 2π units of length, just as $\omega = 2\pi\nu$ is a number of oscillations in a time of 2π units.

A continuous string that is tied down at $x = 0$ and $x = L$, as was the particulate string treated in §11.3, is restricted to amplitudes conforming to the *boundary conditions*:

$$q(0, t) = 0 \qquad \text{and} \qquad q(L, t) = 0. \tag{12.12}$$

The first of these requires choosing the phase $\varphi = 0$ in the sine wave (12.11), and the second restricts the admissible wavelengths to such as lead to a node at $x = L$, as well as at $x = 0$:

$$\sin(kL) = \sin(2\pi L/\lambda) = 0 \qquad \text{or} \qquad 2\pi L/\lambda = n\pi, \tag{12.13}$$

with $n = 1, 2, \ldots$. The corresponding frequencies are

$$\omega_n = ck_n = 2\pi c/\lambda_n = n\pi c/L, \tag{12.14}$$

exactly the eigenfrequencies (12.6) found by the limit process. The resultant amplitudes,

$$A_n(x) = C \sin(n\pi x/L), \tag{12.15}$$

are just the ones already incorporated in the general solution (12.7).

The individual "eigenstates" of steady, simple-harmonic oscillation may be studied by forming real special solutions analogous to (11.69):

$$q_n(x, t) = \sin\left(\frac{n\pi x}{L}\right)\cos\left(\frac{n\pi c}{L}t + \gamma_n\right). \tag{12.16}$$

At given instants, they can be represented as in the generalization shown in Figure 12.1 of the modes exhibited in Figure 11.6. With the infinite number

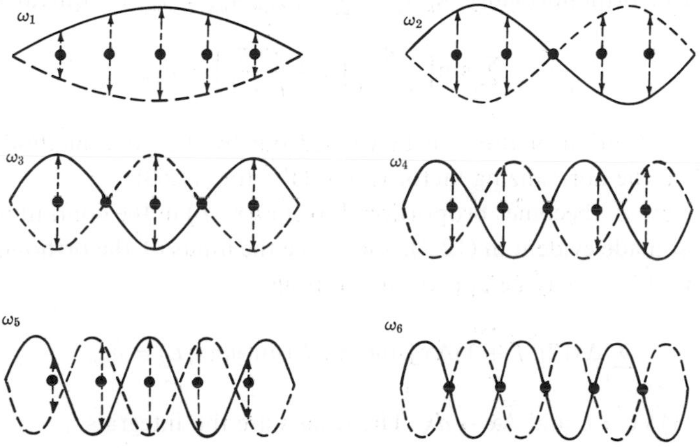

Figure 12.1

of particles characteristic of a continuous string, the number of possible eigenstates is infinite: $n = 1, 2, 3 \ldots, \infty$. The relative displacements at the positions of the five particles of Figure 11.6 are the same in both Figures 11.6 and 12.1. Thus if the motions in those positions only were to be observed, the five-particle model of the preceding section would have been adequate (see the discussion in §6.1).

An arbitrary start of the motion will generally excite a superposition (12.7) of the simple harmonic modes. Finding the amplitudes C_n and phases γ_n with which the various modes are excited introduces a problem of an important type, to be discussed in the next section.

12.2 ORTHOGONAL FUNCTION SETS

The amplitude functions $A_n(x)$ of (12.15) form what is called an "orthogonal set." Such sets serve analytic purposes comparable to those of orthogonal frames in vector spaces. Indeed, the set of functions $A_n(x)$, $n = 1, 2, 3 \ldots$, is the generalization, to a continuous enumeration of "vector" components by x, of the orthonormal eigenvector axes $\mathbf{A}_n$, $n = 1, 2, 3, \ldots$, used in the preceding chapter.

Orthonormality Conditions

General states of motion were presented in (11.24) as resolutions on an orthonormal frame of eigenvectors, $\mathbf{u}_n$ with $\mathbf{u}_n^\dagger \mathbf{u}_m = \delta_{nm}$. In the case of the string of equal masses, the real amplitude eigenvectors $\mathbf{A}_n$, with the components given by (11.68), become orthogonal, and the resolution of the general motion into those components was exhibited in (11.70). The statement of orthonormality $\tilde{\mathbf{A}}_n \mathbf{A}_m = \sum_{s=1}^{f} A_{ns} A_{ms} = \delta_{nm}$ is equivalent to

$$\frac{2}{f+1} \sum_{s=1}^{f} \sin\left(\frac{sn\pi}{f+1}\right) \sin\left(\frac{sm\pi}{f+1}\right) = \delta_{nm}. \tag{12.17}$$

A direct verification of this can be carried out by the same method as was used to find the normalizing factor $(2/f+1)^{1/2}$ for (11.68).

Now A_{ns} becomes proportional to $\sin(n\pi x/L)$ in the continuum limit, as was first made evident in (12.7), and hence the limits of the orthonormality conditions (12.17) may be approached through

$$\sum_{s=1}^{f} \Delta x[2/(f+1)\Delta x] \sin(n\pi x/L) \sin(m\pi x/L) = \delta_{nm},$$

with $(f+1)\Delta x \to L$ and $\Delta x \to dx$. The results are the integrals

$$(2/L) \int_0^L dx \sin(n\pi x/L) \sin(m\pi x/L) = \delta_{nm}. \tag{12.18}$$

These integrals are easily checked by direct integration. In effect, the normalizing factors $(2/f+1)^{1/2}$ have been replaced by $(2\,dx/L)^{1/2}$, and summation by integration. The set $A_n(x)$ of (12.15), when normalizing factors $C = (2/L)^{1/2}$ are adopted, forms an ortho*normalized* set,

$$(A_{ns} \to) \qquad A_n(x) = (2/L)^{1/2} \sin(n\pi x/L), \tag{12.19a}$$

over the domain $0 < x < L$, in the sense that

$$\left(\sum_{s=1}^{f} A_{ns} A_{ms} \to\right) \qquad \int_0^L dx A_n^*(x) A_m(x) = \delta_{nm}. \tag{12.19b}$$

The complex conjugation indicated here has no effect on the real functions; it is included merely for conformity to more general expressions of orthonormality, applying also to sets of complex functions, a generalization of the same type as that to $\mathbf{u}_n^\dagger \mathbf{u}_m = \delta_{nm}$ for basis vectors with discrete components. It is the vanishing of integrals like (12.19*b*), over some domain, which characterizes orthogonality of function sets.

It is helpful to visualize just how the orthogonality of the set (12.19) comes about. Figure 12.2 makes clear that the product of $A_2 \sim \sin(2\pi x/L)$ and $A_4 \sim \sin(4\pi x/L)$ will contribute a canceling negative value for every positive value, to the integrand of (12.19*b*). A similar picture can be formed for every pair of the set.

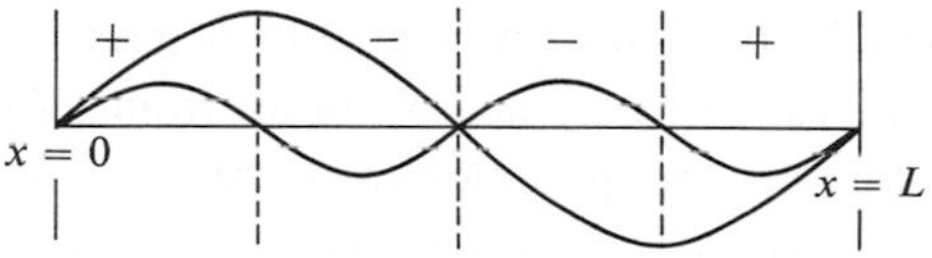

Figure 12.2

Analysis into Orthogonal Modes

Components of any vector on an orthogonal basis (frame) of unit vectors are presentable as projections of the vector on the unit vectors, as in (1.16) or (9.21) or (9.104) or (10.80). The components C_n of the resolution (11.24), giving $\boldsymbol{\mu}^{+1/2}\mathbf{q}$ on the basis $\mathbf{u}_n$, may be represented as

$$C_n = \mathbf{u}_n^\dagger (\boldsymbol{\mu}^{+1/2}\mathbf{q}). \tag{12.20}$$

This follows directly from multiplying (11.24) by $\mathbf{u}_n^\dagger$ and using the orthonormality conditions $\mathbf{u}_m^\dagger \mathbf{u}_n = \delta_{mn}$. A special case of this is the decomposition (11.70), giving the motions of the particulate string on the basis $\mathbf{u}_n$ of (11.68); the component motions are proportional to

$$\begin{aligned}\tilde{\mathbf{A}}_m \mathbf{q}(t) &= \sum_s \left(\frac{2}{f+1}\right)^{1/2} \sin\left(\frac{sm\pi}{f+1}\right) q_s(t) \\ &= [(f+1)/2]^{1/2} C_m \cos(\omega_m t + \gamma_m).\end{aligned} \tag{12.21}$$

In the limit of the continuous string, the summation is replaced by the integration

$$\int_0^L dx (2/L)^{1/2} \sin(m\pi x/L) q(x, t), \tag{12.22}$$

in presenting the components of $q(x, t)$ "along $A_m(x)$."

Any sufficiently well-behaved function $f(x)$, defined over the same domain as that of the orthogonal set $A_n(x)$, may be resolved on this set according to

$$f(x) = \sum_n a_n A_n(x). \tag{12.23}$$

The components a_n here are to be obtained from integrations like

$$a_n = \int_0^L dx A_n^*(x) f(x). \tag{12.24}$$

This can be seen without detailed resort to the preceding considerations simply by multiplying the series (12.23) by $A_m^*(x)$ and integrating with the help of the orthonormality conditions (12.19*b*) that characterize the set. Of course, the representability of any given $f(x)$ by a series like (12.23) depends on, among other things, the existence of finite results for the integrations (12.24). Functions that are too pathological to be thus integrable are simply excluded, as not being needed to represent actual physical circumstances.

The expression for the general solution (12.7) is an example of such a decomposition as (12.23). In this case, it is more appropriate (11.25) to designate the components

$$a_n \to Q_n(t) \equiv (L/2)^{1/2} C_n \cos(\omega_n t + \gamma_n). \tag{12.25}$$

These are just the projections (12.22),

$$\begin{aligned} Q_n(t) &= \int_0^L dx A_n^*(x) q(x, t) \\ &= (2/L)^{1/2} \int_0^L dx \sin(n\pi x/L) q(x, t), \end{aligned} \tag{12.26}$$

and serve as the "normal coordinates" of the continuous string. A *discrete* infinity of degrees of freedom, $Q_1, Q_2, \ldots, Q_n(t), \ldots$ may thus be substituted for the *continuous* infinity, $q(x, t)$; whereas the continuous label x is bounded, $0 < x < L$, the integer label n is not. For this case of the continuous string, the transformation (11.28), which substitutes normal for initial coordinates according to $\mathbf{Q} = S\mathbf{q}$, becomes

$$Q_n = \sum_s S_{ns} q_s \to \int_0^L dx A_n^*(x) q(x, t). \tag{12.27}$$

There is a replacement of discrete elements of a transformation matrix by a continuity of them,

$$S_{ns} \to dx A_n^*(x). \tag{12.28}$$

The proportionality to A_n^* here is the analogue of the connection $S_{sn}^{-1} = A_{ns}$ (11.27); for unitary rotations, $\mathbf{S}^{-1} = \mathbf{S}^\dagger$, the connection becomes $S_{ns} = A_{ns}^*$, as compared to (12.28).

The Excitation Problem

The preceding development systematizes the procedures needed to find what modes of oscillation will be excited when the motion is started in some given way. According to the general solution form (12.7), the problem is a matter of finding amplitudes C_n and phases γ_n for the various modes $A_n(x)$ of Figure 12.1, or, equivalently, the various normal oscillations $Q_n(t)$ of (12.25):

$$q(x, t) = \sum_n Q_n(t) A_n(x), \qquad (12.29)$$

according to (12.7) and the definitions (12.19) and (12.25). This form is just the continuum limit of (11.25).

How a "start" of the motion may be "given" should be considered next. There is involved a generalization of the fact that giving $q_s(0)$ and $\dot{q}_s(0)$ at $t = 0$ suffices to determine the motion $q_s(t)$ that will ensue in given circumstances. The corresponding initial conditions in the case of a continuous string of mass elements labeled by x are $q(x, 0)$ and* $\mathring{q}(x, 0) \equiv (\partial q/\partial t)_0$; hence an initial distortion and "applied impulse" (per unit mass),

$$q(x, 0) = q_0(x) \qquad \text{and} \qquad \mathring{q}(x, 0) = v_0(x), \qquad (12.30)$$

will be presumed given. Each of these functions must be ones which vanish at $x = 0$ and $x = L$, if they are to describe possible starts of a string tied down at those end points.

The motion (12.29) must now be such that

$$\begin{aligned} q(x, 0) &= \sum_n Q_n(0) A_n(x) = q_0(x), \\ \mathring{q}(x, 0) &= \sum_n \dot{Q}_n(0) A_n(x) = v_0(x). \end{aligned} \qquad (12.31)$$

The problem is thus reduced to a matter of finding two series representations of the type (12.23), for given functions $f(x) = q_0(x)$ and $f(x) = v_0(x)$. The coefficients $a_n = Q_n(0)$ and $a_n = \dot{Q}_n(0)$, which will make the respective series

---

* *Partial* time-derivatives will be denoted $\mathring{q}(x, t)$, versus the total time-derivatives $\dot{q}_s(t)$ or $\dot{Q}_n(t)$.

represent the given functions, are to be found from integrations like (12.24):

$$Q_n(0) = \left(\frac{2}{L}\right)^{1/2} \int_0^L dx \sin\left(\frac{n\pi x}{L}\right) \cdot q_0(x),$$
$$\dot{Q}_n(0) = \left(\frac{2}{L}\right)^{1/2} \int_0^L dx \sin\left(\frac{n\pi x}{L}\right) \cdot v_0(x), \tag{12.32}$$

for each mode of oscillation. These in effect replace the giving of two continuous *functions*, $q_0(x)$ and $v_0(x)$, with the discrete starting values $Q_n(0)$ and $\dot{Q}_n(0)$ for each normal oscillation. The amplitudes C_n and phases γ_n of the normal oscillations (12.25) will be determined by

$$Q_n(0) = (L/2)^{1/2} C_n \cos \gamma_n,$$
$$\dot{Q}_n(0) = -\omega_n (L/2)^{1/2} C_n \sin \gamma_n. \tag{12.33}$$

The ensuing motion $q(x, t)$ is thus obtained in the form of a specific superposition (12.7).

As an illustration of how the series representation of the motion works out, consider the following: the motion is started from the straight-line equilibrium configuration by whacking the middle third of the string with a board of suitable width, and thus giving the middle section an initial velocity, v_0. The starting functions then are $q_0(x) \equiv 0$ and

$$v_0(\tfrac{1}{3}L < x < \tfrac{2}{3}L) = v_0, \qquad v_0(x < \tfrac{1}{3}L) = v_0(x > \tfrac{2}{3}L) = 0. \tag{12.34}$$

The first of the integrals (12.32) vanishes when $q_0(x) \equiv 0$, and then (12.33) shows that $\gamma_n = -\pi/2$ may be chosen for every n if C_n is determined from

$$\omega_n C_n = (2/L)^{1/2} \dot{Q}_n(0).$$

The second of the integrations (12.32) now yields

$$\begin{aligned}\omega_n C_n &= (2/L) v_0 \int_{(1/3)L}^{(2/3)L} dx \sin\left(\frac{n\pi x}{L}\right) \\ &= (2v_0/L)(L/n\pi)[\cos(\tfrac{1}{3}n\pi) - \cos(\tfrac{2}{3}n\pi)] \\ &\equiv (2v_0/n\pi)\left[\cos\frac{(n-6)\pi}{3} - \cos\frac{2(n-6)\pi}{3}\right].\end{aligned} \tag{12.35}$$

The object of displaying the equivalent last line is to show that the bracket has the same value for $n + 6$ as it has for n, and so needs to be evaluated separately only for $n = 1, 2, \ldots, 6$. The results are

$$\begin{aligned}\omega_n C_n &= (2v_0/n\pi) && \text{for } n = 1, 5, 7, 11, 13, 17 \ldots, \\ \omega_n C_n &= -2(2v_0/n\pi) && \text{for } n = 3, 9, 15, 21, \ldots, \\ \omega_n C_n &= 0 && \text{for \textit{all even n}.}\end{aligned} \tag{12.36}$$

The last of these—that is, the failure to excite any of the modes with even n—could have been foreseen. The given exciting function, $v_0(x)$ of (12.34) is an *even* function in the sense that it has equal values at points symmetrically placed about the middle of the string: $v_0(L - x) = v_0(x)$. Then only the "even modes," with the property $A_n(L - x) = A_n(x)$, can be excited. Since

$$A_n(L - x) = (2/L)^{1/2} \sin[n(L - x)\pi/L] = (-)^{n+1} A_n(x),$$

the even modes are those which have been labeled by odd values of the integer n. The outcome is a rudimentary manifestation of what is called "parity conservation," by which the "parities" of the modes excited are restricted to be the same as the "parity" of the exciting "source" (the initial exciting function), whenever that *has* a definite parity. The basis for the assertions here will be clarified in a more general context below (page 363).

An explicit series representation of the type (12.23) has now been found for the function $v_0(x)$ of (12.34):

$$\begin{aligned} v_0(x) &= \sum_n \omega_n C_n \sin(n\pi x/L) \\ &= \frac{2v_0}{\pi}\left\{\sin\left(\frac{\pi x}{L}\right) - \frac{2}{3}\sin\left(\frac{3\pi x}{L}\right) + \frac{1}{5}\sin\left(\frac{5\pi x}{L}\right) + \frac{1}{7}\sin\left(\frac{7\pi x}{L}\right) - \cdots\right\}. \end{aligned} \tag{12.37}$$

The ensuing motion (12.29) is then

$$\begin{aligned} q(x, t) &= \sum_n C_n \sin(n\pi x/L)\sin(n\pi ct/L) \\ &= \frac{2L}{\pi^2}\left(\frac{v_0}{c}\right)\left\{\sin\left(\frac{\pi x}{L}\right)\sin\left(\frac{\pi ct}{L}\right) - \frac{2}{9}\sin\left(\frac{3\pi x}{L}\right)\sin\left(\frac{3\pi ct}{L}\right)\right. \\ &\qquad \left. + \frac{1}{25}\sin\left(\frac{5\pi x}{L}\right)\sin\left(\frac{5\pi ct}{L}\right) + \frac{1}{49}\sin\left(\frac{7\pi x}{L}\right)\sin\left(\frac{7\pi ct}{L}\right) - \cdots\right\}. \end{aligned} \tag{12.38}$$

The smaller amplitudes of the high-frequency modes make them mere ripples on the over-all oscillation. The initial conditions are re-established at the end of each fundamental period $2\pi/\omega_1 = 2L/c$; the straight-line configuration $q \equiv 0$ is resumed in each half of this period, but with alternate reversals of $\mathring{q} = \pm v_0(x)$.

An idea of how adding terms to the series (12.37) makes it approach representation of the given function $v_0(x)$ of (12.34) can be gained from Figure 12.3. Curves showing sums of the terms through $n = N$ are plotted, for $N = 1$, 3, and 15.

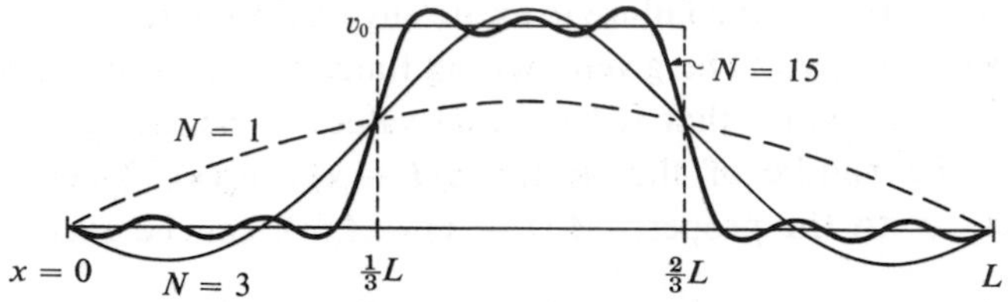

Figure 12.3

The Modal Energies

When the motion is analyzed into the orthogonal modes $A_n(x)$ as in (12.29), the kinetic energy (12.5) becomes

$$\tau = \tfrac{1}{2}\rho \sum_{n,\,n'} \dot{Q}_n \dot{Q}_{n'} \int_0^L dx A_n(x) A_{n'}(x),$$

having, for each n, terms labeled by n' which vanish except for $n' = n$, because of the orthonormality (12.19*b*). Thus

$$\tau = \sum_n \tfrac{1}{2}\rho \dot{Q}_n^2, \tag{12.39}$$

a sum of individual kinetic energies in each mode.

Evaluation of the potential energy (12.4) requires considering

$$\partial q/\partial x = \sum_n Q_n(t) A'_n(x)(n\pi/L),$$

where

$$A'_n(x) \equiv (2/L)^{1/2} \cos(n\pi x/L). \tag{12.40}$$

This is another orthogonal function set* on the interval $0 < x < L$, satisfying the same orthonormality conditions (12.19*b*) as does the set of sines, $A_n(x)$ of (12.19*a*). Making use of this property leads to

$$V = \tfrac{1}{2}\rho c^2 \sum_n (n\pi/L)^2 Q_n^2. \tag{12.41}$$

Since $n\pi c/L = \omega_n$, the total energy may be expressed as

$$E = \sum_n \tfrac{1}{2}\rho[\dot{Q}_n^2 + \omega_n^2 Q_n^2], \tag{12.42}$$

with an individual energy for each mode which differs from (11.37) as to the scale adopted for the definition of the normal coordinates.

---

* It will be "complete" if supplemented with $A'_0 \equiv 1/L^{1/2}$.

Each "modal" energy E_n is individually conserved, as substitution of (12.25) will confirm:

$$E_n \equiv \tfrac{1}{2}\rho[\dot{Q}_n^2 + \omega_n^2 Q_n^2] = \tfrac{1}{4}\rho L \omega_n^2 C_n^2, \qquad \text{constant.} \tag{12.43}$$

In the example of (12.36), this amounts to $(\rho v_0^2 L/n^2\pi^2)$ in each of the modes $n = 1, 5, 7, 11, \ldots$, and is multiplied by four in the modes with $n = 3, 9, 15, \ldots$. To sum these contributions, it is convenient to relabel the modes by $\nu = \frac{1}{2}(n-1) = 0, 1, 2, 3 \ldots$ when $n = 1, 3, 5, 7, \ldots$. Then the total may be evaluated* as

$$E = (\rho v_0^2 L/\pi^2)(4/3)\sum_{\nu=0}^{\infty}(2\nu+1)^{-2} = \tfrac{1}{2}\rho v_0^2(\tfrac{1}{3}L), \tag{12.44}$$

as to be expected, considering the start (12.34) that was given to the motion.

Fourier Analysis

Every member of the orthogonal set $A_n(x)$ of (12.19) vanishes at $x = 0$ and $x = L$. This makes the set suitable for series representations of displacements in a string tied down at those end points, but representations of less special functions are sometimes also called for. A more generally useful set may be formed by supplementing the sine functions $A_n(x)$ with the cosine functions $A_n'(x)$ of (12.40). The cosines have their maximum magnitudes just where the corresponding sines vanish; hence the joint set of A_n and A_n' is better suited for series representations of functions which may have non-vanishing values anywhere.

The famous "Fourier series" representations make use of a set containing both sines and cosines. The explicit fourier set to be introduced here is indicated in the following series for an arbitrary $f(x)$:

$$f(x) = \sum_{n=1}^{\infty}\left[a_n \sin\frac{2\pi n x}{L} + b_n \cos\frac{2\pi n x}{L}\right] + b_0 \tag{12.45a}$$

for $-\frac{1}{2}L < x < \frac{1}{2}L$. That the members of the set here are mutually orthogonal over the specified interval is readily checked; for example,

$$\int_{-(1/2)L}^{+(1/2)L} dx \sin\left(\frac{2\pi n x}{L}\right) \cdot \cos\left(\frac{2\pi n' x}{L}\right) = 0$$

because the integrand is odd in $x \leftrightarrow -x$, and the interval is symmetric in positive and negative x's. As a consequence, no term of the series (12.45*a*)

---

* For example, see Exercise 12.2.

can be represented as a linear combination of the others, and is said to be "linearly independent." Moreover, no function $f(x) \not\equiv 0$ can be found for which the amplitudes

$$a_n = (2/L) \int_{-(1/2)L}^{+(1/2)L} dx \sin\left(\frac{2\pi nx}{L}\right) \cdot f(x),$$

$$b_n = (2/L) \int_{-(1/2)L}^{+(1/2)L} dx \cos\left(\frac{2\pi nx}{L}\right) \cdot f(x) \qquad (12.45b)$$

$$b_0 = \int_{-(1/2)L}^{+(1/2)L} (dx/L) f(x)$$

all vanish, and hence no terms can be added to the series which are not expressible as linear combinations of terms already present. The Fourier set is actually "complete" in that respect. It may be noticed that a term b_0, the coefficient of the cosine with $n = 0$, has been added for completeness; it represents the average of $f(x)$ around which each other term oscillates. No proof of the completeness is being offered here; that reduces to a tedious question of how broad a class of functions may be represented with what accuracy.

The Fourier set introduced here differs from the sets $A_n(x)$ and $A_n'(x)$ discussed above in that the definition of the end position L has been changed by replacing it with $\frac{1}{2}L$. Moreover, the interval of orthogonality has been retained at an over-all length L by an extension of it to negative x-values. A qualitative discussion of why these steps were taken will lend insight into the workings of Fourier representations. If the interval $0 < x < L$ were retained for the joint set of the original sines, $A_n(x)$, and cosines, $A_n'(x)$, then the cosines would *not* be orthogonal to the sines; integrals like (12.24), with $f(x) = A_n'(x)$, do not in general vanish. Any one of the cosines can be expressed as a linear combination of just the sines (and vice versa), despite the fact that $A_n'(0) \neq 0$ is then being represented by a sum of terms which vanish at $x = 0$ (Exercise 12.3). This representation still works in the sense that properly finite values of A_n' are obtained for x-values as close to $x = 0$ as desired, by including enough sine terms in the series. However, obtaining these finite values close to $x = 0$ requires rapid transition from the vanishing values at exactly $x = 0$, attainable only by giving great weight to very short wavelengths, $2L/n$, in the series of the sines. The series must be very prolonged (to high n) before it begins to yield satisfactory approximations. This is really why it becomes desirable to include cosines as additional members of the set, available for representing any finite values wherever the sines vanish (and vice versa). On the other hand, it is also desirable to retain the orthogonality property; it is important for unambiguous determinations, like (12.24), of the coefficients in the series representations, and also for such conclusions, about the sufficient generality (completeness) of the representations provided, as were discussed in connection with the coupled-oscillator solution (11.18). Orthogonality is recaptured by extending the interval of evaluation from $0 < x < L$ to $-L < x < L$, since the sines and cosines have opposite parity in $x \leftrightarrow -x$. A redefinition $2L \to L$ was also adopted for the Fourier set in (12.45*a*), merely to have L rather than $2L$ have the meaning of the full interval length. Unlike the individual sets $A_n(x)$ or $A_n'(x)$, the Fourier set possesses orthogonality on *any* span of length L, whatever its starting point (Exercise 12.4).

An important property of the Fourier set is its subclassifiability into odd and even members (the sines and cosines). It is clear from (12.45*b*) that when $f(x)$ is even, $f(-x) = f(x)$, then all the a_n's vanish and only the cosines appear in the series representation. For odd $f(x) = -f(-x)$, $b_n = 0$ and only the sines appear. Of course every function is expressible as a sum of odd and even functions, equal to $\frac{1}{2}[f(x) \pm f(-x)]$ respectively. Such properties are at the basis of the "parity conservation" alluded to in the discussion of the example (12.36).

More compact Fourier representations are obtained by replacing the sines and cosines with differences and sums of complex exponentials (3.32). Then the Fourier series (12.45) may be written

$$f(x) = \sum_{n=-\infty}^{\infty} \alpha_n e_n(x), \tag{12.46a}$$

containing the "complex Fourier set"

$$e_n(x) \equiv L^{-1/2} e^{2\pi i n x/L} \qquad \begin{cases} -\frac{1}{2}L < x < +\frac{1}{2}L, \\ n = 0, \pm 1, \pm 2, \ldots . \end{cases} \tag{12.47a}$$

This set has the orthonormality properties

$$\int_{-(1/2)L}^{+(1/2)L} dx e_n^*(x) e_{n'}(x) = \delta_{nn'}, \tag{12.47b}$$

as can be very easily checked. The coefficients needed for the representation (12.46*a*) can be obtained from the integrations

$$\alpha_n = \int_{-(1/2)L}^{+(1/2)L} dx e_n^*(x) f(x), \tag{12.46b}$$

as follows from multiplying (12.46*a*) by any one member of $e_n^*(x)$ and integrating with the help of the orthonormality properties. If a *real* function $f(x)$ is to be represented, $\alpha_{-n} = \alpha_n^*$ because $e_{-n}(x) = e_n^*(x)$. Notice that $\alpha_0/L^{1/2}$ is equivalent to b_0 of (12.45). The complex Fourier series (12.46*a*) can always be converted into the "trigonometric" one (12.45*a*) by putting each exponential, exp: $2\pi inx/L$, equal to $\cos(2\pi nx/L) + i\sin(2\pi nx/L)$. Like the real Fourier set, the complex one actually possesses orthogonality on *any* span of length L (Exercise 12.4).

12.3 RUNNING WAVES

In the foregoing, motions in a linear continuum were analyzed into "stationary eigenstate," or "standing-wave," components. For example, the general motion (12.29) is presented as, at every moment, a superposition of steady motions like those of Figure 12.1, each proportional to some $A_n(x) \sim \sin(n\pi x/L)$. Each eigenmode $A_n(x)$ fits the description of a "stationary state" in that it is expressible independently of time, and only the amplitude $Q_n(t)$, with which it is excited, may vary in time. The mode is called a "standing wave" because the position in space of any one of its phases, such as its nodes, remains stationary as time goes on. Such an analysis is particularly suited for describing the "bound" motions in a string tied down at two points, because these then become stationary nodes in a selected set of standing waves like the $A_n(x)$.

An alternative type of analysis is possible, one particularly suited for following the time evolution of initially localized disturbances which may be started in the continuum. It may be introduced by considering afresh the solution of the equation (12.2) for disturbances (displacements) in a continuum,

$$\frac{\partial^2 q}{\partial x^2} = \frac{\partial^2 q}{c^2\, \partial t^2}. \tag{12.48}$$

In the new contexts this will be referred to as the "wave (-propagation) equation." The constant c, characteristic of the continuum, has the dimensions of a speed and is called the "propagation velocity" in that continuum.

The General Running Waves

The "wave equation" obviously demands that the disturbance depend on the lengths x and ct in a similar way. It is promising to try satisfying it with a functional form $f(\xi)$ in which the argument ξ is simply a linear combination of the lengths: $\xi = ax + b(ct)$ with some a, b. Then $\partial f/\partial x = f'(\xi) \cdot a$, $\partial f/c\,\partial t = bf'(\xi)$,

$$\frac{\partial^2 f}{\partial x^2} = a^2 f''(\xi) \qquad \text{and} \qquad \frac{\partial^2 f}{c^2\,\partial t^2} = b^2 f''(\xi),$$

where each "prime" stands for a derivative of the function with respect to its argument. It is now obvious that the wave equation will be satisfied by f if $b^2 = a^2$, or $b = \pm a$. Thus *any* twice-differentiable function may be taken for $q(x, t)$ if x, t appear in it only in the combination $x - ct$ or as $x + ct$. As a result, the general solution of the wave equation may be written

$$q(x, t) = f(x - ct) + g(x + ct), \tag{12.49}$$

with $f(\xi)$ and $g(\eta)$ each a quite arbitrary functional form (it need only be twice-differentiable). The sufficient generality of the solution is attested by its adaptability to arbitrary starting conditions, $q(x, 0) \equiv q_0(x)$ and $\mathring{q}(x, 0) \equiv v_0(x)$. It is only necessary that the forms $f(\xi)$ and $g(\eta)$ be chosen so that

$$\begin{aligned} q(x, 0) &= f(x) + g(x) = q_0(x), \\ \mathring{q}(x, 0) &= -cf'(x) + cg'(x) = v_0(x). \end{aligned}$$

Combining the first of these with an integral of the second from any point of x will yield

$$q(x, t) = \tfrac{1}{2}\left[q_0(x - ct) + q_0(x + ct) + c^{-1}\int_{x-ct}^{x+ct} v_0(\xi)\,d\xi\right] \tag{12.50}$$

for the expression (12.49).

The functions $f(x - ct)$ and $g(x + ct)$ may respectively be called a "right-running wave" and a "left-running wave." The aptness of these names may be clarified by reference to Figure 12.4. Figure 12.4(a) merely defines a possible wave form $f(\xi)$. Figure 12.4(b) shows $f(x - ct)$ at $t = 0$ and at a time $t > 0$. Clearly, the wave form is being propagated to increasing values of x (to the "right") as time goes on. Corresponding "phases" of the wave (such as the node for $\xi = \xi_1$) occur at distances ct apart, upon a time lapse t, and hence c is the "phase velocity" with which the wave is propagated.

A wave may be ascribed a definite wavelength and frequency only when it has a harmonic (trigonometric) form—for example, when $f(x - ct)$ is the real or imaginary part of

$$e^{ik(x-ct)} \equiv e^{i(kx-\omega t)}. \tag{12.51}$$

A constant k, having the dimensions of an inverse length, has had to be

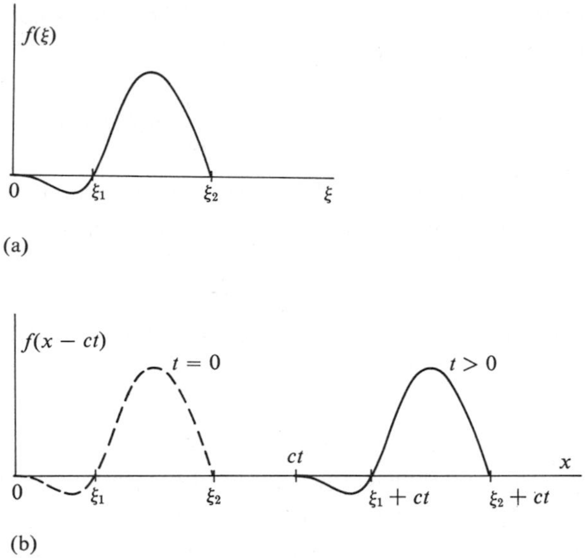

Figures 12.4 (a and b)

introduced here in order to make the exponent a properly dimensionless number. Then, at any given moment t, the phase of the harmonic wave will be effectively the same at two positions x and x' such that $kx' - kx = 2\pi$, since exp: $2\pi i = +1$. The distance between the two positions, $2\pi/|k| = \lambda$, is just the wavelength (12.11). Similarly, at any given position x, effectively the same phase is repeated at times that are apart by $2\pi/|\omega| = 2\pi/c\,|k|$; hence this is the period, and $\nu = |\omega|/2\pi$ is the frequency, of the wave. The great importance of the harmonic wave forms stems from the "Fourier-analyzability" of arbitrary forms, as in (12.46), and as will be explored further below. Various physical operations are natural "Fourier-analyzers," such as the various devices that break up "white light" into component "colors" (frequencies).

An Example of Impulse Propagation

The form (12.50) is convenient to use without preliminary Fourier analysis for disturbances that meet no boundaries, of the continuum characterized by a given c. As an example, consider a disturbance started as in the problem (12.34) to (12.38)—that is, by supplying an impulse, v_0 per unit mass, to a certain span, l, of the string in its equilibrium configuration. With $q_0(x) \equiv 0$, (12.50) reduces to

$$q(x, t) = (2c)^{-1} \int_{x-ct}^{x+ct} v_0(\xi)\, d\xi. \tag{12.52}$$

It will be convenient to put the origin $x = 0$ at the middle of the initially disturbed span, and then $v_0(-\frac{1}{2}l < \xi < \frac{1}{2}l) = v_0$ and $v_0(|\xi| > \frac{1}{2}l) = 0$.

Depending on whether the limits $x \pm ct$ of the integral (12.52) fall within, or outside of, the span in which $v_0(\xi) \neq 0$, several expressions will result, each valid in a different region of x at a given moment t, as indicated in Figure 12.5(a). Thus, for $x + ct < -\frac{1}{2}l$, or for $x - ct > +\frac{1}{2}l$, no nonvanishing

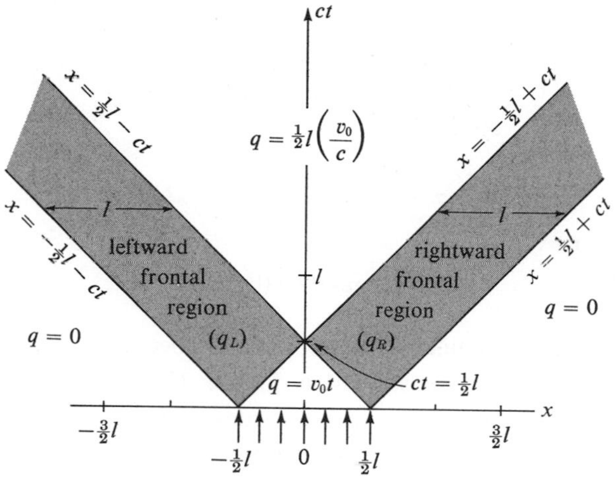

Figure 12.5(a)

portion of $v_0(\xi)$ is enclosed by the integral and $q \equiv 0$; this applies to the regions $x < -\frac{1}{2}l - ct$ and $x > +\frac{1}{2}l + ct$, which the disturbance could not yet have reached at the finite propagation velocity c. When $-\frac{1}{2}l < x - ct < x + ct < \frac{1}{2}l$, possible only at times $t < l/2c$, the integrand is $v_0(\xi) = v_0$ over the entire integration range, and $q = v_0 t$; this result, a displacement of a central portion with the initially imparted velocity v_0, is to be expected in the earliest moments. Because energy is propagated into neighboring portions, the early flat-topped central part $q = v_0 t$, shown in Figure 12.5(b), is narrowed to a span $(l - 2ct)$ by the time $t(<l/2c)$; it has a decreasing kinetic energy, $\frac{1}{2}\rho v_0^2(l - 2ct)$, which finally vanishes at $t = l/2c$. The rest of the initial energy $\frac{1}{2}\rho v_0^2 l$ is carried by the sloped border regions corresponding to the shaded frontal parts marked L (left) and R (right) in Figure 12.5(a). The integral (12.52) gives

$$q_L(x, t) = (v_0/2c)[x + ct + \tfrac{1}{2}l] = q_R(-x, t) \qquad (12.53)$$

for those regions. Within them $\partial q/\partial t = \frac{1}{2}v_0$ and $|\partial q/\partial x| = v_0/2c$, and so they possess the energy (see (12.4) and (12.5)) $\frac{1}{2}\rho(\frac{1}{2}v_0)^2 + \frac{1}{2}\rho c^2(v_0/2c)^2 = \frac{1}{4}\rho v_0^2$ per unit length, divided equally between the kinetic and potential forms. The

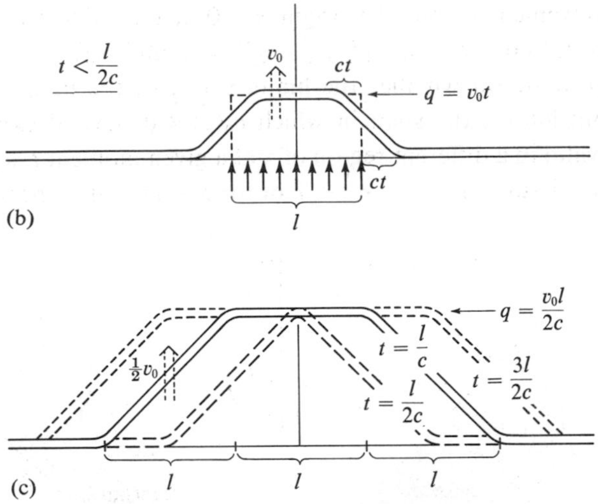

Figures 12.5 (b and c)

central displacement comes to a stop after it becomes equal to $q = lv_0/2c$, at $t = l/2c$. Thereafter, as shown in Figure 12.5(c), the region of static, maximal displacement spreads to both sides, and continues to be bordered by the sloped strips, each of breadth l, which contain all the energy (after $t = l/2c$). The ultimate outcome of the initial impulse is thus seen to be a static displacement by $q = lv_0/2c$ of a section broadening as $2ct - l$; there is a continued flow of the energy to both sides, carried in the outwardly moving, frontal strips of width l.

The foregoing applies so long as no obstacle to the propagation of the impulse is met. In the problem of (12.34) to (12.38), tied-down positions are encountered when the disturbed region has spread to three times its initial span (there designated $\frac{1}{3}L < x < \frac{2}{3}L$ instead of $-\frac{1}{2}l < x < \frac{1}{2}l$). Thus the configuration attained in Figure 12.5(c) at $t = l/c$, spanning $-\frac{3}{2}l < x < +\frac{3}{2}l$, is the latest one unaffected by the obstacles to further spread. Up to the time $t = l/c = L/3c$, the piecewise representations developed in this section should be entirely equivalent to the Fourier series (12.38). At the moment $t = L/3c$ itself, the series gives

$$q = \frac{2L}{\pi^2}\left(\frac{v_0}{c}\right)\frac{3^{1/2}}{2}\left\{\sin\frac{\pi x}{L} - \frac{1}{25}\sin\frac{5\pi x}{L} + \frac{1}{49}\sin\frac{7\pi x}{L} - \cdots\right\}. \quad (12.54)$$

The first term alone should provide a fair approximation since the first correction term has only a 4% amplitude. It is easy to check that there is close agreement between the first term and the exact piecewise representation: $q(0 < x < \frac{1}{3}L) = v_0 x/2c$, $q(\frac{1}{3}L < x < \frac{2}{3}L) = v_0 L/6c$ and $q(\frac{2}{3}L < x < L) = (v_0/2c)(L - x)$.

The Flux of Energy

The preceding example called attention to the propagation of energy associated with running waves. An "energy density" (that is, per unit of *length*) may be defined as

$$U(x, t) = \tfrac{1}{2}\rho[(\partial q/\partial t)^2 + c^2(\partial q/\partial x)^2], \tag{12.55}$$

according to (12.4) and (12.5). Consider how this energy at a point of the continuum may vary in time:

$$\partial U/\partial t = \rho[(\partial q/\partial t)(\partial^2 q/\partial t^2) + c^2(\partial q/\partial x)(\partial^2 q/\partial x\,\partial t)].$$

The acceleration $\partial^2 q/\partial t^2$ may be replaced by $c^2\,\partial^2 q/\partial x^2$, according to the wave-propagation equation, and this enables writing

$$\partial U/\partial t = \rho c^2(\partial/\partial x)[(\partial q/\partial t)(\partial q/\partial x)] \equiv -\partial S/\partial x. \tag{12.56}$$

The quantity thus defined,

$$S(x, t) \equiv -\rho c^2(\partial q/\partial t)(\partial q/\partial x), \tag{12.57}$$

is called the *energy flux*. The suitability of this name becomes evident when the rate of change of the energy within a finite interval, $\Delta x = x_2 - x_1$, is considered:

$$\frac{dE}{dt} = \frac{d}{dt}\int_{x_1}^{x_2} dx U = -\int_{x_1}^{x_2} dx\left(\frac{\partial S}{\partial x}\right) = S(x_1) - S(x_2). \tag{12.58}$$

This result is interpreted to mean that the energy within Δx increases when its "influx" $S(x_1)$, at one end, exceeds its "outflux" $S(x_2)$, at the other.

Suppose that the disturbance has a definite direction of propagation to the right: $q(x, t) = f(x - ct)$. The energy density (12.55) in this case is

$$U = \rho c^2(f')^2. \tag{12.59a}$$

The flux is

$$S = \rho c^3(f')^2 = Uc. \tag{12.59b}$$

It should be expected that the flow of energy in a time Δt should equal the amount resident in a length $c\Delta t$, that is $U \cdot c\Delta t = S\Delta t$, when it is propagated with velocity c.

In the example of Figure 12.5, there is flux only in the frontal regions where both $\partial q/\partial t$ and $\partial q/\partial x$ are nonvanishing, according to $q_{L,R}(x, t)$ of (12.53). The corresponding fluxes are $S_L = -(\frac{1}{4}\rho v_0^2)c$ and $S_R = -S_L$, respectively. These are flows of equal magnitude, to the left and to the right, each carrying half the initial energy density, $\frac{1}{2}\rho v_0^2$.

An absolutely monochromatic disturbance like

$$q(x, t) = \mathrm{Re}\{a(k)e^{i(kx-\omega t)}\} \tag{12.60}$$

has

$$U(x, t) = \tfrac{1}{2}\rho\omega^2[|a|^2 - \mathrm{Re}\{a^2 e^{2i(kx-\omega t)}\}], \qquad S = Uc. \tag{12.61}$$

Aside from the fluctuations of zero average within each wavelength,* these are

$$\langle U\rangle = \tfrac{1}{2}\rho\omega^2|a|^2, \qquad \langle S\rangle = \langle U\rangle c. \tag{12.62}$$

The angular brackets stand for averages over a wavelength (or a period).

12.4 FOURIER TRANSFORMS

When the interest is in propagations through a continuum of indefinite extent, it is desirable to extend the Fourier analysis (12.46) and (12.47) to the infinite domain: $-\infty < x < \infty$. Each Fourier series that results from the analyses over the *finite* interval of length L merely repeats itself when evaluated outside that interval. It has the same value at $x + NL$ as it has at x because $\exp\colon 2\pi in(x + NL)/L = \exp\colon 2\pi inx/L$; it can only represent functions that are periodic in lengths L when applied in the infinite domain. On the other hand, the function represented may be quite arbitrary within the domain $-\frac{1}{2}L < x < +\frac{1}{2}L$. To gain this generality for the infinite domain, the procedure should be adapted to $L \to \infty$.

In the analysis (12.46) and (12.47), successive wave numbers $k = 2\pi n/L$, differing by $\Delta k = 2\pi/L$, were utilized. As $L \to \infty$, these differences tend to vanish, and the succession of wave numbers that is needed approaches continuity, with $\Delta k \to dk$. The individual members of the orthogonal function set (12.47*a*) become

$$e_n(x) \to (\Delta k/2\pi)^{1/2} e^{ikx}, \tag{12.63}$$

distinguished from each other by values of a continuous wave number k, rather than by the discrete integers n. Each member enters the representation (12.46*a*) with a coefficient (12.46*b*):

$$\alpha_n \to (\Delta k/2\pi)^{1/2} \int_{-\infty}^{\infty} dx f(x) e^{-ikx}.$$

* See page 379.

It becomes appropriate to introduce the function of the continuous variable k,

$$F(k) \equiv (2\pi)^{-1/2} \int_{-\infty}^{\infty} dx f(x) e^{-ikx}, \tag{12.64}$$

and to replace the coefficients by $\alpha_n \to F(k)(\Delta k)^{1/2}$. Now the series representation (12.46a) becomes

$$f(x) = \lim_{L \to \infty} \sum \Delta k (2\pi)^{-1/2} F(k) e^{ikx}.$$

Since $\Delta k = 2\pi/L \to 0$ as $L \to \infty$, the summation here, over $-\infty < n = kL/2\pi < \infty$, approaches integration over the continuous variable k:

$$f(x) = (2\pi)^{-1/2} \int_{-\infty}^{\infty} dk F(k) e^{+ikx}. \tag{12.65}$$

The arbitrary function $f(x)$ is thus analyzed into members of the set

$$e(kx) = (2\pi)^{-1/2} e^{ikx}, \tag{12.66}$$

which are enumerated by the continuum of values for $-\infty < k < \infty$, rather than by integers, as are the members of the set $e_n(x)$. The obvious symmetry between the expressions for $f(x)$ and $F(k)$ has led to calling these each other's "Fourier transforms." Just as (12.65) decomposes $f(x)$ into components $F(k)$ "along $e(kx)$," so (12.64) decomposes $F(k)$ into components $f(x)$ "along $e(-xk)$," now regarded as functions of k enumerated by $-x$.

How the orthonormality of the set $e(kx)$ may be expressed follows from the orthonormality properties (12.47b) of the discrete set $e_n(x)$:

$$\int_{-(1/2)L}^{(1/2)L} dx e_n^* e_{n'} \equiv \int_{-(1/2)L}^{+(1/2)L} (dx/L) e^{2\pi i(n'-n)x/L} = \delta_{n'n}.$$

When the replacements $k = 2\pi n/L$, $k' = 2\pi n'/L$, and $L = 2\pi/\Delta k$ are made,

$$\int_{-(1/2)L}^{+(1/2)L} (dx/2\pi) e^{i(k'-k)x} = \delta_{n'n}/\Delta k. \tag{12.67a}$$

The right side vanishes for $k' \neq k$ ($\to n' \neq n$), and for $k' = k$ it becomes $1/\Delta k = L/2\pi$, which approaches infinity in proportion to $L \to \infty$. If its limit is treated as a function of the continuous variable k, it can be said to have the integral

$$\int dk \lim(1/\Delta k) = \lim \Delta k (1/\Delta k) = 1. \tag{12.67b}$$

A singular "function" having all these properties is sometimes called a "needle function," but more often a "Dirac delta function", and is denoted $\delta(k - k')$ when it is the "limit" of (12.67*a*).

Much maneuvering of limiting processes is circumvented simply by treating $\delta(\xi)$ as a real, even function of its argument which vanishes everywhere except where its argument vanishes. Here it is attributed an infinite "value" just such that

$$\int d\xi \delta(\xi) = 1 \tag{12.68}$$

upon integration over any range which encloses the singular point $\xi = 0$.

The limit of the expression (12.67*a*) may now be written as

$$\int_{-\infty}^{\infty} dx e^*(kx) e(k'x) = \int_{-\infty}^{\infty} (dx/2\pi) e^{i(k'-k)x} = \delta(k' - k). \tag{12.69}$$

It has the role of orthonormality conditions on the set $e(kx)$. Comparison to the expression (12.64) for a general Fourier transform shows that the delta function $\delta(k - k')$ is just the Fourier transform of the exponential function $e(k'x) = (2\pi)^{-1/2}$ exp: $ik'x$.

Delta Functions and Completeness

When the Fourier coefficients $F(k)$ in the decomposition (12.65) of $f(x)$ are replaced by the expressions (12.64), the result may be written

$$f(x) = \int_{-\infty}^{\infty} dx' f(x') \int_{-\infty}^{\infty} (dk/2\pi) e^{ik(x-x')} \tag{12.70}$$

upon interchange of the order in which the integrations are performed. This is justified for an arbitrary $f(x)$ only if the integral over k multiplies by zero every value taken on by $f(x')$ except the one, $f(x)$, appearing on the left. Such a behavior is just what can be expected of a delta function $\delta(x' - x)$ analogous to (12.69):

$$\int_{-\infty}^{\infty} (dk/2\pi) e^{ik(x-x')} = \delta(x - x'). \tag{12.71}$$

Moreover, the integration over x' in (12.70) now leads properly to

$$\int_{-\infty}^{\infty} dx' f(x') \delta(x' - x) = f(x) \int_{-\infty}^{\infty} dx' \delta(x' - x) = f(x) \cdot 1. \tag{12.72}$$

It is this integral operation, applicable to any not too singular function $f(x)$, which is often taken to be the defining property of a delta function.

The expression (12.71) amounts to a Fourier-integral decomposition of the delta function. It makes it easy to understand how the characteristic delta-function behavior is acquired. The integrand exp: $ik(x - x')$ plainly fluctuates about a zero average, and integration averages it to zero, except when $x = x'$, and then $\int_{-\infty}^{\infty} dk/2\pi = \infty$.

When the delta function (12.71) is expressed in terms of the orthonormal set $e(kx)$,

$$\int_{-\infty}^{\infty} dke(kx)e^*(kx') = \delta(x - x'), \tag{12.73}$$

it becomes obvious that it is an analogue of the projection operators $\sum_k \mathbf{e}_k \mathbf{e}_k^\dagger = \mathbf{1}$, of (9.109) or (10.82). An arbitrary vector $\mathbf{r}$, in the space spanned by the orthonormal basis vectors $\mathbf{e}_k$, could be resolved on these axes by "inserting the basis" in the form of the unit projection operator: $\mathbf{r} = \mathbf{e}_k \mathbf{e}_k^\dagger \mathbf{r} = \mathbf{e}_k x_k$ if $x_k \equiv \mathbf{e}_k^\dagger \mathbf{r}$ is the kth component of $\mathbf{r}$. Similarly, an arbitrary function $f(x)$ can be resolved on the orthonormal "basis vectors" $e(kx)$ by giving it the equivalent expression (12.72) and then inserting the "projection operator" (12.73):

$$f(x) = \int_{-\infty}^{\infty} dke(kx) \int_{-\infty}^{\infty} dx'e^*(kx')f(x').$$

The integral over x' here is just the Fourier transform $F(k)$ of (12.64), and the result is the Fourier decomposition (12.65).

In the case of the space spanned by the discretely numbered orthogonal unit vectors $\mathbf{e}_k$ ($k = 1, 2, 3, \ldots$), the equivalence to unity of the operation $\mathbf{e}_k \mathbf{e}_k^\dagger = \mathbf{1}$ was at bottom an expression of the *completeness* of the set $\mathbf{e}_k$. This means that every vector $\mathbf{r}$ in the space has no components aside from ones expressable as $x_k \equiv \mathbf{e}_k^\dagger \mathbf{r}$; it also means that any vector that is not completely specifiable by $\mathbf{r} = \mathbf{e}_k x_k$ is not accepted as entirely "in the space." Similarly, the equivalence to the delta function (12.73) is a "completeness relation" in the space spanned by the $e(kx)$, having a continuously infinite number of dimensions.

A completeness property of any orthogonal set $u_1(x)$, $u_2(x)$, ..., $u_n(x)$, ... should be similarly expressible. If the set is supposed to be able to reproduce arbitrarily various values

$$f(x) = \sum_n a_n u_n(x),$$

with $a_n = \oint dx' u_n^*(x')f(x')$, that a function may have at various x's in the domain of orthogonality, $\oint dx$, then the summation in

$$f(x) = \oint dx'f(x') \sum_n u_n(x)u_n^*(x')$$

must give unit weight to the value $f(x)$ of $f(x')$, and multiply by zero every other value $f(x')$. Thus the summation must behave like

$$\sum_n u_n(x)u_n^*(x') = \delta(x - x'), \tag{12.74}$$

or else the set is after all not complete enough to reproduce all the values $f(x)$ desired.

The use of the relation (12.74) should be confined to evaluations like $\oint dx' f(x')\delta(x' - x) = f(x)$ at points x where $\sum_n a_n u_n(x)$ is regarded as an acceptable, "physical" value, if this should differ from $f(x)$, as in the way discussed on page 362. It should be recognized that physicists use functions like $f(x)$ to represent particular physical circumstance in some idealized way, and a difference of $f(x)$ from $\sum a_n u_n(x)$ at some points may mean that the description by $f(x)$ is "overidealized" at those points. This can be maintained when each $u_n(x)$ represents *some* situation without overidealization, so that $\sum a_n u_n(x)$ with every "convergent" choice of the a_n's represents *some* physical circumstances without overidealization (that is, adequately for the purposes of the description). It can then also be maintained that the u_n's form a *complete* enough set to represent all $f(x)$'s that are not overidealized.

Wave Packets

Fourier decompositions of waves are important because they provide classifications by wavelength and frequency. When the harmonic (Fourier) constitution of any wave is known, then its behavior may become predictable from experience with quite different superpositions. Sounds are sometimes reconstructed on the basis of their compositions out of certain frequencies.

When an arbitrary disturbance $q(x, t)$ is decomposed as is $f(x)$ in (12.65), or

$$q(x, t) = (2\pi)^{-1/2} \int_{-\infty}^{\infty} dk F(k, t) e^{ikx}, \tag{12.75a}$$

then the Fourier amplitudes,

$$F(k, t) = (2\pi)^{-1/2} \int_{-\infty}^{\infty} dx q(x, t) e^{-ikx}, \tag{12.75b}$$

must depend on the time in a particular way determined by the wave equation

$$(\partial^2/\partial x^2 - \partial^2/c^2\partial t^2) q(x, t) = 0.$$

Substitution of the form (12.75*a*) into this yields

$$\int_{-\infty}^{\infty} dk e^{ikx}[-k^2 F - c^{-2}\ddot{F}] = 0. \tag{12.76}$$

For this to be satisfied, the coefficient of each independent Fourier component must vanish,

$$\ddot{F} = -c^2k^2F(k, t), \tag{12.77a}$$

and so

$$F(k, t) = (2\pi)^{1/2}[a(k)e^{-ickt} + b(k)e^{+ickt}] \tag{12.77b}$$

in general. The "independence" of the individual Fourier components refers to their orthogonality (12.69). That it has the stated consequence follows mathematically from multiplying (12.76) by $e^{-ik'x}$ and integrating over x—that is, projecting the expression on $e(k'x)$ to obtain

$$\int_{-\infty}^{\infty} dk\delta(k - k')[-k^2F - c^{-2}\ddot{F}] = -c^{-2}[(ck')^2F + \ddot{F}] = 0.$$

This shows that F must obey the equation (12.77a) for every k, since k' could be chosen arbitrarily.

It has now been found that any disturbance $q(x, t)$ will be propagated in accordance with the wave equation if it has the composition

$$q(x, t) = \int_{-\infty}^{\infty} dk[a(k)e^{ik(x-ct)} + b(k)e^{ik(x+ct)}], \tag{12.78}$$

with arbitrary amplitudes $a(k)$ and $b(k)$. This shows that, as already found in (12.49), the general disturbance is an arbitrary superposition of right- and left-running waves. Indeed, (12.78) could have been obtained from the Fourier decompositions of the forms

$$f(\xi) = \int_{-\infty}^{\infty} dka(k)e^{ik\xi}, \qquad g(\eta) = \int_{-\infty}^{\infty} dkb(k)e^{ik\eta}, \tag{12.79}$$

of $f(x - ct)$ and $g(x + ct)$. The $a(k)$ and $b(k)$ have gained meaning as Fourier amplitudes of the right- and left-running parts of the general disturbance.

Plainly, the general expression (12.78) could also have been obtained by starting from a Fourier decomposition of the *time*-dependence:

$$q(x, t) = \int_{-\infty}^{\infty} d\omega A(\omega, x)e^{-i\omega t}, \tag{12.80a}$$

where

$$A(\omega, x) = \int_{-\infty}^{\infty} (dt/2\pi)q(x, t)e^{+i\omega t} \tag{12.80b}$$

must obey the equation (12.8). This emphasizes that terms of different frequency, such as $\omega = \pm ck$, are as independent (orthogonal) as are terms of different wave number.

Sign conventions may be adopted such that the frequency $\omega = c|k|$ is restricted to positive values and $k \lessgtr 0$ label the right- and left-running waves, respectively. These may be introduced by substituting $k' = -k$ into parts of (12.78),

$$\int_0^{\infty} dkb(k)e^{i(kx+\omega t)} = \int_0^{-\infty} (-dk')b(-k')e^{-i(k'x-\omega t)}$$
$$\equiv \int_{-\infty}^{0} dkb(-k)e^{-i(kx-\omega t)}$$

and

$$\int_{-\infty}^{0} dka(k)e^{i(kx+\omega t)} = \int_0^{\infty} dka(-k)e^{-i(kx-\omega t)}.$$

Then

$$q(x, t) = \tfrac{1}{2}\int_{-\infty}^{\infty} dk[C(k)e^{i(kx-\omega t)} + C'(k)e^{-i(kx-\omega t)}],$$

where $\frac{1}{2}C(k > 0) \equiv a(k), \frac{1}{2}C(k < 0) \equiv b(k), \frac{1}{2}C'(k > 0) \equiv a(-k), \frac{1}{2}C'(k < 0) \equiv b(-k)$. For $q(x, t)$ to be a *real* disturbance it is necessary to restrict the superposition to $C'(k) = C^*(k)$ (see the discussion of (12.46*b*)), and so

$$q(x, t) = \text{Re}\left\{\int_{-\infty}^{\infty} dkC(k)e^{i(kx-\omega t)}\right\}, \tag{12.81}$$

with $\omega \equiv c|k| > 0$. Here each element $C(k > 0)$ exp: $i(kx - \omega t)$ is a right-running, and each $C(k < 0)$ exp: $-i(|k|x + \omega t)$ is a left-running, wave.

Any disturbance may be represented by such a superposition of directed harmonic waves as (12.81), and it may next be checked that this is consistent with an always-conserved total energy in the disturbance. The energy may be evaluated by substituting an integral

$$\partial q/\partial x = \tfrac{1}{2}\int_{-\infty}^{\infty} dkC(k)(ik)e^{i(kx-\omega t)} + \text{c.c.}$$

for each factor $\partial q/\partial x$ of a potential energy expression like (12.4), and a similar integral, with $(-i\omega)$ replacing (ik), for each factor $\partial q/\partial t$ in a kinetic energy like expression (12.5). After an integration over $-\infty < x < \infty$ is carried out with the help of (12.69), the results are

$$V = \frac{\pi}{4}\rho c^2 \iint_{-\infty}^{\infty} dk\, dk'kk'$$
$$\times \{C^*(k)C(k')e^{i(\omega-\omega')t}\,\delta(k'-k) - C(k)C(k')e^{-i(\omega+\omega')t}\,\delta(k'+k) + \text{c.c.}\}$$
$$= \frac{\pi}{2}\rho c^2 \int_{-\infty}^{\infty} dkk^2[|C(k)|^2 + \text{Re}\{C(k)C(-k)e^{-2i\omega t}\}] \tag{12.82a}$$

since $\omega' = \omega$ for $k' = \pm k$. The kinetic energy integral has $\omega\omega'$ in place of kk', and $\omega\omega' = +\omega^2 = (ck)^2$ for both $k' = +k$ and $k' = -k$, and hence

$$T = \frac{\pi}{2}\rho c^2 \int_{-\infty}^{\infty} dk k^2 [|C(k)|^2 - \text{Re}\{C(k)C(-k)e^{-2i\omega t}\}]. \quad (12.82b)$$

The total energy is

$$E = T + V = \pi\rho c^2 \int_{-\infty}^{\infty} dk k^2 |C(k)|^2, \quad (12.83)$$

a positive constant, as is proper. Superpositions of harmonic waves like (12.81) are frequently called "wave-packets" or "pulses," especially when their extensions in space and/or time are limited.

Complementarity in Wave Packets

What characterizes a wave-packet is a lack of "monochromaticity," or restriction to any unique, discrete frequency or wavelength. Even an individual harmonic like

$$e^{i(k_0 x - \omega_0 t)} \quad (12.84)$$

becomes a wave-packet if its duration is limited to some finite time interval Δt, or its spatial extent to some Δx. It then lacks monochromaticity in the sense that its Fourier decomposition will contain a range of frequencies different from ω_0, and wavelengths different from $2\pi/|k_0|$.

Consider an element of disturbance which can be represented by (12.84) during the time interval $-\frac{1}{2}\Delta t < t < \frac{1}{2}\Delta t$, but vanishes outside this time interval. The Fourier decomposition of the time-dependence will contain various frequencies with amplitudes which are calculable as in (12.80):

$$A(\omega) = e^{ik_0 x} \int_{-(1/2)\Delta t}^{(1/2)\Delta t} (dt/2\pi) e^{i(\omega - \omega_0)t}. \quad (12.85)$$

The integral here is the function of $\omega - \omega_0$,

$$\frac{\sin(\omega - \omega_0)\Delta t/2}{\pi(\omega - \omega_0)} \to \delta(\omega - \omega_0) \qquad \text{for} \qquad \Delta t \to \infty. \quad (12.86)$$

That it approaches the indicated delta function can be seen from comparison with (12.69) or (12.71). The result shows that the "wave train" (12.84) approaches having a monochromatic frequency, $\omega = \omega_0$, only if it has a long enough duration. For any given finite duration Δt, it contains a distribution of frequencies with amplitudes as shown in Figure 12.6. The indicated rectangle has a unit area like that of the delta function (12.86) and its breadth,

$$\boxed{\Delta\omega = 2\pi/\Delta t,} \tag{12.87}$$

may be used as a measure of the range of frequencies it is essential to include in order that the Fourier superposition have negligible values outside Δt. The small-amplitude frequencies outside the range $\Delta\omega$ arise from the rather artificially sharp cut-offs at $t = \pm\frac{1}{2}\Delta t$ which were presumed for (12.85).

In an exactly analogous fashion, it can be found that a superposition of harmonics having the range of wave numbers

$$\boxed{\Delta k = 2\pi/\Delta x} \tag{12.88}$$

is needed to represent an element of disturbance which is essentially restricted to a span Δx in space. For example,

$$\int_{k_0-(1/2)\Delta k}^{k_0+(1/2)\Delta k} dk e^{ikx} = e^{ik_0x}\frac{\sin \Delta k(x/2)}{(x/2)}$$

is concentrated in the interval $\Delta x = 2\pi/\Delta k$ around $x = 0$.

The relations (12.87) and (12.88) may be called "complementarity" properties of wave representations. It is such properties which have made wave pictures useful in describing quantum phenomena; they provide one way of realizing the famous "Uncertainty Relations."

A consequence of the findings here is that absolutely monochromatic

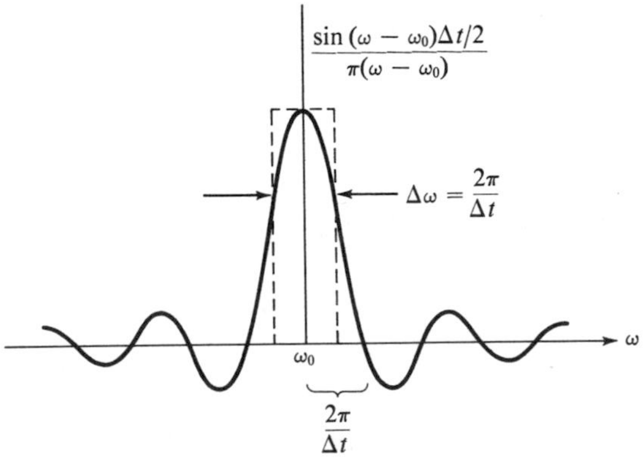

Figure 12.6

waves like (12.60) are not physically realizable in any finite portion of space, or restricted to any finite time. "Near-monochromatic" wave-packets *can* be realized if given sufficient room, and if their duration is longer than any time-span relevant to the situation in which they are to be treated as essentially monochromatic. This is the basis for the greater interest in the time-averaged energies of (12.62), than in the fluctuation within a wave exhibited by (12.61). The absolutely monochromatic representation (12.60) is usually employed when it is an adequate idealization of some wave-packet which is "near-monochromatic," and such wave-packets do not contain the fluctuations in time, as the result (12.83) has attested.

12.5 TRANSMISSION AND REFLECTION

Problems arise in connection with the propagation of waves from one medium into another, contiguous one. A change of medium is represented by a change of the propagation velocity constant, c, in the wave equation. For example, the two media may consist of strings having different linear mass-densities, tied together and stretched under a common tension. The propagation velocities are then different because each is inversely proportional to the square root of the mass-density, according to (12.3). An investigation of the problems may be initiated by considering a system in which the "half-infinite" span $-\infty < x < 0$ is occupied by a medium with one propagation velocity, c_1, while the rest of the space, $0 < x < +\infty$, contains a medium with a different propagation velocity, c_2.

Partial Reflections

Propagations in the composite system may be studied by considering a harmonic element of disturbance, a "representative"

$$q_i(x < 0, t) = a_i e^{i(k_1 x - \omega t)}, \qquad k_1 > 0, \omega = c_1 k_1, \tag{12.89}$$

to be given as "incident" from the remote left ($x = -\infty$). This may be an element of some general superposition like (12.81), with $a_i \equiv C(k_1)dk$, or its real part may be taken to be a complete, real monochromatic wave. In the latter case, a wave-averaged flux of energy, $S_i = \frac{1}{2}\rho_1\omega^2|a_i|^2 c_1$, is being "sent in," according to (12.62).

The incident disturbance will impart a time variation of frequency ω to the boundary point, $x = 0$, and must be expected to start a disturbance *of the same frequency* in the second medium, a "transmitted wave":

$$q(x > 0, t) = a_t e^{i(k_2 x - \omega t)}, \qquad k_2 = \omega/c_2 \quad (>0). \tag{12.90}$$

With a matching frequency, the transmitted wave will have a wavelength $2\pi/k_2 = 2\pi c_2/\omega$, which is different from the incident one, $2\pi c_1/\omega$.

To examine how the penetration into the second medium is started, consider the Taylor series expansion near the boundary point, $x = 0$:

$$q(x, t) = q(0, t) + x\left(\frac{\partial q}{\partial x}\right)_0 + \tfrac{1}{2}x^2\left(\frac{\partial^2 q}{\partial x^2}\right)_0 + \tfrac{1}{6}x^3\left(\frac{\partial^3 q}{\partial x^3}\right)_0 + \cdots.$$

For $x > 0$, it is guided by the wave equation for the given frequency,

$$\partial^2 q/\partial x^2 = -k_2^2 q$$

and hence

$$\begin{aligned} q(x > 0, t) &\approx q(0, t)[1 - \tfrac{1}{2}k_2^2 x^2 + \tfrac{1}{24}k_2^4 x^4 - \cdots] \\ &\quad + (\partial q/\partial x)_0[x - \tfrac{1}{6}k_2^2 x^3 + \tfrac{1}{120}k_2^4 x^5 - \cdots] \\ &= q(0, t)\cos k_2 x + k_2^{-1}(\partial q/\partial x)_0 \sin k_2 x. \end{aligned} \tag{12.91}$$

This is plainly a general solution, with the boundary values of the displacement and its slope as the two arbitrary "integration constants" (at any given t). Both must generally be supplied before the transmitted propagation can be uniquely determined. The matchings

$$q_1(0, t) = q_2(0, t) \qquad (\partial q_1/\partial x)_0 = (\partial q_2/\partial x)_0 \tag{12.92}$$

will furnish the necessary boundary conditions. For convenience, the notations $q_1 \equiv q(x < 0, t)$ and $q_2 \equiv q(x > 0, t)$ are being used for the disturbances in the two media. The boundary conditions adopted here correspond to a "kinkless" tying together of two strings at $x = 0$.

It can now be seen that the two propagations, (12.89) and (12.90), so far supposed to exist in the two media, are "mismatched" under the boundary conditions (12.92). If the boundary displacements are matched, $a_i = a_t$, then the slopes are not: $ik_1 a_i \neq ik_2 a_t$ unless $c_1 = c_2$ and there is no change of medium. This shows that the solution that is supposed to exist must be more generalized. It will not help to introduce new frequencies, since linear conditions like (12.92), to hold at all times, would have to be satisfied independently for each frequency. However, it may be supposed that, besides the right-going incident wave (12.89), there exists a left-going *reflected* wave of the same frequency in the first medium:

$$q(x < 0, t) = a_i e^{i(k_1 x - \omega t)} + a_r e^{-i(k_1 x + \omega t)}. \tag{12.93}$$

Now the boundary conditions can be satisfied with any incident amplitude a_i, if the reflected and transmitted amplitudes are such that

$$\begin{aligned} q(0, t) &= (a_i + a_r)e^{-i\omega t} = a_t e^{-i\omega t}, \\ (\partial q/\partial x)_0 &= ik_1(a_i - a_r)e^{-i\omega t} = ik_2 a_t e^{-i\omega t}. \end{aligned} \tag{12.94}$$

These expressions confirm that the conditions must be satisfied independently for every frequency present, if the time is to cancel out so that they can hold at all times.

The results show that the response to the sending in of an incident wave with amplitude a_i is not only a transmitted wave of amplitude

$$a_t = \frac{2k_1}{k_1 + k_2} a_i = \frac{2c_2}{c_2 + c_1} a_i , \tag{12.95}$$

but also a reflected wave of amplitude

$$a_r = \frac{k_1 - k_2}{k_1 + k_2} a_i = \frac{c_2 - c_1}{c_2 + c_1} a_i . \tag{12.96}$$

The reflected wave would disappear, and $a_t = a_i$ if there were no change of medium.

The energy flow ought to be conserved if the media are coupled to each other in an energy-conserving way; that is,

$$S_t + |S_r| = S_i = \tfrac{1}{2}\rho_1 \omega^2 |a_i|^2 c_1 , \tag{12.97}$$

where

$$S_t = \tfrac{1}{2}\rho_2 \omega^2 |a_t|^2 c_2 \qquad \text{and} \qquad S_r = -\tfrac{1}{2}\rho_1 \omega^2 |a_r|^2 c_1 \tag{12.98}$$

are the transmitted and reflected energy fluxes, according to (12.62). Substitution of the results for a_t and a_r into (12.97) yields

$$S_i = \tfrac{1}{2}\rho_1 \omega^2 |a_i|^2 c_1 \frac{(\rho_2 c_2/\rho_1 c_1)(2c_2)^2 + (c_1 - c_2)^2}{(c_1 + c_2)^2} .$$

The ratio on the right is appropriately equal to one if

$$\rho_2/\rho_1 = c_1^2/c_2^2 , \tag{12.99}$$

in agreement with the special case (12.3) for two strings joined together under a common tension. The ratio has now been found necessary for a consistent description of energy conservation when the energy is given by (12.4) and (12.5) in each medium, and when the two media are coupled by the boundary conditions (12.92). The latter become the "energy-conserving couplings" in the case of the relationship (12.99).

It may not have escaped notice that the solution developed here might have been generalized further even for a given frequency. An added presence of a left-going wave in the second ($x > 0$) medium could have been considered. However, that would merely correspond to a simultaneous sending in of an independent wave from the remote right; the responses to it would merely be superposed on the ones found here.

Total Reflections

Suppose that the new medium upon which a wave is incident consists of elements so heavy as to be practically immovable. That results in a vanishing propagation velocity, $c_2 \to 0$, according to the inverse proportionality to mass, (12.99). The transmitted amplitude (12.95) disappears and the first of the boundary conditions (12.92) reduces to $q(0, t) = 0$. Quite unsurprisingly, this is just the boundary conditions for a string tied down at $x = 0$, as in (12.12).

With the transmitted amplitude vanishing, there is a reflected amplitude (12.96):

$$a_r = -a_i \equiv a_i e^{i\pi}. \tag{12.100}$$

The last of these expressions is written merely to indicate why a "phase shift by π" is sometimes said to characterize reflections from an immovable wall. The result follows from the first of the boundary conditions (12.92), $q(0, t) = 0$ alone, and is also consistent with the condition on the slope in that the second of the equations (12.94) reduces to

$$k_1(a_i - a_r) = \lim k_2 a_t = \lim[2k_1k_2/(k_1 + k_2)]a_i = 2k_1 a_i,$$

which yields $a_r = -a_i$ again. Inasmuch as the incident amplitude a_i can be chosen arbitrarily, the boundary slope can have any value. These results show that no separate attention need be paid to boundary conditions on the slope in the case of reflections from an immovable wall, and none was paid in presenting the conditions (12.12).

Because there is "total reflection," with the reflected amplitude equal to the incident one in absolute value, the disturbance (12.93) develops into a *standing* wave:

$$\begin{aligned} q(x, t) &= a_i e^{-i\omega t}(e^{ik_1x} - e^{-ik_1x}) \\ &= 2ia_i e^{-i\omega t} \sin k_1 x, \end{aligned} \tag{12.101a}$$

having the real part

$$\begin{aligned} \text{Re}[q(x, t)] &= 2a_i \sin k_1 x \sin \omega t \\ &= a_i[\cos(k_1 x - \omega t) - \cos(k_1 x + \omega t)]. \end{aligned} \tag{12.101b}$$

Any standing wave can thus be regarded as a superposition of right- and left-running waves of equal absolute magnitude.

As a contrast with the reflections from an immovable wall, or a fixed point, characterized by a propagation velocity $c_2/c_1 \rightarrow 0$ beyond it, the reflections at a boundary beyond which $c_2/c_1 \rightarrow \infty$ may be considered. Now (12.95) gives $a_t \rightarrow 2a_i$ for the transmitted amplitude, and (12.96) gives

$$a_r \rightarrow +a_i \tag{12.102}$$

for the reflected wave, again a "total" reflection. It may come as a surprise that there need be any reflection at all, in view of the ease of displacement and its ready propagation, in the new medium. However, the mismatch of an incident, purely right-going, wave is quite as thorough for $c_2/c_1 \gg 1$ as it is for $c_2/c_1 \ll 1$; it is $c_2/c_1 = 1$ which is required for a matching which would cause no reflections, as noted in connection with (12.96). The physical reason is that the propagation transmitted into a medium with $c_2 \gg c_1$ is unable to carry away any appreciable part of the incident energy, $S_i = \frac{1}{2}\rho_1\omega^2 |a_i|^2 c_1$. It can be seen that $S_t = \frac{1}{2}\rho_2\omega^2 |2a_i|^2 c_2$ becomes $S_t = S_i(4c_1/c_2) \rightarrow 0$ for $c_2/c_1 \rightarrow \infty$, because of the property (12.99). The situation is somewhat like that met in (5.25), when a massive particle collides with a very light one at rest; the light particle chases ahead with twice the collision velocity but with no appreciable part of the comparatively great collision energy.

With effectively all the energy reflected ($a_r = a_i$) a standing wave is again formed; the total disturbance (12.93) becomes

$$\begin{aligned} q(x, t) &= a_i e^{-i\omega t}(e^{ik_1x} + e^{-ik_1x}) \\ &= 2a_i e^{-i\omega t} \cos k_1 x, \end{aligned} \tag{12.103}$$

accounting for the amplitude $a_t = 2a_i$ needed for the matching transmitted wave. The displacement has a maximum at the interface, $x = 0$, in contrast to its vanishing at the fixed boundary point, as in (12.101). A like contrast is known in the comparison of the pressure waves in open- and closed-end organ pipes.

Development of Reflections

When attention is confined to a monochromatic disturbance, as in the foregoing, then it is an established, steady situation that is perforce being treated. For instance, it had to be supposed in (12.93) that during the entire time of incidence of the wave "sent in" there is a steady reflected wave train in existence, extending back to the source of the incident wave. This situation arises because a disturbance of unique frequency can be claimed to exist only if the time span allowed it is unlimited, as required by the "complementarity" between frequency and time derived in (12.87).

To investigate the time-evolution of a reflection, the incidence of a wave-packet having a definite, moving front should be considered. The investigation here will be restricted to a reflection from a fixed boundary point at $x = 0$, and it will be convenient to set the zero of time as the moment at which the front reaches this boundary. It may also be least confusing to put the region of disturbance at points $x > 0$, and so to consider the incidence of a *left*-running wave from the remote right ($x = +\infty$):

$$q_i(x > 0, t < 0) = g(x + ct). \tag{12.104}$$

This will have a definite front which arrives at $x = 0$ at $t = 0$, if $g(\eta)$ is some functional form defined to vanish for $\eta < 0$ (for example, such a form as is shown in Figure 12.7(a)). At $t = 0$, $\eta \to x$, and so the right half of the figure can be taken to be showing the distribution in space of the wave form as it arrives at the boundary from the right.

Whatever the disturbance which develops after the reflections begin ($t > 0$), it can be given the general form $q(x > 0, t) = F(x - ct) + G(x + ct)$. It must also conform to the boundary condition $q(0, t) = F(-ct) + G(+ct) = 0$ at all times, hence the right- and left-running wave forms must be related as $F(\xi) = -G(-\xi)$. Finally, it will reduce to the given incident wave (12.104)

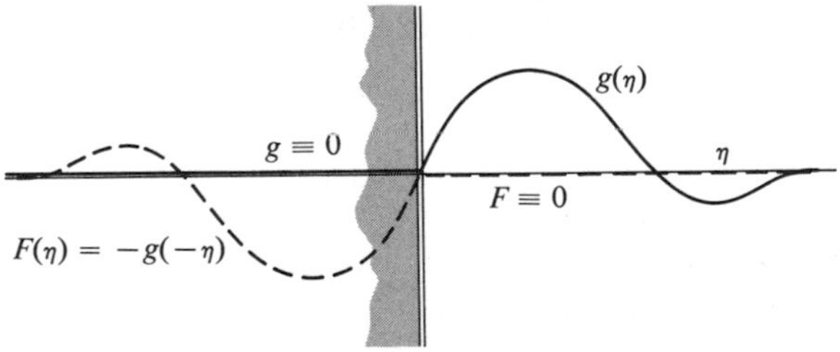

Figure 12.7(a)

for $t < 0$ if $G(\eta) \equiv g(\eta)$ is chosen, with the result

$$q(x > 0, t) = g(x + ct) - g(-x + ct). \tag{12.105}$$

This reduces to (12.104) for $t < 0$ when $g(\eta < 0) \equiv 0$ by definition. The different case (12.101) is also consistent with such a form as (12.105).

The result (12.105) superposes a reflected wave on the incident one at times $t > 0$. The reflected wave form is just an inverted mirror image, in the boundary, of the incident form, as indicated in Figure 12.7(a). The reflected wave behaves like a right-going wave emerging from the "mirror," beginning at $t = 0$, the moment of frontal incidence. The superposition of reflected and incident waves which develops in the "physical region,"

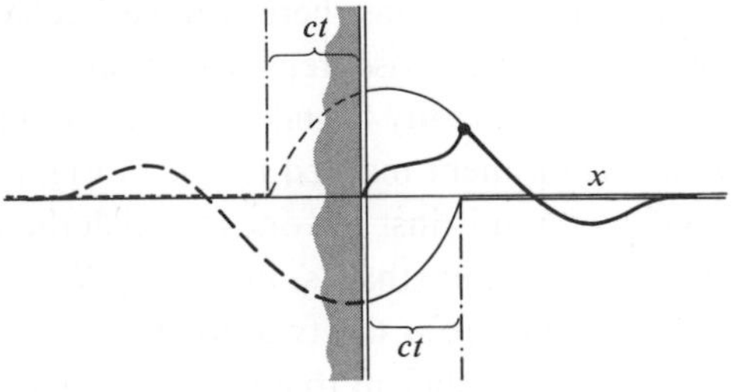

Figure 12.7(b)

$x > 0$, by some time $t > 0$, is plotted in Figure 12.7(b) on the basis of the forms shown in Figure 12.7(a).

Despite the added term in (12.105), each of the forms (12.104) and (12.105) contains the same energy. Substituting the forms derived from (12.105),

$$\frac{\partial q}{\partial x} = g'(x + ct) + g'(-x + ct),$$

$$\frac{\partial q}{\partial t} = cg'(x + ct) - cg'(-x + ct),$$

into (12.4), and (12.5) yields

$$E = \rho c^2 \int_0^\infty dx\{[g'(x + ct)]^2 + [g'(-x + ct)]^2\}.$$

For $t < 0$, when (12.105) reduces to (12.104), this becomes

$$E = \rho c^2 \int_{-c|t|}^\infty d\eta[g'(\eta)]^2 = \rho c^2 \int_0^\infty d\eta[g'(\eta)]^2 \tag{12.106}$$

because $g'(\eta < 0) \equiv 0$. For $t > 0$, the expression for E may be written

$$E = \rho c^2 \left\{\int_{ct>0}^\infty d\eta[g'(\eta)]^2 + \int_{+ct}^{-\infty} (-d\xi)[g'(\xi)]^2\right\}$$

$$= \rho c^2 \int_{-\infty}^\infty d\eta[g'(\eta)]^2 = \rho c^2 \int_0^\infty d\eta[g'(\eta)]^2,$$

which is, as is proper, the same as (12.106).

The findings here may be used to investigate what ensues from trapping the disturbances depicted in Figure 12.5(c) between fixed points at $x = \pm\frac{3}{2}l$. That would correspond to having started the disturbance in a string tied down at two end points, just the motion studied in connection with Figure 12.3. The considerations here enable replacing the infinite Fourier series (12.38) with "piecewise representations" at various stages of the motion, such as the one already found for the series (12.54), representing the configuration at the moment marked $t = l/c$ in Figure 12.5(c).

The latter configuration is just the one at which the reflections begin. Up to a time $3l/2c$ after they begin (that is, up to $t = 5l/2c$ on the time scale of Figures 12.5), the reflections from the two boundary points cannot interfere with each other and so it is sufficient to give attention to just one of them. Figure 12.8 indicates a series of configurations assumed during the reflections from the left-hand boundary, between the times $t = l/2c$ and $t = 5l/2c$ of the above scale. The effect of the reflections is to make the string go through the same sequence of configurations as it did before reflection, but in opposite order. The "turning-point" configuration is the one at $t = 3l/2c$; it is easily checked that it has all the energy in potential, none in the kinetic, form. By the time ($t = 5l/2c$) at which the reflection effects from the two boundary points begin to interfere with each other, an inward propagation has been established, one consisting of the same configurations as those of the initial outward propagation, described in Figures 12.5, but with their sequence "time-reversed." With this result it should be entirely clear just how to construct a "piecewise representation," like any of the ones in Figures 12.5 and 12.8, for any stage of the continuing motion. Any one may be compared to the Fourier representation (12.38), evaluated at the suitable time, and agreement will be found. With the configuration at $t = 3l/2c = L/2c$ identified as the "turning point" of the reflections, it must be expected that $q(x, L/c - t) = q(x, t)$ and $\dot{q}(x, L/c - t) = -\dot{q}(x, t)$; the Fourier representation (12.38) will be found to be in agreement with this expectation.

The foregoing makes it clear that, as the motion continues to the time $t = 3l/c = L/c$, the conditions $q \equiv 0$ and $\dot{q} = -v_0(x)$ arise; by the time $t = 2L/c$, the initial conditions for the disturbances of Figures 12.3 and 12.5 are re-established: $q \equiv 0$ and $\dot{q} = +v_0(x)$. These are just the behaviors pointed out for the Fourier representation (12.38). Here they are obtained by superposing reflected waves emerging from the two boundary points. The final effect of their interference is thus the introduction of a *periodicity* with the fundamental period $T = 2L/c$. It is this that is responsible for suppressing all frequencies in the Fourier representation except the discrete ones $\omega_n = n\pi c/L$. The motions with each of the subperiods must coincide with the fundamental in phase, both at the beginning of a fundamental period and at its end, and this makes possible the existence of only the discrete subperiods.

The "trapping" of a disturbance between fixed end points may also be considered more generally. The result (12.105) represents the effect of just one of the boundary points on a general disturbance. If this superposition

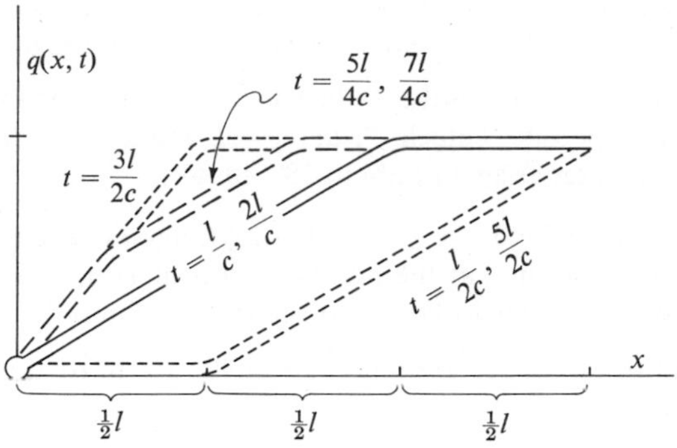

Figure 12.8

of incident and reflected wave forms is Fourier-analyzed as in (12.79), the result may be written

$$q(x > 0, t) = \int_{-\infty}^{\infty} dk b(k) e^{ickt} 2i \sin kx. \tag{12.107}$$

This is little more than an reiteration of the finding (12.101) that total reflections at any boundary will develop standing waves. If now a second fixed point is introduced at $x = L$, reflections from it will interfere with those from $x = 0$, and the definite periodicity ($T = 2L/c$) of the total motion, such as the one found in the above special example, will again develop. This may be seen from considering the superpositions of the two reflected waves, but the powerful method of Fourier analysis makes it unnecessary. It is sufficient to see that the effect on the decomposition (12.107) of the additional requirement $q(L, t) = 0$ is to make $b(k) = 0$ for every k not in the discrete set for which $\sin kL = 0$ (see (12.13)).

EXERCISES

12.1. (a) Suppose that the $n = 2$ mode of the string motion (12.7) is to be excited as efficiently as is possible by "plucking" some *one* point, x_0, of the string. By "plucking" here is meant a start from rest, with an initial configuration which forms sloped straight-line segments meeting in a triangular vertex at x_0. Find the best $x_0 < \frac{1}{2}L$ for exciting $n = 2$, if the "efficiency" is to be gauged by the energy fraction, of the total supplied, which is invested in the desired mode. [The requisite $\xi_0 \equiv 2\pi x_0/L$ will have to be found from $\tan \xi_0 = \xi_0(2\pi - \xi_0)/(\pi - \xi_0)$. Graphical investigation of this expression will show that $x_0 < \frac{1}{4}L$, and series expansions of it, treating $(\frac{1}{4}L - x_0)$ as small, will yield $x_0 \approx \frac{1}{4}L(1 - 4/3\pi^2)$, roughly.]

(b) How much greater is the "efficiency" of a start with two, simultaneous, equal and opposite "plucks", at $x_0 = L/4$ and $x_0 = 3L/4$ respectively? [About 81% here as against about 28% in (a).]

12.2. The straight line $f(0 \leq x \leq L) = 1 - 2x/L$ agrees exactly with $\cos(\pi x/L)$ at the end points and at the middle of the interval, and so the set (12.40) should provide a particularly good representation of it.

(a) Find the series representation, and show that it conforms to "parity conservation."

(b) Use the result to evaluate $\sum_{\nu=0}^{\infty} (2\nu + 1)^{-2}$. [$\pi^2/8$.]

12.3. To complement the discussion on page 362, construct the representation of $\cos(\pi x/L)$ by the series of sines (12.23).

12.4. Show by direct integrations that the members of the real Fourier set in (12.45) are mutually orthogonal on any interval of length L (between x_0 and $x_0 + L$, say). Do the same for the complex Fourier set (12.47).

12.5. For the "bound" string motion started as given by (12.34), identify the entire "turning-point" configuration (see Figure 12.8), and decompose it on the set $A_n(x)$ of (12.19). Check your result against (12.38), evaluated at the appropriate time ($=L/2c$).

12.6. Show that the impulse, $v_0(|x| < \frac{1}{2}l) = v_0$ per unit mass, treated in Figures 12.5, will be propagated solely rightward, with all the leftward propagations eliminated, if it is applied to a string with a suitable initial distortion, $q_0(x)$, and find this.

12.7. The result (12.105) is easy to generalize to the case of an impulse incident on a medium with $c_2 \neq 0$ and $c_2 \neq c_1 \equiv c$.

(a) Show that the reflected wave is again an inverted mirror image of the incident one, but now multiplied by $(c_1 - c_2)/(c_1 + c_2)$ [<0 possible!].

(b) Show that if the impulse $q_i(x - c_1 t)$ is incident from the left at $t = 0$, on a boundary at $x = 0$, then there is a transmitted wave

$$[2c_2/(c_1 + c_2)]q_i((c_1/c_2)x - c_1 t)$$

in the new medium ($x > 0$) for $t > 0$.

(c) With the help of a diagram modeled on Figure 12.5(a), investigate the reflections and transmissions after incidence, at $x = t = 0$, of a simple wave front like $q_i(x < c_1 t - l) = q_0$, $q_i(x > c_1 t) = 0$, and $q_i = (q_0/l)(c_1 t - x)$ in the span between. Detailed diagrams of the displacements, for such cases as $c_2 = \frac{1}{2}c_1$, and $c_2 = 2c_1$, provide instructive generalizations of Figure 12.8.

12.8. Suppose that the two media of §12.5 are separated by a span of width d in which the propagation velocity c is different from both c_1 and c_2.

(a) Find expressions for the reflected and transmitted amplitudes, and check your results by showing that they properly yield energy flux conservation.

(b) For the case in which the span d merely serves as a "window" ($c_2 = c_1$), find the smallest $d/\lambda \neq 0$ for which there will be no reflection.

(c) Suppose that the span d serves as a transition layer between media of $c_2 \neq c_1$. Find the one choice of $c(c_1, c_2)$ for which a nonreflecting d/λ is possible. [$c = \sqrt{c_1 c_2}$ in a "quarter-wave plate," $d = \frac{1}{4}\lambda$.]

CHAPTER 13

SURFACE VIBRATIONS

Mass elements dm which form a plane surface will be considered next. Their motions may represent vibrations of a drumhead,* for example.

The various mass elements may be labeled by their cartesian coordinates in the equilibrium plane x, y, and the mass may be represented by defining a surface mass-density, ρ, such that $dm = \rho\, dx\, dy$. For a uniform medium, the displacements $q(x, y, t)$ of the mass elements at the various positions x, y may be expected to obey equations of oscillational motion which reduce to

$$\left(\frac{\partial^2}{\partial x^2} + \frac{\partial^2}{\partial y^2}\right) q = \frac{\partial^2 q}{c^2\, \partial t^2}, \tag{13.1}$$

as the appropriate generalization of the one-dimensional wave equation (12.2).

Like the one-dimensional equation, the two-dimensional wave equation (13.1) expresses effects of departures from equilibrium, in a " Hooke's

* The reader may not be interested in drums, but, like Archimedes' (lyre) strings, they provide simple introductions to theoretical schemes of wide importance.

Law" approximation of the restoring forces. To see this, consider a restoring force proportional to $-(q - \bar{q})$, where $q(x, y, t)$ is the displacement of the mass element labeled x, y, and $\bar{q}$ is an average displacement of its neighbors. In a uniform medium, the averaging should give equal weight to neighbors in all directions, and this may be done by taking the mass element to be a circular disc of radius $\varepsilon \to 0$. Then the neighbors in all the directions φ, at the edge of this disc, are given equal weight in the average defined by

$$\begin{aligned}
\bar{q} &= \int_0^{2\pi} \left(\frac{d\varphi}{2\pi}\right) q(x + \varepsilon \cos \varphi, y + \varepsilon \sin \varphi, t) \\
&\approx \int_0^{2\pi} \frac{d\varphi}{2\pi} \left[q(x, y, t) + \varepsilon \cos \varphi \frac{\partial q}{\partial x} + \varepsilon \sin \varphi \frac{\partial q}{\partial y} + \tfrac{1}{2}\varepsilon^2 \cos^2 \varphi \frac{\partial^2 q}{\partial x^2} \right. \\
&\qquad \left. + \varepsilon^2 \cos \varphi \sin \varphi \frac{\partial^2 q}{\partial x \, \partial y} + \tfrac{1}{2}\varepsilon^2 \sin^2 \varphi \frac{\partial^2 q}{\partial y^2} + \varepsilon^3 \cdots \right] \\
&= q(x, y, t) + \tfrac{1}{4}\varepsilon^2(\partial^2 q/\partial x^2 + \partial^2 q/\partial y^2) + \varepsilon^3 \cdots .
\end{aligned}$$

With this, the equation of motion for the mass element $dm = \pi\varepsilon^2 \rho$ leads to the proportionality

$$\pi\varepsilon^2 \rho \frac{\partial^2 q}{\partial t^2} \sim \tfrac{1}{4}\varepsilon^2 \left(\frac{\partial^2 q}{\partial x^2} + \frac{\partial^2 q}{\partial y^2} \right) + \varepsilon^3 \cdots,$$

which yields (13.1) in the limit $\varepsilon \to 0$, with some $c^2 \sim 1/\rho$. A comparable basis could also have been promulgated for the one-dimensional wave equation (12.2).

As in the one-dimensional case, "monochromatic" solutions of definite frequency,

$$q(x, y, t) = u(x, y)e^{-i\omega t} \tag{13.2}$$

may be sought first. Then

$$(\partial^2/\partial x^2 + \partial^2/\partial y^2)u = -k^2 u, \tag{13.3}$$

with $k \equiv \omega/c$, is the two-dimensional generalization of (12.8); equations of this type are sometimes called "Helmholtz equations." The approach is obviously equivalent to seeking amplitudes for a Fourier decomposition of the time-dependence, as in (12.80); the general solution can then be expected to be a superposition of the monochromatic ones (13.2), and their complex conjugates.

13.1 THE RECTANGULAR MEMBRANE

The approach represented by (13.2) is sometimes said to be a search for "separable" solutions—that is, consisting of products like $q(x, y, t) = u(x, y)T(t)$, with the dependence on some one of the independent variables in a separate factor. The approach may be carried a step further by seeking solutions of the Helmoltz equation (13.3) in the separable form

$$u(x, y) = X(x) \cdot Y(y). \tag{13.4}$$

Substitution into the equation and then division through by $X \cdot Y$ yields

$$\frac{X''(x)}{X(x)} + \frac{Y''(y)}{Y(y)} = -k^2, \tag{13.5}$$

to be satisfied at all x, y. Here, a term which varies with x at most is being required to produce a constant when added to a term which can vary only with y, if at all. Clearly, each of the two terms must itself be a constant* independent of both x and y; the ratio X''/X will be denoted $-k_x^2$ and and $Y''/Y = -k_y^2$, these being arbitrary constants save that their sum must equal $-k^2$. Thus the product of any solutions of the two equations

$$X'' = -k_x^2 X \qquad \text{and} \qquad Y'' = -k_y^2 Y \tag{13.6}$$

will satisfy the Helmholtz equation if only

$$k_x^2 + k_y^2 = k^2. \tag{13.7}$$

An arbitrary superposition of solutions which are products of

$$X \sim e^{\pm ik_x x} \qquad \text{and} \qquad Y \sim e^{\pm ik_y y}, \tag{13.8}$$

with various *real* $k_{x,y}$, will suffice to yield a general solution, since it will then constitute mere Fourier analyses of the x and y dependences.

Solutions like (13.4), separable in the cartesian coordinates, are particularly suited to fitting situations of rectangular symmetry. To illustrate their use, a rectangular membrane pinned down at edges of lengths L_x and L_y, respectively, will be considered. Its motions will then be subject to the boundary conditions

$$\begin{aligned} q(0, y, t) &= 0 \qquad \text{and} \qquad q(L_x, y, t) = 0, \\ q(x, 0, t) &= 0 \qquad \text{and} \qquad q(x, L_y, t) = 0, \end{aligned} \tag{13.9}$$

* Compare the arguments used in connection with (8.75), when obtaining a solution of the Hamilton–Jacobi equation.

if the x and y axes are chosen appropriately. To vanish on the boundary $x = 0$, the general solution must be a superposition of terms depending on x in proportion to the solution $X \sim \sin k_x x$, of (13.6). They must depend on y in proportion to $Y \sim \sin k_y y$ in order to vanish for $y = 0$. Moreover, the boundary conditions at $x = L_x$ and $y = L_y$ require restricting the choices of k_x and k_y to such that $\sin k_x L_x = 0$ and $\sin k_y L_y = 0$, so that

$$k_x = n_x \pi / L_x, \qquad k_y = n_y \pi / L_y, \tag{13.10}$$

for $n_x, n_y = 1, 2, 3, \ldots$. Consequently, (13.7) restricts the frequencies to a discrete spectrum,

$$\omega_{n_x n_y} = [(n_x/L_x)^2 + (n_y/L_y)^2]^{1/2} \pi c, \tag{13.11}$$

and the general motion is described by

$$q(x, y, t) = \sum_{n_x n_y} C_{n_x n_y} \sin\left(\frac{n_x \pi x}{L_x}\right) \sin\left(\frac{n_y \pi y}{L_y}\right) \cos(\omega_{n_x n_y} t + \gamma_{n_x n_y}), \tag{13.12}$$

a simple generalization of the string motion (12.7).

The Standing Wave Modes

Each of the terms of different frequency, in (13.12) by itself represents a standing wave. The mode $n_x = 1, n_y = 2$, having the spatial distribution

$$q \sim \sin(\pi x / L_x) \cdot \sin(2\pi y / L_y),$$

is pictured in Figure 13.1(a). In addition to the vanishing on the boundaries, it has a stationary node line at $y = \frac{1}{2} L_y$. The mode $n_x = 2$, $n_y = 1$ has a standing node line at $x = \frac{1}{2} L_x$ instead, as pictured in Figure 13.1(b). Plainly, the general standing wave mode n_x, n_y has motionless node lines which divide

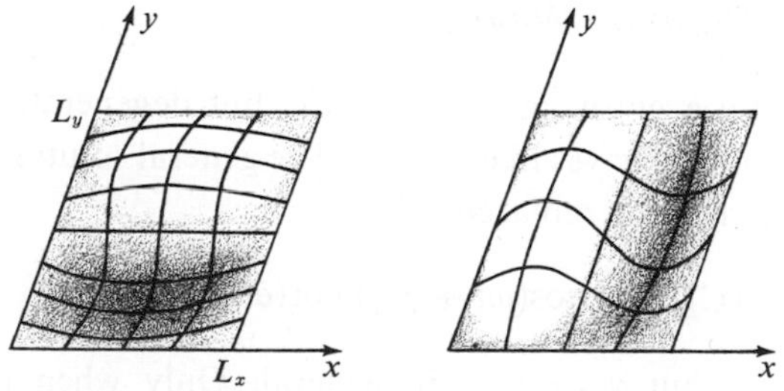

(a) $n_x = 1, n_y = 2$ (b) $n_x = 2, n_y = 1$

Figure 13.1

the membrane into $n_x \times n_y$ rectangles, alternate ones of which enclose displacements of opposite signs at any given moment, in a checkerboard pattern.

A greater variety of standing wave patterns can occur in *square* membranes, when $L_x = L_y = L$. Then the two frequencies for $n_x \neq n_y$,

$$\omega_{n_x n_y} = (n_x^2 + n_y^2)^{1/2} \pi c/L = \omega_{n_y n_x}, \tag{13.13}$$

degenerate to the same value. The word "degenerate" is being used here in the same sense as in the eigenvalue problem of (10.37) and (10.38), and in the eigenfrequency problem discussed following (11.18). It is typical of *two*-fold degeneracy that two orthogonal eigenmodes have the same eigenvalue. Here this results in waves which can "stand" even as arbitrary combinations of two eigenmodes; for example, the two terms of

$$\begin{aligned} q &\sim C_{12} \sin(\pi x/L)\sin(2\pi y/L) + C_{21} \sin(2\pi x/L)\sin(\pi y/L) \\ &= 2C_{12} \sin(\pi x/L)\sin(\pi y/L)[\cos(\pi y/L) + (C_{21}/C_{12})\cos(\pi x/L)] \end{aligned} \tag{13.14}$$

can have a time-independent ratio because they can have the same time factors. This leads to an infinite variety of standing wave patterns, characterized by the continuity of values $-\infty < C_{21}/C_{12} < \infty$, for the ratio as a parameter. The node curves of the patterns with $C_{21}/C_{12} = -1, -\frac{1}{2}, +\frac{1}{2}, +1$, and $+2$ are shown in Figure 13.2. The cases $C_{21}/C_{12} 0 =$ and $C_{21}/C_{12} =$

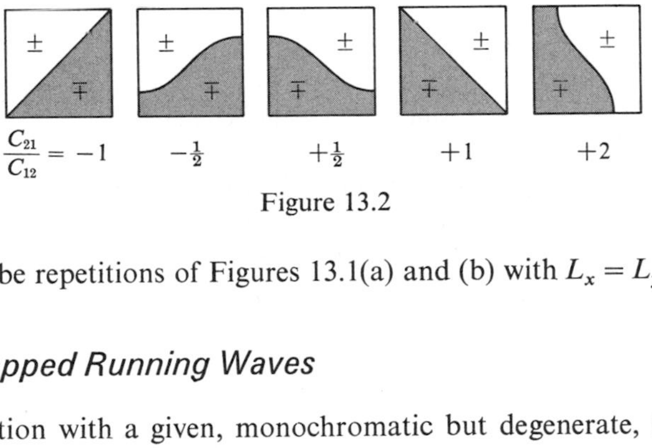

Figure 13.2

$\pm\infty$ would be repetitions of Figures 13.1(a) and (b) with $L_x = L_y = L$.

Trapped Running Waves

Motion with a given, monochromatic but degenerate, frequency is not restricted to standing waves like (13.14). The general solution form (13.12) shows that C_{21}/C_{12} may be replaced by

$$(C_{21}/C_{12})\cos(\omega t + \gamma_{21})/\cos(\omega t + \gamma_{12}), \tag{13.15}$$

with $\omega \equiv \omega_{12} = \omega_{21}$ but $\gamma_{12} \neq \gamma_{21}$, in general. Only when the motion is started in such a way that the phase difference $\gamma_{12} - \gamma_{21}$ is a multiple of π will the time-dependence cancel out and a standing wave pattern result. More generally the ratio (13.15), which determines the node curves of Figure 13.2, varies with time, and then a continuous succession of such node curves is formed as time goes on. The result is a running wave, trapped within the square boundaries and having wave fronts that always intersect the boundary normally except at the instants of turning the corners.

13.2 THE CIRCULAR DRUMHEAD

Suppose that the retangular boundary of the preceding problem is replaced by a circular one. Then the boundary condition on $q(x, y, t)$ becomes $q(x, \pm\sqrt{a^2 - x^2}, t) = 0$, if a is the radius of the boundary circle. Imposing this on the cartesian form (13.12) of the general solution would require solving awkward transcendental equations. Far neater expressions should result from rewriting the general solution in terms of polar coordinates r, φ as substitutes for x, y. Then the mass elements are labeled by r, φ in $q(r, \varphi, t)$, and the "circular drumhead" boundary condition becomes.

$$q(a, \varphi, t) = 0, \tag{13.16}$$

to hold at all angles $0 < \varphi < 2\pi$ and all times. The problem now is to find the expression for the general solution in terms of the polar variables.

The Cylindrical Wave Equation

To remove the awkwardness of describing a situation of circular symmetry on a cartesian "map," the replacement of the cartesian by polar coordinates should begin with the wave equation itself. If the disturbance is treated as a function $q(r, \varphi, t)$ of the polar variables $r = (x^2 + y^2)^{1/2}$, $\varphi = \tan^{-1}(y/x)$, then the cartesian derivatives in the wave equation should be replaced by

$$\begin{aligned}\frac{\partial q}{\partial x} &= \frac{\partial q}{\partial r}\frac{\partial r}{\partial x} + \frac{\partial q}{\partial \varphi}\frac{\partial \varphi}{\partial x} = \frac{x}{r}\frac{\partial q}{\partial r} - \frac{y}{r^2}\frac{\partial q}{\partial \varphi} \\ &\equiv [\cos\varphi(\partial/\partial r) - \sin\varphi(\partial/r\,\partial\varphi)]q(r, \varphi, t), \\ \partial q/\partial y &\equiv [\sin\varphi(\partial/\partial r) + \cos\varphi(\partial/r\,\partial\varphi)]q(r, \varphi, t),\end{aligned} \tag{13.17}$$

and

$$\frac{\partial^2 q}{\partial x^2} + \frac{\partial^2 q}{\partial y^2} = \left(\cos\varphi\frac{\partial}{\partial r} - \frac{\sin\varphi}{r}\frac{\partial}{\partial\varphi}\right)^2 q + \left(\sin\varphi\frac{\partial}{\partial r} + \frac{\cos\varphi}{r}\frac{\partial}{\partial\varphi}\right)^2 q.$$

Reduction of the last expression leads to

$$\frac{\partial^2 q}{\partial r^2} + \frac{1}{r}\frac{\partial q}{\partial r} + \frac{\partial^2 q}{r^2\,\partial\varphi^2} = \frac{\partial^2 q}{c^2\,\partial t^2} \tag{13.18}$$

as the representation of the wave equation in polar coordinates. It is sometimes called the "cylindrical" wave equation because, interpreted in three dimensions, its solutions would have cylindrical wave fronts, with no variations along the (z-) dimension normal to the plane of the polar coordinates.

The Helmholtz equation (13.3) for the monochromatic representatives $u(r, \varphi)$ now becomes

$$-(\partial^2/\partial r^2 + \partial/r\partial r + \partial^2/r^2 d\varphi^2)u = k^2 u. \tag{13.19}$$

Solutions which are separable in the polar coordinates

$$u(r, \varphi) = Z(r), \Phi(\varphi) \tag{13.20}$$

will be sought this time; the notation $Z(r)$ is adopted for the radial functions because they are known as "Zylinderfunktionen" in German works. Substitution into (13.19), followed by multiplication with (r^2/u), yields

$$-\frac{r^2}{Z}\left(Z'' + \frac{1}{r}Z'\right) - \frac{\Phi''}{\Phi} = k^2 r^2.$$

This requires the ratio Φ''/Φ to be some constant, independent of φ as well as r, and so the Helmholtz equation is separated into the two ordinary differential equations

$$\begin{aligned} \Phi'' &= -m^2\Phi, \\ -\left(\frac{d^2}{dr^2} + \frac{d}{rdr} - \frac{m^2}{r^2}\right)Z &= k^2 Z, \end{aligned} \tag{13.21}$$

where m^2 is an arbitrary constant (not necessarily positive, or even real, so far).

The solutions of the first of the equations (13.21) can be expressed very simply: $\Phi \sim \exp\colon \pm im\varphi$. However, a serious restriction on them arises from the circumstance that only a limited range, $0 \leq \varphi < 2\pi$, of the independent variable is needed for the complete description, yet whatever points are taken as the end points of the range are not to be given any special distinction (the description should be invariant to which orientation is labeled $\varphi = 0$). The matter may be settled by considering the effect of removing the limits on φ; then the values φ and $\varphi + 2\pi$ must be regarded as alternative labels for the same point, and only solutions with the property

$$e^{\pm im(\varphi + 2\pi)} = e^{\pm im\varphi}$$

can be admitted for representing disturbances with some unique value at each point. Thus

$$e^{2\pi i m} = 1 \qquad \text{and} \qquad m = 0, \pm 1, \pm 2, \ldots. \tag{13.22}$$

As a result, the general solution formed by superposing the separable solutions will comprise an analysis of the angular dependence into a *discrete* Fourier series like (12.46),

$$u(r, \varphi) = \sum_{m=-\infty}^{\infty} Z_m(r) e^{im\varphi}. \tag{13.23}$$

There is here a resolution on the orthonormal set

$$\begin{aligned} \Phi_m(\varphi) &\equiv e^{im\varphi}/(2\pi)^{1/2}, \\ \oint d\varphi \Phi_m^* \Phi_{m'} &= \delta_{mm'}, \end{aligned} \tag{13.24}$$

with the integration range $0 \leq \varphi < 2\pi$.

The modes (13.24), of dependence on φ, are discretely enumerable for much the same reason as are the modes of a bounded string. As in (12.13), a whole number of wavelengths must fit into a limited domain. When $r\varphi$ is an arc length along some ring of arbitrary radius r, then

$$e^{im\varphi} = e^{\pm i(|m|/r)(r\varphi)}$$

repeats itself in wavelengths, $\lambda_m = 2\pi r/|m|$, which fit exactly $|m|$ times into the ring circumference $2\pi r$.

Cylindrical Harmonics

Dividing the radial equation of (13.21) through by k^2 makes it evident that its solutions will contain k and r only as products, $\zeta \equiv kr$. These products are measures of the radial distance in units of $1/k = \lambda/2\pi$. The equation may be written as

$$Z_m''(\zeta) + \frac{1}{\zeta} Z_m' + \left(1 - \frac{m^2}{\zeta^2}\right) Z_m = 0, \tag{13.25}$$

and has long been known as the "Bessel equation." For its application here, its solutions need be known only for real positive values of ζ and integer values of m. However, solutions defined on the whole complex plane of ζ, and with m not an integer, have also been found useful in physical descriptions.

The various solutions of Bessel's equation are known generically as "cylindrical harmonics" and have been classified into four types: the "Bessel functions," $J_m(\zeta)$; the "Neumann functions," $N_m(\zeta)$; the "Hankel functions of the first and second kinds," $H_m^{(1)}(\zeta)$ and $H_m^{(2)}(\zeta)$. Their interrelationships parallel those among the familiar solutions of the simple oscillator equations, such as the first of the equations (13.21). Just as the oscillator equation, $\Phi_m'' = -m^2\Phi_m$, may be expressed as an arbitrary linear combination of the trigonometric functions cos $m\varphi$ and sin $m\varphi$, so

$$Z_m = (\text{const.})_1 J_m + (\text{const.})_2 N_m \tag{13.26}$$

in general; the Bessel and Neumann functions share with the trigonometric functions the property of having real values for positive real values of their arguments. The general solution of the oscillator equation may instead be expressed as an arbitrary linear combination of the complex exponentials, exp: $\pm im\varphi$. The corresponding alternative to (13.26) is

$$Z_m = (\text{const.})_1' H_m^{(1)} + (\text{const.})_2' H_m^{(2)}, \tag{13.27}$$

for the Hankel functions have been defined as

$$H_m^{(1)} = J_m + iN_m, \qquad H_m^{(2)} = J_m - iN_m, \tag{13.28}$$

in analogy to exp: $\pm\, im\varphi = \cos m\varphi \pm i \sin m\varphi$. Conversely,

$$J_m = \tfrac{1}{2}(H_m^{(1)} + H_m^{(2)}), \qquad N_m = \frac{1}{2i}(H_m^{(1)} - H_m^{(2)}). \tag{13.29}$$

The complex conjugate of $H_m^{(1)}(\zeta)$ is $H_m^{(2)}(\zeta^*)$, just as [exp: $+im\varphi]^* =$ exp: $-im\varphi^*$ when φ is complex.

As for the trigonometric functions, numerical tabulations of the more useful, low-order cylindrical harmonics exist.* Higher-order functions may be derived from low-order ones by means of such relations as

$$Z_{m+1} = (2m/\zeta)Z_m - Z_{m-1}, \tag{13.30}$$

$$Z_{m+1} = -\zeta^m (d/d\zeta)[\zeta^{-m} Z_m]. \tag{13.31}$$

These, and many other relations, which have been found and recorded* facilitate the analytical handling of the functions. Their derivations will not be repeated here. (After all, the mathematicians who found them tried hard to preclude all necessity for further analytical thought in those connections. However, see Exercise 13.4.)

* For example, in E. Jahnke and F. Emde's *Funktionen-tafeln*, Teubner (Leipzig), 1933.

One derivation will be instructive to review because it furnishes an example of a widely applicable technique: the solution of differential equations in the form of power series in the independent variable. Applied to the simple oscillator equation $\Phi_m'' = -m^2\Phi_m$, it would yield the ordinary Taylor expansions:

$$\begin{aligned} \cos m\varphi &= \sum_{n=0}^{\infty} (-)^n (m\varphi)^{2n}/(2n)!, \\ \sin m\varphi &= \sum_{n=0}^{\infty} (-)^n (m\varphi)^{2n+1}/(2n+1)!, \\ e^{im\varphi} &= \sum_{n=0}^{\infty} (im\varphi)^n/n!. \end{aligned} \tag{13.32}$$

Here, $n! \equiv n(n-1)\cdots 3\cdot 2\cdot 1$ for n a positive integer, and $0! \equiv 1$, by definition. The comparable series representations of the Bessel functions may be found by seeking coefficients a_n which will make

$$Z(\zeta) = \zeta^s \sum_{n=0}^{\infty} a_n \zeta^n \tag{13.33}$$

satisfy Bessel's equation. Whatever the lowest, "indicial," power may turn out to be, it is factored out so that a_0 is the coefficient of the first term in the series ($a_0 \neq 0$).

The first two derivatives of the series (13.33) are

$$Z' = \sum_{n=0}^{\infty} a_n(n+s)\zeta^{n+s-1},$$

$$Z'' = \sum_{n=0}^{\infty} a_n(n+s)(n+s-1)\zeta^{n+s-2},$$

and so the Bessel equation yields

$$\sum_{n=0}^{\infty} a_n\{[(n+s)^2 - m^2]\zeta^{n+s-2} + \zeta^{n+s}\} = 0.$$

By writing the two lowest-powered terms separately,

$$\begin{aligned} a_0[s^2 - m^2]\zeta^{s-2} &+ a_1[(1+s)^2 - m^2]\zeta^{s-1} \\ &+ \sum_{n=0}^{\infty} \{a_{n+2}[(n+s+2)^2 - m^2] + a_n\}\zeta^{n+s} = 0, \end{aligned} \tag{13.34}$$

the coefficient of each different power is exhibited. Each of these coefficients must vanish for the balance to zero to be produced at every ζ; cancellations among different powers cannot be counted on, since each power varies with ζ at a different rate.

For the coefficient $a_0[s^2 - m^2]$ to vanish, the series must start with the power of ζ^m, *or* of ζ^{-m}. Next, $a_1[(s+1)^2 - s^2] = 0$, and this requires $a_1 = 0$ (except* possibly for $s = -\frac{1}{2}$). Finally, each higher-powered term will properly cancel out if alternate coefficients have the ratios

$$a_{n+2}/a_n = -1/(n+2)(n+2s+2). \tag{13.35}$$

As a consequence, $a_1 = 0$ will lead to there being no odd powers at all, in the series multiplying ζ^s in (13.33). The even-powered $\zeta^{n \equiv 2N}$ will have the coefficient†

$$\begin{aligned} a_{2N} &= -[2^2 N(N+s)]^{-1} a_{2(N-1)} \\ &= +[2^4 N(N-1) \cdot (N+s)(N+s-1)]^{-1} a_{2(N-2)} \\ &= -[2^6 N(N-1)(N-2) \cdot (N+s)(N+s-1)(N+s-2)]^{-1} a_{2(N-3)} \\ &\;\;\vdots \\ &= (-)^N [2^{2N} N!\,(N+s)!/s!]^{-1} a_0 . \end{aligned}$$

Thus any solution representable by a power series like (13.33) is proportional to

$$J_s(\zeta) = \left(\frac{\zeta}{2}\right)^s \sum_{N=0}^{\infty} \frac{(-)^N}{N!\,(N+s)!} \left(\frac{\zeta}{2}\right)^{2N}, \tag{13.36}$$

in which $s = +m$ *or* $s = -m$. A specific normalization, implemented by the choice $a_0 = 1/2^s s!$, has been adopted here, in order to conform to the definition of the Bessel functions as tabulated.

Since two ostensibly different solutions have been found for each $m \neq 0$, it may seem that the general solution may be written as an arbitrary linear combination of J_m and J_{-m}, except when $m = 0$. This is indeed so‡

---

* This possibility will henceforth be ignored, since it is easy to find that retaining it leads only to a repetition of a solution ($J_{+1/2}$) which will also be found by choosing $s = +\frac{1}{2}$.

† It should be noted that only the property $\zeta! = \zeta \cdot (\zeta - 1)!$, and not the absolute value of $\zeta!$, is needed here. This leaves freedom (afforded by the arbitrary multiplier the solution may have) to adopt the standard definition

$$\zeta! \equiv \int_0^{\infty} x^{\zeta} e^{-x}\, dx$$

for factorials of any number, real or complex. Sometimes $\zeta!$ is called the "gamma function" and is denoted $\Gamma(\zeta + 1) \equiv \zeta!$.

‡ In consistency with (13.26), because

$$N_m(\zeta) = [\cos m\pi \cdot J_m(\zeta) - J_{-m}(\zeta)]/\sin m\pi$$

for *m not* an integer.

when m is not an integer, but the two functions are actually identical within a sign for integer m:

$$J_{-m} = (-)^m J_m \qquad \text{for} \qquad m = 0 \quad \pm 1, \pm 2, \ldots . \tag{13.37}$$

This property is incorporated in the form (13.36) through the definition of the factorial of any negative integer as being infinite; then all powers smaller than $2N + s = +|m|$ vanish from the series for $J_{-|m|}$, and the starting power is $\zeta^{+|m|}$ in both $J_{\pm|m|}$. Quite aside from how factorials are defined, (13.35) shows that a negative integer value for $s = -|m|$, with $a_0 \neq 0$, would lead to $a_{2|m|} \to \infty$. Thus, no linearly independent solution consisting just of a series of powers, except J_m, exists for integer m.

Yet a second solution should exist for integer m, as it does for non-integer m. The two arbitrary constants which a general solution of a second-order differential equation must contain can be introduced only when a second, independent solution can be added to J_m. A second solution can indeed be found, for example by* admitting the possibility that its series representation contains terms proportional to $ln\zeta$, which cannot itself be expressed as a power series in ζ. The Neumann solutions have such a character; in particular, $N_0(\zeta)$ has a term $-(2/\pi)\ln \zeta$ as the dominant one for $\zeta \to 0$. The functions $N_1, N_2, N_3, \ldots$ likewise contain logarithmic terms, but their largest ones near the origin are the inverse powers†

$$N_m \to -\frac{(m-1)!}{\pi}\left(\frac{2}{\zeta}\right)^m \qquad \text{for} \quad m > 0, \zeta \to 0. \tag{13.38}$$

These solutions, "second" to J_m, all become infinite at the origin and are said to be "irregular" there. The Bessel functions J_m of (13.36) provide the only solutions that are finite, "regular," at the origin.

Bessel Waves

Series representations, like (13.36) for the Bessel functions, make it easy to see how a function behaves near the origin, but they are unrevealing about the behavior farther out. For that, it helps to investigate approximations that are valid for large values of the argument, $|\zeta| \to \infty$.

To find the greatest simplification which can arise from assuming $|\zeta| \to \infty$, a power ζ^p is sought such that its separation from $Z_m = \zeta^p f_m(\zeta)$

---

* Or by a method suggested in Exercise 3.16.

† The peculiar normalization of N_m exhibited here is a consequence of choosing it so that, at $|\zeta| \to \infty$, N_m differs from J_m only in phase and not in amplitude (see (13.41)).

makes the equation for $f_m(\zeta)$ particularly simple in the limit $|\zeta| \to \infty$. Substitution into the Bessel equation yields

$$f''_m + \frac{2p+1}{\zeta} f'_m + \left(1 - \frac{m^2 - p^2}{\zeta^2}\right) f_m = 0. \tag{13.39a}$$

This makes it evident that the choice $p = -\frac{1}{2}$ will eliminate the first-derivative term, leaving

$$f''_m + [1 - (m^2 - \tfrac{1}{4})/\zeta^2] f_m = 0, \tag{13.39b}$$

and this approaches the particularly simple form $f''_m \approx -f_m$ for $|\zeta|^2 \gg m^2$. The result thus obtained is that every solution of Bessel's equation must approach proportionality to a trigonometric harmonic, modulated by an amplitude decreasing as $|\zeta|^{-1/2}$; that is,

$$Z_m \sim \zeta^{-1/2} \cos(\zeta + \gamma_m) \tag{13.40}$$

with some phase γ_m.

The expressions like (13.40) which are approached by the Bessel and Neumann functions as standardized in the literature are

$$\begin{aligned} J_m(|\zeta| \to \infty) &\to (2/\pi\zeta)^{1/2} \cos[\zeta - \tfrac{1}{2}(m + \tfrac{1}{2})\pi], \\ N_m(|\zeta| \to \infty) &\to (2/\pi\zeta)^{1/2} \sin[\zeta - \tfrac{1}{2}(m + \tfrac{1}{2})\pi]. \end{aligned} \tag{13.41}$$

The corresponding Hankel functions attain the complex exponential forms that follow from substituting (13.41) into (13.28).

It is now clear that the solutions of Bessel's equations describe oscillations for real values of their arguments. This behavior continues to infinity with a slowly decreasing amplitude, $\sim\zeta^{-1/2}$. Detailed evaluations of J_0, J_1, and J_2 are exhibited in Figure 13.3.

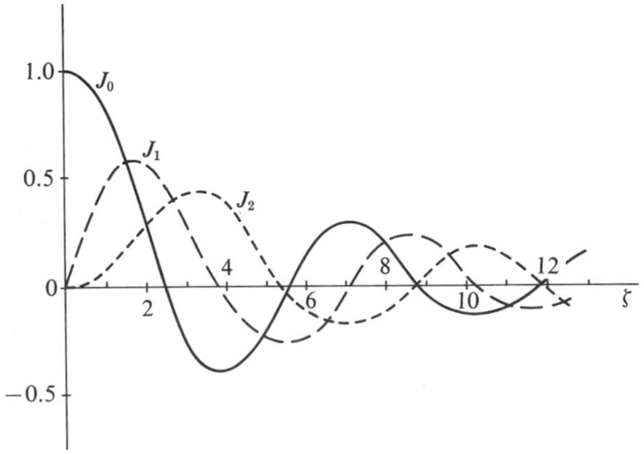

Figure 13.3

The positions of the nodes in the Bessel waves will have interest for the application to be made here. Some are listed in Table 13.1 (see the

Table 13.1 Nodes ζ_{mn} of $J_m(\zeta_{mn}) = 0$.

m \ n	1	2	3	4	5	...
0	2.40_5	5.52	8.65	11.79	14.93	...
± 1	3.83	7.02	10.17	13.32	16.47	...
± 2	5.13_5	8.42	11.62	14.80	17.96	...
± 3	6.38	9.76	13.02	16.22	19.41	...
± 4	7.59	11.06	14.37	17.62	20.83	...
± 5	8.77	12.34	15.70	18.98	22.22	...
⋮	⋮	⋮	⋮	⋮	⋮	

reference on page 398). The nth node, ζ_{mn}, of the Bessel function J_m ought to approach $\zeta_{mn} \to (2n + |m| - \frac{1}{2})\pi/2$ for $\zeta_{mn}^2 \gg m^2$, according to (13.41). It can be seen that the largest values of ζ_{mn} appearing in the table do tend to the expected behavior, being apart by about $\pi \approx 3$ in neighboring columns, and by about $\frac{1}{2}\pi \approx 1.5$ in successive rows.

The Circular Eigenmodes

The foregoing has shown that superpositions of the monochromatic "representatives"

$$ue^{-i\omega t} = Z_m(r)\Phi_m(\varphi)e^{-i\omega t}, \tag{13.42}$$

(13.2) and (13.20), will describe the possible surface vibrations which are finite everywhere, including the point chosen for the center $r = 0$, if the radial waves Z_m are taken to be the regular Bessel functions,* $J_m(kr)$.

In the case of the drumhead of finite radius, a, there is still a boundary condition, $q(a, \varphi, t) = 0$ for every φ and t, to satisfy. That can be done only by such frequencies, $\omega = \omega_{mn}$, as make

$$J_m\left(\frac{\omega_{mn}\, a}{c}\right) = 0. \tag{13.43}$$

* The complementary, irregular Neumann functions of (13.26) can be used only for regions which exclude $r = 0$, and may then be necessary, to help satisfy conditions on whatever boundary is used to exclude $r = 0$.

This defines a discrete spectrum of possible (eigen)frequencies,

$$\omega_{mn} = \zeta_{mn}\, c/a \quad \begin{cases} m = 0, \pm 1, \pm 2, \ldots, \\ n = 1, 2, 3, \ldots, \end{cases} \tag{13.44}$$

where the numbers ζ_{mn} are just the roots of the Bessel functions, $J_m(\zeta_{mn}) = 0$, as tabulated in Table 13.1.

The lowest frequency, $\omega_0 = 2.4c/a$ is that of the "fundamental" vibration proportional to

$$q_0 \sim \text{Re}\{J_0(2.4r/a)e^{-i\omega_0 t}\},$$

vanishing at $r = a$ because $J_0(2.4) = 0$. All the other $m = 0$ modes have such circular symmetry (independence of φ),

$$q_{0n} \sim \text{Re}\left\{J_0\left(\frac{\zeta_{0n} r}{a}\right)e^{-i\omega_{0n} t}\right\}.$$

and have node circles at the radii

$$r = (\zeta_{01}/\zeta_{0n})a,\ (\zeta_{02}/\zeta_{0n})a,\ \ldots,\ a.$$

Profiles are sketched in Figure 13.4; all these circularly symmetric modes

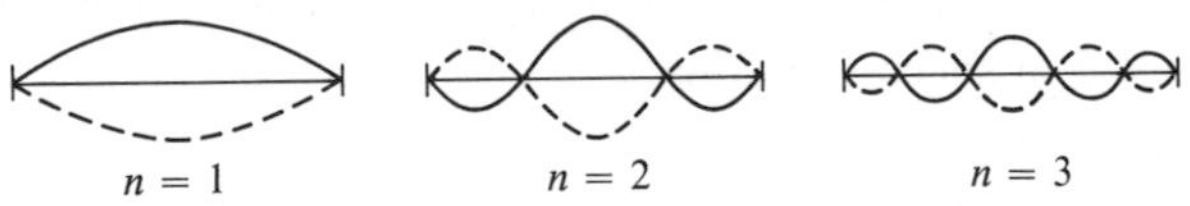

Figure 13.4

consist of standing waves which give nonvanishing displacements to the center, since $J_0(0) = 1$.

The general mode with $m \neq 0$,

$$\begin{aligned} q_{mn} &\sim \text{Re}\{J_m(\zeta_{mn} r/a)e^{i(m\varphi - \omega_{mn} t)}\} \\ &= J_m \cos(m\varphi - \omega_{mn} t) \end{aligned} \tag{13.45}$$

forms a running wave circulating about the center, in a sense determined by the sign of m. These modes all vanish as $J_m \sim r^{|m|}$ at the center, an effect which can be understood as a consequence of centrifugal force arising from the circulation (see §15.3).

The oppositely circulating modes with $m = \pm|m|$ have the same (degenerate) frequency, as reference to (13.44) and Table 13.1 shows, and hence can combine into a standing wave like

$$J_m(\zeta_{mn} r/a) \cdot \sin m(\varphi - \varphi_0) \cdot \cos(\omega_{mn} t + \gamma_{mn}), \tag{13.46}$$

where φ_0 is an arbitrary constant giving the position of a diameter forming some one node line. Node diameters occur at angular intervals $\Delta\varphi = \pi/m$ if

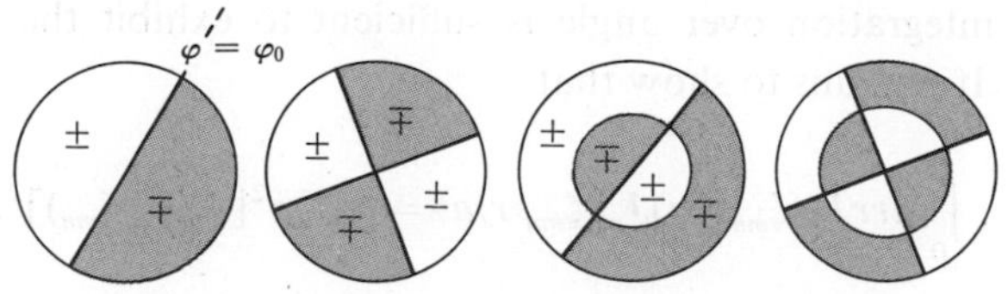

$|m| = 1, n = 1$ $|m| = 2, n = 1$ $|m| = 1, n = 2$ $|m| = 2, n = 2$

Figure 13.5

$m \neq 0$. A few examples of standing waves with $m \neq 0$ are indicated in Figure 13.5, by means of node lines separating regions of opposite displacement. The circulating, "trapped," running waves of the preceding paragraph will have nodes just like these, but the diametrical ones will circulate continuously in one sense or the other ($m \gtrless 0$).

The Orthogonal Set of Circular Eigenfunctions

Whereas the cartesian mapping of planar vibrations led to a classification into modes characterized by real values for k_x, k_y, the polar representation classifies into modes labeled by positive $k = (k_x^2 + k_y^2)^{1/2}$ and an integer m. Pinning the vibrations down on a rectangular boundary restricts the cartesian modes to ones labeled by positive integers $n_x = |k_x| L_x/\pi$, $n_y = |k_y| L_y/\pi$, while a circular boundary restricts the circular eigenmodes to discrete ones labeled by integers m and n, with $k = \zeta_{mn}/a$. Whereas the general vibration of a rectangular membrane can be represented by the superposition (sum over n_x, n_y) in (13.12), the general motion of the circular drumhead may be represented by

$$q(r, \varphi, t) = \sum_{m,\, n} C_{mn} J_m(\omega_{mn} r/c) e^{im\varphi} \cos(\omega_{mn} t + \gamma_{mn}). \tag{13.47}$$

As in the string problem of §12.2, the constants C_{mn} and γ_{mn} are determined by the start given to the motion. The problem is like the one of §12.2 in that again the spatial dependence is being resolved onto an orthogonal set.

The set of functions

$$J_m(\omega_{mn} r/c) e^{im\varphi} \quad \begin{cases} m = 0, \pm 1, \pm 2, \ldots, \\ n = 1, 2, 3, \ldots, \end{cases} \tag{13.48}$$

form an orthogonal set because

$$\int_0^a drr \int_0^{2\pi} d\varphi J_m(\omega_{mn} r/c) J_{m'}(\omega_{m'n'} r/c) e^{i(m-m')\varphi}$$

$$= \delta_{mm'} \delta_{nn'} \cdot \pi a^2 [J_{m+1}(\omega_{mn} a/c)]^2. \tag{13.49}$$

The trivial integration over angle is sufficient to exhibit the orthogonality for $m \neq m'$. It remains to show that

$$\int_0^a drrJ_m(\zeta_{mn} r/a)J_m(\zeta_{mn'} r/a) = \delta_{nn'}\tfrac{1}{2}a^2[J_{m+1}(\zeta_{mn})]^2, \tag{13.50}$$

and the orthogonality for $n \neq n'$ here will now be found to follow directly from the Bessel equation of (13.21). The abbreviated notations $k_1 = \zeta_{mn}/a$ and $k_2 = \zeta_{mn'}/a$ are adopted and the identity

$$d^2/dr^2 + d/rdr \equiv r^{-1}\, d/dr(rd/dr), \tag{13.51}$$

as applied to any function of r, is recognized. Then the two Bessel equations for the two Bessel functions being integrated in (13.50) may be written

$$\begin{aligned} \left[-\frac{1}{r}\frac{d}{dr}\left(r\frac{d}{dr}\right) + \frac{m^2}{r^2}\right]J_m(k_1 r) &= k_1^2 J_m(k_1 r), \\ \left[-\frac{1}{r}\frac{d}{dr}\left(r\frac{d}{dr}\right) + \frac{m^2}{r^2}\right]J_m(k_2 r) &= k_2^2 J_m(k_2 r). \end{aligned} \tag{13.52}$$

These are respectively subjected to the integral operations

$$\int_0^a drrJ_m(k_2 r)\cdots \qquad \text{and} \qquad \int_0^a drrJ_m(k_1 r)\cdots,$$

and then the difference of the results yields

$$\begin{aligned} (k_1^2 - k_2^2)\int_0^a drrJ_m(k_1 r)J_m(k_2 r) &= \int_0^a dr\left\{J_m(k_1 r)\frac{d}{dr}\left[r\frac{d}{dr}J_m(k_2 r)\right] - (k_1 \leftrightarrow k_2)\right\}, \\ &= \int_0^a dr\frac{d}{dr}r\left\{J_m(k_1 r)\frac{d}{dr}J_m(k_2 r) - (k_1 \leftrightarrow k_2)\right\}, \\ &= a\{J_m(k_1 a)[dJ_m(k_2 r)/dr]_a - (k_1 \leftrightarrow k_2)\}, \end{aligned}$$

where each of the symbols $(k_1 \leftrightarrow k_2)$ designates a repetition of the term preceding it, but with k_1 and k_2 interchanged. Now $k_1 a = \zeta_{mn}$ and $k_2 a = \zeta_{mn'}$ are each a root of J_m and so the last line vanishes. This leads to the conclusion that the integral of (13.50) must vanish unless $n = n'$. It should be pointed out that the procedure here has been an adaptation of the procedure represented by (10.32), in which the orthogonality of different eigenvectors of a symmetric operator was demonstrated. The preceding chapter showed how eigenvector equations go over into "eigenfunction" equations like

(13.52). The latter aspect of the formalism will be considered further in the next chapter (§14.3).

There remains the evaluation of the integral (13.50) when it does not vanish. Its particular value is of course a result of the normalization which happens to have been chosen for the Bessel functions. The evaluation is not difficult but does not warrant space here since its result can be obtained from the extant collections of formulas* such as

$$\int d\zeta\zeta Z_m^2(\zeta) = \tfrac{1}{2}\zeta^2\left\{Z_{m+1}^2 + \left(1 - \frac{2m}{\zeta}\right)Z_m^2\right\} \tag{13.53}$$

for the indefinite integral.

13.3 VIBRATIONS OF A SPHERICAL SURFACE

Considered next are surface vibrations which have a limited extent because they are confined to the surface of a sphere. Circumstances in which a spherical surface can be treated as the equilibrium configuration were encountered by Rayleigh when he considered vibrations in the surface shapes of incompressible, but deformable, water droplets. Bohr much later found a like model useful for representing certain excitations of atomic nuclei, in which the nucleus behaves like a drop of incompressible "nuclear fluid."

The wave equation (13.1) relates quantities evaluated at a point, and the considerations that led to it can be expected to hold at each point of a smoothly† curved surface as well as they do on a planar surface. However, it is no longer convenient to use the special cartesian axes having an (x, y) plane in which lies the element of surface at the point of application. It is simple to re-express the equation in an arbitrarily oriented frame, since the axes had been used only for presenting a scalar product of gradient derivatives, $\nabla \cdot \nabla \equiv \nabla^2$. Whereas the special axes had been chosen so that $\nabla_z q \equiv \partial q/\partial z = 0$, in general

$$\nabla^2 = (\partial/\partial x)^2 + (\partial/\partial y)^2 + (\partial/\partial z)^2. \tag{13.54}$$

This is frequently called the "Laplace operator" or the "laplacian."

The wave equation can now be given the frame-independent representation

$$\nabla^2 q(r, t) = \partial^2 q/c^2\, \partial t^2. \tag{13.55}$$

---

* See the reference on page 398, for example, and Exercise 13.5.

† So that the derivatives in the equation exist.

When applying it to surface vibrations, it is necessary to confine the mass-element label, $\mathbf{r}$, to the equilibrium surface. Thus, the cartesian components of $\mathbf{r}$ were restricted to having $z = 0$ in the preceding sections, and should be restricted by $(x^2 + y^2 + z^2)^{1/2} = a$ for a surface which is a sphere of radius a, centered on the origin of $\mathbf{r}$. The restriction to a sphere is awkward to impose on cartesian coordinates but becomes simple in terms of spherical coordinates r, ϑ, φ. The transformation to these may be carried out in the same way as was the change to polar coordinates in (13.17), and the result is (Exercise 13.6)

$$\boxed{\nabla^2 = \frac{1}{r^2}\frac{\partial}{\partial r} r^2 \frac{\partial}{\partial r} + \frac{1}{r^2 \sin\vartheta}\frac{\partial}{\partial\vartheta}\sin\vartheta\frac{\partial}{\partial\vartheta} + \left(\frac{\partial}{r\sin\vartheta\,\partial\varphi}\right)^2}. \tag{13.56}$$

When this is applied to a disturbance $q(\vartheta, \varphi, t)$ of mass elements with equilibrium positions on $r = a$, the radial derivatives vanish, and the wave equation becomes

$$\left(\frac{1}{\sin\vartheta}\frac{\partial}{\partial\vartheta}\sin\vartheta\frac{\partial}{\partial\vartheta} + \frac{1}{\sin^2\vartheta}\frac{\partial^2}{\partial\varphi^2}\right)q(\vartheta, \varphi, t) = \frac{a^2}{c^2}\frac{\partial^2 q}{\partial t^2}. \tag{13.57}$$

The corresponding Helmholtz equation for the monochromatic representative $u(\vartheta, \varphi)$ is

$$-\left(\frac{1}{\sin\vartheta}\frac{\partial}{\partial\vartheta}\sin\vartheta\frac{\partial}{\partial\vartheta} + \frac{1}{\sin^2\vartheta}\frac{\partial^2}{\partial\varphi^2}\right)u = (ka)^2 u, \tag{13.58}$$

if $q = u \exp: - i\omega t$ with $\omega \equiv ck$.

Separable solutions,

$$u(\vartheta, \varphi) = \Theta(\vartheta)\cdot\Phi(\varphi), \tag{13.59}$$

may be sought first. Substitution into (13.58), followed by multiplication with $(\sin^2\vartheta/u)$, yields

$$-\frac{\sin\vartheta}{\Theta}\frac{d}{d\vartheta}\sin\vartheta\frac{d\Theta}{d\vartheta} - \frac{\Phi''(\varphi)}{\Phi(\varphi)} = (ka)^2\sin^2\vartheta.$$

A constancy of $-\Phi''/\Phi = m^2$ must be demanded just as in (13.21), and so the φ-dependence will again be described with the help of the orthogonal set (13.24) of complex exponentials. The corresponding ϑ-dependences must then be described by solutions of

$$\left[-\frac{1}{\sin\vartheta}\frac{d}{d\vartheta}\sin\vartheta\frac{d}{d\vartheta} + \frac{m^2}{\sin^2\vartheta}\right]\Theta = (ka)^2\Theta, \tag{13.60}$$

to be investigated next.

The Eigenfrequencies

Axially symmetric solutions, independent of φ, may be sought first. With $u(\vartheta)$ to be a function of ϑ only, the Helmholtz equation (13.58) reduces to the special case of (13.60) with $m = 0$. A change of independent variable to $\mu \equiv \cos\vartheta$ will prove convenient, with the help of $d/d\vartheta = -\sin\vartheta\, d/d\mu$, it transforms the equation into

$$\left[\frac{d}{d\mu}(1-\mu^2)\frac{d}{d\mu} + (ka)^2\right]u = 0. \tag{13.61}$$

This is a form of what is known as the "Legendre equation." Since the whole sphere is covered by values of ϑ in the range $0 \leq \vartheta \leq \pi$, it is solutions for real μ in the range $+1 \geq \mu \geq -1$ that will be needed.

A power series representation of the solution, as in the case (13.33) for the Bessel equation, may be sought:

$$u = \mu^s \sum_{n=0}^{\infty} a_n \mu^n. \tag{13.62}$$

This time, the first conclusions from substitution into the differential equation are

$$s(s-1)a_0 = 0 \qquad \text{and} \qquad (s+1)sa_1 = 0.$$

Both conditions can be satisfied, with arbitrary a_0 and a_1, if the series is started with $s = 0$; $u = a_0 + a_1\mu + \cdots$. Then the recurrence relation

$$\frac{a_{n+2}}{a_n} = \frac{n(n+1) - (ka)^2}{(n+1)(n+2)} \tag{13.63}$$

emerges for the coefficients of the higher powers. It enables the determination of $a_2, a_4, a_6, \ldots$ in ratio to a_0, and $a_3, a_5, a_7, \ldots$ in ratio to a_1, to yield

$$\begin{aligned} u = a_0&\left[1 - \frac{(ka)^2}{2}\mu^2 - \frac{(ka)^2}{2}\cdot\frac{6-(ka)^2}{12}\mu^4 - \cdots\right] \\ +a_1&\left[\mu + \frac{2-(ka)^2}{6}\mu^3 + \frac{2-(ka)^2}{6}\cdot\frac{12-(ka)^2}{20}\mu^5 + \cdots\right]. \end{aligned} \tag{13.64}$$

With two constants, a_0 and a_1, left arbitrary, the solution is already a general one. It is easy to check that the choices $s = 1$, $a_1 = 0$ or $s = -1$, $a_0 = 0$, each of which also satisfy the "indicial conditions" above, merely repeat special cases of (13.64).

There is an important difference between the recurrence relations (13.63) and those found for the Bessel series, (13.35). The latter gave successive coefficients in rapidly diminishing ratios, $\sim 1/n^2$ as n grew large; consequently the infinitely long Bessel series "converged" to finite results (13.41) even for indefinitely large values of their arguments. On the other hand, (13.63) has a ratio a_{n+2}/a_n which approaches unity for $n^2 \gg (ka)^2$; as a consequence each of the independent series solutions (13.64) has terms proportional to

$$\cdots \mu^N(1 + \mu + \mu^2 + \cdots) = \cdots \frac{\mu^N}{1 - \mu^2},$$

for sufficiently large $N \gg (ka)$. The last form follows from (11.67). Thus the solutions found here "explode" at the poles of the sphere, where $\mu = \pm 1$, and cannot be used to describe disturbances that are finite everywhere on the sphere. Yet any function which is thus finite should be representable by *some* superposition of powers.*

A properly finite superposition of the powers can be obtained only for special values of the wave number, k, in (13.64). Thus, if $a_1 = 0$ is chosen, and k such that $(ka)^2 = 0$ or 6 or 20 or ..., then all terms of power higher than that of $a_0 \mu^0$ or $a_2 \mu^2$ or $a_4 \mu^4$ or ..., respectively, will have vanishing coefficients; the result will be an even-powered polynomial. Odd-powered polynomial solutions can be obtained by choosing $a_0 = 0$ and k such that $(ka)^2 = 2$ or 12 or 30 or .... In general, the wave number which yields a polynomial of degree l is such that

$$(ka)^2 = l(l + 1) \qquad \text{with} \qquad l = 0, 1, 2, 3, \ldots. \tag{13.65a}$$

It has thus been found that stably finite, axially symmetric oscillations of the entire spherical surface, inclusive of the poles, can exist only with the discrete eigenfrequencies

$$\omega_l = [l(l + 1)]^{1/2} c/a. \tag{13.65b}$$

The discreteness here is again really a consequence of fitting "wavelengths" into a finite span. The noninteger square root factor can be ascribed to the nonuniformity† of the "wavelengths" at various latitudes on the sphere. For those which are so short as to span only a nearly flat segment of arc, arising for large values of l,

$$2\pi a \approx l(2\pi c/\omega_l) = \text{“}l\lambda\text{”}$$

* For example, such a solution must possess finite derivatives and so a power series should be obtainable by Taylor expansion.

† Compare the variable "wavelengths", $\lambda = 2\pi r/|m|$, found on rings of various radii, r, in the example on page 397.

follows from (13.65). This corresponds to the fitting of a number l of waves into the circumference of a great circle on the sphere.

The Legendre Polynomials

The "eigenfunctions" $u_l(\vartheta)$ corresponding to the eigenfrequencies ω_l are to be obtained from (13.64), as described above. They are proportional to what are called the "Legendre Polynomials," and will coincide with these as standardized in the literature (see the reference on page 398, for example) if the choices

$$a_0 = l!/(-4)^{(1/2)l}[(\tfrac{1}{2}l)!]^2, \qquad a_1 = 0 \qquad \text{for even } l,$$

$$a_0 = 0, \qquad a_1 = l!/(-4)^{(1/2)(l-1)}\left[\left(\frac{l-1}{2}\right)!\right]^2 \qquad \text{for odd } l,$$

are made for the arbitrary constant multipliers. Then a development somewhat more intricate than the one leading to the expression (13.36) for the Bessel series will yield for the lth polynomial:

$$P_l(\mu) = \sum_s \frac{(-)^s}{2^l s!} \frac{(2l-2s)!}{(l-s)!\,(l-2s)!} \mu^{l-2s}, \tag{13.66}$$

where $s = 0, 1, 2, \ldots, \frac{1}{2}l$ for even l and $s = 0, 1, 2, \ldots, \frac{1}{2}(l-1)$ for odd l. For integers s outside these spans, factorials* in the denominators of (13.66) approach infinity and the terms vanish. The expression (13.66) may be checked by showing that its coefficients satisfy the recurrence relations (13.63) with $(ka)^2 = l(l+1)$.

A so-called "Rodrigues Formula" for "generating" the Legendre polynomials may be derived by substituting

$$\left(\frac{d}{d\mu}\right)^l \mu^{2l-2s} = \frac{(2l-2s)!}{(l-2s)!} \mu^{l-2s}$$

into (13.66), with the result

$$\begin{aligned} P_l(\mu) &= \frac{1}{2^l l!}\left(\frac{d}{d\mu}\right)^l \sum_{s=0}^{l} \frac{(-)^s l!}{s!(l-s)!} \mu^{2(l-s)} \\ &= \frac{1}{2^l l!}\left(\frac{d}{d\mu}\right)^l (\mu^2 - 1)^l. \end{aligned} \tag{13.67}$$

The last step incorporates the binomial theorem (9.131).

---

* See the middle footnote on page 400.

Another formula for generating the Legendre polynomials arises (§14.2) as the power series expansion of $|\mathbf{r}_f - \mathbf{r}_s|^{-1}$:

$$\boxed{\frac{1}{(r_f^2 - 2r_f r_s \cos\vartheta + r_s^2)^{1/2}} = \sum_{l=0}^{\infty} \frac{r_<^l}{r_>^{l+1}} P_l(\cos\vartheta)}\,, \qquad (13.68)$$

where $r_<$ is the lesser of the magnitudes r_f, r_s, and $r_>$ is the greater one. This expression plays an important role in physics, providing what is called a "multipole analysis" of such fields as the potential

$$V = -K/r \equiv -K/|\mathbf{r}_f - \mathbf{r}_s|$$

of (4.26), when the force center is not at the origin of the vectors $\mathbf{r}_f$ and $\mathbf{r}_s$, but rather at either one of the latter positions. It is the angle-dependent coefficients in the power series (13.68) that were originally *defined* as the Legendre polynomials, and this is responsible for the normalization adopted for them. The circumstance that the coefficients in the series expansion of $1/r$ also turn out to be solutions of $-\nabla^2 u(\vartheta) = k^2 u$ stems from the fact that $\nabla^2(1/r) = 0$, as will be discussed in §14.2.

The expression (13.68) makes it easy to find what values the Legendre polynomials have at the poles of the sphere: $\mu \equiv \cos\vartheta = \pm 1$. Take $r_s = 1$ as the length unit, and then for $r_f < 1$

$$\sum_{l=0}^{\infty} r_f^l P_l(\pm 1) = (1 \mp r_f)^{-1} = 1 \pm r_f + r_f^2 \pm r_f^3 + r_f^4 \pm \cdots.$$

Thus

$$P_l(+1) = +1 \qquad \text{and} \qquad P_l(-1) = (-)^l. \qquad (13.69)$$

These values could also have been obtained from the Rodrigues formula (13.67), by substituting $\mu = \pm(1 - \varepsilon)$ and approximating $(\mu^2 - 1)$ by (-2ε). The values at the equator, $\mu = 0$, are most directly given by the sum (13.66):

$$P_l(0) = 0 \qquad \text{for odd } l \qquad (13.70a)$$

because the polynomial then contains only odd powers of μ, whereas there is a term $\sim\mu^0$, with the coefficient

$$P_l(0) = (l-1)!!/(-2)^{(1/2)l}(\tfrac{1}{2}l)! \qquad \text{for even } l. \qquad (13.70b)$$

The "double factorial" here is defined by

$$(2n+1)!! \equiv 1 \cdot 3 \cdot 5 \cdot \cdots \cdot (2n-1) \cdot (2n+1), \qquad (13.71a)$$

for odd integers $(2n + 1)$ only. It is easy to see that

$$(2n + 1)!! = (2n + 1)!/2^n n!, \tag{13.71b}$$

by dividing out the even factors in $(2n + 1)!$.

The expansion (13.68) can also be regarded as a resolution on an orthogonal set, since

$$\int_{-1}^{1} d(\cos \vartheta) P_l P_{l'} = \delta_{ll'} \cdot 2/(2l + 1). \tag{13.72}$$

The orthogonality for $l \neq l'$ can be shown to follow directly from the differential equation (13.61), in much the way such a result was obtained for the Bessel functions of (13.50). However, a method using the Rodrigues formula (13.67) will also allow evaluation of the integral for $l = l'$. Let

$$D_{kl}(\mu) \equiv \left(\frac{d}{d\mu}\right)^k (\mu^2 - 1)^l$$

so that

$$P_l = \frac{D_{ll}}{2^l l!}.$$

Notice that $D_{kl}(\pm 1) = 0$ for $k < l$. In

$$\int_{-1}^{1} d\mu P_l P_{l'} = [2^{l+l'} l! (l')!]^{-1} \int_{-1}^{1} d\mu D_{ll} D_{l'l'},$$

which is invariant to interchanging l and l', let l be the greater order $(l > l')$ and carry out partial integrations which yield

$$\begin{aligned}\int_{-1}^{1} d\mu D_{ll} D_{l'l'} &= -\int_{-1}^{1} d\mu D_{l-1,l} D_{l'+1,l'} \\ &= +\int_{-1}^{1} d\mu D_{l-2,l} D_{l'+2,l'} \\ &\;\vdots \\ &= (-)^{l'} \int_{-1}^{1} d\mu D_{l-l',l} D_{2l',l'}.\end{aligned}$$

If $l > l'$ by at least one unit, onc more such step can be taken, to yield an integral containing

$$D_{2l'+1,l'}(\mu) = \left(\frac{d}{d\mu}\right)^{2l'+1} [\mu^{2l'} - l'\mu^{2l'-2} + \cdots] \equiv 0,$$

and so the orthogonality for $l \neq l'$ is proved. For $l = l'$,

$$\int_{-1}^{1} d\mu P_l^2 = [2^l l!]^{-2} \cdot (-)^l \int_{-1}^{1} d\mu D_{0,l} D_{2l,l},$$

where $D_{0,l} = (\mu^2 - 1)^l$ and $D_{2l,l} = (2l)!$. The integrations by parts,

$$\begin{aligned}\int_{-1}^{1} d\mu(\mu+1)^l(\mu-1)^l &= -[l/(l+1)]\int_{-1}^{1} d\mu(\mu+1)^{l+1}(\mu-1)^{l-1}\\ &= +[l(l-1)/(l+1)(l+2)]\int_{-1}^{1} d\mu(\mu+1)^{l+2}(\mu-1)^{l-2}\\ &\;\;\vdots\\ &= (-)^l[l!/(l+1)\cdots(l+l)]\int_{-1}^{1} d\mu(\mu+1)^{2l},\end{aligned}$$

yields a final integral with the value $2^{2l+1}/(2l+1)$, and then the result (13.72) follows.

The second-order Legendre equation (13.61), with $(ka)^2 = l(l+1)$, must for generality have a second solution besides P_l. The series (13.64) already presents a general solution. Thus the second solution for even l will consist of the terms proportional to a_1, and the terms proportional to a_0 provide second solutions for odd l. For example, the "Legendre function of the second kind" of order $l = 0$ is given by the terms of (13.64) proportional to a_1 and with $(ka)^2 = 0$:

$$Q_0 = \mu + \tfrac{1}{3}\mu^3 + \tfrac{1}{5}\mu^5 + \cdots = \tfrac{1}{2}\ln\frac{1+\mu}{1-\mu}. \tag{13.73}$$

The power series here is just the Taylor expansion of the indicated logarithmic function. Obviously, Q_0 is "irregular" at the poles $\mu = \pm 1$, and this is true of all the solutions Q_l "second to P_l," as expected in view of the considerations that led to the introduction of the integers l into (13.64). The Legendre polynomials P_l remain the only axially symmetric solutions which are finite everywhere on the sphere.

Legendre Modes

The simplest Legendre polynomials are

$$\begin{aligned} P_0 &= 1, & P_2 &= \tfrac{1}{2}(3\mu^2 - 1),\\ P_1 &= \mu, & P_3 &= \tfrac{1}{2}(5\mu^3 - 3\mu). \end{aligned} \tag{13.74}$$

These can be found either from the recursion relation (13.63), or the sum

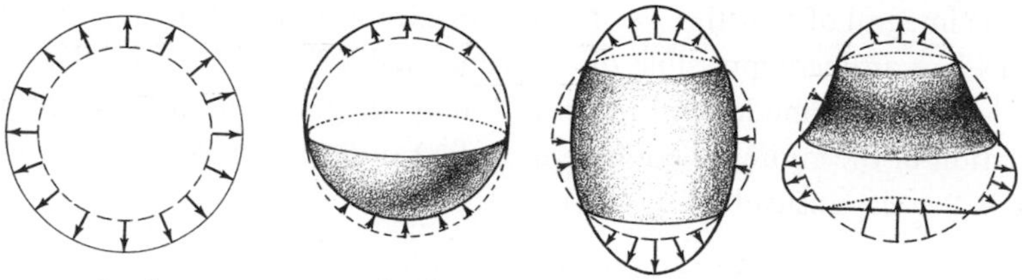

Figure 13.6

(13.66), or the Rodrigues formula (13.67), or the expansion (13.68). They may be pictured in terms of wave patterns they can help describe. Figure 13.6 shows polar diagrams of surfaces $R(\vartheta) = a[1 + \varepsilon_l(t)P_l(\cos\vartheta)]$ for $l = 0$, 1, 2, 3; these represent possible disturbances, $q = R - a$, which satisfy the wave equation if $\varepsilon_l(t)$ has the frequency ω_l. The shaded and unshaded regions represent areas of opposite sign for $R - a$, at some given phase of the standing wave motion; they are separated by motionless node circles, at various latitudes on the sphere.

The case $l = 0$ gives no motion at all, since $\omega_0 = 0$ in (13.65). A uniform distention (or contraction, if $\varepsilon_0 < 0$) is being described, and there are circumstances in which such a pulsation could be set up with some finite period. However, the surface wave equation does not encompass such circumstances, since it allows only for restoring forces that arise from *relative* displacements of neighboring *surface* elements. The absence of an $l = 0$ motion is obviously to be expected when the equilibrium sphere is the envelope of an incompressible fluid; the volume cannot remain properly unchanged unless $\varepsilon_0 \equiv 0$. On the other hand, incompressibility would not hinder any other modes if the oscillations do not grow too large, since the volume is independent of ε_1, ε_2, ε_3, ... for $\varepsilon_l \ll 1$. This can be seen by computing the volume enclosed by

$$R(\vartheta) = a\left[1 + \sum_l \varepsilon_l(t)P_l(\cos\vartheta)\right], \tag{13.75}$$

which will be

$$\begin{aligned}2\pi\int_{-1}^{1} d(\cos\vartheta)\int_0^{R(\vartheta)} dr r^2 &= \left(\frac{2\pi}{3}\right)\int_{-1}^{1} d\mu a^3\left[1 + \sum_l \varepsilon_l P_l\right]^3 \\ &\approx \left(\frac{4\pi}{3}\right)a^3\left[1 + 3\sum_l \varepsilon_l \cdot \tfrac{1}{2}\int_{-1}^{1} d\mu(P_0)P_l\right] \qquad \text{for} \quad \varepsilon_l \ll 1 \\ &= \left(\frac{4\pi}{3}\right)a^3[1 + 3\varepsilon_0],\end{aligned}$$

since the integral of P_l with $P_0 \equiv 1$ vanishes except for $l = 0$. Thus the volume behaves like an incompressible one during small enough $l \geq 1$ oscillations.*

The $l = 1$ mode, as Figure 13.6 may suggest, amounts merely to a translation of the entire sphere by $\varepsilon_1 a$ (to first order in ε_1). A sphere centered on $r \cos \vartheta = a\varepsilon_1$ instead of $r = 0$ has

$$R^2 - 2\varepsilon_1 aR \cos \vartheta + \varepsilon_1^2 a^2 = a^2,$$

or

$$R = \varepsilon_1 a \cos \vartheta + [a^2 - (\varepsilon_1 a \sin \vartheta)^2]^{1/2}$$
$$\approx a(1 + \varepsilon_1 \cos \vartheta) = a(1 + \varepsilon_1 P_1) \text{ for } \varepsilon_1 \ll 1,$$

in place of $R = a$. Such a displacement leads to an oscillation of frequency $\omega_1 = 2^{1/2}c/a$, according to the above procedure, because it was presumed that the equilibrium configuration was a sphere with a fixed center. Usually, isolated bodies, with mass centers which can move *freely* without arousing restoring forces, are of greater interest. The above solutions can still be used, subject to appropriate restrictions. They can represent oscillations relative to the mass center, arising from "internal" forces alone, if they are modified so that they describe a steady $r = 0$ mass-center position. The conditions appropriate to an incompressible enclosed volume can be found by computing

$$\bar{z} = (4\pi a^3/3)^{-1} \cdot 2\pi \int_{-1}^{1} d(\cos \vartheta) \int_0^{R(\vartheta)} dr r^2 (r \cos \vartheta)$$
$$\approx (3a/2) \sum_l \varepsilon_l \int_{-1}^{1} d(\cos \vartheta) P_1 P_l = \varepsilon_1 a.$$

Thus it is only necessary to put $\varepsilon_1 = 0$ to have descriptions of motions relative to the center of mass.

It has thus transpired that the lowest-frequency oscillation which can be induced in an *isolated*, incompressible sphere is

$$R(\vartheta) = a[1 + \varepsilon_2(t) P_2(\cos \vartheta)],$$

with $\omega_2 = 6^{1/2}c/a$. This can be viewed as a sequence of distortions into ellipsoids of revolution, an oscillation between the extended, "prolate" shape indicated in Figure 13.6 and a squashed, "oblate" shape that is distended around the equatorial belt and flattened at the poles. The cartesian description of the ellipsoid is given by

$$\frac{x^2 + y^2}{a^2(1 - \frac{1}{2}\varepsilon_2)^2} + \frac{z^2}{a^2(1 + \varepsilon_2)^2} = 1,$$

* The fact that the incompressibility thus represented is only approximate means merely that exact incompressibility requires small additions of higher-order solutions.

since $P_2(0) = -\frac{1}{2}$ and $P_2(\pm 1) = 1$. With $R = z/\cos\vartheta$, this yields

$$R \approx a/[1 + \varepsilon_2 - 3\varepsilon_2 \cos^2 \vartheta]^{1/2} = a[1 - 2\varepsilon_2 P_2]^{-1/2},$$

which is the same as $a(1 + \varepsilon_2 P_2)$ to lowest order in ε_2.

In the $l = 3$ mode, the oscillations of the upper and lower hemispheres are like two $l = 2$ modes vibrating in opposite phase; that is, as the upper half elongates, the lower is flattened, and vice vèrsa.

The profiles pictured in Figure 13.6 show that each mode may be regarded as a fitting of "wavelengths" into great circles on the sphere. The lth mode will have l crests and l troughs forming latitude bands encircling the polar axis. This band structure has led to the name "zonal harmonics" for the Legendre polynomials.

The Associated Legendre Functions

As indicated in arriving at equation (13.60), a more general disturbance than an axially symmetric one will be some superposition of representatives

$$\Theta_m(\vartheta)e^{i(m\varphi - \omega t)}, \qquad m = 0, \pm 1, \pm 2, \ldots, \tag{13.76}$$

with Θ_m a solution of

$$\left[\frac{d}{d\mu}(1 - \mu^2)\frac{d}{d\mu} + (ka)^2 - \frac{m^2}{1 - \mu^2}\right]\Theta_m = 0. \tag{13.77}$$

This equation is just (13.60) with $\mu \equiv \cos\vartheta$ substituted for ϑ.

The generalization from the axially symmetric case (13.61) to $m \neq 0$ introduces a term into the balance (13.77) which becomes infinite at the poles $\mu = \pm 1$. It will prove convenient to introduce a substitute for the dependent variable, by means of

$$\Theta_m = (1 - \mu^2)^\alpha f(\mu), \tag{13.78}$$

with such an α that the singularity is eliminated* from the consequent differential equation for $f(\mu)$:

$$(1 - \mu^2)f'' - 2\mu(2\alpha + 1)f' + \left[(ka)^2 - 2\alpha - \frac{m^2 - 4\alpha^2\mu^2}{1 - \mu^2}\right]f = 0.$$

This shows that the singular term can be reduced to a finite constant by

---

* An analogous elimination of singularities at poles was found convenient in producing Figure 10.7.

choosing $4\alpha^2 = m^2$. More specifically, $\alpha = +\frac{1}{2}|m|$ will be adopted for both $m \gtrless 0$, because the $f(\mu)$ thus defined obeys an equation,

$$(1 - \mu^2)f'' - 2\mu(|m| + 1)f' + [(ka)^2 - |m|(|m| + 1)]f = 0 \quad (13.79)$$

which can be solved somewhat more directly than the one with the sign of α reversed.

Equations like (13.79) can be generated from the equation (13.61) for the $m = 0$ case,

$$(1 - \mu^2)\Theta_0'' - 2\mu\Theta_0' + (ka)^2\Theta_0 = 0,$$

by successive differentiations:

$$(1 - \mu^2)\Theta_0''' - 2\mu(1 + 1)\Theta_0'' + [(ka)^2 - 2]\Theta_0' = 0,$$
$$(1 - \mu^2)\Theta_0^{(\mathrm{IV})} - 2\mu(2 + 1)\Theta_0''' + [(ka)^2 - 6]\Theta_0'' = 0,$$
$$(1 - \mu^2)\Theta_0^{(\mathrm{V})} - 2\mu(3 + 1)\Theta_0^{(\mathrm{IV})} + [(ka)^2 - 12]\Theta_0''' = 0,$$

and so on. After $|m|$ derivatives,

$$(1 - \mu^2)\Theta_0^{(|m|+2)} - 2\mu(|m| + 1)\Theta_0^{(|m|+1)} + [(ka)^2 - |m|(|m| + 1)]\Theta_0^{(|m|)} = 0.$$

Comparison with (13.79) shows that it will be satisfied by

$$f(\mu) = (d/d\mu)^{|m|}\Theta_0$$

if Θ_0 is any solution of the $m = 0$ equation.

It was found that the general solution Θ_0 for arbitrary k, given in (13.64), diverges to infinity at the poles of the sphere. These singularities only grow worse for derivatives of Θ_0. On the other hand, derivatives of the special Legendre polynomial solutions plainly remain finite everywhere, and provide the only regular solutions. These exist only for the discrete eigenfrequencies ω_l of (13.65), and thus it transpires that the ω_l's constitute the *entire* spectrum of frequencies, of all possible finite oscillations of the whole spherical surface, and not only for the axially symmetric motions.

The regular solutions of equation (13.77),

$$P_l^{|m|} \equiv (1 - \mu^2)^{\frac{1}{2}|m|}(d/d\mu)^{|m|}P_l(\mu)$$
$$= \frac{(1 - \mu^2)^{\frac{1}{2}|m|}}{2^l l!}\left(\frac{d}{d\mu}\right)^{l+|m|}(\mu^2 - 1)^l, \quad (13.80)$$

are known as the "Associated Legendre Functions." A point of first importance is that these everywhere finite solutions exist only for values of m less than, or equal to, l in absolute value:

$$m = 0, \pm 1, \pm 2, \ldots, \pm l. \tag{13.81}$$

This can be seen from the fact that the highest power of μ occurring in the binomial expansion of $(\mu^2 - 1)^l$ is μ^{2l}, and so just $l + |m| = 2l$ derivatives reduce it to a constant, $(2l)!$. The $(2l + 1)$st derivative vanishes identically.

The solution (13.80) is the only independent regular one for both $m = \pm|m|$, since only m^2 appears in the equation (13.77); however, a $P_l^{-|m|}$ differing from $P_l^{|m|}$ by just a constant factor will be convenient to define:

$$\begin{aligned} P_l^{-|m|} &\equiv \frac{(1 - \mu^2)^{-\frac{1}{2}|m|}}{2^l l!} \left(\frac{d}{d\mu}\right)^{l-|m|} (\mu^2 - 1)^l \\ &= (-)^m[(l - |m|)!/(l + |m|)!]P_l^{|m|}. \end{aligned} \tag{13.82}$$

One way to prove that this is consistent with the expression (13.80) for $P_l^{|m|}$ is to work in terms of the derivatives $D_{kl}(\mu)$ of $(\mu^2 - 1)^l$, already employed to prove (13.72). Here it will suffice to abbreviate the notation for the kth derivative to $D_k(\mu) \equiv D_{kl}$, since the same l will be considered throughout. Then the Legendre polynomial (13.67) is just $P_l = D_l/2^l l!$ and D_l obeys the Legendre equation (13.61):

$$l(l + 1)D_l(\mu) = (d/d\mu)(\mu^2 - 1)D_{l+1}(\mu).$$

An indefinite integration, extending from either pole, converts this into

$$l(l + 1)D_{l-1}(\mu) = (\mu^2 - 1)D_{l+1}(\mu),$$

since $D_k(\pm 1) = 0$ for $k < l$. A succession of such integrations, together with a dexterous use of the differential equations known to be satisfied by each $f(\mu) \sim D_{l+|m|}(\mu)$, generates a succession of expressions

$$[(l + m)!/(l - m)!]D_{l-m} = (\mu^2 - 1)^m D_{l+m},$$

but it will suffice to show that if this is true for any given value of m, then it is also true for $m + 1$ or $m - 1$. That only requires differentiation of the last expression rewritten in terms of $|m|$:

$$\begin{aligned} &[(l + |m|)!/(l - |m|)!]D_{l-|m|+1} \\ &\quad = (\mu^2 - 1)^{|m|-1}\{(\mu^2 - 1)D_{l+|m|+1} + 2|m|\mu D_{l+|m|}\}. \end{aligned}$$

The first term in the curly bracket may be replaced by

$$-2\mu|m|D_{l+|m|} + [l(l + 1) - (|m| - 1)|m|]D_{l+|m|-1}$$

according to the differential equation (13.79), and thus a proportionality of $D_{l-(|m|-1)}$ and $D_{l+(|m|-1)}$ emerges. The correct proportionality constant also emerges, since the last square bracket can be rewritten as $(l+|m|) \times (l-|m|+1)$. With the results of this paragraph, obtaining (13.82) is elementary. Notice that the last of the relations (13.82) remains valid if $|m|$ is replaced by $m \gtrless 0$.

The Spherical Harmonics

The "eigenfunctions" (13.76) which will represent finite oscillations of an entire spherical surface form the discrete set

$$P_l^m(\vartheta)e^{i(m\varphi-\omega_l t)} \quad \begin{cases} \omega_l = \dfrac{[l(l+1)]^{1/2}c}{a}, \\ l = 0, 1, 2, 3, \ldots, \\ m = 0, \pm 1, \pm 2, \ldots, \pm l. \end{cases} \tag{13.83}$$

The angular dependence of each is proportional to a member of the *orthonormal* set

$$\begin{aligned} Y_{lm}(\vartheta, \varphi) &= C_{lm} P_l^m(\vartheta)\Phi_m(\varphi) \\ &\equiv \frac{(-)^{l+m}}{2^l l!}\left[\frac{2l+1}{4\pi}\cdot\frac{(l-m)!}{(l+m)!}\right]^{1/2} \\ &\quad \times \sin^m \vartheta[d/d(\cos\vartheta)]^{l+m} \sin^{2l}\vartheta \cdot e^{im\varphi}. \end{aligned} \tag{13.84}$$

These are known as the "Spherical Harmonics" and their orthonormality property may be represented as

$$\oint d\Omega Y_{lm}^*(\Omega) Y_{l'm'}(\Omega) = \delta_{ll'}\,\delta_{mm'}, \tag{13.85}$$

where

$$\oint d\Omega \equiv \int_0^{2\pi} d\varphi \int_0^{\pi} d\vartheta \sin\vartheta = \int_0^{2\pi} d\varphi \int_{-1}^{1} d(\cos\vartheta) = 4\pi \tag{13.86}$$

is the complete solid angle. Proof of the orthogonality, and the determination of the normalization constant C_{lm}, is considered next.

Suppose the smaller of the l's in (13.85) is called l'—that is, $l > l'$ in

$$\begin{aligned} \oint d\Omega Y_{lm}^* Y_{l'm'} &= C_{lm}^* C_{l'm'} \int_{-1}^{1} d\mu P_l^m P_{l'}^{m'} \cdot \delta_{mm'} \\ &= \delta_{mm'} C_{lm}^* C_{l'm}(-)^m \frac{(l+m)!}{(l-m)!} \int_{-1}^{1} d\mu P_l^m P_{l'}^{-m}. \end{aligned}$$

The motive for using (13.82) to replace $P_{l'}^m$ by $P_{l'}^{-m}$ is that there will follow a cancellation of the factor $(1-\mu^2)^{\pm(1/2)m}$ in front of P_l^m and $P_{l'}^{-m}$, respectively, as given in (13.80) and (13.82). In terms of the derivatives $D_{kl}(\mu)$ used in the proof of (13.72),

$$\int_{-1}^{1} d\mu P_l^m P_{l'}^{-m} = [2^{l+l'} l!\,(l')!]^{-1} \int_{-1}^{1} d\mu D_{l+m,\,l} D_{l'-m,\,l'}.$$

A succession of $(l'+m)$ partial integrations will reduce the last integral to one already encountered in the proof of (13.72):

$$(-)^{l'+m} \int_{-1}^{1} d\mu D_{l-l',\,l} D_{2l',\,l'} = \delta_{ll'} (-)^m 2^{2l+1} (l!)^2/(2l+1).$$

This shows that the unit value for the integral (13.85) with $l=l'$ and $m=m'$ will be obtained if the choice

$$|C_{lm}|^2 = [(2l+1)/2][(l-m)!/(l+m)!]$$

is made. The square root of this, multiplied by the normalization factor $(2\pi)^{-1/2}$ from $\Phi_m(\varphi)$, is exhibited in (13.84). The sign adopted is immaterial for the orthonormality; the standardization of sign shown here is a result of certain accidents of its history.

The behavior (13.82) of the Associated Legendre Functions under a sign change of m results in the property

$$Y_{l,\,-m} = (-)^m Y_{lm}^* \tag{13.87}$$

of the spherical harmonics. Whereas the two functions $P_l^{\pm m}$ are not linearly independent because of (13.82), the two functions $Y_{l,\,\pm m} \sim \Phi_{\pm m}$ are linearly independent because of their orthogonal φ-dependences.

For $m=0$, the spherical harmonics become axially symmetric, and differ from the Legendre polynomials only in normalization:

$$Y_{l0} = [(2l+1)/4\pi]^{1/2} P_l(\cos\vartheta) \tag{13.88}$$

as a comparison of (13.84) with (13.67) shows. For $l=0$,

$$Y_{00} = (4\pi)^{-1/2}, \tag{13.89}$$

just $P_0 = 1$ renormalized per unit solid angle. At the extreme $m=l$,

$$Y_{l,\,l} = \frac{(-)^l}{2^l l!} \left[\frac{(2l+1)!}{4\pi}\right]^{1/2} \sin^l \vartheta e^{il\varphi}, \tag{13.90}$$

while $Y_{l,\,-l} = (-)^l Y_{ll}^*$ for $m=-l$.

Degeneracies in the Oscillations

The oscillations labeled by a specific l-value—the ones proportional to any of the orthogonal eigenfunctions $Y_{lm}(\vartheta, \varphi)$ with the same l but having $m = 0, \pm 1, \ldots, \pm l$—all have the same frequency: $\omega_l = [l(l+1)]^{1/2} c/a$. Arbitrary linear combinations of the $(2l+1)$ independent eigenmodes can form standing waves,

$$q_l(\vartheta, \varphi, t) = \left[\sum_m C_m Y_{lm}(\vartheta, \varphi)\right] \cos(\omega_l t + \gamma).$$

There is thus a $(2l+1)$-fold *degeneracy* of the eigenfrequency ω_l.

A reason that such a variety of motions can all have the same frequency is most transparent in the $l = 1$ case. The spherical harmonics with $l = 1$ are

$$Y_{10} = (3/4\pi)^{1/2} \cos \vartheta, \qquad Y_{1,\pm 1} = \mp (3/8\pi)^{1/2} \sin \vartheta e^{\pm i\varphi}. \tag{13.91}$$

An arbitrary linear combination of these will have opposite signs on two halves of the spherical surface, separated by a node curve which is some great circle of the sphere, just as in the special case $q \sim Y_{10}$ shown in Figure 13.6. This is easiest to see in terms of the cartesian coordinates, $x = r \sin \vartheta \cos \varphi$, $y = r \sin \vartheta \sin \varphi$, $z = r \cos \vartheta$:

$$Y_{10} = \left(\frac{3}{4\pi}\right)^{1/2} \frac{z}{r}, \qquad Y_{1,\pm 1} = \mp \left(\frac{3}{8\pi}\right)^{1/2} \frac{x \pm iy}{r}. \tag{13.92}$$

Thus an arbitrary linear combination can always be written

$$q_1 = \alpha x + \beta y + \gamma z, \tag{13.93}$$

and this has nodes on a plane through the center, hence halving the sphere. Various such linear combinations will differ only as to the orientation of the plane containing the node circle. They must all be expected to have the same frequency under forces which have been supposed uniform on the sphere.

Mere differences of orientation can also be seen to characterize various of the $l = 2$ motions described with the help of

$$\begin{aligned} Y_{20} &= (5/16\pi)^{1/2}(3 \cos^2 \vartheta - 1), \\ Y_{2,\pm 1} &= \mp (15/8\pi)^{1/2} \sin \vartheta \cos \vartheta e^{\pm i\varphi}, \\ Y_{2,\pm 2} &= (15/32\pi)^{1/2} \sin^2 \vartheta e^{\pm 2i\varphi}. \end{aligned} \tag{13.94}$$

Standing wave patterns proportional to

$$\begin{aligned}
\mathrm{Re}(Y_{21}) &\sim Y_{21} - Y_{2,-1} \sim \sin\vartheta\cos\vartheta\cos\varphi,\\
\mathrm{Im}(Y_{21}) &\sim Y_{21} + Y_{2,-1} \sim \sin\vartheta\cos\vartheta\sin\varphi,\\
\mathrm{Re}(Y_{22}) &\sim Y_{22} + Y_{2,-2} \sim \sin^2\vartheta\cos 2\varphi,\\
\mathrm{Im}(Y_{22}) &\sim Y_{22} - Y_{2,-2} \sim \sin^2\vartheta\sin 2\varphi,
\end{aligned}$$

are compared in Figure 13.7. Shaded versus unshaded areas represent regions in which the displacements have opposite signs, and are separated by node circles. That the cases shown differ only in orientation is quite apparent. However, the case of displacements proportional to Y_{20}, shown in Figure 13.6, does not differ from those of Figure 13.7 merely by reorienta-

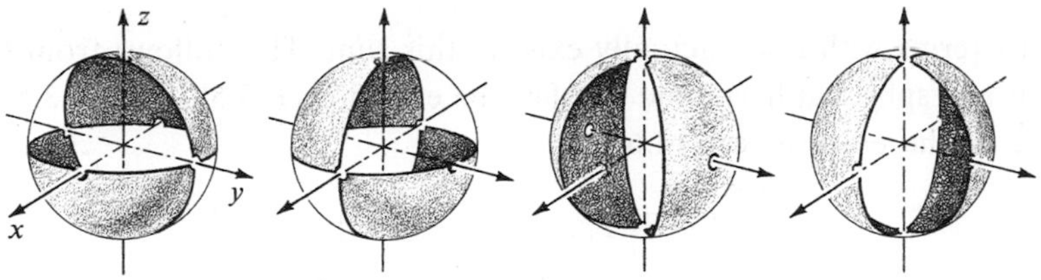

Figure 13.7

tion. It does have in common with the latter cases the occurrence of two node circles. Moreover, it will be seen in the next subsection that any reorientation of the Y_{20} pattern produces a linear combination of the five $l = 2$ patterns referred to the initial frame; hence the five-fold degeneracy remains an orientational one in this sense.

Rotational Properties

A more general examination of the orientational degeneracies amounts to investigating the behavior of the spherical harmonics under frame rotations. For the investigation here, it will suffice to specify a rotation of just the polar (z-)axis as in Figure 13.8. The amount of rotation is measured by the angles ϑ', φ' giving the direction of the new polar (z'-) axis, and Θ, Φ are the new coordinates of an arbitrary point ϑ, φ on the sphere.

The new coordinates are definite functions of ϑ, φ, for a given frame rotation, and so the spherical harmonic $Y_{lm}(\Theta, \Phi)$ may be treated as a function of ϑ, φ resolvable on the orthonormal set of spherical harmonics (13.84):

$$Y_{lm}(\Theta, \Phi) = \sum_{l'm'} C_{l'm'}\, Y_{l'm'}(\vartheta, \varphi). \tag{13.95}$$

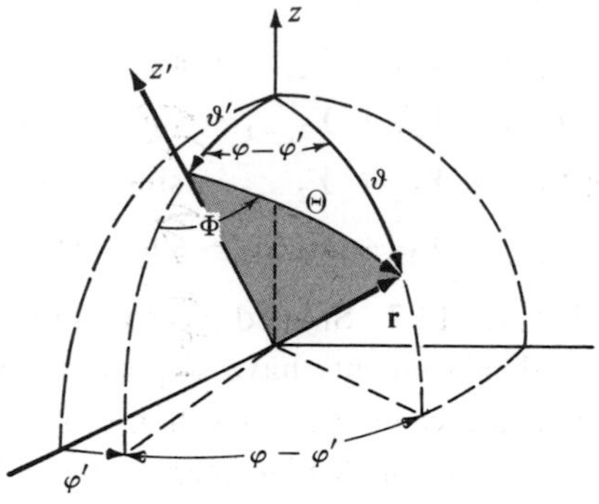

Figure 13.8

Only terms with $l' = l$ actually exist in this sum. This follows from the fact that any spherical harmonic satisfies the equations (13.58), and (13.56), with one of the eigenvalues (13.65),

$$-\nabla^2 Y_{lm} = [l(l+1)/a^2]Y_{lm}, \tag{13.96}$$

irrespective of the frame used ($\nabla^2 \equiv \nabla \cdot \nabla$ is an invariant!). As a consequence, applying the operation $\nabla^2 + l(l+1)/a^2$ to (13.95) makes it vanish, producing

$$\sum_{l'm'} C_{l'm'}[-l'(l'+1)/a^2 + l(l+1)/a^2]Y_{l'm'} = 0.$$

The independent (orthogonal) terms of this must each vanish individually, and so $C_{l'm'} = 0$ unless $l' = l$. The result

$$Y_{lm}(\Theta, \Phi) = \sum_{m'} C_{lm'}(\vartheta', \varphi')Y_{lm'}(\vartheta, \varphi) \tag{13.97}$$

shows that a reorientation of any spherical harmonic wave-pattern produces a linear combination of patterns, referred to the initial polar axis, all with the same frequency, $\omega_l = [l(l+1)]^{1/2}c/a$. Of course the frequency cannot have been changed merely by viewing the motion from a new angle.

The coefficients in (13.97) naturally depend on the amount of rotation, as indicated. Their specific values are perhaps most interesting when the motion being described is such that the new polar axis becomes an axis of symmetry and the relation becomes independent of Φ:

$$Y_{lm}(\Theta, \Phi) \to Y_{l0} = [(2l+1)/4\pi]^{1/2}P_l(\cos\Theta).$$

The relation (13.97), now independent of Φ, should become completely symmetrical between ϑ, φ and ϑ', φ', as reference to Figure 13.8 may help make obvious. It should then be unsurprising, in that respect at least, that the symmetrical expression

$$\boxed{P_l(\cos\Theta) = [4\pi/(2l+1)] \sum_m Y_{lm}^*(\vartheta', \varphi') Y_{lm}(\vartheta, \varphi)}, \qquad (13.98)$$

known as the spherical harmonic "addition theorem," can be derived. The complex conjugation operation (*) can rest on either of the factors, since the sum must have a real result. This operation is needed if the result is to depend only on the *difference* of the angles φ and φ', as should be expected.

The expression (13.98) can perhaps be derived most easily by taking advantage of the fact that summing it over l yields the "completeness" relation (see (12.74)),

$$\sum_{lm} Y_{lm}(\Omega) Y_{lm}^*(\Omega') = \delta(\Omega - \Omega'), \qquad (13.99)$$

complementary to the orthonormality (13.85) in a way analogous to the relation of (12.73) to (12.69), or of the projection operator $\mathbf{e}_k \mathbf{e}_k^\dagger = \mathbf{1}$ (9.109) to the orthonormal basis, $\mathbf{e}_k^\dagger \mathbf{e}_l = \delta_{kl}$. The completeness relation is an ex-expression of the resolvability of an arbitrary function $f(\vartheta, \varphi)$ according to

$$\begin{aligned} f(\Omega) &= \sum_{lm} C_{lm} Y_{lm}(\Omega) \\ &= \oint d\Omega' f(\Omega') \sum_{lm} Y_{lm}^*(\Omega') Y_{lm}(\Omega), \end{aligned} \qquad (13.100)$$

with the last line the result of substituting for C_{lm} the integration by which it is determined for a given $f(\Omega)$. If the final integral is to yield just the value $f(\Omega)$, whatever this function may be, the summation (13.99) must be equivalent to a delta function which vanishes unless $\Omega' = \Omega$, and has the property

$$\oint d\Omega' \, \delta(\Omega' - \Omega) = 1. \qquad (13.101)$$

Now the difference of directions $\Omega'(\vartheta', \varphi') - \Omega(\vartheta, \varphi)$ can depend only on Θ, not on Φ, and so the delta function should be resolvable according to

$$\delta(\Omega' - \Omega) = \sum_l A_l P_l(\cos\Theta).$$

The coefficients here may be determined from

$$A_l = \tfrac{1}{2}(2l+1) \int_{-1}^{1} d(\cos\Theta)\, \delta(\Omega' - \Omega) P_l(\cos\Theta),$$

as reference to the orthogonality condition (13.72) shows, and since $d\Omega \equiv d(\cos\Theta)\, d\Phi$,

$$A_l = [(2l+1)/4\pi] \oint d\Omega\, \delta(\Omega - \Omega')\, P_l(\cos\Theta),$$
$$= [(2l+1)/4\pi] P_l(1) = (2l+1)/4\pi.$$

This converts the completeness relation (13.99) into

$$\sum_l [(2l+1)/4\pi] P_l(\cos\Theta) = \sum_{lm} Y^*_{lm}(\Omega')\, Y_{lm}(\Omega).$$

From the result (13.97) it is known that the terms from the two sides with a given l must be independently equal, and so the "addition theorem" (13.98) is proved on the grounds outlined here.

The case of (13.98) with $l = 1$ yields the relation

$$\cos\Theta = \cos\vartheta \cos\vartheta' + \sin\vartheta \sin\vartheta' \cos(\varphi - \varphi'), \qquad (13.102)$$

well known in spherical trigonometry. Various examples of it occur as elements of the rotation matrix (9.49), giving direction cosines between basis unit vectors (also see Exercise 1.9).

The three spherical harmonics of first order ($l = 1$) are just renormalized "spherical components" of a unit vector $\hat{\mathbf{r}}$ in the direction, ϑ, φ (denoted Ω above), as can be seen from comparing (13.92) to (10.79):

$$Y_{1m}(\hat{\mathbf{r}}) \equiv (3/4\pi)^{1/2} \hat{r}_m. \qquad (13.103)$$

These are just projections on the rotation matrix eigenvectors, $\mathbf{u}_m$; that is,

$$\hat{r}_m = (\mathbf{u}_m^\dagger \hat{\mathbf{r}})^* \qquad (13.104)$$

according to (10.80). The expression (13.102) amounts to a scalar product,

$$\hat{\mathbf{r}} \cdot \hat{\mathbf{r}}' = \sum_m \hat{r}_m (\hat{r}'_m)^* = (4\pi/3) \sum_m Y_{1m}(\hat{\mathbf{r}}) Y^*_{1m}(\hat{\mathbf{r}}') \qquad (13.105)$$

like (10.87), of two unit-vector directions as resolved on the rotation eigenvector basis $\mathbf{u}_m$.

A higher-order spherical harmonic, Y_{lm}, is just one of the $(2l + 1)$ components of a "reduced" tensor of the lth rank. For example, $r^2 Y_{2m}(\hat{\mathbf{r}})$ is proportional to one of the five independent components of the reduced second-rank tensor

$$\mathsf{T} = \mathbf{r}\tilde{\mathbf{r}} - \tfrac{1}{3} r^2 \mathbf{1}, \qquad (13.106)$$

constructed to have a vanishing trace, as is (9.88). If

$$\mathsf{T}_{mm'} = r^*_m r_{m'} - \tfrac{1}{3} r^2\, \delta_{mm'} \qquad (13.107)$$

stands for a component on the rotation eigenvector basis (m, $m' = 0, \pm 1$), then it can be found that

$$
\begin{aligned}
(16\pi/45)^{1/2} r^2 Y_{20} &= T_{00} = -2T_{+1,+1} = -2T_{-1,-1}, \\
(4\pi/15)^{1/2} r^2 Y_{21} &= T_{0,+1} = T^*_{+1,0} = -T_{-1,0} = -T^*_{0,-1}, \\
(8\pi/15)^{1/2} r^2 Y_{22} &= -T_{-1,+1} = -T^*_{+1,-1}.
\end{aligned}
\qquad (13.108)
$$

The first of the lines shows proper consistency with the vanishing trace, $T_{mm} = 0$.

EXERCISES

13.1. For a square membrane, find the lowest eigenfrequency which allows standing waves with a *three*-fold degeneracy. [Five times the fundamental.]

13.2. A blow at the center of a rectangular membrane at rest, which gives it a total impulse $(\rho L_x L_y)v_0$, may be represented by the initial velocity distribution

$$v_0(x, y) = v_0 L_x L_y \,\delta(x - \tfrac{1}{2}L_x)\,\delta(y - \tfrac{1}{2}L_y),$$

vanishing except at the center point. Find the intensity ratio of the two lowest modes that will be excited, as a function of $L_y/L_x > 1$. ("Intensity" is a measure of energy proportional to the square of the amplitude. See (12.43) or (12.62).) Can you see why harmonics with even n_x or n_y are not excited?

13.3. (a) A blow at the center of a circular drum, giving it a total impulse $\rho \cdot \pi a^2 v_0$, may be represented by

$$v_0(r) = 2\pi a^2 v_0 \,\delta(r^2) \quad \text{with} \quad \int_0^{>0} d(r^2)\,\delta(r^2) = 1.$$

Find the intensity ratio with which the two lowest modes are excited. [For this, values of $J_1(\zeta_{0n})$ at roots ζ_{0n} of $J_0(\zeta_{0n}) = 0$ are needed. A table in the reference of the footnote on page 398 gives $J_1(\zeta_{01}) = 0.519$ and $J_1(\zeta_{02}) = -0.340$. Also see Figure 13.3].

(b) Compare the ratio of the two lowest *frequencies* excited here to the corresponding ratio of Exercise 13.2, specialized to a square drum.

13.4. The relations (13.30) and (13.31) are easy to prove for the special case of the regular Bessel functions $J_m(\zeta)$, from the series representation (13.36).

(a) Show that the term-by-term derivative of the series for $(\zeta^{-m} J_m)$ is just the series for J_{m+1} multiplied by $-\zeta^{-m}$, thus yielding (13.31).

(b) Prove similarly that

$$(d/d\zeta)(\zeta^m J_m) = \zeta^m J_{m-1}.$$

(c) Combine the results in (a) and (b) to demonstrate (13.30), and also

$$\frac{dJ_m}{d\zeta} = \tfrac{1}{2}(J_{m-1} - J_{m+1}).$$

(d) By using (13.31) in the form

$$\frac{Z_m}{\zeta^m} = -\frac{d}{\zeta d\zeta}\left(\frac{Z_{m-1}}{\zeta^{m-1}}\right),$$

show that

$$Z_m = \zeta^m(-d/\zeta d\zeta)^m Z_0.$$

13.5. (a) By replacing $k_2 a$ in (13.52) by a general ka which is *not* a root of $J_m(ka)$, show that

$$(k^2 - k_1^2)\int_0^a drr\, J_m(kr)J_m(k_1 r) = k_1 a J_m'(k_1 a)J_m(ka),$$

where $J_m'(\zeta) \equiv dJ_m/d\zeta$.

(b) By taking the derivative (d/dk) of this expression, then allowing $k \to k_1$, show that the integral in (13.50) equals $\frac{1}{2}a^2[J_m'(\zeta_{mn})]^2$ for $n' = n$.

(c) Now the result (13.50) can be obtained by utilizing the expression found in Exercise 13.4(c), and (13.30).

13.6. (a) From the relations between the cartesian and spherical coordinates, show that

$$\partial/\partial z = \cos\vartheta\, \partial/\partial r - \sin\vartheta\, \partial/r\partial\vartheta,$$

$$\frac{\partial}{\partial x} \pm i\frac{\partial}{\partial y} = e^{\pm i\varphi}\left[\sin\vartheta\frac{\partial}{\partial r} + \cos\vartheta\frac{\partial}{r\partial\vartheta} \pm i\frac{\partial}{r\sin\vartheta\partial\varphi}\right].$$

Check these with (2.30) using $(\mathbf{i}\cdot\mathbf{e}_r) = \sin\vartheta\cos\varphi$, and so on.

(b) Derive the spherical coordinate form (13.56) for $\nabla^2 \equiv \boldsymbol{\nabla}\cdot\boldsymbol{\nabla}$. [This *may* be done by starting from the results in (a), but it can be briefer to square the gradient operator (2.30), duly recognizing that $\mathbf{e}_{r,\vartheta,\varphi}$ change with ϑ, φ as implied in (1.33). A more conventional method makes use of the "Gauss Theorem," to be quoted in (15.29), as applied to a volume consisting of just one element $r^2 dr\, d(\cos\vartheta)\, d\varphi$.]

13.7. Useful relations among the Legendre polynomials can be found from the "generating" expression (13.68), by taking derivatives of it with respect to $\rho \equiv r_</r_>$, or with respect to the argument $\mu \equiv \cos\vartheta$ itself.

(a) From the derivative by $d/d\rho$, show that

$$(\mu - \rho)\sum_{l=0}^{\infty}\rho^l P_l = (1 - 2\mu\rho + \rho^2)\sum_{l=0}^{\infty} l\rho^{l-1}P_l.$$

Then, by equating coefficients of ρ^l, show that the so-called "recurrence relation,"

$$(2l+1)\mu P_l = (l+1)P_{l+1} + lP_{l-1},$$

follows. Use this to determine P_4 from P_2 and P_3.

(b) From the derivative by $d/d\mu$, show that

$$dP_l/d(\cos\vartheta) = l(P_{l-1} - \cos\vartheta P_l)/\sin^2\vartheta$$

can be obtained if a derivative of the above recurrence relation is also used. Check that such results for the derivatives of the P_l's are consistent with the Legendre differential equation (13.61).

13.8. Expand the power μ^n into a series of Legendre polynomials. This will require obtaining the useful results that the integral $\int_{-1}^{1} d\mu\, \mu^n P_l$ vanishes unless $l \leq n$ and has the same parity as n (both even or both odd). Show that, when it does not vanish,

$$\int_{-1}^{1} d\mu\, \mu^n P_l(\mu) = 2^{l+1} n! \left(\frac{n+l}{2}\right)! \bigg/ \left(\frac{n-l}{2}\right)! (l+n+1)!.$$

[The procedure used to obtain (13.72) provides one way.]

13.9. Suppose that oscillations of an elastic sphere are started from rest by "poking in" opposite poles, and simultaneously constricting an equatorial belt, before letting go. What is the principal frequency you may expect to excite in this way? [$2\sqrt{5}\, c/a$.] How large should the initial equatorial displacement have been made, in ratio to each of the (equal) displacements at the poles, to make this eigenmode of the motion as "pure" as possible? Find the latitudes of the nodes in the eigenmode here.

13.10. Show that the values of Y_{lm} at any two diametrically opposite points on a sphere are related as

$$Y_{lm}(\pi - \vartheta, \varphi + \pi) = (-)^l Y_{lm}(\vartheta, \varphi).$$

(Then a distribution proportional to a Y_{lm} with an odd l is said to have odd "parity." Compare the allusions to parity in one dimension, on pages 359 and 363 and Exercise 12.2(a).)

13.11. Suppose that the initial distributions of displacements and velocities on an elastic sphere are given as $q_0(\vartheta, \varphi)$ and $v_0(\vartheta, \varphi)$, respectively.

(a) Show that the ensuing motion may be represented by

$$q(\vartheta, \varphi, t) = \text{Re} \sum_{lm} C_{lm} Y_{lm} e^{-i\omega_l t},$$

with the (generally complex) coefficients

$$C_{lm} = \oint d\Omega\, Y_{lm}^* (q_0 + iv_0/\omega_l).$$

(b) For the special case of an axially symmetric start, $q_0(\vartheta)$ and $v_0(\vartheta)$, show that only standing waves can be excited. [The reader may want to check that this is so even relative to a polar axis perversely chosen at some angle to the symmetry axis. He may do this by transforming his result through the use of (13.98), then showing that the running wave component of each $m = +|m|$ term combines to a standing wave with the $m = -|m|$ term.]

(c) Show that a start of definite parity, q_0, $v_0(\pi - \vartheta, \varphi + \pi) = (+ \text{ or } -)q_0$, $v_0(\vartheta, \varphi)$, excites a motion of the same parity (l even only, or l odd only).

CHAPTER 14

SPATIAL WAVES

Mechanical oscillations may be propagated through any space containing a suitable "medium." The propagations that will be treated here are ones governed by the wave equation (13.55),

$$\boxed{\nabla^2 q(\mathbf{r}, t) = \partial^2 q / c^2 \partial t^2} \tag{14.1}$$

for some generalized coordinate $q(\mathbf{r}, t)$, used as a measure of the disturbance at various points, $\mathbf{r}$. This equation is found to apply in a wide variety of circumstances basically because it can represent first effects of deviations from equilibrium, as suggested in connection with the two-dimensional case (13.1) (and see Exercise 14.1).

For oscillations in a three-dimensional continuum, questions about the directions of oscillational displacements naturally arise; a displacement *vector*, $\mathbf{q}(\mathbf{r}, t)$, rather than a scalar quantity, should perhaps be used to represent the disturbance at each point, $\mathbf{r}$. However, the degrees of freedom at a point can usually be analyzed into uncoupled ("normal") coordinates, each independently satisfying a wave equation. This permits restricting attention to some one degree of freedom, representable by a scalar $q(\mathbf{r}, t)$. Simultaneous excitations of other degrees of freedom may be ignored as long as no coupling to them is explicitly introduced.

An important example of the concurrent, but independent, propagation of uncoupled oscillations is provided by the longitudinal and transverse waves in near-rigid media. Longitudinal displacements, parallel to a propagation direction, transfer energy from element to element by means of forces of compression and decompression (expansion), whereas it is shearing forces that are involved in transverse relative displacements. Consequently, the two types of disturbance usually have different propagation velocities even in an isotropic medium. Each satisfies a wave equation like (14.1), but containing a different c. Seismic disturbances of the earth exhibit such a behavior.

Both light and sound (when this is not so intense as to form "shocks" or "sonic booms") are likewise propagated according to the wave equation (14.1), with appropriate choices of velocity, c. In the case of light, q has no necessary association with any "material" medium, but has to do with electromagnetic fields in the space. Sound is propagated through the atmosphere in longitudinal waves, since shearing forces are negligible in air; the $q(\mathbf{r}, t)$ may then stand for the compressible air-density distribution, or for deviations from an average air pressure. Sonic vibrations may be transmitted through solid media in both longitudinal and transverse waves.

14.1 PLANE WAVES

The disturbance $q(\mathbf{r}, t)$ may be treated as a function $q(x, y, z, t)$ of cartesian position-vector components, and then the x-, y-, and z-dependences may be successively Fourier-analyzed as in (12.75):

$$q(x, y, z, t) = (2\pi)^{-1/2} \int_{-\infty}^{\infty} dk_x\, a(k_x, y, z, t) e^{ik_x x},$$

if

$$a(k_x, y, z, t) = (2\pi)^{-1/2} \int_{-\infty}^{\infty} dx\, q(x, y, z, t) e^{-ik_x x};$$

$$a(k_x, y, z, t) = (2\pi)^{-1/2} \int_{-\infty}^{\infty} dk_y\, b(k_x, k_y, z, t) e^{ik_y y},$$

if

$$b(k_x, k_y, z, t) = (2\pi)^{-1/2} \int_{-\infty}^{\infty} dy\, a(k_x, y, z, t)e^{-ik_y y}$$

$$= (2\pi)^{-1} \iint_{-\infty}^{\infty} dx\, dy\, q(x, y, z, t)e^{-i(k_x x + k_y y)};$$

$$b(k_x, k_y, z, t) = (2\pi)^{-1/2} \int_{-\infty}^{\infty} dk_z\, c(k_x, k_y, k_z, t)e^{ik_z z},$$

if

$$c(k_x, k_y, k_z, t) = (2\pi)^{-1/2} \int_{-\infty}^{\infty} dz\, b(k_x, k_y, z, t)e^{-ik_z z}$$

$$= (2\pi)^{-3/2} \iiint_{-\infty}^{\infty} dx\, dy\, dz\, q(x, y, z, t)e^{-i(k_x x + k_y y + k_z z)}. \quad (14.2)$$

Meanwhile, the first of these expressions has become

$$q(x, y, z, t) = (2\pi)^{-3/2} \iiint_{-\infty}^{\infty} dk_x\, dk_y\, dk_z\, c(k_x, k_y, k_z, t)e^{i(k_x x + k_y y + k_z z)}. \quad (14.3)$$

It is obviously appropriate to define vectors **k**, with cartesian components $k_{x,y,z}$, so that the Fourier transforms (14.3) and (14.2) may be written

$$q(\mathbf{r}, t) = (2\pi)^{-3/2} \oint dV(\mathbf{k})c(\mathbf{k}, t)e^{i\mathbf{k}\cdot\mathbf{r}} \quad (14.4a)$$

$$c(\mathbf{k}, t) = (2\pi)^{-3/2} \oint dV(\mathbf{r})q(\mathbf{r}, t)e^{-i\mathbf{k}\cdot\mathbf{r}} \quad (14.4b)$$

where $c(\mathbf{k}, t) \equiv c(k_x, k_y, k_z, t)$. Whereas $dV(\mathbf{r}) \equiv dx\, dy\, dz$ is an ordinary spatial volume element, $dV(\mathbf{k}) \equiv dk_x\, dk_y\, dk_z$ is a "volume element" in a "**k**-space." The symbols $\oint$ stand for integrations over the entire domains of the integration variables.

The Fourier decomposition (14.4*a*) may be regarded as a resolution of the **r**-dependence onto the basis

$$u(\mathbf{k}, \mathbf{r}) = e^{i\mathbf{k}\cdot\mathbf{r}}/(2\pi)^{3/2}, \quad (14.5)$$

generalized from the one-dimensional basis (12.66), and having the orthonormality property

$$\oint [dV(\mathbf{r})/(2\pi)^3]e^{i(\mathbf{k}-\mathbf{k}')\cdot\mathbf{r}} = \delta(\mathbf{k} - \mathbf{k}'). \quad (14.6)$$

Here, the delta function of the vector variable may be treated as

$$\delta(\mathbf{k} - \mathbf{k}') = \delta(k_x - k_x')\,\delta(k_y - k_y')\,\delta(k_z - k_z'), \tag{14.7a}$$

and its essential property is represented in

$$\oint dV(\mathbf{k})\,\delta(\mathbf{k} - \mathbf{k}')f(\mathbf{k}) = f(\mathbf{k}') \tag{14.7b}$$

for any function $f(\mathbf{k})$ having a definite value at $\mathbf{k} = \mathbf{k}'$. All these are rather obvious generalizations of formalisms developed in Chapter 12.

Any function of $\mathbf{r}$ and t may be decomposed into Fourier components as in (14.4). For the function to satisfy the wave equation, its Fourier transform must be restricted in its time-dependence to

$$c(\mathbf{k}, t) = \tfrac{1}{2}(2\pi)^{3/2}[C(\mathbf{k})e^{-i\omega t} + D(\mathbf{k})e^{+i\omega t}],$$

with $\omega \equiv c\,|\mathbf{k}|$ and C, D arbitrary. Then

$$q(\mathbf{r}, t) = \tfrac{1}{2}\oint dV(\mathbf{k})[C(\mathbf{k})e^{i(\mathbf{k}\cdot\mathbf{r}-\omega t)} + D(-\mathbf{k})e^{-i(\mathbf{k}\cdot\mathbf{r}-\omega t)}]$$

is properly real only if $D(-\mathbf{k}) = [C(\mathbf{k})]^*$, and

$$\boxed{q(\mathbf{r}, t) = \text{Re}\oint dV(\mathbf{k})C(\mathbf{k})e^{i(\mathbf{k}\cdot\mathbf{r}-ckt)}}\,. \tag{14.8}$$

This is the generalization to three dimensions of the wave-packet (12.81).

An individual element proportional to

$$\exp[i(\mathbf{k}\cdot\mathbf{r} - \omega t)] \equiv \exp[ik(\hat{\mathbf{k}}\cdot\mathbf{r} - ct)] \tag{14.9}$$

describes a wave propagated in the direction $\hat{\mathbf{k}} = \mathbf{k}/|\mathbf{k}|$ of the "wave-vector," $\mathbf{k}$, having the wavelength $\lambda = 2\pi/|\mathbf{k}| \equiv 2\pi/k$. Figure 14.1 may help clarify such points. Each plane normal to $\mathbf{k}$ is a locus of positions at which the phase has some uniform value, $\mathbf{k}\cdot\mathbf{r} - \omega t$, at a given moment, t. Plainly, a "wave front" on which the phase continues to have the constant value $k(\hat{\mathbf{k}}\cdot\mathbf{r} - ct)$, as time goes on, must advance with velocity, c, to steadily greater values of the distance $\hat{\mathbf{k}}\cdot\mathbf{r}$ along $\mathbf{k}$. The two wave fronts indicated in the figure are a wavelength, λ, apart if $\hat{\mathbf{k}}\cdot(\mathbf{r}' - \mathbf{r})$ is such that $\mathbf{k}\cdot(\mathbf{r}' - \mathbf{r}) = 2\pi = k\lambda$. Each front of given phase is planar; hence the name "plane waves" for the elements (14.9), and "plane wave packets" for such superpositions as (14.8).

Oblique Incidences of Plane Waves

Plane wave representatives are useful for exploring what happens at an interface between two media characterized by different propagation velocities, c_1 and c_2. The results obtained in §12.5, in the one-dimensional

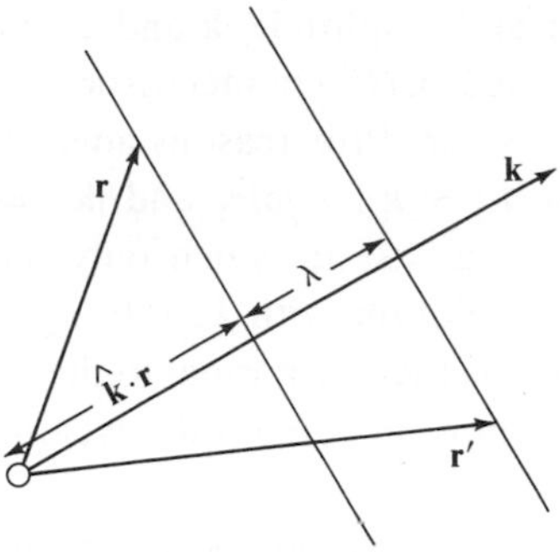

Figure 14.1

exploration, will still apply in three dimensions for normal incidences of plane waves on plane surfaces, since nothing varies in two of the three dimensions in such circumstances. It is for oblique incidences that new findings will now be forthcoming.

The specific circumstances which will be treated are indicated in Figure 14.2. The surface separating the two media is taken to be a plane;

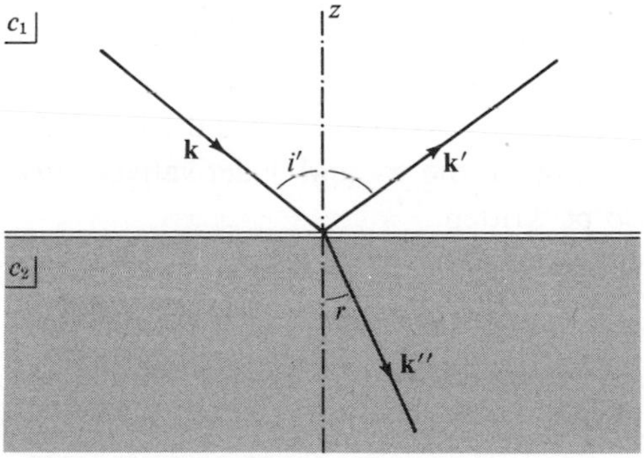

Figure 14.2

actually the results will also be valid as contributions from an infinitesimal element of a curved surface, if this only has deviations from planarity which are negligibly small as compared to a wavelength of the disturbance over spans of a wavelength or less (cases of so-called "specular reflection"). Then a solution will be sought which, in the first medium, has the form of a superposition of incident and reflected plane waves,

$$q_1 = (a_i e^{i\mathbf{k}\cdot\mathbf{r}} + a_r e^{i\mathbf{k}'\cdot\mathbf{r}})e^{-i\omega t}, \tag{14.10a}$$

and in the second medium it is

$$q_2 = a_t e^{i(\mathbf{k}''\cdot\mathbf{r} - \omega t)}. \tag{14.10b}$$

The incident wave-vector and amplitude, $\mathbf{k}$ and a_i, are to be prescribed, but the reflected and transmitted wave characteristics are to be found. Only a single frequency, ω, is considered for reasons adequately discussed in §12.5. Then it is necessary that $|\mathbf{k}| = |\mathbf{k}'| = \omega/c_1$ and $|\mathbf{k}''| = \omega/c_2 = (c_1/c_2)\,|\mathbf{k}|$. For the description to apply to a case in which only the plane wave with $\mathbf{k}$ is "sent in" (prescribed), a solution with $k_z < 0$, $k_z' > 0$, and $k_z'' < 0$ will be sought when the z-direction is the one normal to the interface, as shown in the diagram. It will also be important to consider whether $\mathbf{k}'$ and $\mathbf{k}''$ are unique for a given $\mathbf{k}$.

The solution (14.10) has already been specified sufficiently so that its satisfaction of the wave equation in both media is guaranteed. Boundary conditions coupling the two media remain to be satisfied. For the special case of normal incidence, when all circumstances are independent of x and y, the boundary conditions introduced in (12.92) should still apply. Rewritten for the present context, they are

$$\begin{aligned} q_1(x, y, 0, t) &= q_2(x, y, 0, t), \\ (\partial q_1/\partial z)_{z=0} &= (\partial q_2/\partial z)_{z=0}\,. \end{aligned} \tag{14.11}$$

The only effective changes to be expected for oblique incidences are variations with x and y, and then the conditions must be satisfied at every x, y, just as for every t. The matching $q_1 = q_2$ at every x and y is sufficient to guarantee a corresponding matching of the x- and y-derivatives; hence the boundary conditions may also be written

$$q_1(\mathbf{r}_S, t) = q_2(\mathbf{r}_S, t), \qquad \nabla q_1\Big|_{\mathbf{r}_S} = \nabla q_2\Big|_{\mathbf{r}_S} \tag{14.12}$$

where $\mathbf{r}_S$ is any position on the boundary surface. Formal proof is possible that prescribing q and the normal component of its gradient on a boundary surface is sufficient to determine a uniquely specific solution within the boundary (a so-called "Green Theorem"), in generalization of the one-dimensional result (12.91). It will not be presented here, but saved for contexts of field theory in which it can be given greater generality.

The solutions (14.10) will satisfy the boundary condition (14.11) only if

$$a_i \exp[i(k_x x + k_y y)] + a_r \exp[i(k_x' x + k_y' y)] = a_t \exp[i(k_x'' x + k_y'' y)]$$

for every x, y. The factors containing x, and also the ones containing y, must cancel out to achieve the independence of x and y; hence

$$k_x = k_x' = k_x'' \qquad \text{and} \qquad k_y = k_y' = k_y'' \tag{14.13}$$

(the same conclusions also follow from the linear independence of Fourier elements having unequal wave numbers). With (14.13) the two conditions (14.11) reduce to

$$a_i + a_r = a_t \qquad \text{and} \qquad k_z a_i + k_z' a_r = k_z'' a_t$$

or

$$\frac{a_r}{a_i} = -\frac{k_z'' - k_z}{k_z'' - k_z'}, \qquad \frac{a_t}{a_i} = \frac{k_z - k_z'}{k_z'' - k_z'}. \tag{14.14}$$

More explicit consequences of these findings follow next.

Laws of Reflection and Refraction

It has been found that the reflected wave-vector, $\mathbf{k}'$, can differ from the incident one, $\mathbf{k}$, only in the sign of its normal (z-)component; that is, because of (14.13) and $|\mathbf{k}'| = \omega/c_1 = |\mathbf{k}|$, there can only be a $k_z' > 0$ equal to

$$k_z' = -k_z.$$

This amounts to the "Law of Reflection," usually stated as: the angle of reflection equals the angle of incidence.

The relation of the transmitted wave-vector, $\mathbf{k}''$, to the incident one may be expressed in terms of the "angle of refraction," r, as well as the incidence angle, i', both shown in Figure 14.2:

$$\sin i' = (k_x^2 + k_y^2)^{1/2}/|\mathbf{k}|, \qquad \sin r = [(k_x'')^2 + (k_y'')^2]^{1/2}/|\mathbf{k}''|.$$

According to the above findings, $k_x'' = k_x$ and $k_y'' = k_y$, but $|\mathbf{k}''| = (c_1/c_2)\,|\mathbf{k}|$, and hence

$$\frac{\sin i'}{\sin r} = \frac{c_1}{c_2}. \tag{14.15}$$

This is well known as "Snell's Law of Refraction," and the ratio of velocities is frequently called an "index of refraction."

The amplitude ratios (14.14) may now be expressed entirely in terms of the incidence angle and the refractive index, c_1/c_2:

$$\frac{a_r}{a_i} = -\frac{[(c_1\, c_2)^2 - \sin^2 i']^{1/2} - \cos i'}{[(c_1/c_2)^2 - \sin^2 i']^{1/2} + \cos i'}, \tag{14.16a}$$

$$\frac{a_t}{a_i} = \frac{2\cos i'}{[(c_1/c_2)^2 - \sin^2 i']^{1/2} + \cos i'}. \tag{14.16b}$$

These properly reduce to $a_r \to 0$ and $a_t \to a_i$, independently of the incidence angle, when there is no change of medium ($c_1 = c_2$). They also reduce to the "one-dimensional" results (12.95) and (12.96) for normal incidence ($i' = 0$), as expected. It can be seen that for $c_2 < c_1$, the transmission decreases monotonically, but reflection increases, with obliquity of incidence.

Total Reflection at Oblique Incidences

When the normal incidences were investigated in §12.5, it was found that total reflection occurs both at a medium so dense that $c_2/c_1 \to 0$, and also at one so rarefied that $c_2/c_1 \to \infty$. The total reflection for $c_2 \to 0$ ($a_r = -a_i$) persists at all angles, as reference to (14.16) shows. When $c_2 > c_1$ instead, it is no longer necessary to have c_2 approach infinity to get total reflection; sufficiently oblique angles of incidence will yield total reflection for any $c_2 > c_1$, as will now be confirmed.

When $c_2 < c_1$, the maximum angle of refraction that can be given by (14.15), reached at "grazing" incidence ($i' \to \frac{1}{2}\pi$), is

$$r_{\max} = \sin^{-1}(c_2/c_1) \ [<\tfrac{1}{2}\pi \text{ for } c_2 < c_1], \tag{14.17}$$

less than 90°. However, when $c_2 > c_1$, then the refraction angle already reaches 90° at an incidence less than grazing, at a "critical angle," i_C:

$$i_C = \sin^{-1}(c_1/c_2) < \tfrac{1}{2}\pi \text{ for } c_1 < c_2 . \tag{14.18}$$

If an incident wave closer to grazing than this is nevertheless "sent in," then no real value exists for

$$\begin{aligned} k_z'' &= -[|\mathbf{k}''|^2 - (k_x'')^2 - (k_y'')^2]^{1/2} \\ &= -[(c_1/c_2)^2 k^2 - k_x^2 - k_y^2]^{1/2} \\ &= -k[(c_1/c_2)^2 - \sin^2 i']^{1/2} \\ &= -k[\sin^2 i_C - \sin^2 i']^{1/2}, \end{aligned} \tag{14.18a}$$

obviously imaginary when $i' > i_C$.

Despite the last result, the solution (14.10) remains valid. It satisfies all conditions with an imaginary value for

$$k_z'' = -i\kappa \qquad \text{with} \qquad \kappa(i') \equiv k[\sin^2 i' - \sin^2 i_C]^{1/2}, \tag{14.18b}$$

in terms of which (14.10*b*) may be rewritten*

$$q_2 = a_t e^{\kappa z} e^{i(k_x x + k_y y - \omega t)} \tag{14.19a}$$

---

* The existence of solutions with imaginary wave-vector components as here may seem to indicate a lack of completeness of the set (14.5) with real **k**. However, it can be found that the set is after all complete enough so that, for example, $e^{\kappa z}$ with $\kappa z < 0$ can be represented by it, as a superposition of exp: $ik_z''' z$ with real k_z'''. Thus, the solutions with complex **k** can be represented as linear combinations of solutions with real **k**'s, over regions in which the solutions with complex **k** remain finite.

This describes a propagation parallel to the boundary $z = 0$, together with a "penetration" in depth, falling off exponentially as $z \to -\infty$. The "mean penetration depth," $1/\kappa$ (the distance in which the amplitude is decreased to a fraction $1/e$) varies with the incidence angle (14.18*b*). The amplitude at the boundary, a_t, should now be rewritten from (14.16*b*) as

$$a_t = 2a_i \cos i' / [\cos i' + i\kappa(i')/k]. \tag{14.19b}$$

Of greater interest is the amplitude of the reflected wave:

$$\frac{a_r}{a_i} = \frac{k \cos i' - i\kappa}{k \cos i' + i\kappa} \qquad \text{with} \qquad \left|\frac{a_r}{a_i}\right|^2 = 1. \tag{14.20}$$

This shows that total reflection occurs for all incidences with $i' > i_C$ (when κ is real).

Even in the total reflections the disturbance penetrates into the reflecting medium (unless its $c_2 = 0$!). This is not inconsistent with conservation principles. The monochromatic plane waves being used in this description correspond to an established, steady situation (see (12.87)!). It must be supposed that the steady flow along the boundary of the reflecting medium was established together with the steady reflections. The point will be discussed further, in connection with energy conservation, in §15.2.

14.2 SPHERICAL WAVES

The plane waves are usually the most suitable for describing disturbances far from their sources. There the wave fronts are spread out over a wide area, and any finite sector, such as is intercepted by a detector, can be treated as planar. However, near a source of limited size the wave fronts are usually curved around it, and effects of the curvature may become noticeable. It is important to gain familiarity with the description of the spherical waves which can be expected to emanate from any one *point* of initial disturbance into an isotropic medium.

It will conform best with the symmetry of the circumstances to use spherical coordinates centered in the source for describing waves emanating from it, and hence solutions $q(r, \vartheta, \varphi, t)$ of the wave equation will now be sought. As in the two-dimensional case (13.2), a monochromatic "representative,"

$$q(\mathbf{r}, t) = u(\mathbf{r})e^{-i\omega t} \qquad \text{with} \qquad \nabla^2 u = -k^2 u, \tag{14.21}$$

will be obtained first. It will make possible the representation of a general time-dependence by serving as a Fourier component of it. Similarly, the representation of a general directional dependence can be prepared by taking advantage of the resolvability of such dependence on the orthonormal basis formed by the spherical harmonics (13.84). Thus, solutions of the form

$$u_{klm}(r) = R(r)\,Y_{lm}(\vartheta, \varphi) \tag{14.22}$$

are to be sought, with radial waves $R(r)$ which are to satisfy the equation obtained upon substitution into the "Helmholtz equation" (14.21).

The spherical coordinate representation of the Laplace operator, ∇^2, was given in (13.56). The part of it which operates on angular dependences was found to have the effect

$$-\left[\frac{1}{\sin\vartheta}\frac{\partial}{\partial\vartheta}\sin\vartheta\frac{\partial}{\partial\vartheta} + \frac{1}{\sin^2\vartheta}\frac{\partial^2}{\partial\varphi^2}\right]Y_{lm} = l(l+1)Y_{lm} \tag{14.23}$$

on the spherical harmonics (13.56) and (13.96). As a consequence, the substitution of the form (14.22) into the Helmholtz equation yields the radial wave equation

$$\left[-\frac{d}{r^2 dr}r^2\frac{d}{dr} + \frac{l(l+1)}{r^2}\right]R_{kl} = k^2 R_{kl}. \tag{14.24}$$

Plainly, its solutions depend on the choices made for $k = \omega/c$ and l, but not on m (and accordingly not on the absolute orientation of the disturbance, as discussed in connection with (13.91), and following it). The radial derivative operations in (14.24) can be given alternative forms,

$$\frac{d}{r^2\,dr}r^2\frac{d}{dr} = \frac{d^2}{dr^2} + \frac{2}{r}\frac{d}{dr} = \frac{d^2}{r\,dr^2}\,r = \left(\frac{d}{dr} + \frac{1}{r}\right)^2, \tag{14.25}$$

one or the other of which may be the more convenient for any given purpose.

The Spherical Bessel Functions

Like the Bessel equation (13.25), the radial equation (14.24) has solutions which contain the radius r only in the combination $\zeta = kr$, since a substitution of ζ for r yields

$$\left[\frac{d^2}{d\zeta^2} + \left(1 - \frac{l(l+1)}{\zeta^2}\right)\right](\zeta R) = 0. \tag{14.26}$$

Thus, solutions for different "wavelengths" $2\pi/k$ will differ only in scale.

Solution of the equation may be approached in the same way that the asymptotic behavior of the Bessel functions was found in (13.39). A power of the variable is separated off through a substitution $R = \zeta^p f(\zeta)$, and this yields the equation

$$f'' + \frac{2(p+1)}{\zeta} f' + \left[1 - \frac{l(l+1) - p(p+1)}{\zeta^2}\right] f = 0$$

for $f(\zeta)$. It becomes evident that the choice $p = -\frac{1}{2}$ will reduce this to a Bessel equation,

$$f'' + \frac{1}{\zeta} f' + \left[1 - \frac{(l+\frac{1}{2})^2}{\zeta^2}\right] f = 0,$$

with solutions $f \sim Z_{l+\frac{1}{2}}$, these being any of the cylindrical harmonics of order $(l + \frac{1}{2})$. Solutions called "Spherical" Bessel, Neumann, or Hankel functions may then be defined:

$$\left.\begin{aligned} j_l(\zeta) &= (\pi/2\zeta)^{1/2} J_{l+\frac{1}{2}}(\zeta), \\ n_l(\zeta) &= (\pi/2\zeta)^{1/2} N_{l+\frac{1}{2}}(\zeta), \\ h_l(\zeta) &= (\pi/2\zeta)^{1/2} H^{(1)}_{l+\frac{1}{2}}(\zeta), \end{aligned}\right\} \tag{14.27}$$

normalized so as to have certain asymptotic values (14.39). The Hankel function of the second kind leads to $h_l^*(\zeta^*)$ in the same way. Because of the interrelationships

$$\left.\begin{aligned} h_l(\zeta) &= j_l + i n_l, \\ j_l(\zeta) &= \frac{1}{2}(h_l + h_l^*), \\ n_l(\zeta) &= \frac{1}{2i}(h_l - h_l^*), \end{aligned}\right\} \tag{14.28}$$

following from (13.28) and (13.29) for real values of the argument, the general solution for a given choice of k and l may be written

$$R_{kl}(r) = A j_l(kr) + B n_l(kr), \tag{14.29a}$$

or

$$R_{kl}(r) = \tfrac{1}{2}(A - iB) h_l + \tfrac{1}{2}(A + iB) h_l^*, \tag{14.29b}$$

with arbitrary constants A, B.

A noteworthy fact about the spherical functions (14.27) is that they may be expressed as linear combinations of trigonometric functions, with coefficients that are polynomials in inverse powers of the argument ((14.34), below). This can be seen by starting from the spherically symmetric $l = 0$ case, for which (14.22) reduces to $u_{k00} = R_{k0}(r)/(4\pi)^{1/2}$ and equation (14.26) to

$$\frac{d^2}{d\zeta^2}(\zeta R_{k0}) = -(\zeta R_{k0}).$$

This obviously has the two independent solutions:

$$\frac{\sin\zeta}{\zeta} \equiv j_0(\zeta), \qquad \frac{\cos\zeta}{\zeta} \equiv -n_0(\zeta). \tag{14.30}$$

The identifications here follow from the asymptotic forms (13.41). The corresponding spherical Hankel function is

$$h_0 = j_0 + in_0 = e^{i\zeta}/i\zeta. \tag{14.31}$$

Next, it can be seen that the property (13.31) of the cylindrical harmonics leads to

$$h_{l+1} = -\zeta^l(d/d\zeta)(\zeta^{-l}h_l). \tag{14.32}$$

Thus

$$h_1 = -\left(1 + \frac{i}{\zeta}\right)\frac{e^{i\zeta}}{\zeta}, \qquad h_2 = -\frac{e^{i\zeta}}{i\zeta}\left(1 + \frac{3i}{\zeta} - \frac{3}{\zeta^2}\right), \ldots, \tag{14.33}$$

and (14.28) yields

$$\left.\begin{aligned} j_1 &= \frac{\sin\zeta}{\zeta^2} - \frac{\cos\zeta}{\zeta}, &\qquad j_2 &= \left(\frac{3}{\zeta^3} - \frac{1}{\zeta}\right)\sin\zeta - \frac{3}{\zeta^2}\cos\zeta, \ldots, \\ n_1 &= -\frac{\cos\zeta}{\zeta^2} - \frac{\sin\zeta}{\zeta}, &\qquad n_2 &= -\left(\frac{3}{\zeta^3} - \frac{1}{\zeta}\right)\cos\zeta - \frac{3}{\zeta^2}\sin\zeta, \ldots. \end{aligned}\right\} \tag{14.34}$$

A general formula

$$h_l = (-\zeta)^l\left(\frac{d}{\zeta\, d\zeta}\right)^l h_0 \tag{14.35}$$

follows from (14.32); it also holds with j_l, j_0, or n_l, n_0 replacing h_l, h_0.

The Static Solutions

It will be instructive to give some attention to the $\omega = ck = 0$ solutions, which obey the "Laplace equation," $\nabla^2 u = 0$, a special case of the

Helmholtz equation (14.21). They may not seem germane to a study of wave motion, since they are time-independent, but they do form an element in the analysis of a general disturbance. They are more important for descriptions of various types of *static* fields, such as electrostatic potentials, or ones arising from gravitational attraction.

With $k = 0$, the radial equation (14.24) reduces to

$$r^2 \frac{d^2}{dr^2}(rR_l) = l(l+1)(rR_l). \tag{14.36}$$

Trying a power like $R \sim r^\alpha$ will show that this equation is satisfied by $\alpha = l$ and also by $\alpha = -(l+1)$, and so it has the two independent solutions respectively proportional to

$$R_l^{\text{reg}} \sim r^l \qquad \text{and} \qquad R_l^{\text{irreg}} \sim 1/r^{l+1}. \tag{14.37}$$

Thus the general static solution (of the Laplace equation) can be written

$$u(\mathbf{r}) = \sum_{lm} (A_{lm} r^l + B_{lm}/r^{l+1}) Y_{lm}(\vartheta, \varphi). \tag{14.38}$$

This evidently becomes infinite at the origin if any $B_{lm} \neq 0$, and also as $r \to \infty$ if any A_{lm} with $l \geq 1$ exists. It can nevertheless be used to represent a field that is finite both at $r = 0$ and as $r \to \infty$, in a "piecewise" manner. A set of coefficients (all $B_{lm} = 0$, some $A_{lm} \neq 0$) may be used to represent $u(\mathbf{r})$ in some bounded region which contains the origin, and then a different set (all $A'_{lm}(l \geq 1) = 0$ and some $B'_{lm} \neq 0$) may be used to represent the same function outside the bounds, in the region extending to $r \to \infty$ (an example will be considered in the next paragraph). On the other hand, the representation of a function which does become infinite at the origin is frequently desired. For example, the potential energy of a point-charge, like $V = zZe^2/r$ of Figure 4.4, is a solution of the Laplace equation which has $B_{00} = (4\pi)^{1/2} zZe^2 \neq 0$ right down to $r = 0$; it is just this singular behavior that is used to indicate the presence of a point source at $r = 0$.

The result (14.38) can help in confirming the assertions of (13.68), that the power series development there generates the Legendre polynomials as coefficients. The expression there is just $u = 1/r$, but referred to a frame-origin shifted from $r = 0$ by $z_f - z = r_s$, as Figure 14.3 may help make clear. Now, the laplacian ∇^2 is plainly invariant to the shift ($z \to z_f = z + r_s$) and so* $u(r_f, \vartheta) \equiv (r_f^2 - 2r_f r_s \cos\vartheta + r_s^2)^{-1/2}$ is a solution of the Laplace equation just as $u = 1/r$ is. It must then be possible to put $u(r_f, \vartheta)$ in the form

* The subscripts f and s are used because r_f and r_s are frequently called "field" and "source" points, respectively.

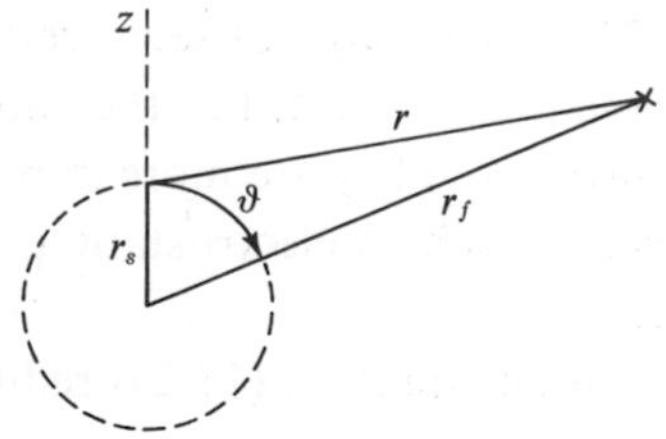

Figure 14.3

of the general solution (14.38), with nonvanishing coefficients for $m = 0$ only, because of the axial symmetry (independence of φ). Since

$$Y_{l0} = [(2l + 1)/4\pi]^{1/2} P_l(\cos \vartheta),$$

according to (13.88),

$$u(r_f, \vartheta) = \sum_l [a_l(r_s) r_f^l + b_l(r_s)/r_f^{l+1}] P_l(\cos \vartheta)$$

with coefficients which depend on the value of r_s. Because of the symmetry to interchanges of r_f and r_s and because the expression must be finite at both $r_f \to \infty$ and $r_f = 0$,

$$u(r_f \geqslant r_s, \vartheta) = \sum_l c_l (r_s^l / r_f^{l+1}) P_l(\cos \vartheta).$$

That is, $a_l(r_s) = 0$ and $b_l(r_s) = c_l r_s^l$ must be chosen to get a representation for r_f "outside the source." The expression for $u(r_f \leq r_s, \vartheta)$ with r_f "inside the source" must be the same except that r_f and r_s are interchanged. What the constants c_l should be depends on the normalization with which the $P_l(\cos \vartheta)$ are defined. It has already been stated, in connection with (13.68), that the conventional normalization was adopted just so $c_l = 1$ for every l. Thus emerges the result (13.68). (It also follows from a normalization such that $P_l(1) = 1$, as can be seen from the way (13.69) was obtained.)

The Radial Waves

Characteristic properties of the spherical functions become evident when their asymptotic behavior is examined, with the help of (13.41) and the definitions (14.27). Thus

$$\begin{aligned} j_l(\zeta \to \infty) &\to (1/\zeta)\cos[\zeta - (l + 1)\pi/2], \\ n_l(\zeta \to \infty) &\to (1/\zeta)\sin[\zeta - (l + 1)\pi/2] \end{aligned} \tag{14.39}$$

for $\zeta \gg l$, when the term $l(l + 1)/\zeta^2$ can be neglected in (14.26). The corresponding behavior of the spherical Hankel functions follows from (14.28):

$$\begin{aligned} h_l(\zeta \to \infty) &\to (1/\zeta) e^{i[\zeta - (l+1)\pi/2]} \\ &= (-i)^l e^{i\zeta}/i\zeta. \end{aligned} \tag{14.40}$$

It can be seen from (14.30) and (14.31) that h_0, j_0, and n_0 begin their "asymptotic" behavior right from the origin!

The role of each function in the description of spherical waves is now clearer. When a representative like (14.21), (14.22) is formed with $R = h_l$,

$$q(\mathbf{r}, t) = h_l(kr)Y_{lm}(\vartheta, \varphi)e^{-i\omega t} \to (-i)^{l+1} Y_{lm} \frac{e^{i(kr-\omega t)}}{kr}, \tag{14.41}$$

it behaves at infinity like a spherical *outgoing* wave, emanating from $r = 0$, and diminishing in "intensity" (squared amplitude, see §15.2) inversely as the square of the radius. The interesting point about the latter property is that it corresponds to the same *total* "intensity" at each radius—that is, multiplied by the entire area, $4\pi r^2$—of every successively larger sphere centered on $r = 0$.

The "outgoing" character of the wave (14.41) is not only the asymptotic behavior, but is retained at all radii. At nearer distances, the wave e^{ikr}/r is merely modulated in amplitude, by polynomials in $(kr)^{-1}$ as reference to (14.35) and (14.31) shows.

To obtain an *ingoing* spherical wave instead, $h_l(kr)$ in (14.41) should be replaced by its complex conjugate, so that

$$q(\mathbf{r}, t) \to i^{l+1} Y_{lm}(\vartheta, \varphi) \frac{e^{-i(kr+\omega t)}}{kr}. \tag{14.42}$$

Note, however, that the complex conjugate of the *entire* expression (14.41) is still an outgoing wave.

The representative

$$\begin{aligned} q(\mathbf{r}, t) &= j_l(kr) Y_{lm} e^{-i\omega t} \\ &\to Y_{lm} \frac{\cos[kr - (l+1)\pi/2]}{kr} e^{-i\omega t} \end{aligned} \tag{14.43}$$

is a radially "standing" wave, having nodes at fixed radii. The functions $n_l(kr)$ likewise describe radially standing waves, asymptotically out of phase by 90° from those proportional to $j_l(kr)$.

The standing radial waves j_l and n_l are sometimes distinguished as respectively "regular" and "irregular" at the origin $r = 0$. The behavior $j_l(kr \to 0)$ is evident from the definition (14.27) and the series representation (13.36) of $J_{l+\frac{1}{2}}$:

$$\boxed{j_l(kr \to 0) \to (kr)^l/(2l+1)!!} \tag{14.44}$$

where $(2l+1)!! \equiv 1 \cdot 3 \cdot 5 \cdots (2l-1) \cdot (2l+1)$ as in (13.71). The behavior of

$n_l(kr \to 0)$ may be found by using the identity $h_l = j_l + in_l$ and applying the operation (14.32) repeatedly to $h_0 \to 1/(ikr)$:

$$\begin{aligned} h_l(kr \to 0) &\to -i \cdot 1 \cdot 1 \cdot 3 \cdot 5 \cdots (2l-1)/(kr)^{l+1}, \\ n_l(kr \to 0) &\to -1 \cdot (2l-1)!!/(kr)^{l+1}, \end{aligned} \tag{14.45}$$

when $kr \equiv \zeta \ll l$, so that 1 can be neglected relative to $l(l+1)/\zeta^2$ in (14.26). The finding of the two independent types of behavior near the origin, $\sim(kr)^l$ and $\sim(kr)^{-(l+1)}$, might have been expected from the static behaviors, $\sim r^l$ and $\sim r^{-(l+1)}$, found in (14.37). After all, when dealing with functions of a variable $\zeta \equiv kr$, a like behavior is to be expected when $\zeta \to 0$ because $r \to 0$, as when $\zeta \to 0$ because $k \to 0$.

A lesson to be learned from the behavior near the origin (14.45) is that in order to have a *purely* outgoing wave (proportional to any linear combination of the h_l's) emanate from a *point*, the disturbance must be singular at the point. This conforms with the finding for the static case in the preceding subsection, that singularities correspond to point *sources* of disturbance. On the other hand, the regular functions $j_l(kr)$ describe standing waves having ingoing, as well as outgoing, parts, and with equal absolute magnitudes; for them, $r = 0$ is merely a point of passage like any other.

The Orthogonality of the Spherical Bessel Functions

The regular standing waves of a given order l, as enumerated by the continuum of values for $k = |\omega|/c$, form an orthogonal set:

$$\int_0^\infty dr r^2 j_l(kr) j_l(k'r) = 0 \qquad \text{if } k \neq k'. \tag{14.46}$$

This can be shown to follow from the radial wave equation (14.24) in a way closely the same as the way that the orthogonality (13.50) of the Bessel functions was shown. That procedure yields the result

$$\left[\frac{r^2}{k^2 - (k')^2}\left\{j_l(kr)\frac{d}{dr}j_l(k'r) - j_l(k'r)\frac{d}{dr}j_l(kr)\right\}\right]_0^\infty$$

for the integral (14.46). The value at the lower limit vanishes* because the expression becomes proportional to a power not smaller than that of r^{2l+1} as $r \to 0$. At the upper limit, $r \to \infty$, the asymptotic expressions (14.39) may be used, and then the integral becomes

$$\lim_{r \to \infty} \frac{1}{2kk'}\left\{\frac{\sin(k-k')r}{k-k'} - (-)^l\,\frac{\sin(k+k')r}{k+k'}\right\}.$$

* This is not so for the irregular functions, n_l and h_l, and hence the restriction of the considerations to just the regular functions.

Comparison with the representation (12.86) of the delta function shows that this may be written as

$$(\pi/2k^2)\{\delta(k-k') + (-)^l\delta(k+k')\}. \tag{14.47}$$

Actually, only functions labeled by positive k's need be admitted to the set, since $j_l(-\zeta) = (-)^l j_l(\zeta)$, as can be confirmed by reference to (14.30) and (14.35). Thus, the second delta function in (14.47) always vanishes in the domain of definition and

$$\int_0^\infty dr r^2 j_l(kr) j_l(k'r) = (\pi/2k^2)\,\delta(k-k'). \tag{14.48}$$

The orthogonal set thus defined is complete enough to represent functions that are finite over the domain of the integration, inclusive of $r = 0$.

The normalization of the radial waves in (14.48) is comparable to the "continuous normalization" (12.69) of the plane waves. In going from the discrete plane wave set (12.47), defined for $|x| < \frac{1}{2}L$, to the continuously enumerated one (12.66), there was a replacement $\frac{1}{2}L \to \pi$ in the normalization. Similarly, the set $j_l(kr)$ may be regarded as the limit for $a \to \infty$ of the discrete set $(\pi/2k_n r)^{1/2} J_{l+\frac{1}{2}}(k_n r)$ of (13.50). The discrete wave numbers k_n are defined by $J_{l+\frac{1}{2}}(k_n a) = 0$, and so, as $a \to \infty$,

$$\begin{aligned} \cos[k_n a - \tfrac{1}{2}(l+1)\pi] &= 0, \\ k_n a &= (n + \tfrac{1}{2}l)\pi. \end{aligned} \tag{14.49}$$

Evaluation of the integral (13.50) then yields $\delta_{nn'}\, a/2k_n^2$ for $a \to \infty$, and so there is a replacement $a \to \pi$ in going over to the continuous normalization in (14.48).

14.3 EIGENFUNCTIONS OF DIFFERENTIAL OPERATORS

The radial functions $j_l(kr)$ furnish yet another example of an orthogonal "eigenfunction" set. Their orthogonality property was expressed in (14.48), and they are called "eigenfunctions" of the operation on the left side of

$$\left[-\left(\frac{d}{dr} + \frac{1}{r}\right)^2 + \frac{l(l+1)}{r^2}\right] j_l(kr) = k^2 j_l(kr) \tag{14.50}$$

because it is equivalent to multiplying them by a constant, an "eigenvalue" k^2, as can be seen from (14.24) and (14.25). Earlier examples of orthogonal eigenfunctions were provided by the Fourier sets (12.47), (12.66), and (14.5), all of which are eigenfunctions of $-\nabla^2(\equiv k^2)$, and by exp: $im\varphi$ of (13.24), an eigenfunction of $-\partial^2/\partial\varphi^2(\equiv m^2)$. The Bessel functions of (13.48) are eigenfunctions of the operator in (13.21), and the spherical harmonics Y_{lm} are eigenfunctions of the operator in (14.23), with eigenvalues $l(l+1)$. It becomes apparent that the connection between orthogonality and "eigenfunctionality" is quite a general one, and that will now be explored.

The connection is not surprising, since a corresponding one has already been found for eigen*vectors* in (10.32). Eigen*functions* may be regarded as generalizations of eigenvectors, in a way illustrated by the passage from the eigenvectors $\mathbf{A}_n$ of (11.12) to the eigenfunctions $A_n(x)$ of (12.8). There was a passage from a space of vectors $\sum c_n \mathbf{A}_n$ to a "function space" containing all possible linear combinations $f(x) = \sum_n c_n A_n(x)$.

Characteristics of function spaces have been alluded to in §12.2, and in connection with the completeness relation (12.74). Here, use will be made of "scalar products" of functions, like the projection $\oint dx\, A_n^*(x) f(x)$ in (12.24), for example, or the orthonormality integral $\oint dx\, e_n^* e_{n'} = \delta_{nn'}$ of (12.47). A widely used symbol for this type of scalar product is

$$(f, g) \equiv \oint dx\, f^*(x) g(x). \tag{14.51}$$

The integration is to extend over the domain, of the continuous variable, in which are defined the functions $f(x), g(x), \ldots$, accepted as part of the function space in use—for example, all the functions representable by $\sum_n c_n A_n(x)$ as above.

Hermitian Operators

The eigenfunction equations of the type (14.50) may all be discussed as examples or generalizations of

$$Qu_\lambda(x) = \lambda u_\lambda(x), \tag{14.52}$$

where Q stands for some operation, usually including differentiations with respect to continuous variables like x. All the operators in the examples cited above are of a type called "hermitian," and the considerations here will be restricted to that most important type.

An operator Q is said to be hermitian if the relation

$$\boxed{(f, Qg) = (g, Qf)^*} \tag{14.53}$$

is valid for any pair of the functions in the space. It is considered a property of the operator because $f(x)$ and $g(x)$ can be chosen with that arbitrariness. The results of the operations, $Qf(x)$ and $Qg(x)$, are also presumed to belong to the function-space manifold to be used.

To attain the hermitianship, some functions may have to be excluded from the manifold, but they are in any case not needed for descriptions of actual physical circumstances. A simple example is provided by the operator $Q \equiv -i d/dx$, for which

$$\begin{aligned}(f, Qg) &= \int_{x_1}^{x_2} dx f^*(x)[-i\, dg/dx] \\ &= -i[f^*g]_{x_1}^{x_2} + i\int_{x_1}^{x_2} dxg\, df^*/dx.\end{aligned} \tag{14.54}$$

The final term here is just $(g, Qf)^*$, and so Q will be hermitian if

$$f^*(x_2)g(x_2) - f^*(x_1)g(x_1) = 0.$$

The implications of this are best seen by Fourier-analyzing each function as in (12.46), after shifting the origin of x so that its domain can be represented as $|x| < \frac{1}{2}L \equiv \frac{1}{2}(x_2 - x_1)$. Then

$$f(x) = \sum_n \alpha_n e^{2\pi inx/L}, \qquad g(x) = \sum_m \beta_m e^{2\pi imx/L}, \tag{14.55}$$

and the condition may be expressed as

$$\sum_{m,n} \alpha_n^* \beta_m [e^{i(m-n)\pi} - e^{-i(m-n)\pi}] = \sum_{m,n} \alpha_n^* \beta_m \cdot 2i \sin(m-n)\pi = 0,$$

automatically satisfied. However, the use of the Fourier analysis has entailed the exclusion of any functions f, g which the Fourier set may not be complete enough to represent. The functions thus excluded are judged too pathological to correspond to any actually realized physical circumstances. It is noteworthy that the Fourier set here is composed of eigenfunctions, in the domain, of the operator in question: $Q \equiv -i\, d/dx$ ($\equiv 2\pi n/L$). Moreover, the inclusion of the imaginary unit in the operation was essential for the hermitianship; without it $(f, dg/dx) = -(g, df/dx)^*$.

The hermitianship of the operation $Q^2 = -d^2/dx^2$ follows from that of $Q = -i\, d/dx$. The product of any two hermitian operators, Q_1, Q_2, is also hermitian if the operators commute: $Q_1 Q_2 = Q_2 Q_1$.

Because hermitianship is a property of the operator, it is often expressed independently of any functions on which it may operate, as

$$Q^\dagger = Q, \tag{14.56}$$

for function-space operators just as for matrix operators (10.97). In the example of $Q = -i\, d/dx$,

$$(f, Q^\dagger g) = (f, Qg) \equiv \oint dx\, f^*[-i\, dg/dx]$$

when $Q^\dagger = Q$, but also

$$(f, Q^\dagger g) \equiv \oint dx f^*[-i\, d/dx]^\dagger g \equiv i \oint dx f^*\widetilde{[d/dx]}g$$

since $Q^\dagger \equiv \tilde{Q}^*$. Comparison to (14.54) shows that the "transpose" of a differential operator must be interpreted as equivalent to its "backward" application: $f^*\widetilde{[d/dx]} \equiv df^*/dx$.

The Orthogonality of the Eigenfunctions

Suppose that the function-space operator Q in (14.52) has another eigenfunction $u_{\lambda'}(x)$, with another eigenvalue λ'. Then it is easy to see that

$$(u_{\lambda'}, Qu_\lambda) - (u_\lambda, Qu_{\lambda'})^* = [\lambda - (\lambda')^*](u_{\lambda'}, u_\lambda) = 0, \qquad (14.57)$$

vanishing for an hermitian operator. The expression is also valid for $u_{\lambda'} \equiv u_\lambda$ and $\lambda' \equiv \lambda$, and hence it shows that the eigenvalues of an hermitian operator must be real, $\lambda^* = \lambda$, for then

$$(u_\lambda, u_\lambda) \equiv \oint dx\, |u_\lambda|^2 \qquad (14.58a)$$

cannot vanish. When $\lambda' \neq \lambda$,

$$(u_{\lambda'}, u_\lambda) \equiv \oint dx u_{\lambda'}^* u_\lambda = 0, \qquad (14.58b)$$

showing that eigenfunctions for distinct eigenvalues must be orthogonal to each other. This is the basis for the orthogonality of the eigenfunctions found in the above examples. Indeed, it is through forming expressions like (14.57) that the orthogonalities of the Bessel functions in (13.50) and of the radial waves in (14.48) were demonstrated.

The expression (14.57) is just the generalization tc a function-space of the expression (10.32), which was used to demonstrate the orthogonality possessed by eigenvectors of real symmetric matrices. An extension to hermitian matrices like (10.97) was left to Exercise 10.24. The comparisons here indicate why such projections as

$$(u_{\lambda'}, Qu_\lambda) \equiv \langle \lambda' | Q | \lambda \rangle \equiv Q_{\lambda'\lambda} \qquad (14.59)$$

are called "matrix elements" on the basis of eigenfunctions $u_{\lambda'}(x)$, $u_{\lambda''}(x)$, ... Compare such matrix projections as (9.106). The last two forms introduced in (14.59) are merely alternative notations that are frequently to be seen in the literature. In terms of the last symbol, hermitianship is represented as $Q^\dagger_{\lambda'\lambda} = Q^*_{\lambda\lambda'}$. The use of eigenfunctions of Q as the basis for the elements $Q_{\lambda'\lambda}$ makes it an element of a "diagonal matrix": $Q_{\lambda'\lambda} = \lambda\delta_{\lambda'\lambda}$, on the presumption that the basis has been normalized to unity, so that the integral (14.58*a*) is equal to one.

Degeneracy and Simultaneous Eigenfunctions

The demonstration of orthogonality through (14.57) does not work for pairs of eigenfunctions $u_{\lambda'}$, u_λ, which are distinguishable but have the same eigenvalue ($\lambda' = \lambda$). The expression then vanishes without requiring that the scalar product $(u_{\lambda'}, u_\lambda)$ do so. This should not be surprising. The two functions exp: ikx and $\sin kx$ are both eigenfunctions of $-\nabla^2$ ($\equiv k^2$), yet the exponential is orthogonal only to the last part of

$$\sin kx = (e^{ikx} - e^{-ikx})/2i.$$

On the other hand, the two exponentials, exp: $\pm ikx$, are mutually orthogonal, and provide a two-element basis complete enough to represent $\sin kx$ as a linear combination.

The occurrences of several different eigenfunctions with the same eigenvalue are cases of "degeneracy." This has been discussed for matrix operators in connection with (10.38) and (11.18) to (11.24), and for the operator $-\nabla^2$ in connection with (13.14), (13.46), and Figures 13.2, 13.5, and 13.7. The degeneracy is "g-fold" when just g *linearly independent* eigenfunctions having the degenerate eigenvalue can be found. There is an infinite variety of ways to choose the g functions, since any linear combination of them is also an eigenfunction with the same eigenvalue. The most convenient way to encompass all these possibilities is to choose some set of g functions which are mutually orthogonal,* and then use it as a subbasis of the representation.

One can arrive at a completely defined set of orthogonal eigenfunctions by finding hermitian operators additional to Q which possess the same eigenfunctions as Q. The additional operators may have *different* eigenvalues for the functions which are degenerate with respect to operation by Q. Thus, the $(2l + 1)$ spherical harmonics Y_{l0}, $Y_{l,\pm 1}$, ..., $Y_{l,\pm l}$ all yield the same,

* That can always be done by what is called a "Schmidt process." This is rarely used by physicists, who find it more meaningful to "remove the degeneracy."

degenerate eigenvalue, $l(l+1)$, upon operation with $-r^2\nabla^2$ of (14.23). On the other hand, Y_{mm}, $Y_{m+1,m}$, $Y_{m+2,m}$, ... are an infinitely degenerate set with respect to operation by $-i\,\partial/\partial\varphi(\equiv m)$. A specific $Y_{l_1 m_1}$ is singled out when it is required to be *simultaneously* an eigenfunction of $-r^2\nabla^2$ with eigenvalue $l_1(l_1+1)$ on the sphere in the space of ϑ, φ, and an eigenfunction of $-i\,\partial/\partial\varphi$ with eigenvalue m_1. This is said to "remove the degeneracy."

When enough operators Q, R, S, ... have been found to remove a degeneracy—that is, to define a simultaneous eigenfunction which is unique apart from an arbitrary constant multiplier—they are said to constitute "a complete set of operators." Guidance in finding such sets comes from the fact that only mutually commuting operators can have simultaneous eigenfunctions. For suppose that $u_{\lambda\mu}$ is a simultaneous eigenfunction of Q and R:

$$Qu_{\lambda\mu} = \lambda u_{\lambda\mu} \qquad \text{and} \qquad Ru_{\lambda\mu} = \mu u_{\lambda\mu}. \tag{14.60}$$

Then

$$QRu_{\lambda\mu} = Q\mu u_{\lambda\mu} = \mu\lambda u_{\lambda\mu} = R\lambda u_{\lambda\mu} = RQu_{\lambda\mu}.$$

Thus it must not matter in which order Q and R are applied: $QR = RQ$.

An instructive example is provided by the general solution of the eigenvalue equation $-d^2u/dx^2 = k^2u$,

$$u = ae^{ikx} + be^{-ikx}, \tag{14.61a}$$

or

$$u = A\cos kx + B\sin kx. \tag{14.61b}$$

The members of the orthogonal pair exp: $\pm ikx$, and of the orthogonal pair $\cos kx$, $\sin kx$, all have a degenerate eigenvalue for $Q = -d^2/dx^2(\equiv k^2)$. The two eigenfunctions exp: ikx and exp: $-ikx$ can be distinguished as having the different eigenvalues $+k$ and $-k$ upon operation with $R = -id/dx$, which obviously commutes with $Q = -d^2/dx^2$. The distinction thus made can be a physically meaningful one, into a "right-" and a "left-running wave." The "standing waves," $\cos kx$ and $\sin kx$, may be distinguished as having opposite "parities"—as being respectively even and odd under the operation of coordinate reversal, $x \to -x$. The latter operation may be formally symbolized as the application of a "parity operator," P. Then

$$\begin{aligned} P\cos kx &\equiv \cos(-kx) = +1\cdot\cos kx, \\ P\sin kx &\equiv \sin(-kx) = -1\cdot\sin kx, \end{aligned} \tag{14.62}$$

and the eigenvalues of P are $+1$ and -1. Notice that, although both $R = -id/dx$ and P commute with $Q = -d^2/dx^2$, they do *not* commute with each other: $PR = -RP$. Thus, exp: $\pm ikx$ are neither of them eigenfunctions of P, and $\cos kx$, $\sin kx$ are not eigenfunctions of R. *Each* pair Q, R and Q, P is a complete set of mutually commuting operators, leading to *alternative* eigenfunction bases. Another example of alternative eigenfunction bases will be considered in the next section.

14.4 PLANE AND SPHERICAL WAVE INTERTRANSFORMATIONS

Alternative bases for resolving functions of position in ordinary, three-dimentional space, "scalar fields" $f(\mathbf{r})$, are provided by the plane wave set (14.5) versus the spherical wave set

$$j_l(kr)\, Y_{lm}(\vartheta, \varphi). \tag{14.63}$$

Both sets contain subsets of eigenfunctions which are degenerate under the operation $-\nabla^2(\equiv k^2)$, with a specific eigenvalue for the magnitude k. The various *plane* waves of definite wave number k can be distinguished from each other as differing in their directions of propagation, $\hat{\mathbf{k}}$; any given member, $\sim \exp\colon i\mathbf{k}\cdot\mathbf{r}$, is simultaneously an eigenfunction of $-\nabla^2\ (\equiv k^2)$ and of the operation

$$-ik^{-1}\nabla \qquad (\equiv \hat{\mathbf{k}}). \tag{14.64}$$

The *spherical* waves are simultaneous eigenfunctions of $-\nabla^2\ (\equiv k^2)$, of $-i\partial/\partial\varphi\ (\equiv m)$ and of

$$-\frac{1}{\sin\vartheta}\frac{\partial}{\partial\vartheta}\sin\vartheta\frac{\partial}{\partial\vartheta} + \frac{m^2}{\sin^2\vartheta} \qquad [\equiv l(l+1)].$$

Thus the alternative bases are defined by using alternative complete sets of mutually commuting operators to remove the degeneracy left by $-\nabla^2\ (\equiv k^2)$.

The alternative resolutions of $f(\mathbf{r})$ are respectively given by the Fourier transforms,

$$f(\mathbf{r}) = (2\pi)^{-3/2} \oint dV(\mathbf{k}) g(\mathbf{k}) e^{i\mathbf{k}\cdot\mathbf{r}},$$

with (14.65)

$$g(\mathbf{k}) = (2\pi)^{-3/2} \oint dV(\mathbf{r}) f(\mathbf{r}) e^{-i\mathbf{k}\cdot\mathbf{r}},$$

and by the analysis into spherical waves

$$f(\mathbf{r}) = \sum_{l,m} \int_0^\infty dk\, g_{lm}(k) j_l(kr) Y_{lm}(\hat{\mathbf{r}}),$$

with (14.66)

$$g_{lm}(k) = (2k^2/\pi) \oint dV(\mathbf{r}) f(\mathbf{r}) j_l(kr) Y^*_{lm}(\hat{\mathbf{r}}).$$

The last expression may be verified by carrying out the integration after substituting the preceding expression for $f(\mathbf{r})$, using the orthogonalities (13.85) and (14.48). Actually, the basis of just the regular spherical waves (14.63) is at most complete enough only for functions $f(\mathbf{r})$ which are regular at $r = 0$. Otherwise, the set must be supplemented with the irregular functions $n_l(kr)Y_{lm}$, even when $r = 0$ is excluded from the region of interest.

The Decomposition of a Plane Wave into Partial Waves

The connection between the plane- and spherical-wave bases can be found by taking $f(\mathbf{r})$ to be $j_l(kr)Y_{lm}(\hat{\mathbf{r}})$ in (14.65) or $f(\mathbf{r}) = \exp\!: i\mathbf{k}\cdot\mathbf{r}$ in (14.66). Unfamiliar integrations can be avoided by starting with the more modest analysis

$$e^{i\mathbf{k}\cdot\mathbf{r}} \equiv e^{ikr\cos\Theta} = \sum_l c_l(kr)P_l(\cos\Theta)$$

of the plane wave, treated as a function of the cosine, $\hat{\mathbf{k}}\cdot\hat{\mathbf{r}}$, into the Legendre set. Then it follows from the orthogonality property (13.72), of the Legendre polynomials, that

$$c_l(kr) = \tfrac{1}{2}(2l+1)\int_{-1}^{1} d\mu P_l(\mu)e^{ikr\mu}.$$

A power series expansion of the exponential yields

$$c_l = \tfrac{1}{2}(2l+1)\sum_{n=0}^{\infty}[(ikr)^n/n!]\int_{-1}^{1} d\mu\mu^n P_l.$$

Of the integrals here, only those will survive in which $n = l$ or n exceeds l by an even integer ($n = l + 2N$ with $N = 0, 1, 2, \ldots$, as found in Exercise 13.8). Moreover,

$$\int_{-1}^{1} d\mu\mu^{l+2N}P_l = 2^{l+1}\frac{(l+N)!(l+2N)!}{N!(2l+2N+1)!},$$

and

$$\begin{aligned}(2l+2N+1)! &\equiv (2l+2N+1)(2l+2N)\cdots 4\cdot 3\cdot 2\cdot 1\\ &= 2^{2l+2N+1}(l+N)![(l+N+\tfrac{1}{2})!/(-\tfrac{1}{2})!],\end{aligned}$$

with $(-\tfrac{1}{2})! = \pi^{1/2}$ (Exercise 14.8). Substitution of all these results into the expression for c_l yields

$$c_l = \tfrac{1}{2}\pi^{1/2}(2l+1)(ikr/2)^l\sum_{N=0}^{\infty}(-)^N(kr/2)^{2N}/N!(l+N+\tfrac{1}{2})!.$$

Now a comparison of this power series to the one (13.36) which defines the Bessel functions shows that

$$c_l = (2l + 1)i^l(\pi/2kr)^{1/2}J_{l+\frac{1}{2}}(kr).$$

With the definition (14.27) of the spherical Bessel functions, the plane wave expansion becomes

$$\boxed{e^{i\mathbf{k}\cdot\mathbf{r}} = \sum_{l=0}^{\infty}(2l + 1)i^l j_l(kr)P_l(\hat{\mathbf{k}}\cdot\hat{\mathbf{r}}).} \tag{14.67}$$

This is usually referred to as the decomposition of a plane wave into "partial waves."

The transformation can finally be brought to the form (14.66) by using the spherical harmonic addition theorem (13.98) to write

$$e^{i\mathbf{k}\cdot\mathbf{r}} = 4\pi\sum_{l,m} i^l j_l(kr)Y^*_{lm}(\hat{\mathbf{k}})Y_{lm}(\hat{\mathbf{r}}). \tag{14.68}$$

The symmetry in **k** and **r** is to be expected, as is the restriction to regular spherical waves, since a plane wave has no singular point.

These analyses of the plane wave may be carried out relative to any point chosen as $\mathbf{r} = 0$, because the plane wave is uniformly regular at all points. Whatever the center chosen, each radial wave component, $j_l(kr)$, can be considered a superposition, $\frac{1}{2}(h_l^* + h_l)$, of ingoing waves converging upon, and outgoing waves diverging from, that center. These reflections from each center are total, the outgoing amplitude being equal to the ingoing one, enabling the equivalence to a plane wave traveling on through $\mathbf{r} = 0$ without incident.

Analyses of Spherical Waves, Huyghens Wavelets

The transformation converse to (14.68), giving the plane wave composition of any one of the spherical waves, is perhaps simplest to obtain by regarding (14.68) as a resolution of the plane wave on the orthogonal set $Y^*_{lm}(\hat{\mathbf{k}})$. Then the projections must be

$$4\pi i^l j_l(kr)Y_{lm}(\hat{\mathbf{r}}) = \oint d\Omega(\hat{\mathbf{k}})Y_{lm}(\hat{\mathbf{k}})e^{i\mathbf{k}\cdot\mathbf{r}}. \tag{14.69}$$

Plane waves into all possible directions, $\hat{\mathbf{k}}$, are superposed here, but their wave number is restricted to the given k. Notice that this implies that $j_0(kr)$ is merely an isotropic average of plane waves into all directions.

Of greater interest is an analysis of the isotropic outgoing wave which is known in various contexts as a "Green's function," a "propagator," a "signal wave," or a "Huyghens wavelet" (14.31):

$$h_0(kr) = \frac{e^{ikr}}{ikr} = \frac{e^{ik|\mathbf{r}_f - \mathbf{r}_s|}}{ik|\mathbf{r}_f - \mathbf{r}_s|}, \tag{14.70}$$

with the last form appropriate to a shift of origin, as in Figure 14.3. The shift of origin is desired as preparation for gauging the influence, at some "field-point" $\mathbf{r}_f$, of emanations from *various* "source-points," $\mathbf{r}_s$. What is now wanted amounts to a generalization to $k \neq 0$ of the static field-distribution (13.68), which was discussed in connection with Figure 14.3.

The conclusion

$$\boxed{\frac{e^{ik|\mathbf{r}_f - \mathbf{r}_s|}}{ik|\mathbf{r}_f - \mathbf{r}_s|} = 4\pi \sum_{l,m} [j_l(kr_<)Y^*_{lm}(\hat{\mathbf{r}}_<)][h_l(kr_>)Y_{lm}(\hat{\mathbf{r}}_>)]} \tag{14.71}$$

can be argued. First, there is directional dependence only on the cosine $\hat{\mathbf{r}}_s \cdot \hat{\mathbf{r}}_f$, representable as an expansion into the Legendre polynomial combinations

$$P_l(\hat{\mathbf{r}}_s \cdot \hat{\mathbf{r}}_f) = [4\pi/(2l+1)] \sum_m Y^*_{lm}(\hat{\mathbf{r}}_s)Y_{lm}(\hat{\mathbf{r}}_f),$$

forming the spherical harmonic addition theorem. Next, it is known that $h_0(kr)$ is an eigenfunction of $-\nabla^2$ ($\equiv k^2$) and ∇^2 is invariant to the shift of origin. This means that, as a function of $\mathbf{r}_s$, the expression is an eigenfunction of $-\nabla_s^2$ ($\equiv k^2$), involving derivatives with respect to the coordinates of $\mathbf{r}_s$, and, as a function of $\mathbf{r}_f$, it must be an eigenfunction of $-\nabla_f^2$ ($\equiv k^2$). Such eigenfunctions, with only the eigenvalue k specified, are degenerate to the degree that they may be any linear combinations of the spherical waves proportional to Y_{lm}, with all possible values of l and m. The products of the waves proportional to $Y_{lm}(\hat{\mathbf{r}}_s)$ and $Y_{l'm'}(\hat{\mathbf{r}}_f)$ respectively, are restricted to $l' = l$ and $m' = m$ by the first of the arguments here. The particular *radial* waves which can appear, together with their coefficients, is deducible from the behavior of $h_0(k|\mathbf{r}_f - \mathbf{r}_s|)$, on the left, for $r_> \to \infty$. Let $\mathbf{r}_< = \mathbf{r}_s$ and $\mathbf{r}_> = \mathbf{r}_f$. Now, for $r_f \gg r_s$,

$$|\mathbf{r}_f - \mathbf{r}_s| = r_f\left[1 - \frac{2\mathbf{r}_f \cdot \mathbf{r}_s}{r_f^2} + \left(\frac{r_s}{r_f}\right)^2\right]^{1/2} \approx r_f - \hat{\mathbf{r}}_f \cdot \mathbf{r}_s, \tag{14.72}$$

so that

$$\frac{e^{ik|\mathbf{r}_f-\mathbf{r}_s|}}{ik|\mathbf{r}_f-\mathbf{r}_s|}\approx\frac{e^{ikr_f}}{ikr_f}\,e^{-i\mathbf{k}\cdot\mathbf{r}_s}\left(1+\frac{\hat{\mathbf{r}}_f\cdot\mathbf{r}_s}{r_f}+\cdots\right), \tag{14.73}$$

with $\mathbf{k}\equiv k\hat{\mathbf{r}}_f$ and the last term negligible for $r_f\to\infty$. Thus, only the outgoing wave solutions $h_{l'}(kr_f)$, with the asymptotic behavior

$$h_{l'}(\infty)\to(-i)^{l'}e^{i\mathbf{k}\cdot\mathbf{r}_f}/ikr_f,$$

are admissible for representing the dependence on r_f. Moreover, the plane-wave factor, $\exp\colon -i\mathbf{k}\cdot\mathbf{r}_s$, can be analyzed into spherical waves as in (14.68) to show that

$$\frac{e^{ik|\mathbf{r}_f-\mathbf{r}_s|}}{ik|\mathbf{r}_f-\mathbf{r}_s|}\approx 4\pi\sum_{l,m}i^{-l}j_l(kr_s)Y^*_{lm}(\hat{\mathbf{k}})Y_{lm}(\hat{\mathbf{r}}_s)i^{l'}h_{l'}(kr_f).$$

Since $\hat{\mathbf{k}}\equiv\hat{\mathbf{r}}_f$ here, and $l'=l$ according to arguments above, the conclusion (14.71) is borne out. It may still be checked that the expression (multiplied by ik) properly reduces to (13.68) for $k\to 0$.

Applications of some of the results obtained in this section will be taken up in the next chapter (§15.3), after the role of the observable intensities and energies has been clarified.

EXERCISES

14.1. (a) An average of displacements $\bar{q}(\mathbf{r},t)$, around a point $\mathbf{r}$ of an isotropic medium, may be defined as

$$\bar{q}(\mathbf{r},t)=\oint\frac{d\Omega(\hat{\boldsymbol{\varepsilon}})}{4\pi}\,q(\mathbf{r}+\boldsymbol{\varepsilon},t),$$

where $\boldsymbol{\varepsilon}$ is a radius vector from $\mathbf{r}$ to points on a sphere of radius $\varepsilon\to 0$, centered at $\mathbf{r}$. Show that

$$q(\mathbf{r}+\boldsymbol{\varepsilon},t)\approx q(\mathbf{r},t)+\boldsymbol{\varepsilon}\cdot\nabla q+\tfrac{1}{2}\sum_{i,j}\varepsilon_i\varepsilon_j\nabla_i\nabla_j q+\cdots,$$

and average this over all directions of $\boldsymbol{\varepsilon}$.

(b) Now suppose that there is a restoring force at $\mathbf{r}$, on the mass-element $(4\pi\varepsilon^3/3)\rho$, which is proportional to the deviation from equilibrium, $-(q-\bar{q})$, with a force-constant proportional to ε. Show that the wave-equation (14.1) follows, with a $c^2\sim 1/\rho$ (that is, inversely proportional to ρ).

14.2. The "Doppler effect," in which the frequency of sound changes according to $\Delta\nu/\nu = \pm v/c$, when the source approaches or recedes with a velocity v, is familiar from elementary physics.

(a) Generalize the expression to the continuum of frequency changes that are heard when the source follows a straight track past a listener at some distance from the track. [$\Delta\nu/\nu = (v/c) \cos\vartheta$ for sound leaving the source at a point such that the line from it to the listener makes the angle ϑ with the track.]

(b) Show that the relation found in (a) is exactly what is needed for an *invariance* of the plane-wave phase, $\mathbf{k} \cdot \mathbf{r} - \omega t$, to Galilean transformations. (Thus shown is how frequencies transform between frames in uniform relative motion.)

14.3. Consider the possible eigenfrequencies of an elastic medium confined to a cube of side L, having perfectly reflecting, immovable walls. (Vibrations in a crystal, and also in many other circumstances, are often treated in this way.)

(a) Find the ratio to the fundamental of the lowest eigenfrequency corresponding to a motion of *six*-fold degeneracy; also of the one with *four*-fold degeneracy. (The latter has the lowest "accidental" degeneracy —that is, not arising from rotation symmetry.)

(b) The distinct eigenmodes enumerated by n_x, n_y, n_z may be represented by "lattice" points in a space formed by plotting the integers n_x, n_y, n_z along orthogonal directions. Then a "volume" of it enclosed by $0 < n_{x,y,z} < N$ contains about N^3 points each standing for a distinct eigenmode. Reasoning on this basis, argue that

$$4\pi V\nu^2 \Delta\nu/c^3$$

eigenmodes of motion with frequencies between $\nu \equiv \omega/2\pi$ and $\nu + \Delta\nu$ can be fitted into a large volume $V (\equiv L^3$ with $L \gg \lambda \equiv c/\nu)$.

14.4. Show that the results (14.16) may also be written

$$\frac{a_r}{a_i} = \frac{\sin(r - i')}{\sin(r + i')}, \qquad \frac{a_t}{a_i} = \frac{2 \sin r \cos i'}{\sin(r + i')},$$

where r is the refraction angle.

14.5. Generalize the investigation in Exercise 12.8 to the case of oblique incidences.

(a) Show that correct expressions for reflected and transmitted intensity ratios can be obtained from those found for the normal incidence merely by replacing c_1 with $\cos i$, c_2 with $\mu_2 \equiv [(c_1/c_2)^2 - \sin^2 i]^{1/2}$ and c with $\mu \equiv [(c_1/c)^2 - \sin^2 i]^{1/2}$.

(b) For the case $c_1 = c_2$, show that no reflected intensity will reach a given observation point (in the medium of c_1) from points on the interface at reflection angles

$$i_n = \sin^{-1}\{(c_1/c)[1 - (n\lambda/2d)^2]^{1/2}\},$$

where $\lambda = 2\pi c_1/\omega$ and $n = 0, 1, 2, \ldots$. (What is then observed are called "interference fringes" in optics, but corresponding effects also arise in "soundings" through strata.)

(c) Show that interference fringes will also arise when $c_2 \neq c_1$, if only $c_1 c_2 \leq c^2$. Now i_n is the same as in (b) except that $\frac{1}{2}\lambda$ is to be replaced by $\frac{1}{4}\lambda$.

14.6. Suppose that $c > c_1 \geq c_2$ in the system of the preceding exercise, and that the incidence is at an angle greater than the critical one, $i_C = \sin^{-1}(c_1/c)$. Show that even at incidences $i > i_C$, there will be a transmitted wave (into the medium with c_2), having an intensity that is proportional to

$$\left|\frac{a_t}{a_i}\right|^2 \approx \frac{16c^4 c_2^2 \kappa^2 e^{-2\kappa d}}{c_1^2 k^2(c^2 - c_1^2)(c^2 - c_2^2)} \cos^2 i$$

when the thickness d is large enough. ($\kappa d \gg 1$, with $\kappa \equiv k[\sin^2 i - (c_1/c)^2]^{1/2}$; the exponential factor is sometimes called the "penetrability" of the intervening "barrier.")

14.7. Some properties of the heavier atomic nuclei resemble those of an elastic, incompressible "nuclear fluid" (page 407). Interior vibrations might be treated by supposing the medium to be confined by an immovable, perfectly reflecting, spherical surface with the "nuclear radius" R (as for an interior described as in Exercise 5.6, but with $D \to \infty$). Suppose, therefore, that the nuclear excitations may be represented by some

$$q(r, \vartheta, \varphi, t) = \sum_{lmn} C_{lmn} j_l(\zeta_{ln} r/R) Y_{lm}(\vartheta, \varphi) e^{-i\omega_{ln} t},$$

where ζ_{ln} is a root of $j_{ln}(\zeta_{ln}) = 0$. Then the eigenfrequencies will approach

$$\omega_{ln} \to (2n + l)\pi c/2R$$

in value, for large enough R (see page 447).

(a) This asymptotic expression for ω_{ln} may be used for small l, n with only moderate error. On that basis, count the number of distinct eigenmodes with frequencies up to twice the fundamental, and compare the result with the number of eigenmodes in a cubical volume (Exercise 14.3), also with frequencies up to twice the fundamental. [Such systems are sometimes compared as to "level diagrams." A horizontal line is charted, to represent each eigenfrequency, at a position in a vertical scale proportional to its size in units of the fundamental. A count of the eigenmodes is maintained by attaching, to each level, a number equal to the degree of degeneracy at the corresponding frequency.]

(b) Show that the number of eigenmodes which have a given frequency $\omega = N\pi c/2R$ is $\frac{1}{2}N(N-1)$.

14.8. Show that $(-\frac{1}{2})! = \sqrt{\pi}$ follows from the definition in the footnote on page 400. (The integration may be carried out handsomely by first changing to the square of the integration variable, then showing that the square of the entire integral is equivalent to an integration over an appropriately defined area, simple to complete upon a switch to polar variables.)

14.9. Using (13.30), (13.31), and the results of Exercise 13.4, show that

$$j_{l-1}(\zeta) + j_{l+1}(\zeta) = [(2l+1)/\zeta]j_l(\zeta),$$
$$dj_l/d\zeta = [lj_{l-1} - (l+1)j_{l+1}]/(2l+1),$$
$$j_{l+1} = -\zeta^l(d/d\zeta)\zeta^{-l}j_l,$$
$$j_{l-1} = +\zeta^{-(l+1)}(d/d\zeta)\zeta^{l+1}j_l.$$

These relations also hold for n_l's and h_l's. Show how the third of these leads to (14.35).

CHAPTER 15

LAGRANGIANS AND ENERGIES IN CONTINUA

Variational, Lagrangian, and Hamiltonian formulations are supposed to be possible as fundamental descriptions of any mechanical system (Chapters 7 and 8). Making them explicit for continua in which mechanical waves can be propagated will be the first concern of this chapter. An immediate result will be a means for identifying a conservable energy, one which does not depend on a prior association of mass with the degree of freedom chosen for treatment. Representing conservation of the energy will require considering its transfer from point to point of the continuous medium; the energy flux densities consequently defined serve as measures of the "intensities" with which waves are propagated. Their use will be illustrated by propagations around an obstacle, in the final section.

15.1 THE VARIATIONAL FORMULATION OF THE WAVE EQUATION

Every type of motion is supposed to be derivable from the Variational Principle (7.48). This has actually been demonstrated so far only for systems of discrete variables: $q_1(t)$, $q_2(t)$, ..., $q_s(t)$, ... It remains to be seen how it can be done for such motions as are governed by the wave equation. These are described by a *continuum* of variables, $q(\mathbf{r}', t)$, $q(\mathbf{r}' + \mathbf{dr}, t)$, ..., $q(\mathbf{r}'', t)$, ..., one for each of the continuum of positions $\mathbf{r}$, used to label the moving mass-elements of an elastic medium, for example. The problem is to adapt the variational procedure to such continua of degrees of freedom.

For the Linear Continuum

It will be helpful to examine first how a variational formulation fares in the "limit process" of §12.1, by which discrete variables $q_s(t)$ pass over into the continuously enumerated ones $q(x, t)$. To implement a variational formulation, a lagrangian must be constructed, and that will now be done for the example of the particulate string studied in §11.3 and §12.1.

The correct equations of motion will follow from the usual choice, $L = \tau - V$, and it will here be written (see the development of (12.4) and (12.5))

$$L = \sum_{s=0}^{f} \tfrac{1}{2}\rho\, \Delta x \left[\dot{q}_s^2 - c^2 \left(\frac{q_{s+1} - q_s}{\Delta x} \right)^2 \right], \tag{15.1}$$

with the substitutions $l = \Delta x$, $m = \rho\, \Delta x$, $\kappa = \rho c^2/\Delta x$ in preparation for the limit process. The momentum conjugate to the discrete variable q_s is

$$p_s = \partial L/\partial \dot{q}_s = \rho \Delta x \dot{q}_s \equiv m\dot{q}_s,$$

and the Euler-Lagrange equations arising from the Variational Principle are

$$\dot{p}_s = \partial L/\partial q_s = (\rho c^2/\Delta x)[(q_{s+1} - q_s) - (q_s - q_{s-1})],$$

which check with (11.53). The limit to which the lagrangian passes is evidently the integral

$$L = \int_0^L dx \tfrac{1}{2}\rho c^2 \left[\left(\frac{\partial q}{c\, \partial t} \right)^2 - \left(\frac{\partial q}{\partial x} \right)^2 \right], \tag{15.2}$$

just the difference of the kinetic and potential energies (12.5) and (12.4), but it is not nearly so clear how this can be treated as a function like

$$L(q_1, q_2, \ldots, \dot{q}_1, \dot{q}_2, \ldots),$$

for which the derivative operations of the Euler-Lagrange equations have meaning.

The new lagrangian is what is called a "functional," a "function of functions" in that its values depend on what functions $q(x, t)$ are used in evaluating it. It will be convenient to recognize that the derivatives $q_x(x, t) \equiv \partial q/\partial x$ and $q_t \equiv \partial q/\partial t \equiv \mathring{q}(x, t)$ may appear in it explicitly, just as $\dot{q}$'s appear in the lagrangians of discrete variables, and that it may generally be written as

$$L = \oint dx\, l[q(x, t), q_x(x, t), q_t(x, t)] \tag{15.3}$$

in terms of a "lagrangian density," l, such that $L = \oint dx\, l$. In the special example above,

$$l = \tfrac{1}{2}\rho[q_t^2 - c^2 q_x^2], \tag{15.4}$$

and *only* the derivatives appear explicitly.

To see what happens to the Euler-Lagrange equations, the application of the variational principle, $\delta \int L\, dt = 0$, will be considered afresh. Undue repetition will be avoided by at once generalizing to a system of volume elements, $dV(\mathbf{r})$, filling some three-dimensional region, and hence labeled by $\mathbf{r}$ instead of x.

For the Three-Dimensional Continuum

The appropriate generalization of the Variational Principle now to be considered is evidently

$$\delta \int_{t_1}^{t_2} dt L = \delta \int_{t_1}^{t_2} dt \oint dV l[q(\mathbf{r}, t), \nabla q(\mathbf{r}, t), \mathring{q}(\mathbf{r}, t)] = 0. \tag{15.5}$$

A continuum of variables $q(\mathbf{r}, t)$, which specifies a time-varying value for some degree of freedom, at each point $\mathbf{r}$ in a region $\oint dV$, may be said to constitute a "field" or "state" of motion in the continuum.

The dependent variable $q(\mathbf{r}, t)$ contrasts with the variable $\mathbf{r}$, which is used merely as a label, to distinguish various ones of the elements $dV(\mathbf{r})$, and does not itself vary with time as part of the motion. Thus $\mathbf{r}$ is an *independent* variable, as is t itself, varied only as attention is transferred from one moving element to another—for example, in describing a coupling proportional to $q_{s+1} - q_s \rightarrow (\partial q/\partial x)\Delta x$. It is essential to remember such distinctions when considering what constitutes the "configuration space" of the continuous system, and how paths in it are to be described.

In the configuration space of a system of discrete mass-particles, a single system-point is located at moment t by giving a discrete set of numbers: $q_1(t), q_2(t), \ldots, q_s(t), \ldots$. For a system containing a continuum of mass-elements, $\rho\, dV(\mathbf{r})$, the single system-point requires giving a *continuity* of configuration-coordinate values, $q(\mathbf{r}', t), q(\mathbf{r}' + \mathbf{dr}, t), \ldots, q(\mathbf{r}'', t), \ldots$, for its specification at any moment; therefore a configuration space with a continuous infinity of dimensions, one for each value that $\mathbf{r}$ may have, must be employed. Thus an entire field distribution—in—$\mathbf{r}$, $q(\mathbf{r}, t)$, must be given at each moment t, to describe a path in the latter configuration space, with its continuous infinity of dimensions.

The path variations, needed in applying the Variational Principle, were introduced for the discrete systems by choosing an independent variation, $\delta q_s(t) \equiv q_s'(t) - q_s(t)$, for each degree of freedom and moment, t. Now an entire "field of variations" must be introduced,

$$\delta(\mathbf{r}, t) \equiv q'(\mathbf{r}, t) - q(\mathbf{r}, t), \tag{15.6a}$$

with a generally different value at each moment for each of the continuum of elements $dV(\mathbf{r})$. Just as the variations $\delta q_s(t)$ led to variations in the velocities, $\dot{q}_s'(t) - \dot{q}_s(t) = (d/dt)\,\delta q_s(t)$, so the field of variations leads to variations in the derivatives of $q(\mathbf{r}, t)$:

$$\mathring{q}'(\mathbf{r}, t) - \mathring{q}(\mathbf{r}, t) = \partial\delta(\mathbf{r}, t)/\partial t, \tag{15.6b}$$

$$\nabla q'(\mathbf{r}, t) - \nabla q(\mathbf{r}, t) = \nabla\delta(\mathbf{r}, t). \tag{15.6c}$$

Each of these contributes to the variation $\delta \int L\,dt$ of (15.5).

The variations just reviewed affect only the lagrangian density within the integral; that is,

$$\delta \int_{t_1}^{t_2} dt \oint dV l = \int_{t_1}^{t_2} dt \oint dV \delta l,$$

since it is alternative paths for a *given* volume of medium between *fixed* times that are to be compared. When l explicitly contains q, ∇q, and $\mathring{q}$, it undergoes the variation

$$\delta l = \frac{\partial l}{\partial q}\delta + \sum_{x,y,z} \frac{\partial l}{\partial[\nabla_x q]}\frac{\partial\delta}{\partial x} + \frac{\partial l}{\partial\mathring{q}}\frac{\partial\delta}{\partial t} \tag{15.7}$$

in a self-explanatory notation. This is prepared for partial integrations as in (7.45), by making such substitutions as

$$\frac{\partial l}{\partial\mathring{q}}\frac{\partial\delta}{\partial t} = \frac{\partial}{\partial t}\left[\frac{\partial l}{\partial\mathring{q}}\delta\right] - \left[\frac{\partial}{\partial t}\left(\frac{\partial l}{\partial\mathring{q}}\right)\right]\delta, \tag{15.8a}$$

$$\frac{\partial l}{\partial[\nabla_x q]}\frac{\partial\delta}{\partial x} = \frac{\partial}{\partial x}\left[\frac{\partial l}{\partial[\nabla_x q]}\delta\right] - \left[\frac{\partial}{\partial x}\left(\frac{\partial l}{\partial[\nabla_x q]}\right)\right]\delta. \tag{15.8b}$$

In the time-integration required for (15.5),

$$\int_{t_1}^{t_2} dt \oint dV \frac{\partial}{\partial t}\left[\frac{\partial l}{\partial\mathring{q}}\delta\right] = \oint dV \left[\frac{\partial l}{\partial\mathring{q}}\delta(\mathbf{r}, t)\right]_{t_1}^{t_2}. \tag{15.9a}$$

Just as for the systems of discrete variables, no variations will be permitted at the end times, t_1 and t_2, so that $\delta(\mathbf{r}, t_1) = \delta(r, t_2) = 0$ everywhere, and the contribution (15.9*a*), from the first term of (15.8*a*), vanishes.

The required integration $\oint dV$, over the volume of the medium, similarly permits partial integration of the first term in (15.8*b*):

$$\oint dV \frac{\partial}{\partial x}\left[\frac{\partial l}{\partial[\nabla_x q]}\delta\right] = \iint dy\, dz\left[\frac{\partial l}{\partial[\nabla_x q]}\delta(\mathbf{r}, t)\right]_{x_1(y,z)}^{x_2(y,z)}. \qquad (15.9b)$$

It is apparent that, for the purpose of the integration over x here, the whole volume was divided up into thin parallel "prisms" of cross section $dy\, dz$, penetrating the region from one end to the other, parallel to the x-axis. The $x_{1,2}(y, z)$ are the coordinates of the ends of the prism at y, z. They are coordinates of positions $\mathbf{r}_{1,2}$ on the *surface* enclosing the volume $\oint dV$ (a very remote surface if the volume approaches infinity!). No variations of $q(\mathbf{r}, t)$ will be permitted on the boundaries of the medium; $\delta(\mathbf{r}_{1,2}, t) = 0$ will be taken everywhere on the surface, however remote, at all times. This is done for much the same reasons as the ones discussed in §7.4, which have led to taking

$$\delta(\mathbf{r}, t_{1,2}) = 0.$$

The latter amounted to settling for equations of motion which cannot yield a uniquely specified motion until an equivalent of "initial conditions" is supplied. Similarly, equations of motion for $q(\mathbf{r}, t)$ will be settled for, which will not be able to yield a unique solution $q(\mathbf{r}, t)$ unless *boundary* conditions are supplied. These must be the same for the fields of motion, $q(\mathbf{r}, t)$ and $q'(\mathbf{r}, t)$, being compared in the application of the Variational Principle, and hence for boundary points.

With the vanishing of such partial integrals as (15.9*a*) and (15.9*b*), the expression of the Variational Principle is reduced to

$$\int_{t_1}^{t_2} dt \oint dV\left[\frac{\partial l}{\partial q} - \sum_{x,y,z}\frac{\partial}{\partial x}\left(\frac{\partial l}{\partial[\nabla_x q]}\right) - \frac{\partial}{\partial t}\left(\frac{\partial l}{\partial \mathring{q}}\right)\right]\delta(\mathbf{r}, t) = 0. \qquad (15.10)$$

This is supposed to vanish whatever the value given to $\delta(\mathbf{r}, t)$ at each point $\mathbf{r}$ and each moment t, within the domains of integration, and so, for every $\mathbf{r}$, t in the domain,

$$\frac{\partial l}{\partial q} - \sum_{x,y,z}\frac{\partial}{\partial x}\left(\frac{\partial l}{\partial[\nabla_x q]}\right) - \frac{\partial}{\partial t}\left(\frac{\partial l}{\partial \mathring{q}}\right) = 0 \qquad (15.11)$$

serves as the "field equation of motion," replacing the Lagrange equations (7.33) for discrete variables.

In the example

$$l = \tfrac{1}{2}\rho c^2\left[\left(\frac{\partial q}{c\,\partial t}\right)^2 - (\nabla q)^2\right], \qquad (15.12)$$

generalized from (15.4), the derivative $\partial l/\partial q = 0$ and $\partial l/\partial[\nabla_x q] = -\rho c^2\, \partial q/\partial x$. The field equation becomes

$$\rho c^2\left[\frac{\partial}{\partial x}\left(\frac{\partial q}{\partial x}\right) + \frac{\partial}{\partial y}\left(\frac{\partial q}{\partial y}\right) + \frac{\partial}{\partial z}\left(\frac{\partial q}{\partial z}\right) - \frac{\partial}{\partial t}\left(\frac{\partial q}{c^2\,\partial t}\right)\right] = 0,$$

and thus a variational formulation of the wave equation has been found.

The Hamiltonian Formulation

It is desirable, as usual, to find quantities which may be conserved, and for discrete systems such quantities were expresssed with the help of the conjugate momenta, $p_s = \partial L/\partial \dot{q}_s$ of (7.38). It is not immediately clear how the conjugates to a field of variables like $q(\mathbf{r}, t)$ are to be identified.

Identification can be helped by reducing the problem to the familiar one, through treating the continuum as a limit of discrete elements, generalized from the one-dimensional ones, Δx, of (15.1). The medium is divided up into discrete *cells* with volumes $\Delta V_1, \Delta V_2, \ldots, \Delta V_s, \ldots,$ and the lagrangian of (15.5) is regarded as a limit of

$$L = \sum_s \Delta V_s\, l_s[q_s, (\nabla q)_s, \dot{q}_s], \tag{15.13}$$

where $q_s, (\nabla q)_s, \dot{q}_s$ are averages over the volume of the sth cell. The occurrence of $(\nabla q)_s$ serves to represent the dependence of the lagrangian density l_s at the sth cell on the displacements in neighboring cells, a generalization from (15.1). Of course, the replacement of finite differences, like those in (15.1), by the average gradients in (15.13), approaches validity only in the limiting case.

The identification of conjugate momenta is obtainable from considering variations of the lagrangian, like

$$\delta L(q, \dot{q}) = \sum_s [(\partial L/\partial q_s)\, \delta q_s + (\partial L/\partial \dot{q}_s)\, \delta \dot{q}_s]$$

for the discrete systems; the conjugate momentum $p_s = \partial L/\partial \dot{q}_s$ is then identifiable as the coefficient of $\delta \dot{q}_s$. For the lagrangian of (15.5),

$$\begin{aligned}\delta L &= \oint dV\left\{\frac{\partial l}{\partial q}\,\delta + \sum_{x,y,z} \frac{\partial l}{\partial[\nabla_x q]}\frac{\partial \delta}{\partial x} + \frac{\partial l}{\partial \dot{q}}\,\mathring{\delta}\right\} \\ &= \oint dV\left\{\left[\frac{\partial l}{\partial q} - \sum_{x,y,z}\frac{\partial}{\partial x}\left(\frac{\partial l}{\partial[\nabla_x q]}\right)\right]\delta + \frac{\partial l}{\partial \dot{q}}\,\mathring{\delta}\right\},\end{aligned}$$

after partial volume integrations (15.8*b*). The last line is the limit of

$$\delta L = \sum_s \Delta V_s\left\{\left[\frac{\partial l}{\partial q} - \sum_{x,y,z}\frac{\partial}{\partial x}\left(\frac{\partial l}{\partial[\nabla_x q]}\right)\right]_s \delta_s + \frac{\partial l}{\partial \dot{q}_s}\,\mathring{\delta}_s\right\}, \tag{15.14}$$

where the averages over the cell ΔV_s, δ_s and $\mathring{\delta}_s$, are equivalent to the variations denoted δq_s and $\delta \dot{q}_s$ in the usual expressions for a discrete system. It becomes convenient to define a new type of derivative, a "functional derivative" with respect to a function like $q(\mathbf{r}, t)$:

$$\frac{\delta L}{\delta q} \equiv \lim\left(\frac{(\delta L)_s}{\Delta V_s\, \delta_s}\right) \equiv \frac{\partial l}{\partial q} - \sum_{x,y,z} \frac{\partial}{\partial x}\left(\frac{\partial l}{\partial[\nabla_x q]}\right), \tag{15.15}$$

where $(\delta L)_s$ is the contribution to δL from the cell ΔV_s, arising from variations of the function $q(\mathbf{r}, t)$ and its gradient. In such terms, the functional derivative with respect to the function $\mathring{q}(\mathbf{r}, t) \equiv \partial q/\partial t$, in place of $q(\mathbf{r}, t)$, is

$$\delta L/\delta \mathring{q} = \partial l/\partial \mathring{q}, \tag{15.16}$$

when $\nabla \mathring{q}$ is absent from the functional l, as above. The Lagrange field equations (15.11) may now be written

$$\frac{\delta L}{\delta q} - \frac{\partial}{\partial t}\left(\frac{\delta L}{\delta \mathring{q}}\right) = 0, \tag{15.17}$$

more closely the same in appearance as the Lagrange equations for discrete systems.

Before the limit is reached, the momentum conjugate to the discrete q_s is $p_s = (\partial l_s/\partial \dot{q}_s)\Delta V_s$, as follows from (15.14). In the limit, a finite "conjugate momentum *density*" becomes definable:

$$\pi(\mathbf{r}, t) \equiv \lim \frac{(\delta L)_s}{\Delta V_s\, \mathring{\delta}_s} \equiv \frac{\partial l}{\partial \mathring{q}}. \tag{15.18}$$

In the special case (15.12), this is

$$\pi(\mathbf{r}, t) = \rho(\partial q/\partial t) \equiv \rho \mathring{q}(\mathbf{r}, t), \tag{15.19}$$

just the momentum density to be associated with the displacement velocity at $\mathbf{r}$.

For the discrete system at hand before the limit is reached, the hamiltonian is

$$H = \sum_s p_s \dot{q}_s - L = \sum_s \Delta V_s \left[\frac{\partial l_s}{\partial \dot{q}_s} \dot{q}_s - l_s\right],$$

and this has the limit

$$H = \oint dV \left[\frac{\partial l}{\partial \mathring{q}} \mathring{q} - l\right] \equiv \oint dV h(q, \nabla q, \mathring{q}), \tag{15.20a}$$

defining a hamiltonian density

$$\boxed{h = \pi\mathring{q} - l.} \tag{15.20b}$$

In the special example of (15.12) and (15.19),

$$h = \rho\mathring{q}^2 - \tfrac{1}{2}\rho[\mathring{q}^2 - c^2(\nabla q)^2] = \frac{\pi^2}{2\rho} + \tfrac{1}{2}\rho c^2(\nabla q)^2. \tag{15.21}$$

Thus

$$H = \tfrac{1}{2}\oint dV\, \rho[\mathring{q}^2 + c^2(\nabla q)^2], \tag{15.22}$$

just the generalization of the total energy, $\tau + V$, to be expected from (12.4) and (12.5).

It should also be expected that the equations of motion for the continuous system can be given the canonical form:

$$\mathring{q} = \delta H/\delta\pi, \qquad \mathring{\pi} = -\delta H/\delta q. \tag{15.23a}$$

Translated into ordinary partial derivatives, these are

$$\begin{aligned} \frac{\partial q}{\partial t} &= \frac{\partial h}{\partial \pi} - \sum_{x,y,z} \frac{\partial}{\partial x}\left(\frac{\partial h}{\partial[\nabla_x \pi]}\right), \\ \frac{\partial \pi}{\partial t} &= -\frac{\partial h}{\partial q} + \sum_{x,y,z} \frac{\partial}{\partial x}\left(\frac{\partial h}{\partial[\nabla_x q]}\right). \end{aligned} \tag{15.23b}$$

If $h = \pi\mathring{q} - l(q, \nabla q, \mathring{q})$, which is independent of $\mathring{q}$ (for a given π) because $\pi = \partial l/\partial\mathring{q}$, the first of the equations is merely the identity $\mathring{q} = \mathring{q}$, and the second yields just the Lagrange field-equations (15.11). Thus the canonical equations (15.23*a*) are indeed valid.

15.2 THE FLOW OF ENERGY

If $H = \oint h\, dV$ of (15.20) is to be identified as the generalized energy, consider how it may be conserved. A direct examination (as in (8.10)) of how H may change with time yields*

* A total time derivative of a volume integral becomes a partial time derivative within the integral because, in the latter expression, a contribution from each independently chosen volume element $dV(\mathbf{r})$ is counted, and $\mathbf{r}$ is therefore held constant in evaluating each contribution.

$$\frac{dH}{dt} = \oint dV \frac{\partial h}{\partial t} = \oint dV[\mathring{\pi}\mathring{q} + \pi\mathring{\mathring{q}} - \mathring{l}],$$

where

$$\mathring{l} \equiv \frac{\partial l}{\partial t} = \frac{\partial l}{\partial q}\mathring{q} + \sum_{x,y,z} \frac{\partial l}{\partial[\nabla_x q]}\frac{\partial \mathring{q}}{\partial x} + \frac{\partial l}{\partial \mathring{q}}\mathring{\mathring{q}},$$

when l contains q, ∇q, and $\mathring{q}$ as in all the considerations above. Now $\pi = \partial l/\partial \mathring{q}$ (15.18), and so the last term in the expression for $\mathring{l}$ cancels against $\pi\mathring{\mathring{q}}$, and

$$\mathring{\pi}\mathring{q} - \frac{\partial l}{\partial q}\mathring{q} = \mathring{q}\left\{-\sum_{x,y,z}\frac{\partial}{\partial x}\left(\frac{\partial l}{\partial[\nabla_x q]}\right)\right\}$$

follows from the Lagrange equation (15.11). It becomes clear that the entire integrand may be written as a sum of spatial derivatives:

$$\frac{dH}{dt} = \oint dV \frac{\partial h}{\partial t} = -\oint dV \sum_{x,y,z}\frac{\partial}{\partial x}\left\{\mathring{q}\frac{\partial l}{\partial[\nabla_x q]}\right\}. \tag{15.24}$$

The quantity in the curly brackets is the x-component of a vector,

$$S_x \equiv \mathring{q}\frac{\partial l}{\partial[\nabla_x q]}, \tag{15.25}$$

if l depends on ∇q in a suitable way, and on this presumption (of rotational invariance!) the last result may be written

$$\oint dV\{\partial h/\partial t + \nabla \cdot \mathbf{S}\} = 0. \tag{15.26}$$

Here h and $\mathbf{S}$ are being treated as explicit functions of $\mathbf{r}$ and t, as when they are evaluated from some specific motion $q(\mathbf{r}, t)$; to emphasize that $h(\mathbf{r}, t)$ is no longer serving as an hamiltonian $h(q, \pi)$, as in (15.21), but as an energy distribution, a new symbol is substituted,

$$h[q(\mathbf{r}, t), \pi(\mathbf{r}, t)] \equiv U(\mathbf{r}, t), \tag{15.27}$$

in conformity with the one-dimensional example, (12.55). The result (15.26) implies that

$$\boxed{\partial U/\partial t + \nabla \cdot \mathbf{S} = 0,} \tag{15.28}$$

since the integration volume was chosen arbitrarily, and can be reduced to a point.

The Energy Flux-Density

Expressions of the type (15.28), called "continuity equations," occur frequently in theories of mechanical continua (such as fluids) and in field theories. They are used to express the conservation of quantities that are distributed in a continuous density, like $U(\mathbf{r}, t)$, and are transferred from point to point as is the energy during wave motion. In the latter connection $\mathbf{S}(\mathbf{r}, t)$ serves to represent the direction and magnitude of the energy flux density—that is, the amount of energy passing through the point $\mathbf{r}$, per unit time, *and* per unit of cross-sectional area.

The consistency of the interpretations may be seen by carrying out the partial integrations in (15.24), as in (15.9*b*). In terms of the directed area element $\mathbf{dA}$ with components $dA_x = dy\,dz$, $dA_y = dz\,dx$, $dA_z = dx\,dy$, (15.24) becomes*

$$dH/dt = -\oint dV \nabla \cdot \mathbf{S} = -\oint \mathbf{dA} \cdot \mathbf{S}, \tag{15.29}$$

with the area integral extending over outward-facing elements $\mathbf{dA}$, enclosing the volume $\oint dV$. Thus, the energy H enclosed in the volume decreases when there is a net outward flux (positive contributions $\mathbf{S} \cdot \mathbf{dA}$ predominating) escaping through the whole surface; an increase of H requires the incidence of energy from outside (with flux densities $\mathbf{S}$ antiparallel to the outward-facing elements $\mathbf{dA}$ predominating).

Clearly, a nonvanishing flux anywhere on the surface implies the existence of energy transfers between the medium enclosed by the volume and external systems. To describe an isolated volume of medium, the field of motion $q(\mathbf{r}, t)$ within it must be such that $\mathbf{S} \cdot \mathbf{dA} \equiv S_n\,dA = 0$ for every element of the enveloping surface. The components of the flux density normal to the surface are $S_n = \mathring{q}\,\partial l/\partial[\nabla_n q]$ according to (15.25). Thus a solution $q(\mathbf{r}, t)$ must be subjected to the boundary conditions

$$\mathring{q} = 0 \qquad \text{or} \qquad \partial l/\partial[\nabla_n q] = 0 \tag{15.30}$$

on the surface, if it is to describe an isolated,† energy-conserving field of motion.

The interpretations introduced here are generalizations of ones already illustrated by the examples in Chapter 12 for motions in a linear continuum stemming from the lagrangian (15.4). The corresponding energy density expression was given in (12.55) and the flux in (12.57), this now being

* The equation of the volume- to the surface-integral that occurs here is valid for any vector field $\mathbf{S}(\mathbf{r}, t)$ possessing the derivatives and is known as the "Gauss Theorem."

† One which is not even "radiating" energy into vacuo.

derivable from (15.25). The string tied down at both ends provides an example of an isolated system which has a vanishing flux at each end because $q = 0$ there, at all times, and hence $\mathring{q} = 0$. An example of an isolated system which has a vanishing flux at each end because $\partial q/\partial x = 0$ there, rather than $\mathring{q} = 0$, is also easy to construct. It may be an "open-ended" vibrating rod, having each end left free to oscillate "in vacuo." That $\partial q/\partial x = 0$ at the boundary in such a case was shown by (12.103), which describes the reflections taking place without energy loss at a "medium" which simulates vacuo by having $c_2 = \infty$ in it. The example of the reflections and transmissions at a boundary between two media, as discussed in connection with (12.97), is a case in which there is a steady influx of energy at one end ($x = -\infty$!), balanced by effluxes at both $x = \pm\infty$; no finite part of this system is isolated.

The Energy Flow in Monochromatic Waves

For the motions guided by the three-dimensional wave equation, the lagrangian density is (15.12) and the flux density expression (15.25) yields

$$\mathbf{S} = -\rho c^2 \mathring{q} \nabla q, \tag{15.31}$$

a simple generalization of the one-dimensional flux expression (12.57). The proportionality of the flow to the negative gradient (for $\mathring{q} > 0$) was graphically illustrated by the propagation of the impulse pictured in Figures 12.5 and discussed following equations (12.59).

Throughout the descriptions of the wave motions, it has been found most significant to express results for monochromatic, complex "representatives," $q(\mathbf{r}, t) = u(\mathbf{r})$ exp: $-i\omega t$, which can serve as Fourier elements of any general time-dependence. Real motions are then to be represented by real linear combinations of the complex representatives. However, expressions of the energy for a single complex representative are relatively meaningless. They omit an essential interference between the complex conjugate components of a real disturbance. This is evident when the energy expressions are evaluated for the *real* representative,

$$\text{Re}\{q\} \equiv \tfrac{1}{2}(q + q^*) = \tfrac{1}{2}(ue^{-i\omega t} + u^* e^{i\omega t}), \tag{15.32}$$

customarily used whenever representation of the reality seems essential (as in pictures like Figure 13.7). Now such an energy contribution as the density of kinetic energy in (15.22) becomes

$$\begin{aligned}\tfrac{1}{2}\rho \cdot \tfrac{1}{4}(\mathring{q} + \mathring{q}^*)^2 &= -\tfrac{1}{8}\rho\omega^2(ue^{-i\omega t} - u^* e^{i\omega t})^2 \\ &= \tfrac{1}{4}\rho\omega^2(|u|^2 - \text{Re}\{u^2 e^{-2i\omega t}\}).\end{aligned}$$

The steady, positive-definite part of this is expressed by the product $qq^* = uu^*$, for which the addition of q^* to q in (15.32) is essential. Indeed, the remaining time-dependent part, representing fluctuations within a period $2\pi/\omega$, usually goes undetected and ignored.* Usually, energy measurements are made on "incoherent superpositions" of waves, of random phase. The members of such superpositions having frequency ω may be represented as proportional to exp: $-i\omega(t - t_0)$, with t_0 equally likely to have any value within a period. The only results observable are then time-averages over a period, obtainable by isolating and dropping terms having fluctuations of zero average. Even when dealing with "coherent waves," of well-defined phase, the transient fluctuations "within a wave" are rarely of any interest. (A case in which there was interest in the time-fluctuations of a coherent wave packet led to the result (12.83). This was a check of energy conservation, which is supposed to hold throughout every phase of any fluctuations in time.)

The full, time-averaged energy density in (15.22), evaluated for the real representative (15.32), is readily calculated to be

$$\langle U \rangle = \tfrac{1}{4}\rho[\omega^2 |u(\mathbf{r})|^2 + c^2 |\nabla u|^2]. \tag{15.33}$$

Similarly, replacing q with $\frac{1}{2}(q + q^*)$ in the flux density expression (15.31) leads to the time average:

$$\langle \mathbf{S} \rangle = \tfrac{1}{2}\rho c^2\omega \cdot [u^*\nabla u - u\nabla u^*]/2i. \tag{15.34}$$

This is real because the bracket constitutes $2i$ times the imaginary part of $u^*\nabla u$. Notice that u must be complex for the steady directed energy flow to exist, in concordance with the custom of representing running waves by expressions which are complex in their spatial dependence, and representing standing waves by real spatial dependences, $u(\mathbf{r})$.

In the case of a plane wave with propagation vector $\mathbf{k}$, $u(\mathbf{r}) = a$ exp: $i\mathbf{k} \cdot \mathbf{r}$ may be taken. The energy density in the wave, as given by (15.33), is

$$\langle U \rangle = \tfrac{1}{2}\rho\omega^2 |a|^2, \tag{15.35a}$$

and the time-averaged flux density is

$$\langle \mathbf{S} \rangle = \hat{\mathbf{k}}(\langle U \rangle c), \tag{15.35b}$$

in consistency with (12.62). Even without the time averaging, $\mathbf{S} = \hat{\mathbf{k}}(Uc)$, an obvious generalization of (12.61).

Now possible is an exploration of the energy flows during the reflections and refractions treated in §14.1. The flux density in the incident wave is

$$\langle \mathbf{S}_i \rangle = \hat{\mathbf{k}}\tfrac{1}{2}\rho_1\omega^2 |a_i|^2 c_1 = \mathbf{k}\tfrac{1}{2}\rho_1 c_1^2\omega |a_i|^2, \tag{15.36a}$$

---

* See the discussion on page 379.

and in the reflected wave it is

$$\langle \mathbf{S}_r \rangle = \mathbf{k}'\tfrac{1}{2}\rho_1 c_1^2 \omega \, |a_r|^2 . \tag{15.36b}$$

The transmitted flux density is

$$\langle \mathbf{S}_t \rangle = \mathbf{k}''\tfrac{1}{2}\rho_2 c_2^2 \omega \, |a_t|^2 = \mathbf{k}''\tfrac{1}{2}\rho_1 c_1^2 \omega \, |a_t|^2 , \tag{15.36c}$$

since $\rho_2 c_2^2 = \rho_1 c_1^2$ according to (12.99). A comparison of the normal components (in the z-direction of Figure 14.2) can be made by noting that $k_z' = -k_z$ and

$$k_z'' = k_z(a_i - a_r)/a_t \qquad \text{with} \qquad a_t = a_i + a_r ,$$

according to (14.14). Then

$$\langle S_{tz} \rangle = \langle S_{iz} \rangle \frac{a_i - a_r}{a_i + a_r} \left| \frac{a_i + a_r}{a_i} \right|^2 = \langle S_{iz} \rangle \left[1 - \left| \frac{a_r}{a_i} \right|^2 \right], \tag{15.37}$$

just the difference between the values of the incident and reflected normal flux densities to be expected from energy conservation.

The transverse flux densities are obtained by putting $k'_{x,y} = k_{x,y} = k''_{x,y}$, according to (14.13), and there is no loss of generality if the x, y axes are so oriented that $k_y = k_y' = k_y'' = 0$. These transverse fluxes are conserved in a trivial way; each, individually, simply stays unchanged from $x = -\infty$ to $x = +\infty$, since it encounters no changes of medium. Each must have been "sent in" right from $x = -\infty$, to establish the stationary wave picture being described.

There may be some doubt about the validity of speaking about separate incident and reflected fluxes, since both waves coexist in the same part of the space and are "coherent" relative to each other, as attested by the well-defined ratio a_r/a_i of (14.14) and (14.16). When the flux is calculated for the full disturbance,

$$u(\mathbf{r}) = a_i e^{i\mathbf{k}\cdot\mathbf{r}} + a_r e^{i\mathbf{k}'\cdot\mathbf{r}},$$

then not only the sum of $\langle \mathbf{S}_i \rangle$ and $\langle \mathbf{S}_r \rangle$, above, is found, but added to them is an interference term

$$\begin{aligned} &\tfrac{1}{4}\rho_1 c_1^2 \omega(\mathbf{k} + \mathbf{k}')[a_i^* a_r e^{i(\mathbf{k}'-\mathbf{k})\cdot\mathbf{r}} + \text{c.c.}] \\ &\quad = \tfrac{1}{2}\rho_1 c_1^2 \omega(\mathbf{k} + \mathbf{k}')\,|a_i|^2 (a_r/a_i)\cos(\mathbf{k}' - \mathbf{k})\cdot\mathbf{r}, \end{aligned}$$

and $\cos(\mathbf{k}' - \mathbf{k})\cdot\mathbf{r} = \cos 2k_z z$, since $k_x' = k_x$, $k_z' = -k_z$. This interference term does not exist in the normal flux, since $k_z' + k_z = 0$ and so it does not contribute to the conservation relation (15.37). However, it produces a stationary pattern in the transverse flux, even after time-averaging. The pattern will already have to exist in the transverse fluxes that are "sent in" at $x = -\infty$, while establishing the stationary wave picture. It consists of fluctuations within a wavelength which are not readily detected (especially for short wavelengths); when there is no interest in such fluctuations within a wave length, then only "wave averages," in which the interference terms vanish, are expressed.

Invariance to Time Shifts

Energy is supposed to be identifiable as something conserved in an isolated motion because this should be invariant to shifts in the zero of time, as discussed in §7.5 and Chapter 8. This ostensibly more fundamental identification has been delayed to the present stage because it helps to have prior familiarity with the concept of flux, introduced in the preceding sub-sections.

Suppose $q(\mathbf{r}, t)$ describes a state of motion in an isolated volume of medium, $\oint dV$, and so is a solution of the field equations of motion (15.11) subject to suitable boundary conditions (15.30). If the motion is isolated from external events, it should not matter when it is carried out, and so

$$Q(\mathbf{r}, t) = q(\mathbf{r}, t + \delta t), \tag{15.38}$$

the same motion occurring during times earlier* by δt, must be another solution of the same equation of motion under the same boundary conditions. The equation, now to be written

$$\left[\frac{\partial}{\partial Q} - \sum_{x,y,z} \frac{\partial}{\partial x}\left(\frac{\partial}{\partial[\nabla_x Q]}\right) - \frac{\partial}{\partial t}\left(\frac{\partial}{\partial \mathring{Q}}\right)\right] l(Q, \nabla Q, \mathring{Q}) = 0, \tag{15.39}$$

will certainly be the same one if l has the same form† as before, merely with symbols Q substituted for q's.

---

* "Earlier" rather than "later" was chosen here merely so that Figure 8.1 would be appropriate again, now to indicate the motion at some one point, **r**. A "later" time, denoted $\Delta t = -\delta t$, was chosen for the discussion in §7.5.

† Actually, the form can be taken to differ as in

$$l'(q, \nabla q, t) = l + \nabla \cdot \mathbf{f}(q, t) + \frac{\partial f_0(q, t)}{c\partial t}$$

because of a "gauge invariance" generalized from (7.51). It is easy to check that no matter how the $f_{0,1,2,3}(q, t)$ are chosen, they drop out of the equation of motion (15.11).

The consequences of the invariance may be discussed as in §7.5. The time-shifted motion should be possible without a change of the "expenditure," which is

$$\mathscr{I} = \int_{t_1-\delta t}^{t_2-\delta t} dt \oint dV l[Q(\mathbf{r}, t), \nabla Q(\mathbf{r}, t), \mathring{Q}(\mathbf{r}, t)]$$

during the motion $Q(\mathbf{r}, t)$, as compared to the motion $q(\mathbf{r}, t)$ carried out between the times t_1, t_2. Now (15.38) yields

$$Q(\mathbf{r}, t) = q(\mathbf{r}, t) + \mathring{q}\delta t \equiv q(\mathbf{r}, t) + \delta q(\mathbf{r}, t),$$

defining the change $\delta q \equiv Q - q$ induced by the time shift. It is then quite straightforward to find, with the techniques exhibited in (15.7) to (15.9), that because $q(\mathbf{r}, t)$ obeys the equation of motion (15.11),

$$\delta I = \mathscr{I} - I = -\delta t \oint dV[l]_{t_1}^{t_2}$$
$$+ \int_{t_1}^{t_2} dt \oint dV \left\{ \sum_{x, y, z} \frac{\partial}{\partial x}\left[\frac{\partial l}{\partial[\nabla_x q]}\delta q\right] + \frac{\partial}{\partial t}\left[\frac{\partial l}{\partial \mathring{q}}\delta q\right]\right\}.$$

After the final term is integrated over time, $\delta q \equiv \mathring{q}\delta t$ substituted, and the definitions (15.20) and (15.25) of h and $\mathbf{S}$ introduced, the expression becomes

$$\delta I = \delta t\left\{H(t_2) - H(t_1) + \int_{t_1}^{t_2} dt \oint dV \nabla \cdot \mathbf{S}\right\}.$$

This shows that the energy conservation, and the conditions for isolation, can be discussed exactly as in the preceding subsection, but now as a consequence of the fact that the time-shifted motion should be possible to carry out with the same expenditure, and $\delta I = 0$.

Of course, an isolated motion should also be invariant to translations and rotations in space and this should lead to identifications of linear and angular "momentum" densities. It also leads to the introduction of new types of flux (of the "momenta"). The details are better explored in the context of field theory (in "space-time") but one result will be quoted here in order to explain why the word "momentum" is being enclosed in quotation marks.

The result to be quoted will be for the change ΔI which arises in the expression for the expenditure when the entire motion is considered to be displaced bodily by $\Delta\mathbf{r}$, so that $\Delta q = q(\mathbf{r} + \Delta\mathbf{r}, t) - q(\mathbf{r}, t) = \Delta\mathbf{r} \cdot \nabla q$. The specification of the new types of flux, mentioned above, can be avoided by supposing that boundary conditions isolating the medium have already been imposed, such that all fluxes vanish at its outer surface. Then it can be found that

$$\Delta I = \left[\oint dV \pi \, \Delta q \right]_{t_1}^{t_2},$$

where $\pi(\mathbf{r}, t) \equiv \partial l / \partial \mathring{q}$ is just the conjugate field defined in (15.18). Invariance to the uniform translation in which $\Delta q = \nabla q \cdot \Delta \mathbf{r}$ implies that

$$\left[\frac{d}{dt} \oint dV \pi \nabla q \right] \cdot \Delta \mathbf{r} = 0.$$

Thus identified is the generalized "momentum" density

$$\mathbf{g} \equiv -\pi \nabla q. \qquad (15.40)$$

The sign in this definition is chosen to make it the same as that of the energy flux, (15.31), in the special case of motions governed by the wave equation, when $\pi = \rho \mathring{q}$ (15.19). In that special case,

$$\mathbf{g} = -\rho \mathring{q} \nabla q = \mathbf{S}/c^2. \qquad (15.41)$$

differs from the energy flux only by a constant factor.

The last is a curious result. It shows that the *conserved* "momentum" here identified is *not* the same as the ordinary momentum, of density $\pi = \rho \mathring{q}$, associated with the displacement velocities of the mass-elements of the medium. It may have a completely different direction, as in transverse plane wave motions for which (15.35)

$$\mathbf{g} = \mathbf{S}/c^2 = \hat{\mathbf{k}}(U/c), \qquad (15.42)$$

has the direction of the wave propagation, $\hat{\mathbf{k}}$, rather than of the displacement velocity, $\mathring{q}$, transverse to it. The conserved "momentum" is associated with the flow of *energy*, rather than the motion of the mass-elements of medium, and may be called the "wave-momentum." It can exist even when the ordinary mass-momenta average to zero.

According to the discussion in §2.2, mass is essentially *defined* as the coefficient of velocity in expressing a momentum. On this basis, because the energy flow in (15.42) has the propagation velocity c, the energy U can be treated as equivalent to the "mass," $g/c = U/c^2$. Einstein's famous relation between energy and mass may be given a similar basis. However, Einstein's relation is a universal one, stemming from the universality of the *light* velocity, c. Here, such a usage would merely be a mode of expression; for example, an energy of sound, U, could be spoken of as equivalent to mass, equal in amount to U divided by the square of the *sound* velocity, only insofar as the wave equation is applicable, and with a constant c.

It can also be found, as a consequence of invariance to rotations, that

$$\mathbf{r} \times \mathbf{g} \tag{15.43}$$

is a conservable *angular* momentum density, relative to any momental center chosen as the origin of $\mathbf{r}$.

15.3 THE SCATTERING OF WAVES BY AN OBSTACLE

The use of the spherical wave representations, introduced in the preceding chapter, may be illustrated by finding the effect of a small obstacle on the propagation of waves. The problem has fundamental significance for much the same reason (mentioned in the first paragraph of §5.2) that gives similar importance to the scattering of *particles* by an obstacle. The reflection of waves from an object, or force center, serves to detect it or to "feel out" the nature of its force field. Electromagnetic light waves are most often used in observing objects, but mechanical or sound waves are employed in some circumstances. For example, it was observations on longitudinal and transverse ("pressure" and "shear") earth tremors that served to detect the earth's iron core. Sound waves used in "sonar" devices provide another instance.

The prototype problem to be treated here will be concerned with the reflections from a fixed, impenetrable sphere. As in the problem of the reflections from a plane surface (§14.1), they may be explored by sending in an "incident" plane wave from "infinity":

$$q_i(\mathbf{r}, t) = a_i e^{i(\mathbf{k} \cdot \mathbf{r} - \omega t)}. \tag{15.44}$$

Just as in the string problem of (12.93), or in (14.10), this cannot be the entire disturbance when the obstacle is present. However, the full solution

$$q(\mathbf{r}, t) = u(\mathbf{r})e^{-i\omega t} \tag{15.45}$$

is to correspond to q_i insofar as incidences are concerned. No refracted wave need be provided for, as in (14.10), since the sphere is to be impenetrable. The sphere makes itself felt only by requiring that $q = 0$ on its surface, this being the appropriate boundary condition at an impenetrable medium, as discussed in connection with the total reflection (12.100).

The Scattered-Wave Picture

The most general form that the solution (15.45) may have is

$$u(\mathbf{r}) = \sum_{lm} [C_{lm} h_l(kr) + D_{lm} h_l^*(kr)] Y_{lm}(\hat{\mathbf{r}}), \tag{15.46}$$

according to what was learned in arriving at (14.29). The arbitrary constants here must be determined from the boundary conditions at the obstacle, and at infinity. By the "boundary condition at infinity" is here meant the requirement that the solution correspond to "sending in" only the plane wave, q_i, from the direction $\hat{\mathbf{k}}$. The imposition of the latter condition will be considered first, since it is independent of the obstacle and permits conclusions that will be valid for any type of obstacle.

The "running wave" basis was chosen for (15.46), instead of the "standing waves" (j_l, n_l) of (14.29), because this has permitted the isolation of constants, D_{lm}, which must be determined solely by what is incident, independently of the obstacle. The terms proportional to $h_l^*(kr)$ describe purely *ingoing* waves (14.42), and if no waves are to be sent in aside from the plane wave, q_i, the only ingoing waves present must be those contained in the plane wave.

The last statement refers to a fact implicit in the decomposition (14.68), that any plane wave can be treated as a superposition of ingoing waves converging upon, and outgoing waves emanating from, any point chosen as the origin of $\mathbf{r}$. (Naturally, some point within the obstacle is most convenient for this.) With $j_l = \frac{1}{2}(h_l + h_l^*)$, according to (14.28), the incident plane wave can be expressed as

$$a_i\, e^{i\mathbf{k}\cdot\mathbf{r}} = 4\pi a_i \sum_{lm} i^l Y_{lm}^*(\hat{\mathbf{k}}) \cdot \tfrac{1}{2}(h_l + h_l^*) Y_{lm}(\hat{\mathbf{r}}). \tag{15.47}$$

To have only those ingoing waves which are present here, in the final solution $u(\mathbf{r})$, it is necessary to choose

$$D_{lm} = 2\pi a_i\, i^l Y_{lm}^*(\hat{\mathbf{k}}). \tag{15.48}$$

Now the solution may be written

$$\begin{aligned} u(\mathbf{r}) &= 2\pi a_i \sum_{lm} i^l [h_l^*(kr) + S_{lm} h_l(kr)] Y_{lm}^*(\hat{\mathbf{k}}) Y_{lm}(\hat{\mathbf{r}}) \\ &= a_i \left[e^{i\mathbf{k}\cdot\mathbf{r}} + 2\pi \sum_{lm} i^l (S_{lm} - 1) h_l(kr) Y_{lm}^*(\hat{\mathbf{k}}) Y_{lm}(\hat{\mathbf{r}}) \right], \end{aligned} \tag{15.49}$$

where

$$S_{lm} \equiv C_{lm}/2\pi a_i\, i^l Y_{lm}^*(\hat{\mathbf{k}})$$

is a constant measuring the amplitude in the "outgoing l, m channel." It is sometimes referred to as an element of a "scattering matrix." It is so defined that its difference from unity is a measure of the change in the plane wave due

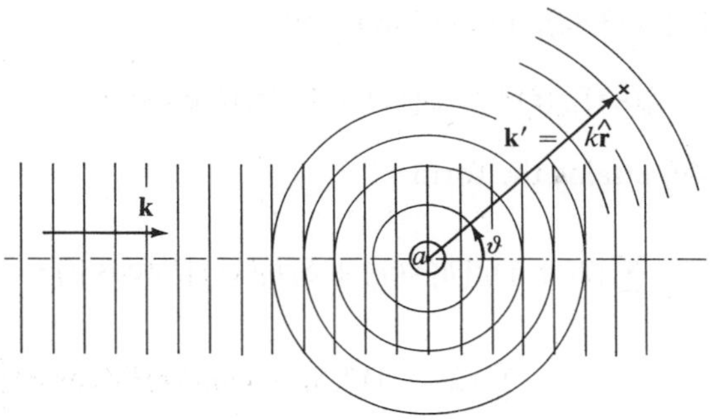

Figure 15.1

to the presence of the obstacle. The remaining "unit" of outgoing waves helps "send on" the plane wave, including those parts of it which have "missed" the obstacle.

The last form given to the solution shows that it may be thought of as consisting of the incident plane wave, undisturbed, and superposed on it a purely outgoing, "scattered" spherical wave emanating from the obstacle, as indicated in Figure 15.1. Of course, this does not mean that the wave sent in marches by undisturbed without even a hole punched into it. It is only that *any* total wave picture can be analyzed as a superposition of plane and/or spherical waves. When a plane wave component with the incidence direction is subtracted from the picture that is appropriate here, what is left are just outgoing spherical waves. It is the interference between the plane and spherical components, at points reached by both, that is to produce the "shadows" and other effects that can be expected.

That a field of motion initiated by a plane wave should finally consist of the same plane wave, and spherical wave-motions in addition, may seem inconsistent with conservation principles. However, as already pointed out in connection with the one- and two-dimensional reflection problems, the restriction to monochromatic waves is tantamount to limiting the description to a "stationary" state of motion, with a steady inflow and outflow of energy, already established before the complete picture is formed. Whatever energy was needed to establish it must have been supplied before the stationary state was reached. It may be done in a type of process already reviewed for the simpler reflections of §12.5.

The picture in Figure 15.1 makes it particularly evident that the incidence direction $\hat{\mathbf{k}}$ will be a convenient choice for the direction of the polar axis. Then

$$Y_{lm}(\hat{\mathbf{k}}) \equiv Y_{lm}(\vartheta_k = 0, \varphi_k) = \delta_{m0}\left[\frac{2l+1}{4\pi}\right]^{1/2}, \tag{15.50}$$

according to (13.84), (13.88), and (13.69), and

$$Y_{lm}^*(\hat{\mathbf{k}})Y_{lm}(\hat{\mathbf{r}}) = \delta_{m0}[(2l+1)/4\pi]P_l(\cos\vartheta).$$

The solution (15.49) attains the form

$$u(\mathbf{r}) = \tfrac{1}{2}a_i \sum_l (2l+1)i^l[h_l^*(kr) + S_l h_l(kr)]P_l(\cos\vartheta)$$
$$= a_i\left[e^{i\mathbf{k}\cdot\mathbf{r}} + \frac{1}{2}\sum_l (2l+1)i^l(S_l - 1)h_l(kr)P_l(\cos\vartheta)\right], \tag{15.51}$$

with $S_l \equiv S_{l0}(\hat{\mathbf{k}})$, in general, as defined in (15.49).

For a spherically symmetric obstacle, like the sphere to be treated here, constants S_l which are independent of the incidence direction, $\hat{\mathbf{k}}$, can be expected to be adequate for satisfying the boundary conditions on the obstacle. This will be confirmed in (15.56), below. Then it is not surprising that the effects of the obstacle, as incorporated in the wave (15.51), are axially symmetric about the incidence direction, depending only on the angle ϑ with that direction.

The Wave-Scattering Cross Section

Simple general results are obtainable for what can be observed many wavelengths away from the object, as a result of sending in the plane wave. "Many wavelengths away" means $r \gg \lambda/2\pi = 1/k$, and the behavior for $kr \gg 1$ of the outgoing waves (14.40) leads to*

$$u(\infty) \to a_i[e^{i\mathbf{k}\cdot\mathbf{r}} + f(\vartheta)e^{ikr}/r], \tag{15.52a}$$

with

$$f(\vartheta) = (2ik)^{-1}\sum_l (2l+1)(S_l - 1)P_l(\cos\vartheta), \tag{15.52b}$$

for the asymptotic behavior of the solution (15.51). The function $f(\vartheta)$, independent of the radial distance, is called the "scattered amplitude"; it is axially symmetric about the incidence direction if the obstacle is spherically symmetric, permitting S_l to be independent of the incidence direction.

Consider now what will be observed in the neighborhood of a finite sector on a sphere of radius $r = R$, remote from the object. The sector area will be denoted $R^2\, d\Omega(\vartheta)$ if it subtends a solid angle $d\Omega$ in a direction making

* This expression may be regarded as the formal representation of the "boundary condition at ∞."

the angle ϑ with the incident wave-vector. If R is large enough, the variations from $1/R$ of the inverse radius, $1/r$, modulating the outgoing waves, may be ignored; that is, the curvature of the wave fronts may be neglected, and the sector of waves simulates the plane wave

$$(a_i/R)f(\vartheta)e^{i\mathbf{k}'\cdot\mathbf{r}} \quad \text{with} \quad \mathbf{k}' = k\hat{\mathbf{r}}. \tag{15.53}$$

This shows that in the neighborhood being observed, the outgoing wave parts of the expressions (15.51) and (15.52) behave as if they were linearly independent of the plane wave part, since they effectively have different wave-vectors, $\mathbf{k}' \neq \mathbf{k}$, and can be *separately* observed. In measurements, a *spatial* separation from the incident wave can also be obtained by using for the latter a "beam" of finite cross section, and placing the detector outside its path. Of course, a wide enough beam must be used, as compared to the size of the object, if treating it as a plane wave is to remain valid to an adequate approximation.

The results may be expressed in terms of "intensities," as measured by wave-averaged energy fluxes in the detected waves. The steady influx of energy, brought in by the incident wave, has the density

$$\mathbf{S}_i = \hat{\mathbf{k}}(Uc) \qquad \text{with } \langle U \rangle = \tfrac{1}{2}\rho\omega^2 \, |a_i|^2, \tag{15.54a}$$

according to (15.35). The corresponding efflux in the effective plane wave (15.53) has the density

$$\mathbf{S}_{\text{out}} = \hat{\mathbf{k}}'(Uc)\,|f(\vartheta)|^2/R^2, \tag{15.54b}$$

since the amplitude differs from the incident one by just the factor $f(\vartheta)/R$. The flux passing out through the sector area element $dA = R^2 \, d\Omega(\vartheta)$ is

$$\mathbf{S}_{\text{out}} \cdot \mathbf{dA} = (Uc)\,|f(\vartheta)|^2 \, d\Omega, \tag{15.54c}$$

independent of the distance R, if only this is large enough to make the curvature negligible.

The ratio of the efflux, per unit solid angle, to the incident flux density, Uc, is called the "wave-scattering differential cross section," and is denoted

$$d\sigma/d\Omega = |f(\vartheta)|^2. \tag{15.55}$$

It plainly has the dimensions of an area, since $f(\vartheta)$ in (15.52) has the dimensions of length. The comparability of the conception to the *particle*-scattering cross section (5.34) is obvious. Indeed, the two types of quantity become indistinguishable in the contexts of quantum mechanics.

The Phase Shifts

More that can be learned will be illustrated by the scattering from an impenetrable sphere. The boundary condition $u = 0$ at $r = a$, on the surface of the sphere, must now be imposed. Since the condition is to hold all around the sphere, and each of the terms in the first expression (15.51) for $u(\mathbf{r})$ is linearly independent in its variations with ϑ, each one must vanish separately at $r = a$. Thus

$$S_l = -h_l^*(ka)/h_l(ka) \equiv e^{2i\delta_l} \tag{15.56}$$

is required. The ratio here is plainly a complex number of absolute value unity, and $2\delta_l$ is standard notation for its phase. The result $|S_l| = 1$ is obtained for every spherically symmetric object. Only the phase $2\delta_l$ is different for a penetrable refracting sphere, or for a region with no sharp boundary but representable by a smoothly varying propagation velocity, $c(r)$, which approaches constancy as the radius increases.

The result $|S_l| = 1$ has such generality because it is necessitated by energy flux conservation. Compare the incident and reflected parts of any one l-wave component of the full solution (15.51). An incident part is proportional to $h_l^*(kr)$, exactly as in the incident plane wave. The corresponding reflected part merely has $S_l h_l(kr)$ replacing $h_l^*(kr)$. Now the flux in each part is to be calculated from

$$\oint d\Omega r^2 S_r = \tfrac{1}{2}\rho c^2 \omega \oint d\Omega r^2 \left[u^* \frac{\partial u}{\partial r} - u \frac{\partial u^*}{\partial r} \right] \Big/ 2i,$$

according to (15.34). It is then plain that the reflected part of the l-wave flux can differ in magnitude from the incident part only by the factor $|S_l|^2$. For the two magnitudes to be the same, $|S_l|^2 = 1$ is necessary.

The quantity δ_l is called a "phase shift" generated by the scatterer. The reason for the name becomes evident when the asymptotic behavior of each radial wave in (15.51) is examined:

$$\begin{aligned} \lim_{kr\to\infty} &\left\{ R_l(kr) \equiv \tfrac{1}{2}[h_l^*(kr) + S_l h_l(kr)] \right\} \\ &= (2kr)^{-1}\{e^{-i[kr - \frac{1}{2}(l+1)\pi]} + e^{2i\delta_l} e^{+i[kr - \frac{1}{2}(l+1)\pi]}\} \\ &= (kr)^{-1} e^{i\delta_l} \cos[kr + \delta_l - \tfrac{1}{2}(l+1)\pi]. \end{aligned} \tag{15.57a}$$

This is to be compared with the corresponding radial wave in the absence of the scatterer, as in the plane wave (15.47) alone:

$$\lim_{kr\to\infty} j_l(kr) = (kr)^{-1} \cos[kr - \tfrac{1}{2}(l+1)\pi]. \tag{15.57b}$$

Thus each of the partial, radial waves composing the incident plane wave has its phase shifted, by δ_l, as the effect of the scatterer, the absolute amplitude remaining unchanged.

A particularly simple picture of the effect on the $l = 0$ wave can be constructed, since this begins the simple "asymptotic" behavior right at the origin, as has already been pointed out following (14.40). The incident wave has (14.30)

$$krj_0(kr) = \sin kr$$

and

$$kr\,|R_0| = \sin(kr + \delta_0).$$

These waves are diagramed in Figure 15.2. The wave R_0 must vanish at $r = a$,

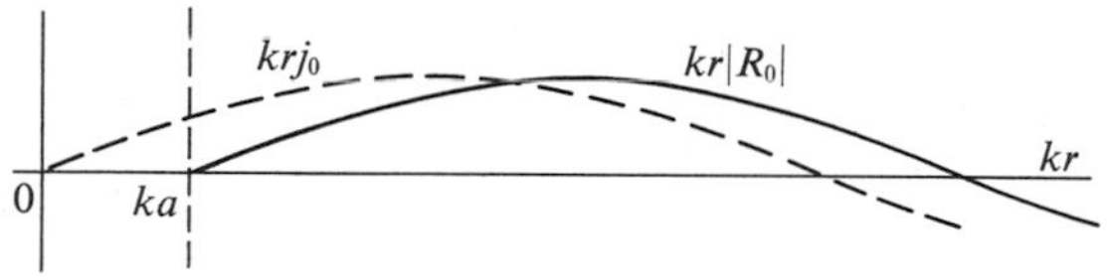

Figure 15.2

and so a phase shift $\delta_0 = -ka$ is to be expected. This is exactly what follows from (15.56), since

$$h_0(ka) = e^{ika}/ika \to e^{2i\delta_0} = e^{-2ika}.$$

It may be noticed that (15.56) defines δ_l only within an additive multiple of π. This ambiguity has no effect on what is observable, but does sometimes necessitate conventions about which span of width π is to be adopted to represent the entire range of effects. Compare the similar situation for the phase of the resonance problem in §3.6.

In the usual derivations, the standing wave basis, j_l, n_l rather than h_l, h_l^* is employed for expressing the radial waves. The formula

$$\tan \delta_l = j_l(ka)/n_l(ka) \tag{15.58}$$

is then found for the phase shifts. It can be easily shown to follow from (15.56) also.

In terms of the phase shifts, the scattering amplitude (15.52*b*) becomes

$$\boxed{f(\vartheta) = k^{-1} \sum_l (2l+1) e^{i\delta_l} \sin \delta_l \, P_l(\cos \vartheta)}\,, \tag{15.59}$$

and the differential cross section (15.55):

$$\boxed{d\sigma/d\Omega = (\lambda/2\pi)^2 \left| \sum_l (2l+1)e^{i\delta_l} \sin \delta_l P_l(\cos \vartheta) \right|^2}. \qquad (15.60a)$$

Because of the orthogonality of the Legendre polynomials, this is easy to integrate to

$$\boxed{\sigma = (4\pi/k^2) \sum_l (2l+1)\sin^2 \delta_l}, \qquad (15.60b)$$

as the total cross section, measuring the flux scattered into all directions.

A result that is obvious from comparing the *forward* scattering amplitude (15.59),

$$\begin{aligned} f(0) &= k^{-1} \sum_l (2l+1)e^{i\delta_l} \sin \delta_l \\ &= k^{-1} \sum_l (2l+1)[\cos \delta_l \sin \delta_l + i \sin^2 \delta_l], \end{aligned}$$

with the total cross section (15.60*b*), is that

$$\boxed{\sigma = (4\pi/k) \operatorname{Im}\{f(0)\}}. \qquad (15.61)$$

This is an instance of what is known as "the optical theorem" in general scattering theory. Like $|S_l| = 1$, it is at bottom necessitated by flux conservation, since it is this which allows expressing the effects of scattering by phase shifts.

The formulas (15.59) and (15.60) hold for all spherically symmetric scatterers, with some set of phase shifts. For the special case of the impenetrable sphere, it is easy to see that

$$d\sigma/d\Omega = (1/k^2) \left| \sum_l (2l+1)[j_l(ka)/h_l(ka)]P_l(\cos \vartheta) \right|^2, \qquad (15.62a)$$

and

$$\sigma = (4\pi/k^2) \sum_l (2l+1) j_l^2(ka)/|h_l(ka)|^2. \qquad (15.62b)$$

The denominators $|h_l|^2$ are frequently to be seen in the equivalent form $|j_l + in_l| = j_l^2 + n_l^2$. The generally nonisotropic angular distribution (15.62*a*) contrasts with the isotropic reflection of *particles* (5.44) by a fixed impenetrable sphere. This is understood as arising from effects on portions of the incident wave front not impinging directly on the sphere but immediately outside its edges. A closer discussion of such effects follows.

The Scattering of the $l = 0$ Partial Wave

Suppose that the incident wave-length is very large as compared to the size of the object; that is, $ka \ll 1$ in the arguments of the spherical functions occuring in the cross section expressions (15.62). Then the functions may be approximated by their values for $ka \ll l(\geq 1)$, as given in (14.44) and (14.45):

$$j_l(ka)/h_l(ka) = i(ka)^{2l+1}/(2l+1)!!(2l-1)!! \qquad (l \geq 1). \qquad (15.63)$$

This is just a first term in an expansion into powers of ka, the terms of higher power being neglected as small in comparison to the lowest power. For consistency, only the lowest occurring power of ka can then be retained in the summations over l, required for the cross sections. Thus only the $l = 0$ waves are scattered appreciably, and since $j_0 \approx 1$ and $h_0 \approx 1/ika$ for $ka \ll 1$,

$$d\sigma/d\Omega \to a^2, \qquad \sigma \to 4\pi a^2. \qquad (15.64)$$

The scattering of the very long wavelengths approaches isotropy, and the total scattered flux the magnitude incident on four times the *cross-sectional* area, πa^2, of the sphere. It is as if the whole *surface* area of the sphere, $4\pi a^2$, were involved in the scattering of the long waves; such effectively "bathe" it on all sides and get around it with only the relatively small amount of disturbance indicated by $\sigma = 4\pi a^2 \ll 4\pi\lambda^2$. The isotropic convergence on the sphere, of the $l = 0$ ingoing partial wave, accounts for the isotropy of the result.

The phase shifts of the very long waves, as given by (15.58), are

$$\tan \delta_l \approx \delta_l \approx -(ka)^{2l+1}/(2l+1)!!(2l-1)!!,$$

with the replacement of the tangent by its argument justified by the smallness of the result. With the $l = 0$ phase shift the most appreciable one, the cross-section expressions (15.60) reduce to

$$d\sigma/d\Omega = (1/k^2)\sin^2 \delta_0, \qquad (15.65a)$$

showing the isotropy again, and

$$\sigma = (4\pi/k^2)\sin^2\delta_0 \qquad (\approx 4\pi\delta_0^2/k^2). \qquad (15.65b)$$

The object of exhibiting such expressions in terms of the phase shifts, δ_0, is that they are actually valid for *any* object which is small as compared to the wavelength, and not only for the impenetrable sphere, when $\delta_0 = -ka$ as shown by Figure 15.2. A conclusion is that very long waves yield very poor "resolution," showing no anisotropies to be correlated with irregularities of the object (it can only look like a round "blob" in the "light" of such waves!).

The Radial Distributions

As the wavelength is decreased, in an attempt to get better "resolution," more of the partial waves might be expected to contribute. Just how many of the l-wave components $\sim j_l(kr)$, present in the incident wave (15.47), will be appreciably disturbed, can be understood from examining their respective radial distributions.

An example ($l = 7$) is shown in Figure 15.3. It shows how the distribu-

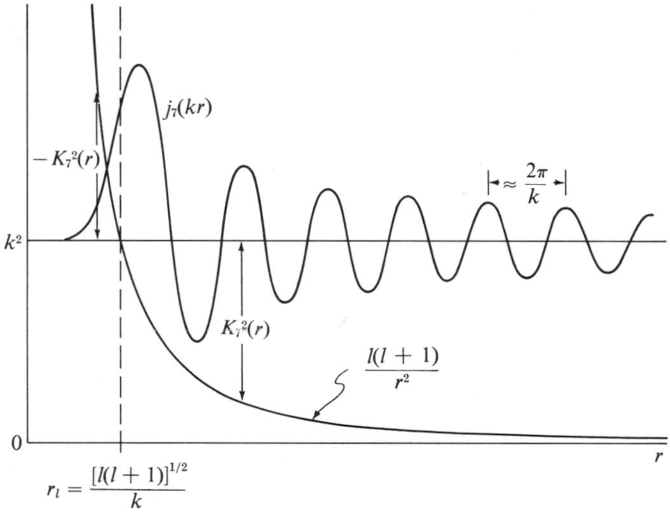

Figure 15.3

tion is pushed out from a central region in accordance with the proportionality to $(kr)^l$ near $r = 0$ (14.44); it also shows how the motion settles down to an oscillation outside that region, in accordance with the asymptotic behavior (14.39). A natural dividing line between the two types of behavior is provided by the radial wave equation (14.26), which $j_l(kr)$ must obey:

$$(rj_l)''/(rj_l) = -[k^2 - l(l+1)/r^2] \equiv -K_l^2(r). \tag{15.66}$$

This shows that

$$r_l \equiv \{l(l+1)\}^{1/2}/k[\approx(l+\tfrac{1}{2})\lambda/2\pi \text{ for } l^2 \gg \tfrac{1}{4}] \tag{15.67}$$

is an inflection point in the curvature of (rj_l). The initial monotonic rise must continue to r_l, since the rate of change of the slope $(rj_l)'$ has the same sign as (rj_l) up to that point. Curvatures that are concave toward the axis instead, as needed for an oscillatory behavior, begin only beyond r_l.

The function $K_l(r > r_l)$ is real, and may be thought of as a variable "wave-number" such that $K_l(\infty) = k$. The role of $K_l(r)$ as a wave number may be made more explicit by trying to put the solutions of (15.66) into the forms, exp: $\pm iw(r)$. Substitution into the equation yields

$$(w')^2 \mp iw'' = K_l^2(r) \tag{15.68a}$$

as the equation to be satisfied by the "phase function," $w(r)$. Far enough out, $K_l(r) \approx k$ and $w \approx kr +$ constant, $w' \approx k$, $w'' \approx 0$. The variable wave member $K_l(r)$ descends from the value k quite gradually (see Figure 15.3) as r approaches r_l from above, and so $|w''|$ should continue to be negligible relative to $|w'|^2$ down to within a few wavelengths of r_l. This invites the approximation*

$$(w')^2 \approx K_l^2(r) \equiv k^2[1 - l(l+1)/(kr)^2], \tag{15.68b}$$

---

* Shown here is a first step in what is called the "WKB approximation" procedure, after G. Wentzel, H. Kramers, and L. Brillouin. The full procedure includes a correction, $w_1(r)$, to be added to the phase-function (15.69). The equation (15.68a) can be seen to yield

$$w_1' \approx \pm iw''/2w' = \pm\tfrac{1}{2}i \ln w' = \pm i \ln[K_l(r)]^{1/2}$$

and hence the WKB approximations proportional to

$$e^{\pm iw(r)}/[K_l(r)]^{1/2}.$$

When these are combined so as to have the asymptotic behaviors characteristic of $j_l(kr)$ and $n_l(kr)$, they yield

$$j_l \approx \frac{\cos[w - \frac{1}{4}\pi]}{kr[K_l(r)/k]^{1/2}}, \qquad n_l \approx \frac{\sin[w - \frac{1}{4}\pi]}{kr[K_l(r)/k]^{1/2}},$$

for $r > r_l$. For $r < r_l$, the correct behaviors (14.44) and (14.45) are shown by the approximations

$$j_l \approx \tfrac{1}{2}e^{-\chi(r)}/kr|K_l(r)/k|^{1/2}, \qquad n_l \approx -e^{\chi(r)}/kr|K_l(r)/k|^{1/2},$$

where

$$\chi(r < r_l) \equiv \int_r^{r_l} |K_l(r)|dr = iw(r).$$

(The $r \to 0$ behaviors here are brought to the forms shown in (14.44) and (14.45) by employing the "Stirling approximation" for factorials.)

The same results can be obtained, using methods beyond the scope of this book, directly from exact expressions for $j_l(kr)$ and $n_l(kr)$. That can be seen in P. Morse and H. Feshbach's *Methods of Theoretical Physics* (McGraw-Hill, 1953), equation (5.3.37).

having solutions proportional to

$$e^{\pm iw(r)} \qquad \text{with} \qquad w(r) = \int_{r_l}^{r} K_l(r)\, dr. \tag{15.69}$$

Of course, a real linear combination of these solutions must be used to approximate $rj_l(kr > kr_l)$. The asymptotic behavior is reproduced exactly, since the approximation procedure started from just that, and so j_l will be best represented at any given r if $k(r - r_l) = 2\pi(r - r_l)/\lambda \gg 1$. Thus the approximation (15.68*b*) is best suited to short wavelengths.

Inside of r_l, $K_l(r)$ is purely imaginary and so is $w(r < r_l)$ of (15.69). The approximate solutions now become real exponentials, exp: $\pm|w(r)|$. It is the solution which *decreases* exponentially, with penetration inward from r_l to $r < r_l$, that must be used to represent $rj_l(kr < kr_l)$, since this descends monotonically to zero at $r = 0$, as pointed out just below (15.67). Whereas $K_l(r > r_l)$ served as a variable wave number, $|K(r < r_l)|^{-1}$ serves as a "penetration depth," like κ^{-1} in the part (14.19*a*) of a totally reflected wave that penetrates into a reflecting medium, as discussed in connection with (14.19*a*). Thus, even though there is no change of medium here, the l-wave is in effect totally reflected at the surface $r = r_l$. The point $r = r_l$ on each radius serves as a "turning point" for the wave motion, analogous to the turning point r_L of Figure 4.4. The latter applies to particles approaching collision against a potential (force) barrier centered on $r = 0$. The situation here (no change of medium at $r = r_l$) is more nearly analogous to the case of *free* particles ($|K| = 0$ in Figure 4.4), which cannot approach the center more closely than $r_L = L/(2mE)^{1/2} = L/mv = b$ when they start in with angular momentum $L = mvb$ about $r = 0$ (initial "lever arm" b). The "centrifugal barrier," $L^2/2mr^2$, of Figure 4.4 is replaced in Figure 15.3 by $l(l + 1)/r^2$.

It should now be clear that a partial wave with a given l penetrates very little within its turning point r_l. If an obstacle of radius $a < r_l$ is introduced, it will disturb the l-wave only negligibly. The condition that the wave vanish at $r = a < r_l$ is practically satisfied by $j_l(kr)$ itself, without appreciable phase-shift. Of all the l-waves contained in an incident plane wave, only those will be substantially scattered which have $kr_l \approx (l + \frac{1}{2}) \lesssim ka$. There arises a practical upper limit, of the order $l \approx ka$, on the number of appreciable terms in the summations over l needed for the cross sections (15.60) to (15.62).

The upper limit $l \approx ka$ is the more sharply defined the larger it is. This is because, for larger l-values, the fractional penetration into the region $r < r_l$ is smaller, as the increased power of $j_l(0) \sim (kr)^l$ suggests. Accompanying this is a greater fractional concentration of the oscillations into the region just outside $r \approx r_l$, as suggested by the greater relative amplitudes of the first oscillations in Figure 15.3. For very large l's, the most appreciable amplitudes are concentrated into a spherical shell which may be taken to be many wavelengths thick, and yet will be very thin in comparison to the radius, $r_l \approx l\lambda/2\pi$, of the shell.

The Ray Approximation

Simple results may be expected not only for the long waves already discussed, but also at the opposite extreme, $ka \gg 1$, when the wavelength, $\lambda = 2\pi/k \ll a$, is very small in comparison to the object.

Very short wavelength propagations may be treated as consisting of "rays," each following a rectilinear path. A ray may be considered a beam which is very broad in comparison to the wavelength, yet is extremely narrow as compared to the object. Such a beam, many wavelengths wide, may be treated as a monodirectional plane wave within its lateral extent, as already pointed out in connection with (15.53). A restriction of it to a transverse breadth $\Delta x \gg \lambda = 2\pi/k$ ought only introduce transverse wave-vector components $\Delta k_x \approx 2\pi/\Delta x \ll |\mathbf{k}|$, according to the complementarity (12.88), and so no appreciable deviations from the propagation direction $\hat{\mathbf{k}}$ will develop. If now the beam, although many wavelengths wide, can yet be narrow in comparison to relevant dimensions in a given set of circumstances, then it can be treated as a straight-line ray in those circumstances. It is, for example, just the very short wavelengths of visible light which make it cast rectilinearly propagated shadows, as ordinarily observed.

Now, consider the "bundle" of rays in the incident plane wave which is intercepted by an area element $dA = a^2\, d\Omega(i)$ on the sphere (Figure 15.4). The bundle forms a sector of the plane wave front through which passes the flux

$$d\sigma_s = [a^2\, d\Omega(i)]\cos i = a^2 \cos i\, d(\cos i)\, d\varphi,$$

for each unit of flux density in the incident wave. The intercepted bundle can be treated as a plane wave incident on a reflecting "plane" of area dA, if only the wavelength is as short as supposed, and so short enough for "specular" reflections as noted in connection with Figure 14.2. The incidence angle is just i, on the area element at i, and the reflection direction makes the equal angle i to the other side of the normal to the sphere. The outcome is a net deflection by an angle $\vartheta = \pi - 2i$, and when the differential cross section is put into terms of this, it becomes

$$d\sigma_s = \tfrac{1}{4}a^2 \sin\vartheta\, d\vartheta\, d\varphi \equiv \tfrac{1}{4}a^2\, d\Omega(\vartheta). \qquad (15.70)$$

Thus, an isotropic scattering into all angles $\pi > \vartheta > 0$ is to be expected from

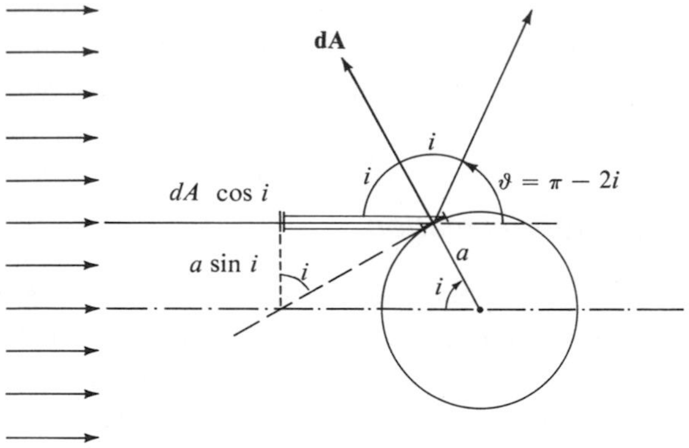

Figure 15.4

the "illuminated" area elements on the front of the sphere, at all the angles $0 < i < \frac{1}{2}\pi$. The total specular reflection cross section is $\sigma_s = \pi a^2$, just equal to the geometrical cross section of the sphere.

This result agrees exactly with the cross section (5.44) for the scattering of *particles* by a fixed impenetrable sphere. Moreover, the rays as indicated in Figure 15.4 coincide with corresponding particle trajectories in their scattering. Such comparability is quite general, and accounts for the efficacy, in ray optics, of Fermat's "Principle of Least Time," constructed like the Least Action Principle (8.119) for particle trajectories. It is a very simple Exercise (15.7) that shows the laws of reflection and refraction, which were obtained in §14.1 from the wave picture, also follow from requiring a least* time of propagation along the rays "joining" any two points. It is equally easy to find that the deflections by just $\pi - 2i$, in the problem of Figure 15.4, are obtainable as consequences of the principle.

Any segment ds of path has some direction which may be denoted $\hat{\mathbf{k}}$ and the Principle of Least Time may be expressed as

$$\delta \int_{s(\mathbf{r}_0)}^{s(\mathbf{r}_1)} ds/c(\mathbf{r}) = \delta \int_{\mathbf{r}_0}^{\mathbf{r}_1} \hat{\mathbf{k}} \cdot \mathbf{dr}/c = 0.$$

In general, $\hat{\mathbf{k}}(\mathbf{r})$ will be variable if $c(\mathbf{r})$ is. It is only while the propagation is uninterruptedly confined to a "single" medium, with a uniform propagation velocity c, that the minima will be obtained for straight line paths and the problem reduce to one of finding constant propagation directions $\hat{\mathbf{k}}$. To play

*As its origin in the Variational Principle indicates, Fermat's principle more generally requires the correct path to provide a "stationary" value for times of traversal, relative to neighboring paths. This may sometimes be a maximum, rather than a minimum.

the role of the action $W(\mathbf{r}, \mathbf{p}_0)$ of (8.118) for a given constant energy E, a phase-function $w(\mathbf{r})$ may be defined, for a given constant frequency ω, such that

$$dw(\mathbf{r}) = \omega ds/c(\mathbf{r}) = \mathbf{k}(\mathbf{r}) \cdot \mathbf{dr}. \qquad (15.71a)$$

The Hamilton-Jacobi equation (8.116) for particle trajectories is replaced by

$$(\nabla w)^2 = k^2(\mathbf{r}) \qquad (15.71b)$$

for rays of a given frequency. The function $w(\mathbf{r})$ provides the phases for very short wavelength waves of which the rays are parts (for example, $w(\mathbf{r}) = \mathbf{k} \cdot \mathbf{r} + \text{constant}$ in a monodirectional plane wave $\sim\exp\colon i\mathbf{k} \cdot \mathbf{r}$) just as $W(\mathbf{r}, \mathbf{p}_0)$ determines the phases of configuration waves associated with the particle trajectories discussed in §8.6. While the configuration waves furnish a way of representing collections of possible trajectories for ensembles of particles, such as are considered whenever there is uncertainty about just which of the possible paths any one particle may have been started on, the rays represent trajectories to be associated with waves of very short wave-length. There is here a certain "correspondence" between the particle and wave motions of classical mechanics. It helps justify Newton's famous conclusion that very short wave-length, rectilinearly propagated, light may be thought of as consisting of particle-like "corpuscles" (called "photons" nowadays, and "phonons" in connection with sound and other mechanical waves).

The wave-vector $\mathbf{k}(\mathbf{r})$ in the phase function (15.71) takes the place of the particle momentum $\mathbf{p}$ in the expression (8.118) for the phase of a particle configuration wave. There is indeed a "momentum," the wave-momentum of density $\mathbf{g} = \hat{\mathbf{k}}(U/c) = \mathbf{k}(U/\omega)$ given by (15.42), to be associated with the energy flow in each ray. Thus the wave-vector $\mathbf{k}$ becomes a certain measure of the wave-momentum to be associated with a monochromatic wave, normalized to some standard energy density U.

The Angular Momenta in Waves

In the scattering problems of Chapter 5, the incident particles are classified according to their "impact parameters," b, which serve as the "lever arms" for the particle angular momenta, $L = mvb$. Plainly, also a ray impinging on the sphere at a distance b from the axis through the sphere center has an "angular momentum" about that center. It is of the wave type (15.43), and is equal to $(U/c)b$ when the ray carries a wave-momentum of magnitude U/c; the direction of this angular wave-momentum is about the axis perpendicular to the plane containing the ray and the sphere center.

The analysis of the incident plane wave into various angular wave-momenta will now be found to correspond to its decomposition into the various spherical l-waves of (15.47). To discuss angular wave-momentum components directed like those considered in the preceding paragraph, it will help to choose that direction to be the polar (z'-)axis, as indicated in Figure 15.5. In the frame shown, $k = k_{x'} = k \sin \vartheta_k' \cos \varphi_k'$ and $\vartheta_k' = \frac{1}{2}\pi$, $\varphi_k' = 0$. The

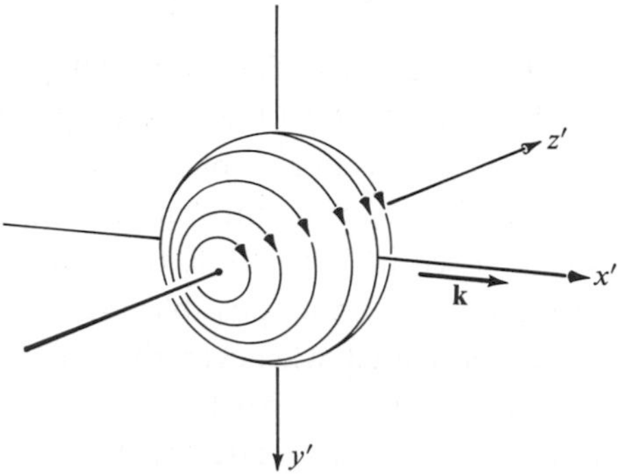

Figure 15.5

plane wave (15.47) becomes a sum of the "(l, m')-wave" components

$$u_{lm'}(\mathbf{r}) = 4\pi a_i\, i^l Y_{lm'}^*(\tfrac{1}{2}\pi, 0) j_l(kr) Y_{lm'}(\vartheta', \varphi'). \tag{15.72}$$

A directed flux of energy, $\mathbf{S}(l, m')$, may be associated with each (l, m')-wave component, and a wave-momentum density, $\mathbf{g}(l, m') = \mathbf{S}/c^2$, which has the distribution of lever arms, $r \sin \vartheta'$, about the z'-axis. The directions of the flux are to be found from (15.34), so that

$$\mathbf{g}(l, m') = \mathbf{S}(l, m')/c^2 = \tfrac{1}{2}\rho\omega \operatorname{Re}\{u_{lm'}^*(-i\boldsymbol{\nabla})u_{lm'}\}, \tag{15.73}$$

and it is the component

$$-i\nabla_{\varphi'} = (r \sin \vartheta')^{-1}(-i\, \partial/\partial\varphi')$$

of the gradient operator (see (2.30)) which contributes to the angular wave-momentum about the z'-axis. This will then be proportional to $(-i\partial/\partial\varphi')\, Y_{lm'} = m' Y_{lm'}$, with m' an eigenvalue of the "angular momentum operation" here: $-i(\mathbf{r} \times \boldsymbol{\nabla})_{z'} = -i\partial/\partial\varphi'$. Of course, the entire plane wave contains a canceling contribution from $m' = -|m'|$ for every one proportional to $m' = +|m'|$. (The opposite circulations of $m' = \pm|m'|$ waves have been pointed out several times, first in connection with Figure 13.5. The arrows in Figure 15.5 correspond to $m' > 0$.) Such cancellations must be expected, since the plane wave as a whole carries no net angular momentum; it can always be divided into a uniform distribution of rectilinear rays, all carrying the same wave-momentum, impinging uniformly on all sides of any momental center.

The maximum angular wave-momentum component around the positive $z' =$ axis will be carried by the $m' = l$ wave, for a given l, according to the preceding discussion. The remaining (l, m')-waves carry z'-components proportional to $+l > m' \geq -l$; hence m' is a measure, in steps a unit apart, of z'-component of an angular wave-momentum having a maximum magnitude of l units, for a given l, on the scale thus introduced. The $m' = 0$ wave of given l is proportional to $Y_{l0}(\vartheta', \varphi') = [(2l+1)/4\pi]^{1/2} P_l(\cos \vartheta')$ and has no z'-component of angular wave-momentum; however, a total magnitude of l units must still be associated with it, since, by mere reorientation of axes, it becomes resolvable into components with $m'' \neq 0$, of angular wave-momenta about a new (z'')-axis. Consequently, such numbers as $|m'| \leq l$ or $|m''| \leq l$ are called "projections" of l on the corresponding polar axes, since they are measures of angular wave-momentum components about those axes.

In the discussion of Figure 15.3 on page 488 it was pointed out that the amount $l(l+1)/r^2$—by which the squared "variable wave-number," $K_l^2(r)$, characteristic of a given l-wave, is diminished by approach toward the center—is analogous to the centrifugal barrier, $L^2/2mr^2$, to particles approaching a center with an angular momentum L. Since it has now been found that l is a measure of angular wave-momentum to be associated with a given l-wave, the quantity $l(l+1)/r^2$ may be regarded as a measure of "centrifugal force," acting on "equivalent mass$(=g/c)$" rather than on actual mass-elements of the medium. This can be said to account for the pushing away, from the center, of the oscillations bearing the main density of energy flux, to the spherical shell at the "turning point" $r_l = [l(l+1)]^{1/2}\lambda/2\pi$.

The phase-function $w(r)$, which gives the phases of a radial l-wave, was found to obey the equation $(w')^2 \approx k^2 - l(l+1)/r^2$ (15.68*b*), in an approximation which is best for short waves (see the remark following (15.69)). The "corresponding" configuration-wave phase-function, for particles of given angular momentum L, receives a radial phase contribution, $W_r(r)$ of (8.109*b*), which is determined by the Hamilton-Jacobi equation (8.112),

$$(W_r')^2 = (mv)^2 - L^2/r^2,$$

when $Z = 0$, as for a free particle. There is implicit here the correspondence $L/mv = b \leftrightarrow [l(l+1)]^{1/2}\lambda/2\pi = r_l$, of the turning points. A free particle follows a straight-line path like a ray carrying a definitely directed linear wave-momentum, $\mathbf{g} = \hat{\mathbf{k}}(U/c) = \mathbf{k}U/\omega$. Viewed as a carrier of angular momentum L, the particle can be said to follow a straight-line *orbit* (from $\vartheta = \pi$ to $\vartheta = 0$) which has the lever arm $b = L/mv$. Corresponding to this is a spherical

l-wave, rather than "ray," component of the incident plane wave, having the "lever arm" $b = r_l = [l(l+1)]^{1/2}\lambda/2\pi$.

The Short-Wave Scattering Cross Section

The practical limit, $l \approx ka$, on the l-waves which will be appreciably scattered, pointed out on page 488, can now be understood as arising because rays of "impact parameter" $b = r_l \approx l/k > a$ will "miss" the scatterer. This limit is best defined for the large $l \approx ka \gg 1$ values characterizing short-wave scattering ($\lambda = 2\pi/k \ll 2\pi a$), when the ray approximation is best. It was found in (15.70) that the "specular reflection" cross section of the rays is just the geometrical one, $\sigma_s = \pi a^2$. This will now be compared to the $ka \gg 1$ limit of the more general wave-scattering cross section expression (15.62*b*).

With $ka \gg 1$, the factors

$$j_l(ka)/h_l(ka) = \tfrac{1}{2}[1 + h_l^*(ka)/h_l(ka)]$$

may be evaluated with the asymptotic value (14.40) of $h_l(ka)$:

$$\begin{aligned} j_l(ka)/h_l(ka) &\approx \tfrac{1}{2}[1 - (-)^l e^{-2ika}], \\ &= e^{-ika} \cdot i \sin ka \qquad \textit{for } l \textit{ even}, \\ &= e^{-ika} \cdot \cos ka \qquad \textit{for } l \textit{ odd}. \end{aligned} \tag{15.74}$$

The terms with l values up to ka become extremely numerous as $ka \to \infty$, the distinction between odd and even l's can be forgotten, and an average used:

$$|j_l/h_l|^2 = \tfrac{1}{2}(\sin^2 ka + \cos^2 ka) = \tfrac{1}{2}$$

The result for the wave-scattering cross section is

$$\sigma \to (4\pi/k^2)\sum_{l=0}^{(ka)}(2l+1)\cdot\tfrac{1}{2} \to 2\pi a^2 \qquad \text{for } ka \to \infty. \tag{15.75}$$

More precise considerations, not resorting to the averaging here, lead to the same conclusion.

The conclusion is that the "wave-scattering cross section," as it is defined, counts *twice* as many deflections as are accounted for by the specular reflections of the rays. A proper understanding of why this happens requires considering the complete wave picture, including the shadow formation that stems from the interference of the plane and outgoing spherical waves of the complete and exact solution (15.51). The partial wave analysis in terms of which (15.51) is presented is not very well suited for investigating this (as attested by the infinite number of terms that must be counted as $ka \to \infty$). The most economical treatments lie outside the scope of this treatise, requiring methods best learned in connection with "wave optics." Accordingly, the remainder of the discussion here will only be qualitative, aimed at plausibility rather than proof. Indeed, most of the provable results lack plausibility until discussed in some such way as here!

The phenomenon which provides the extra deflections, not accounted for by the specular reflections, is known as "Fraunhofer diffraction" in optics. It is responsible for the fact that even very short wavelengths produce a "bright spot" in the middle of the shadow behind an object, at a certain minimum distance. Evidently rays are still being "bent" around the object, although not as completely as very long $l = 0$ waves. Somewhat farther back, the shadow disappears practically completely, and the brightness expected from an uninterrupted wave is essentially restored, although not immediately behind the object, as for the very long waves. The distance L at which restoration occurs is the greater, the larger the obstacle and the shorter the wavelength.

The dimensions associated with the phenomenon may be rendered plausible with the help of the diagram in Figure 15.6. There are indicated

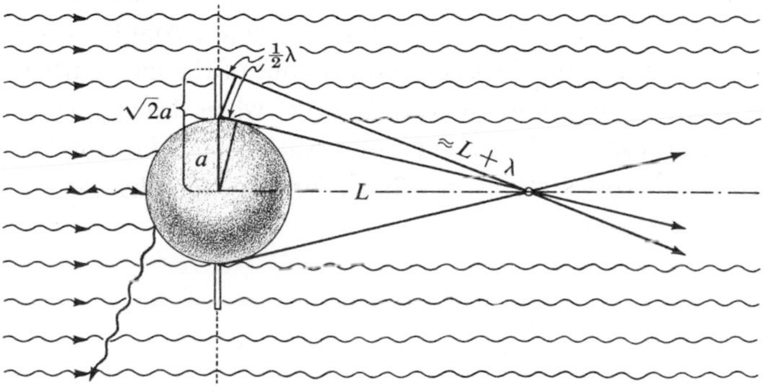

Figure 15.6

outer and inner radii, $\sqrt{2}\,a$ and a, of a circular "first zone" around the edge of the sphere, chosen so as to have just the area πa^2, upon which enough flux is incident to restore an uninterrupted value of intensity to points on the axis of the shadow. This is not meant to imply that all the rays incident on this zone are bent to the point at which the brightness beyond the shadow begins, but only that it can serve as a surrogate for all contributing parts of the wave front. Moreover, there is no sharply definable "point at which the brightness begins"; there is inmost a region of "diffraction rings" which merge to near-uniform brightness only farther out. The nonuniformities in the brightness arise from constructive and destructive interferences of waves arriving with various phase differences.

It is plausible that the distance L is roughly such that the path lengths from the inner and outer edges of the first zone differ by about a half wavelength. A whole wavelength would include contributions which average out to zero by interference. The path length from the inner edge of the zone is roughly $(L^2 + a^2)^{1/2} \approx L(1 + a^2/2L^2)$, and from the outer edge, $(L^2 + 2a^2)^{1/2} \approx L(1 + a^2/L^2)$. The difference is $a^2/2L \approx \lambda/2$, and hence $L \approx a^2/\lambda$ is the order of magnitude of the distance to which the inmost rays are "bent." Their maximum angle of deflection is thus roughly $a/L \approx \lambda/a$. This result is also rendered plausible by the fact that "rays" of as large a breadth as the radius a of the intercepted beam have generated in them deflections of the order $\Delta k_y/k \approx 2\pi/ka = \lambda/a$, according to such complementarity relations as the one already quoted above.

It should now be clear that, added to the isotropically distributed specular reflections, there should be expected an equal flux of small deflections, up to roughly the angle λ/a from the forward direction. These diffraction deflections may be considered ignorable, as too small to be noteworthy, for observations well within the shadow, at $z \ll a^2/\lambda$. However, the wave-scattering cross section measures the flux at "infinity," far beyond even the long (for $ka \gg 1$) distance a^2/λ. At distances so great, even the tiny deflections up to angles λ/a may take the deflected rays out of the incident beam. This is especially true on the atomic or nuclear scale, when dealing with objects of $a \approx 10^{-8}$ cm or $a \approx 10^{-12}$ cm. Then observations as "far out" as 1 cm will detect deflection cross sections $\sigma = 2\pi a^2$, for wave lengths $\lambda \ll a$.

EXERCISES

15.1. Most of the relations of §15.1 and §15.2 may be written economically (and significantly) in a "space-time formulation." A four-dimensional space is employed, with "position vectors," x, having components $x_1 \equiv x$, $x_2 \equiv y$, $x_3 \equiv z$, and $x_4 = ict$. A "volume" element in it is denoted $d^4x (\equiv dx\, dy\, dz.\, icdt)$, and the four-component gradient operator, ∂_μ (see (9.89)), has $\partial_4 \equiv \partial/ic\,\partial t$.

(a) Identify relations in the text which may be written

$$\delta \int_{x(1)}^{x(2)} d^4x\; l(q, \partial_\mu q) = 0,$$

$$\partial_\mu\{\partial l/\partial[\partial_\mu q]\} = \partial l/\partial q,$$

$$l = -\tfrac{1}{2}\rho c^2(\partial_\mu q)^2, \qquad \partial_\mu^2 q = 0,$$

when the summation convention, first introduced in (9.19*b*), is employed.

(b) Identify the equation $\partial_\mu S_\mu = 0$, if $S_4 \equiv icU$. Note that $c^{-2}\sum_\mu S_\mu^2 = c^2g^2 - U^2 = 0$ for the plane-wave quantities of (15.42). This constitutes an invariance to "rotations" in the four dimensional space.

15.2. Show that the outcome of evaluating the energy (15.22) for the arbitrary wavepacket (14.8) is time-independent, and review the implications of this for idealizations in which wave packets, containing component amplitudes strongly "peaked" at some one frequency ω_0, are treated as *mono*chromatic waves.

15.3. For the system of Exercise 14.6, show that the reflection falls short of being total by just the right amount to account for the normal energy flux found to be "leaking" out of the other side of the intervening "barrier" medium. (It may be most instructive to restrict the investigation to $\kappa d \gg 1$, yet not destroy the matching of the fluxes.)

15.4. Suppose that spherical wavelets of equal amplitude but opposite phase, as represented by

$$q(\mathbf{r}, t) = \text{Re}\, A\left\{\frac{e^{ikR}}{R} - \frac{e^{ikR'}}{R'}\right\}e^{-i\omega t},$$

emanate from two source-points a distance $\varepsilon \ll \lambda$ apart. Here $R \equiv |\mathbf{r} - \mathbf{r}_s|$ and $R' = |\mathbf{r} - \mathbf{r}_s'|$, where $r_s = z_s = +\frac{1}{2}\varepsilon$ and $r_s' = -z_s' = \frac{1}{2}\varepsilon$. Use (14.71) to show that radial wave components of odd parity only are generated (see Exercise 13.10). Also show that, for $k\varepsilon \ll 1$, the fraction of the outward energy flow emerging at "infinity" per unit solid angle in the direction ϑ is

$$(3/4\pi)\cos^2\vartheta.$$

(This is a characteristic "dipole radiation pattern.")

15.5. Show that the effect of a "complex conjugation operation" on the solution $u(\mathbf{r})$ of (15.51) is to produce a description of the "time-reversed situation," that is $q(\mathbf{r}, t) = u^*(\mathbf{r}) \exp{:}\ -i\omega t$ is to be interpreted as the emergence of a plane wave at infinity, in consequence of having "sent in" a spherical wave to converge on a scatterer. [Consider what replaces (15.52*a*). The "time-reversed situation" would be difficult to arrange physically, but the result has important implications. It shows what part of any more general disturbance is responsible for any one plane-wave component which might be found to be emerging "at infinity."]

15.6. Consider the scattering of very long waves by a small refracting sphere (c_2 in $r < a$).

(a) Argue that a disturbance proportional to

$$u(r < a) \sim \sin(k_2 r)/k_2 r, \qquad k_2 = (c/c_2)k,$$

as its entire spatial dependence, is the only one appreciably generated inside the sphere.

(b) Find the phase shift, $\delta_0(c, c_2, ka)$, in the incident wave.

(c) Show that in the limit $k_2 a \ll 1$ (and $ka \ll 1$), the (isotropic) scattering cross section will be

$$\sigma \approx 4\pi a^2\left(\frac{c^2 - c_2^2}{3c_2^2}\right)^2\left(\frac{2\pi a}{\lambda}\right)^4.$$

[This represents so-called "Rayleigh scattering." The characteristically greater magnitude for shorter waves is supposed to account for the blue color of the sky. It also accounts for certain acoustical effects.]

15.7. An idea of how the "Principle of Least Time" operates may be gained about as simply as possible from the following.

(a) Derive the law of reflection by comparing all possible ray paths which consist of pairs of straight-line segments, one of the pair extending from any fixed starting point to a variable point on a plane reflecting surface, the other from the reflection point to any suitably fixed end point.

(b) Derive the law of refraction in a similar way, using a starting point in a medium of propagation speed c_1, and an end point in a medium with c_2.

(c) Finally, consider a *mass-particle* which follows the refracted path of (b). If this path is to be derivable from the Principle of Least Time, the particle's "configuration wave" (§8.6) speeds must have the ratio c_1/c_2. Show that the particle's actual speeds $v_{1,2}$ must have the *inverse* ratio, if they differ in the two media only because the particle acquires a constant potential energy in crossing the interface (thus receiving a normal component of impulse). Find the potential difference needed. [$\frac{1}{2}mv_1^2(1 - c_1^2/c_2^2)$.]

15.8. From (8.71) and (8.116), it follows that

$$-\partial S/\partial t = (1/2m)(\nabla S)^2 + V(\mathbf{r})$$

is the Hamilton-Jacobi equation for a mass-particle in a potential field $V(\mathbf{r})$.

(a) Show that the substitution $\psi = e^{iS}$ yields the equation

$$i\partial\psi/\partial t = -(1/2m)\nabla^2\psi + V\psi - (i/2m)\psi\nabla^2 S$$

for ψ. If the curvature, $\sim \nabla^2 S$, of the phase-function is neglected, this amounts to the famous "Schrödinger equation" of wave mechanics. [It was seen in connection with (15.68) and (15.71) that the curvature of a phase function becomes most nearly negligible when the "correspondence" between particle-paths and wave-rays is best.]

(b) Show that the "Schrödinger equation," and its complex conjugate, follow from a variational principle with

$$l = i\psi^*\mathring{\psi} - (1/2m)(\nabla\psi^* \cdot \nabla\psi) - V\psi^*\psi,$$

if ψ and ψ^* are varied independently.

A POSTSCRIPT

A final grand exercise which the reader would find profitable to perform is a summary, or series of summaries according to his interests, of the entire contents of this book. It can be very brief, once the concepts, and the mathematical languages developed for their expression, are familiar. It can be greatly abbreviated if the implications of every statement, obtained by deduction *and* by generalization (induction), can be taken for granted.

The last point is a significant one to enlarge upon. It should always be borne in mind that a properly physical, "open-ended" theory gives at best a *starting basis* for applications of it. Its implications cannot be restricted to mere deductions as from a purely mathematical set of axioms. To meet new situations, creative acts of induction are called for. The experimentalist, in particular, performs a crucial act whenever he adopts correspondences between his operations and some set of theoretical concepts. The reader himself can contribute to the growth of the theory by finding new problems that he can solve with its help. He should be encouraged to do so.

INDEX